Global Perspectives on Literary Tourism and Film-Induced Tourism

Rita Baleiro
University of the Algarve, Portugal & CiTUR, Portugal

Rosária Pereira
University of the Algarve, Portugal & CinTurs, Portugal

IGI Global
PUBLISHER of TIMELY KNOWLEDGE

A volume in the Advances in Hospitality, Tourism, and the Services Industry (AHTSI) Book Series

Published in the United States of America by
IGI Global
Business Science Reference (an imprint of IGI Global)
701 E. Chocolate Avenue
Hershey PA, USA 17033
Tel: 717-533-8845
Fax: 717-533-8661
E-mail: cust@igi-global.com
Web site: http://www.igi-global.com

Library of Congress Cataloging-in-Publication Data

Names: Baleiro, Rita, editor. | Pereira, Rosária, 1970- editor.
Title: Global perspectives on literary tourism and film-induced tourism / Rita Baleiro and Rosária Pereira, editors.
Description: Hershey, PA : Business Science Reference, 2022. | Includes bibliographical references and index. | Summary: "This book aims at fulfilling a research gap acknowledged in recent studies on literary tourism and on aspects of film-induced tourism by offering contributions on both practical and academic approaches to offer advances in research on the intersections of literature, film, and the act of travelling"-- Provided by publisher.
Identifiers: LCCN 2021024346 (print) | LCCN 2021024347 (ebook) | ISBN 9781799882626 (hardcover) | ISBN 9781799882633 (paperback) | ISBN 9781799882640 (ebook)
Subjects: LCSH: Heritage tourism. | Literary journeys. | Film tourism.
Classification: LCC G156.5.H47 G558 2022 (print) | LCC G156.5.H47 (ebook) | DDC 338.4/791--dc23
LC record available at https://lccn.loc.gov/2021024346
LC ebook record available at https://lccn.loc.gov/2021024347

This book is published in the IGI Global book series Advances in Hospitality, Tourism, and the Services Industry (AHTSI) (ISSN: 2475-6547; eISSN: 2475-6555)

British Cataloguing in Publication Data
A Cataloguing in Publication record for this book is available from the British Library.

For electronic access to this publication, please contact: eresources@igi-global.com.

Advances in Hospitality, Tourism, and the Services Industry (AHTSI) Book Series

Maximiliano Korstanje
University of Palermo, Argentina

ISSN:2475-6547
EISSN:2475-6555

Mission

Globally, the hospitality, travel, tourism, and services industries generate a significant percentage of revenue and represent a large portion of the business world. Even in tough economic times, these industries thrive as individuals continue to spend on leisure and recreation activities as well as services.

The Advances in Hospitality, Tourism, and the Services Industry (AHTSI) book series offers diverse publications relating to the management, promotion, and profitability of the leisure, recreation, and services industries. Highlighting current research pertaining to various topics within the realm of hospitality, travel, tourism, and services management, the titles found within the AHTSI book series are pertinent to the research and professional needs of managers, business practitioners, researchers, and upper-level students studying in the field.

Coverage

- Hotel Management
- Tourism and the Environment
- Casino Management
- Health and Wellness Tourism
- Customer Service Issues
- Sustainable Tourism
- International Tourism
- Service Design
- Destination Marketing and Management
- Travel Agency Management

IGI Global is currently accepting manuscripts for publication within this series. To submit a proposal for a volume in this series, please contact our Acquisition Editors at Acquisitions@igi-global.com or visit: http://www.igi-global.com/publish/.

The Advances in Hospitality, Tourism, and the Services Industry (AHTSI) Book Series (ISSN 2475-6547) is published by IGI Global, 701 E. Chocolate Avenue, Hershey, PA 17033-1240, USA, www.igi-global.com. This series is composed of titles available for purchase individually; each title is edited to be contextually exclusive from any other title within the series. For pricing and ordering information please visit http://www.igi-global.com/book-series/advances-hospitality-tourism-services-industry/121014. Postmaster: Send all address changes to above address.

Titles in this Series

For a list of additional titles in this series, please visit: http://www.igi-global.com/book-series/advances-hospitality-tourism-services-industry/121014

Prospects and Challenges of Community-Based Tourism and Changing Deographics
Ishmael Mensah (University of Cape Coast, Ghana) and Ewoenam Afenyo-Agbe (University of Cape Coast, Ghana)
Business Science Reference • © 2022 • 300pp • H/C (ISBN: 9781799873358) • US $195.00

Challenges and New Opportunities for Tourism in Inland Territories Ecocultural Resources and Sustainable Initiatives
Gonçalo Poeta Fernandes (CITUR, Polytechnic Institute of Guarda, Portugal)
Business Science Reference • © 2022 • 295pp • H/C (ISBN: 9781799873396) • US $195.00

Rebuilding and Restructuring the Tourism Industry Infusion of Happiness and Quality of Life
André Riani Costa Perinotto (Universidade Federal do Delta do Parnaíba, Brazil) Verônica Feder Mayer (Federal Fluminense University, Brazil) and Jakson Renner Rodrigues Soares (Universidade da Coruña, Spain & Universidade Estadual do Ceará, Brazil)
Business Science Reference • © 2021 • 330pp • H/C (ISBN: 9781799872399) • US $195.00

Handbook of Research on the Impacts and Implications of COVID-19 on the Tourism Industry
Mahmut Demir (Isparta University of Applied Sciences, Turkey) Ali Dalgıç (Isparta University of Applied Sciences, Turkey) and Fatma Doğanay Ergen (Isparta University of Applied Sciences, Turkey)
Business Science Reference • © 2021 • 906pp • H/C (ISBN: 9781799882312) • US $415.00

Socio-Economic Effects and Recovery Efforts for the Rental Industry Post-COVID-19 Strategies
Maximiliano Korstanje (University of Palermo, Argentina)
Business Science Reference • © 2021 • 281pp • H/C (ISBN: 9781799872870) • US $195.00

Risk, Crisis, and Disaster Management in Small and Medium-Sized Tourism Enterprises
Diego R. Toubes (University of Vigo, Spain) and Noelia Araújo-Vila (University of Vigo, Spain)
Business Science Reference • © 2021 • 308pp • H/C (ISBN: 9781799869962) • US $195.00

Resiliency Models and Addressing Future Risks for Family Firms in the Tourism Industry
Anita Zehrer (MCI The Entrepreneurial School, Austria) Gundula Glowka (MCI The Entrepreneurial School, Austria) Katrin Magdalena Schwaiger (MCI The Entrepreneurial School, Austria) and Victoria Ranacher-Lackner (MCI The Entrepreneurial School, Austria)
Business Science Reference • © 2021 • 361pp • H/C (ISBN: 9781799873525) • US $195.00

IGI Global
PUBLISHER of TIMELY KNOWLEDGE

701 East Chocolate Avenue, Hershey, PA 17033, USA
Tel: 717-533-8845 x100 • Fax: 717-533-8661
E-Mail: cust@igi-global.com • www.igi-global.com

Editorial Advisory Board

Table of Contents

Detailed Table of Contents

Departing from the assumption that cinema has always had the capacity to represent social structures and movements and provide valuable sources of information about societal phenomena, this chapter employs representation as a research approach to offer contributions to understanding realities "outside the film" regarding literary tourists' motivations, experiences, and literary places. The authors analyse cinematic representations of literary tourism in feature films and take the perspective of literary tourism studies, reviewing the literature on cultural tourism, special interest tourism, niche tourism, literary tourism, and literary sites and landscape. The analysis and interpretation of the cinematic sequences reveal two opposed ideas of what literary tourism experiences might be: a shallow, disappointing, and inauthentic experience or a meaningful and authentic event.

Author-related literary tourism places can create attractive elements for visitors who have an inner interest in the life stories of authors. Literary pilgrims place particular emphasis on authenticity level in literary sites. One of the attractive literary places in terms of authenticity, which is also Turkey's first literary museum, is Asiyan Museum where Tevfik Fikret lived between the years 1906-1915. In 2010, the museum underwent a comprehensive restoration. During this restoration, serious archive research was carried out and the house approached the original condition in the period of the 1910s. This study aims to examine the relationships between the authenticity of the place and the museum management mentality, research, and tourists' expectations by pointing out the prominence of authenticity in literary places. Using the interview technique, data were collected from the museum official in March 2018 on subjects such as the history of the museum, interpretation techniques, restoration works.

This chapter aims at understanding travelers' literary experience in "The Museum of Innocence," which is known as a museum based on a book that was written by Orhan Pamuk. To reach the purpose, a qualitative research approach was adopted in this study, and online reviews posted by visitors of the museum were used to explore the main components of their experience in a literary context. The collected data was examined through content analysis. As a result of analysis, visualization, a sense of nostalgia, awe, and memorable components were revealed. These components were interpreted in different ways by visitors. Theoretical and practical implications were provided as well as limitations for future research lines.

It is the purpose of this chapter to investigate the opportunities that lie with drawing on the personal letters of an iconic historical figure so as to implant into the process of telling his story through the medium of tourism a sensitive interpretation and meaning of historical events in which he was involved. The claim will be that input from the letters could be used to conjure a particular 'spirit' of and 'personality' for any place at which significant events occurred, and in so doing give to that place a special identity, its genius loci. It is a process of enrichment and value-adding. Using a story embedded in the history of 19th-century Australia, the study uses a custom-built form of narrative analysis to disaggregate six selected personal letters authored by the selected historical figure and to interpret their potential for enriching the telling of his story through tourism.

The castle of If is one of the most popular tourist attractions in Marseille. This 16th-century fortress was a prison from 1540 until the First World War. Tourists from all over the world are drawn by the presence of a prisoner who has never been there because he never existed: Edmond Dantès, the Count of Monte-Cristo. This chapter analyses how this castle became a confluence space between fiction and reality and a venue for various film adaptations of Dumas' novel. It also shows how this romantic pamphleteer managed to overcome the paradox of giving authenticity to a tourist spot without paying any attention to historical facts or verisimilitude. The chapter draws on the concepts of literary and dark tourism, the concept of aura linked to authenticity in the work of art, and the concept of literary tourist.

Turkish folk narratives formed around the Gallipoli Campaign, which reflect the mental, psychological, and cultural attitude of Turks towards this war and hold an important place in Turkish folklore, also

literary heritage. The authors present an evaluation tool (validated by experts), whose aim is to analyze the different dimensions and elements that should be incorporated in digital maps. The intention is to enable the analysis of existing digital literary tourism maps and to encourage the future use of the many options offered by digital cartography in maps of this nature.

This chapter aims to analyze the novels Mrs, Dalloway (1925) and Quarenta Dias (2014) in the perspective of elucidating the view of the woman writer-character-traveller on the city, showing continuities and ruptures between the modern city and the contemporary city. Therefore, three paths of analysis are proposed: (1) the understanding of urban territories as a way of elaborating subjectivities and experiences; (2) the link between city and memory, place and identity; (3) the link between city and memory, place, identity, and gender. It was possible to verify that both in Mrs. Dalloway and in Forty Days, women have a central role in the construction of narratives about the city and that this is placed in a centrality-character.

This chapter analyses two travel narratives within the scope of literature and tourism studies, aiming to explore the motivations to undertake journeying and the experience of (literary) pilgrims. The first is the novel Flights, by Olga Tokarczuk (2007), and the second is "How the Mind Works," by Patti Smith (2017). This chapter defines the umbrella concept of "tourist literature" and takes a cross-disciplinary perspective combining the hermeneutics process with findings from the literature review on tourism studies. The analysis of Flights reveals the touring identity and experience of a pilgrim and reflections about airports, travel guides, tourists, and their syndromes. The analysis of Patti Smith's short story uncovers the touring identity and experience of a literary pilgrim who is strongly motivated to undertake literary-inspired trips towards the authors' places.

By focusing on television film-induced tourists, this chapter will contribute to a better understanding of tourist behaviour in relation to motivations for travel to filming locations. The chapter combines, analyses, and critiques the main debates raised by key authors in relation to identifying the motivational factors that prompt site-specific film tourism as well as providing contributions from this author's 2016 research on television film-induced tourism in Ireland. The chapter is divided into the following sections: a brief history of film and television and a review and discussion on film tourism, a breakdown of the different categories of film tourist and a summation of the various motivations for television induced film tourism, followed by an insight into the concept of authenticity in film tourism studies and the use

analyses interviews with local travel companies and information on their social media platforms. The clear aim of the research is to find out how a successful film may affect the tourism of a location. Using the results of the research, the study discusses how a site previously associated with disaster tourism utilizes the opportunities of film tourism, as well as the opportunities provided by a possible change of profile.

Avid readers find in literature real or imaginary scenarios, which are written in such a way that make them visit the portrayed places. The same goes for films, or documentary series, where the narrator plays a fundamental role in captivating the viewer. Both means are inductors of cultural and/or nature tourism, regardless of everyone's personal motivations. The transmission of values is important and varies over time. In the past some marine animals were perceived negatively by humans, which has faded over time. Ecotourism with activities related to the sea is a relatively recent tourism niche. Nature films or series are often boosters of impulses in tourists for the search for natural, unexplored places or places with little human pressure. Some tourist activities in the nature tourism line developed all over the world—from this induction—such as the cases of diving and observation of marine animals.

Preface

Literature, cinema and television series are fascinating creative expressions and media through which art exists. At the same time, they meet crucial human needs by translating profound and abstract concepts and thoughts that help make meaning of life and the world and encourage much needed leisurely moments and imagination. Although the influence of literature on cinema is more evident than the influence of cinematographic work on literature, as literature is by far ancienter than cinema and many novels have been adapted into screenplays, there is a mutual influence between these two media as they share narrative and storytelling (Alqadi, 2015). They both motivate people to enter a creative universe where interpretation becomes an impulse through mental or visual perception.

In the scope of tourism, these two art forms play an essential role in inducing people to travel to places they have read about on the pages of a book or seen on the screen, to travel to destinations where the author lived, worked or died, where the director shot the films, and to places where literary art and motion pictures are celebrated. In this regard, art is communication as it allows people from diverse backgrounds to communicate via stories, images, and sounds. Both these art forms – literature and cinema – share the potential for inspiring new forms of representations of places, for giving meaning to places and although *representation* is the keyword when we talk about art, literary cartography and cinematic cartography illustrate how literature and cinema can function as maps and transform those implicit and explicit representations of places into significant places for tourists and, thus, into attractive tourist destinations.

The influence of art, together with a transformation of the traditional forms of tourism, which occurred in the last decades of the 20th century, propelled tourist destinations to develop forms of special interest tourism to meet the needs of visitors, but also to present the "micro" as a solution to the "macro" issue (Liu, 2003). Two of these niches are literary tourism and film-induced tourism, both falling under the umbrella of cultural tourism and fast-growing and commercially significant phenomena (Beeton, 2016; Watson, 2006). The former refers to travelling in the footsteps of authors and the maps created by novels (Squire, 1994), and the latter serves to describe those tourism visits made to a destination or attraction resulting from its featuring in cinema films and television programmes (Beeton, 2016).

Tourist destinations have also identified both niches as strategies to take advantage of their cultural assets to promote the uniqueness of tourist experiences and sustainable development. In this context, literature and cinematographic art have become highly relevant attracting factors and can provide a sense of place to visitors, even before they visit the destinations.

Global Perspectives on Literary Tourism and Film-Induced Tourism assembles contributions from an international body of scholars interested in the multi and interdisciplinary study of literary tourism and film-induced tourism and provides theoretical and practical insights on these two niches of cultural tourism that have expanded significantly over the last forty years. Thus, the purpose of this book is to contribute to the literature on literary and film-induced tourism by being a platform for critical discussion, examining concepts, trends and methods, and documenting the advances in research on the intersections of literature, film, and the act of travelling. Besides, both literary texts and films can convey plenty of information about the act of travelling, tourism practices, tourists and travellers, across the world and times, and this volume also offers these contributions.

Encompassing a wide range of topics from literary museums, literary tours, motivations of film-induced tourists and literary tourists, women travellers, film tourism destinations, literary places to digital literary tourism, this book covers literature and film as drivers of sustainable tourism development, cultural identity, preservation of heritage, community relations and development, literary and cultural literacy, and destination marketing.

The first chapter of this collection is "Reading Between the Scenes: Cinematic Representations of Literary Tourism," and the choice of opening the volume with this text was motivated by the fact that the authors bring together films, literature, and tourism. Departing from the assumption that cinema always could represent social structures and movements and provide valuable sources of information about societal phenomena, this chapter employs representation as a research approach to offer contributions to understanding realities "outside the film" regarding literary tourists' motivations, experiences and literary places. To this end, the authors analyse literary-inspired tourism experiences depicted in two sequences of two feature films: *The leisure seeker* (2017, dir. Paolo Virzi) and *Paterson* (2016, dir. Jim Jarmusch). The analysis and interpretation of the cinematic sequences reveal two opposing ideas of a literary tourism experience: a meaningful and authentic event or a shallow, disappointing and inauthentic experience.

In the following chapter, "Authenticity in Literary Tourism Places: The Case of 'Asiyan Museum'," researcher Samet Çevik focuses on the concepts of literary tourists' expectations, author-related literary place and authenticity, after the analysis of the first literary museum in Turkey: the Asiyan Museum. This museum was the home of the Turkish poet Tevfik Fikret between 1906-1915, and on its first floor, visitors may see the bedroom, along with other rooms, where the author's objects are exhibited. The plans, architectural drawings and decorations of the house also belong entirely to him. The museum adopted a research-based approach and made identical replicas using resources such as photographs and books to aim for greater authenticity. Thus, the cultural heritage value and this literary place have increased the sense of authenticity among its visitors. This type of literary place is an "astonishingly persistent" place of memory and one of the first in the history of literary tourism (Young, 2015, p. 229), but new ones continue to emerge every year.

As such, in the third chapter, authors Ozan Atsiz e Selman Temiz study the international visitors' experience in another literary place: the Museum of Innocence, which celebrates a book by Orhan Pamuk. This qualitative study reveals four main themes in the visitors' experience in this literary museum: visualisation, nostalgia, awe and memorable. The first is the motivation to visit a site to gather information about the setting of the novel and its characters; the second shows a wish to return to the past when they first read the book and met the characters, but also to the past of the city – Istanbul – and see how people lived and behaved. The third theme is linked to the amazement of seeing the collection of objects mentioned in the novel, and the fourth brings together the whole experience as most visitors are sure they will never forget this literary-inspired visit.

The chapter "The Jerilderie Letter and Others: Using Personal Letters to Enrich Heritage-Based Tourism" is authored by Michael Fagence and researches the opportunities that lie with drawing on the personal letters of an iconic historical figure – Ned Kelly – to tell his story through the medium of tourism. The author intends to show that the result of the analysis of these letters may be used to conjure a particular "spirit" of and "personality" for any place at which significant events occurred, giving a unique identity to that place: its *genius loci*. It is a process of enrichment and value-adding. Using a story embedded in the history of nineteenth-century Australia, the study uses a custom-built form of narrative analysis to disaggregate six selected personal letters and to interpret their potential for enriching the telling of his story through tourism.

The two following chapters focus on dark literary tourism: one of the micro-niches of literary tourism. Departing from literary cartography as a means by which a literary text functions as a map, guiding the visitors and helping them making sense of what they see, Adriana Coelho-Florent focuses on *The Count of Monte-Cristo* by Alexandre Dumas and the castle of If, one of the most popular tourist attractions in Marseille. The castle of If is an in-between place, a literary place where fiction and reality meet, and a cinematic place as it was the setting for numerous adaptations of the famous novel by Alexandre Dumas. The chapter also discusses how this romantic pamphleteer overcame the paradox of giving authenticity to a tourist spot without paying any attention to historical facts or verisimilitude. As such, the author also addresses the concept of aura linked to authenticity in the work of art.

"Warfare, Oral Tradition, and Tourism: Valorisation of the Folk Narratives About the Gallipoli Campaign" is written by Erol Gülüm, and it also explores the topic of dark literary tourism, associated with battlefield tourism, focusing on the Turkish folk narratives on the Gallipoli Campaign, and how they illustrate the mental, psychological, and cultural attitude of Turks towards this event. The author aims to discuss authenticity, creativity and innovation in tourist attractions through the valorisation and re-enactment of intangible war heritage via the example of folk narratives that may effectively convey the mystical, supernatural, and miraculous events believed to have taken place in this war.

Chapters 7 and 8 focus on literary tourism products and experiences. The first is by Zeliha Öztürk, and it is entitled "The Literary Geography of Istanbul and the Ahmet Hamdi Tanpinar in the Framework of the Narration and Tourism Relationship." Departing from the assumption that writers and literary works play a vital role in constructing the identity of a city, the author understands that literary tourism impacts not only the tourism industry but also artistic activities. To this end, the chapter focuses on the experience of literary tours organised by the Ahmet Hamdi Tanpinar Research and Application Centre to celebrate the iconic Turkish novel *Huzur* and its author: Ahmet Hamdi Tanpinar (1949). The second chapter, "Porto, Gaia, and the Sant'Ana Arch: Places in Almeida Garrett's Literary Work," co-written by Isabel Vaz de Freitas e Helena Albuquerque, suggests a new literary tourism product and experience: a tour inspired on the *O Arco de Sant'Ana*, a novel by Almeida Garrett, one the most important 19th-century Portuguese writers. Following a geographical information system as a methodological tool, the authors of the chapter conducted a literary cartographic analysis, identifying places, tangible and intangible heritage mentioned in the historical novel, intending to create a literary itinerary that promotes a journey to medieval times and a journey into the writer's literary work.

"Evaluating the Use of Digital Cartography to Showcase the Intangible Heritage: The Case of Literary Tourism" is the ninth chapter in this volume, and Jordi Arcos-Pumarola and colleagues co-write it. In it, the authors underline the idea that new narratives are essential for ensuring the sustainability of tourist destinations and improving the tourist experience. To this effect, destinations can draw on a critical resource: intangible heritage, which digital cartography can help visitors interpret. The aim of this chapter is to analyse the opportunities digital cartography offers for the use of literary heritage. The authors present an evaluation tool that analyses the different dimensions and elements that may be incorporated in digital maps. The study outcomes enable the analysis of existing digital literary tourism maps and encourage the future use of digital cartography's various options for tourism purposes.

Cities are one of the elements of interaction between tourism and culture. The following chapter focuses on the concept of the city and women writers-characters-travellers. Drawing from the analysis of two literary texts: *Mrs Dalloway* (1925), by Virginia Woolf (1925) and *Quarenta Dias*, by Maria Valéria Rezende (2014), the chapter reveals that women writers-characters-travellers play a crucial role in the construction of narratives about the city. Therefore, the analysis is threefold: (a) understanding of urban territories as a way of elaborating subjectivities and experiences; (b) a link between city and memory, place, and identity; (c) a link between city and memory, place, identity, and gender.

Chapter 11 also examines two travel narratives by women writers-characters-travellers. The chapter is written in the scope of literature and tourism studies, aiming to explore the motivations to undertake journeying and the experience of (literary) pilgrims. The first is the novel *Flights* by Olga Tokarczuk (2007), and the second is "How the mind works" by Patti Smith (2017). This chapter defines the concept of "tourist literature" and takes a cross-disciplinary perspective combining the hermeneutics process with findings from the literature review on tourism studies. The analysis of the novel *Flights* reveals the touring identity and experience of a pilgrim, and reflections about airports, travel guides, tourists and their syndromes. The analysis of Patti Smith's short story uncovers the touring identity and experience of a literary pilgrim, who is strongly motivated to undertake literary-inspired trips towards the authors' places.

Focusing on television film-induced tourists, the chapter "Could Television Streaming Sites Lead the Charge for Film-Induced Tourism Post COVID? Seeking the 'Sites/Sights' Seen on the Small Screen!" is authored by Jennifer Stewart and contributes to a better understanding of tourists' behaviour and motivations. The chapter analyses studies and critiques on the topic of motivation, identifying the authors who have participated in this debate and the motivational factors that trigger site-specific film tourism. Besides presenting the author's 2016 research on television film-induced tourism, the chapter also addresses the concept of authenticity in film tourism studies and the use of Virtual Reality and Augmented Reality as a means to provide a more immersive experience.

Chapter 13, "Experiencing the Story: The Role of Destination Image in Film-Induced Tourism," by Eleni Michopoulou and colleagues, provides insights into how movies influence the featured destination's image by focusing on specific film tourists' perceptions and motivations and their emotional relation to the movies. The chapter begins by offering a definition of film-induced tourism followed by a typology of film tourists within the context of film fans. Then, factors influencing film tourism destination image are studied: destination marketing activities, film-specific factors, and destination attributes. Two case studies are provided to better understand the findings from the literature review.

The following chapter is co-authored by Árpád Ferenc Papp-Váry and Timea Zsófia Tóth and is entitled "Analysis of Budapest as a Film Tourism Destination." The researchers claim that the Hungarian film industry has undergone a considerable change recently, becoming a pull sector and repositioning the country in Europe. The study seeks to highlight factors that can boost film tourism in Budapest and make the city even more prominent on the film map of Europe. The city's diversity is clearly outlined, as it primarily serves as a backdrop for various film productions. Using primary and secondary sources, this research focuses on the resources of the city for film tourism and the changes in the economic environment affecting the film industry. The analysis explores different types of film-induced tourism through interviews with the performers of this niche of cultural tourism.

Chapter 15 analyses another destination of film tourism: Chernobyl, which after the HBO mini-series registered a record number of tourists. In this chapter, Timea Zsófia Tóth and Árpád Ferenc Papp-Váry offer a complex tourism analysis of this destination, after the examination of the demand and supply sides of local film tourism. The study analyses interviews with local travel companies and information on their social media platforms to discuss how a site, previously associated with disaster tourism, utilises the opportunities of film-induced tourism, as well as the opportunities provided by a possible change of profile.

The last chapter, "Marine Life Features and Their Transmissibility Through Broadcasting: Tourism Capitalisation," by Jorge Ramos, reflects on how nature films or series are often boosters of tourists' impulses for searching natural, unexplored places or places with little human pressure. These films and documentary series, in which narrators play a fundamental role in captivating the viewer, are inductors of cultural and nature tourism and promote the creation of several tourism experiences, such as diving and marine animals' observation.

Global Perspectives on Literary Tourism and Film-Induced Tourism strengthens the bridge between literary tourism studies and film-induced research, as they share methodological and theoretical principles and deal with imaginary cartography that impulses people to travel, and it contributes a varied and inspiring collection of chapters to travel agents, tourism agencies, tour operators, government officials, postgraduate students, researchers, academicians, cultural development councils and associations, and policymakers.

We are very grateful to all the researchers who participate in this volume and Katie McLoughlin, the Assistant Development Editor at IGI Global, for all the help in the process.

Rita Baleiro
University of the Algarve, Portugal & CiTUR, Portugal

Rosária Pereira
University of the Algarve, Portugal & CinTurs, Portugal

REFERENCES

Alqadi, K. (2015). Literature and cinema. *International Journal of Language and Literature*, *3*(1), 42–48. doi:10.15640/ijll.v3n1a6

Beeton, S. (2016). *Film-induced tourism* (2nd ed.). Channel View Publications. doi:10.21832/9781845415853

Liu, Z. (2003). Sustainable tourism development: A critique. *Journal of Sustainable Tourism*, *11*(6), 459–475. doi:10.1080/09669580308667216

Squire, S. J. (1994). The cultural values of literary tourism. *Annals of Tourism Research*, *21*(1), 103–120. doi:10.1016/0160-7383(94)90007-8

Watson, N. J. (2006). *The literary tourist. Readers and places in Romantic and Victorian Britain*. Palgrave.

Young, L. (2015). Literature, museums, and national identity; or why are there so many writers' house museums in Britain? *Museum History Journal*, *8*(2), 229–246. doi:10.1179/1936981615Z.00000000052

Chapter 1

Reading Between the Scenes: Cinematic Representations of Literary Tourism

Rita Baleiro
https://orcid.org/0000-0002-3188-5150
University of the Algarve, Portugal & CiTUR, Portugal

Rosária Pereira
https://orcid.org/0000-0002-2335-5694
University of the Algarve, Portugal & CinTurs, Portugal

ABSTRACT

Departing from the assumption that cinema has always had the capacity to represent social structures and movements and provide valuable sources of information about societal phenomena, this chapter employs representation as a research approach to offer contributions to understanding realities "outside the film" regarding literary tourists' motivations, experiences, and literary places. The authors analyse cinematic representations of literary tourism in feature films and take the perspective of literary tourism studies, reviewing the literature on cultural tourism, special interest tourism, niche tourism, literary tourism, and literary sites and landscape. The analysis and interpretation of the cinematic sequences reveal two opposed ideas of what literary tourism experiences might be: a shallow, disappointing, and inauthentic experience or a meaningful and authentic event.

INTRODUCTION

At the end of the 20th century, the traditional forms of tourism underwent a transformation in the sense that they expanded by the introduction of new postmodern tourism forms, bringing innovative offers to the marketplace. This transformation had multiple causes, but one of the most impactful was the acknowledgement that the growth of mass tourism after the 1950s did more than favouring economic benefits. It has also led to new and severe environmental degradation, adverse social and cultural impacts, and uneven distribution of benefits. Therefore, in the 1980s, the concept of sustainable development emerged,

DOI: 10.4018/978-1-7998-8262-6.ch001

and destinations started to develop forms of special interest tourism, presenting the "micro" as a solution to the "macro" issue (Liu, 2003). One of these new forms was literary tourism, which has grown into a commercially significant phenomenon (Watson, 2006). This fast-growing niche of cultural tourism refers to travelling to places connected to a literary text or writer (Baleiro & Quinteiro, 2018, p. 31).

This chapter complements the authors' previous studies where literary fiction and narrative interpretation were a means to understanding and providing new perspectives on travel and tourism motivations and experiences and aims to analyse literary-inspired tourism experiences depicted in two sequences of two feature films: The *leisure seeker* (2017, dir. Paolo Virzi) and *Paterson* (2016, dir. Jim Jarmusch). Two films that have not yet been examined in relation to tourism.

As far as the authors know, no previous studies focus on the cinematographic representations of literary tourism. Hence, this chapter does not draw on similar background research. Instead, this chapter brings forward an innovative approach to contribute information on the representation of literary touring and literary tourists. In studies such as this one, one of the main challenges is finding films featuring literary tourists. After thorough research on the International Movie Database, via analysis of keywords and the authors' personal history of film viewing, two feature films were selected.

The assumption underpinning this work on literary tourism and cinema is that films are deeply embedded in other social practices. As such, they help us make sense of and communicate about the world (Sampaio, 2014, p. 185). Thus, this chapter wishes to contribute to discussing types of literary sites and landscapes, literary tourists' motivations and experiences, and literary tourism products. To that end, the authors start by focusing on the concepts of cultural tourism, special interest tourism and niche tourism, then on the definition of literary tourism, and, lastly, on the content analysis of the two cinematic sequences in the scope of literary tourism.

THE CONCEPTS OF CULTURAL TOURISM, NICHE TOURISM AND LITERARY TOURISM

Cultural Tourism

From the first touring experiences, tourism and culture go hand in hand, and cultural tourism definitions are now almost as multiple and diverse as the definitions of culture (Pereiro, 2009, pp. 108-120). Hence, each research study operationalises the concept to fit their research goals. Some include the critical notion that this form of touring often complements several recreation or business travels (McIntosh et al., 1995). Others allow technical comparative statistical surveys, such as "cultural tourism is all movements of persons to specific cultural attractions, such as heritage sites, artistic and cultural manifestations, arts and drama outside their normal place of residence" (Richards, 2001, p. 37). Still, others define cultural tourism as a conceptual umbrella of a group of activities that contribute to differentiation in a time when capitalist marketing instruments have segmented the tourism market for mercantilist purposes (Pereiro, 2009, p. 111). In this perspective, cultural tourism is a synonym to the commodification process of culture and, as such, a commercial label to sell culture and cultural heritage as merchandise.

In this chapter, the authors depart from the UNWTO (2017) definition, according to which cultural tourism is driven by the travellers' primary motivation "to learn, discover, experience and consume the tangible and intangible cultural attractions/products". In this view, cultural tourism stems from intrinsic human motivations and features, *i.e.*, nostalgia, curiosity, escapism, search for happiness, search for the sacred, leisure, knowledge and information, and the wish to understand the "other". Such tourists' motivations have had different emphasis throughout the history of tourism (Richards, 1996, p. 264), and, overall, they are defined as an intricate system of biological and cultural influences which steers travel choices, behaviour, and experience (Pearce et al., 1998). To meet the tourists' motivations, the tourism industry has produced cultural tourism experiences and products and commodified culture. Therefore, the end of the 20th century witnessed increased segmentation and the creation of several tourism niches tailored to specific market demands (*e.g.*, dance tourism, photographic tourism, wine tourism, music tourism, food tourism). The following section of the chapter addresses this issue, focusing on the concepts of special interest tourism and niche tourism.

Special Interest Tourism and Niche Tourism

The concept of "niche" was introduced by zoologist G. Evelyn Hutchinson in 1957, suggesting an optimum location that an entity may exploit against its competitors (Qiao et al., 2016, p. 805). At that time, as it is now, niche meant "small" and "specialised", but it still had not crossed the borders of zoology studies. In 1992, marketeers Keegan *et al.* presented the concept as a smaller market not yet served by competing products; and slowly but steadily, the term 'niche' was embraced by the business literature. However, according to Ali-Knight (2011), in the late 1980s, the tourism industry was still using the term "special interest tourism" as the umbrella term for an alternative to mass tourism, which catered for niche markets, motivated by special interests and needs of individuals and groups. At that time, creating products and experiences which targeted a specific audience with specific motivations was a "prime force associated with the expansion of tourism and the motivation around which tourist activity was planned and advanced." (Read, 1980 cited in Hall & Weiler, 1992, p. 5). This has set the move from mainstream tourism offer to one that was more specialised and unique, and the terms "special interest tourism" and "niche tourism" became synonyms referring to emergent tourist markets that derive from the needs of clusters of tourists and from the needs of the tourism industry which aims at diversifying its product base (Derrett, 2001). At present, the concept is widely used "by the producers of tourism, rather than the consumers" (Robinson & Novelli, 2005, p. 5), but more and more consumers are aware of the existence of these niches which cater for their interests. One such niche is literary tourism, and the next section in this chapter will address this concept.

Literary Tourism

Literary tourism is a niche of cultural tourism (Robinson & Andersen, 2002), and although people have been travelling 'under the influence of a favourite author, book or poem' (MacLeod et al., 2018, p. 388) for the last five hundred years, only in the last three decades, research on literary tourism has flourished, and the tourism industry has started to cater for literary tourists. It is an example of a niche that drew scholars' and tourism industry interest much later than the tourists' motivation to undertake literary-inspired trips.

Literary tourism results "from an interest in the personal life histories of writers or their works of literature" (Agarwal & Shaw, 2017, p. 16), and it is an expression of the material search for places and environments that have been significant in the lives of authors (from the beginning to the end of their lifetimes) and the construction of literary imaginaries. This type of touring is primarily motivated by the consumption of literary sites, events, or products, and contrary to most special interest forms that have emerged in reaction to mass tourism (Richards, 1996), its genesis dates to the 15th and 16th centuries, when literary pilgrims visited the Canterbury Cathedral after the reading of William Chaucer's *The Canterbury Tales* (Westover, 2012) and writers' houses, such as that of Francesco Petrarca (Moryson, 1617, pp. 373-375). Visitors were motivated by the admiration for the author, seeking and celebrating literary creativity in the places where the source of the genius was believed to be found (Hendrix, 2008, p. 23).

Over the following three hundred years (*i.e.*, up to the nineteenth century), the motivations to undertake such trips expanded, and although some persisted in the act of travelling to writers' places (namely their homes and graves), visitors began to visit the places depicted in the literary texts (Hendrix, 2014, p. 27). To this end, the rise of the popular novel played a key role (Thomas, 2012), especially in the European setting (Westover, 2012), as it propelled the fascination for the places of fictional characters, along with a continuous interest in the places of the authors. In Europe, the experience of the Grand Tour also had an impact in fostering literary tourism (Thomas, 2012; Watson, 2006; Watson, 2009) and the nineteenth century witnessed the consolidation of the attraction for literary places and the preservation and construction of literary destinations (Taylor & Pouliot, 2020), *e.g.*, the Shakespeare Birthplace Museum in Stratford-upon-Avon and the Lake District, in England, the Robert Burns Birthplace Museum, in Scotland, and the home of Harriet Beecher Stowe, Florida, in the United States of America.

Recently, this segment of cultural tourism has expanded: in academia, where the number of researchers and case studies is increasingly higher, and in the tourism industry, where projects, campaigns, and programmes such as the UNESCO's cities of literature have had an impact on the number of new literary destinations, products, and experiences. As a consequence, many more literary tourism places – ultimately literary tourism destinations – were created: Verona (Italy), where tourists go to visit Juliet's balcony, one of the main elements in the setting of William Shakespeare's *Romeo and Juliet;* the West Yorkshire Moors (England), the countryside setting of the novels by the Brontë sisters; London through the words of Charles Dickens or Virginia Woolf; Paris where Honoré de Balzac, the founder of French realism, lived and wrote in a five-room garden apartment, in Rue Raynouard; Dublin through James Joyce's life and work; Venice through the words of Thomas Mann and so many other literary destinations.

Such literary destinations were created after the visitors' desire to see the places depicted in the writings of fictions, but also "from the desire to see aspects of the author's real-life", such as the places closely connected with their favourite writers, the buildings where they lived and their artefacts or memorabilia (Agarwal & Shaw, 2017, p. 17). As stated by David Herbert, this type of setting enhances the experiential quality of these contacts. To illustrate this point, Herbert quotes from a guide to former British and Irish writers' homes: "In these, a visitor can still today walk out of a house and into landscapes which have barely changed since the writer drew breath from them and breathed literature into them . . . We walk in our writers' footsteps and see through their eyes when we enter these spaces" (1996, p. 77). Such literary places are sold as a portal through which tourists may access a time before their own time, a portal straight to the lives of their favourite authors and the possibility of walking in the writers' footsteps. Another motivation to undertake literary-inspired visits is that literary places often evoke distant memories and impact self-identity, a finding revealed by Squire's research (1994) on the visitors to Hill Top Farm, in the Lake District, where Beatrix Potter was born. Squire concluded that the experience was about a return to childhood memories and feelings of Englishness, rurality, and seemingly simpler lifestyles for many of the visitors. In this situation, the literary place "acted as a medium through which a range of cultural meanings and values are communicated" (Agarwal & Shaw, 2017, p. 17). Gradually, the tourism industry became aware of the several attractive features of literary places, and currently, there is an array of literary tourism products that include many manufactured literary places, *i.e.* places where the connection to the writers and literature is not constructed via authors' biography or the sites depicted in literary texts as it happens in literary festivals and literary dinners (Baleiro & Quinteiro, 2018).

CINEMATIC REPRESENTATIONS AND TOURISM STUDIES

The attention to feature films in tourism studies has been mostly steered by promotional aims, under the assumption that they may influence how places are perceived, transformed, and experienced, structuring the viewers' sense of reality (Bonarou, 2021). As such, many researchers have studied the concept of film-induced tourism (Beeton, 2005; Buchmann et al, 2010), analysing the impact of this medium on traveller motivation (Macionis, 2004), desire (Griffin, 2002), expectation (Coshall, 2000), behaviour (Schofield, 2000) and influence on destination image (Pike, 2002), destination choice (Laws et al, 2002), and constructing and reinforcing images of people and place (Andsager & Drzewiecka, 2002).

Considering that cinema often adopts tourism and travel as its subject matters, cinematic representations – as cultural productions – may provide insights into the analysis of tourism experiences. As Johannes Ehrat states, representation is one of the "fundamental functions of cinema" (2005, p. 6), and it can be defined and criticised from numerous approaches. Despite those variations, a representation is a portrayal that conveys the director or the writer's message and view and a way of reflecting identities, issues, and events to the audiences. As such, cinematic representations are an opportunity for academic debate to "move from purely abstract analyses to a more active engagement with kinds of representation – modalities of image, forms of speech and address to viewers" (Datta, 2000, p. 71).

Therefore, in this chapter, the authors depart from the analysis of representations of literary tourism to provide insights on the consumption of places and their meaning to tourists.

In this chapter, representation is employed as a research approach (Gupta & Gupta, 2013) which means that this instrument will help the authors access unique empirical findings regarding literary tourism. As such, after reviewing the literature on the core concepts of cultural and literary tourism, the authors proceed to the examination of two sequences in the selected feature films focusing on the content analysis of images and dialogues to build the links between the reviewed concepts and the two portrayals of literary tourism experiences, motivations and sites. In the context of this analysis, authors will move "outside the text" to reflect on the current trends in this tourism niche, as cinema can act as a mediator of social realities and collective concerns, and it has the potential to redirect the cultural and material fabric of life (Jain & Rai, 2002).

In the following sections, the authors present the analysis and interpretation of these two filmic sequences in the scope of literary tourism studies.

The Leisure Seeker (2017): Literary Tourism, Writers' Homes and Expectations

The leisure seeker was directed by Paolo Virzi and released in 2017. The main characters are John and Ella Spencer (performed by Donald Sutherland and Helen Mirren), a couple in their seventies, who decide to embark on one last road trip to visit the Ernest Hemingway Home and Museum, in Key West, after learning about their fatal illness's diagnosis. Ella is the driving force of this literary-inspired tour, although the literary expert is John, who has always wanted to visit Hemingway's home. As such, they rent a motorhome and drive more than 2000 km from Boston to Florida. The arrival to this destination coincides with the final sequences of the film.

Writers' homes are "the apogee of literary tourism sites" (Watson, 2006, p. 90), and one of the first products and experiences of literary tourism, since the first recorded trips to Petrarca's home (Hendrix, 2007, pp. 15-30). Most of the writers' residences "developed into museums at the end of the eighteenth and beginning of nineteenth centuries, when houses inhabited by Walpole, Rosseau and Petrarch, as well as Canova's birthplace, were turned into tourist destinations" (Hendrix, 2020, p. 8). The trend expanded in the twenty-first century, and currently, there are more than four hundred authorial homes in France (Hendrix, 2014, p. 27), fifty in the United Kingdom and seventy in the United States of America (MacLeod, 2020, p. 2). Writers' homes are heritage sites that satisfy most common literary tourists' motivations: to see the places associated with the revered writer's life (MacLeod et al., 2018); to commemorate the favourite author (MacLeod et al., 2018); to learn about the authors' lives and texts (Herbert, 2001; MacLeod et al, 2018) and to feel close to the author, a sense that is stimulated by the curated real-life items that he or she used.

Apart from being one of the most famous literary tourism products, authors' homes and museums are an example of a "top-down" literary place resulting from an imposition from the tourism industry (Robinson, 2002, p. 13). This means the place was prepared by tourism promoters, mapped and pre-interpreted to the public. In this sense, they are an example of the commodification of literature and authors, a trend which started in the nineteenth century propelled by guides as the likes of William Howitt's *Homes and haunts of the most eminent British poets* (1847), which encouraged readers to visit writers' homes and other literary landscapes (Macleod, 2020). This genre subsisted and expanded, and since then, several books devoted to the topic have been published, *e.g.*, *American writers at home* (2004), by J. D. McClatchy and Erica Lennards, Anne Trubek's *A sceptic's guide to writers' houses* (2010), and Nick Channer's *Writers' houses: Where great books begin* (2015).

In Paolo Virzi's film, when the two main characters get to Hemingway's house, they feel utterly disappointed with the site: the writer's house has been transformed into the worst of the tourism industry. In such a way that the female character apologises to her husband for seeing that place converted into a "party town" (as another visitor says). In this 3-minute-long sequence, the soundtrack features loud instrumental Latin music synchronised to the images of busy tourists taking photographs of Hemingway's bed with mobile phones. Their experience is mediated by the images on their mobile, most likely in an attempt to authenticate it: "I was there, and I saw it with my own eyes (Sontag, 1979, p. 7). As Harald Hendrix underlines (2001, p. 313), not all visitors share a single mind of dedication that impels them to go on literary-inspired trips, paying homage to their favourite authors; instead, many visit such places out of curiosity and general interest, and that is what the viewer witnesses in this cinematic representation of literary tourism.

The expectation is a crucial component in the experience of place (Pocock, 1992), and at the sight of what has become of Hemingway's sanctuary, the female main character laments to her husband, "this is unbelievable…you have always wanted to come here and now look at it…" and she apologises to him. Instead of the place's sacred aura (MacLeod, 2020), they find a wedding party, loud music, images of the six-toed cats and numerous persons in the rooms taking pictures with their mobiles. This part of the film illustrates how tourists' expectations can be dashed. It clearly shows that the character's expectations, based on her mental representation of the place (*i.e.*, image), in turn generated by the knowledge about the author and his work, were neither surpassed nor even met.

Image is a form of expectations facilitator (Grönroos, 1990). In a context with high levels of uncertainty, individuals consider their preconceived images as a signal of the true competencies of the provider (Weigelt & Camerer, 1988). In tourism, destination image is a mental representation of the place that helps individuals anticipate their destination experiences (Jenkins, 1999). Therefore, image affects individuals' expectations of a tourist destination before the visit (Bigné et al., 2001). The more positive the preconceived image of a destination, the higher the tourist expectations. Bosque et al. (2009) concluded that the preconceived image of a tourist destination is the most critical factor generating tourist expectations. As a result, individuals have tremendous confidence in this psychological factor to form their expectations of a future destination experience.

Aware that the film is an artistic simulacrum resulting from the directors' and the screenwriters' choices, a visit to the website of the Ernest Hemingway Home and Museum confirms some of the features portrayed in the film: there are contests for Hemingway's lookalikes, wedding balls, and other venues, and a paraphernalia of souvenirs at the gift shops: key chains, earrings, cufflinks, Christmas ornaments, can openers, cups and bowls, calendars, caps, coupons, t-shirts, among many other objects. Although the current visitors to writers' homes are more autonomous and have departed from the late-eighteenth-century paradigm "when visitors were considered primarily as welcome guests that appreciated and indeed expected being accompanied and instructed during their visit in order to immerse themselves in the intellectual and artistic universe of the illustrious personage commemorated" (Hendrix, 2020, p. 19), they still expect to "gain insight and more intimate knowledge" of the author (Morris, 2017, p. 93) through a combination of biographical details and the recreated home interiors, in the sense that "the home allows the visitor to see what may have remained hidden, getting closer to the subject, or getting to know them" (Morris, 2017, p. 92). In this frame, "the cult of genius and the fetishisation of the process of creativity" are not experienced as they were in the late 1700 and early 1800 (Morris, 2017, p. 91), but research shows that the reason for visiting is both intellectual and emotional (Busby & Shetliffe, 2013).

The representation of the writer's home in *The leisure seeker* depicts a superficial, repetitive and standardised literary tourism product, and that may be seen as a form of heritage commodification or staged authenticity (MacCannell, 1973). Something that is fostered by the tourism promoters but also by the tourists as both parties tend to focus on simplified and marketable experiences, even though they often claim the opposite (Goulding, 2000; Kirshenblatt-Gimblett, 1998). This contradiction can have significant consequences on heritage preservation and tourists' and local communities' behaviours, as it demands a continuous staging of reality. John Urry and Jonas Larsen (1990/2011) describe most tourists as "gazers" who have a superficial relation with the places they visit, swiftly registering the sites "with a camera draped around their neck" (p. 18). Although the concept of authenticity is subjective, varying according to the visitors' expectations and other indicators, in the context of literary tourism, it can be defined as a coincidence between the reality of the place with the reality of the mind, *i.e.*, the expectations the visitors are met (Watson & Saunders, 2004).

In writers' homes, "authenticity" is very much determined by the element "home" and biographical objects, as the place is paramount in achieving closeness (Brown, 2016). As such, most tourists would feel satisfied with a standardised experience and choose to live the experience positively (Neal et al., 1999). However, that is not the case of the female tourist portrayed in *The leisure seeker*: the place's superficial staging destroys all projections of authenticity, and it does not satisfy her psychological or emotional needs for a sense of achievement as it does not meet the quest to be close to a sacred site. As Urry and Larsen (1990/2011, p. 9) put it gazing is a performance that is conditioned not only by circulating, exterior elements but also framed by the tourists' "personal experiences and memories", "resources", and "cultural lenses" that enable them "to see the physical forms and material spaces before their eyes as 'interesting, good or beautiful'". In this cinematic representation, the female tourist's gaze upon the home of Ernest Hemingway is filtered by her expectations and motivations, namely nostalgia: a common internal drive in most literary tourists (Gentile & Brown, 2015). Nostalgia is linked to emotional involvement with the past to escape a less meaningful present (Schofield, 1996, p. 335), a yearning for the past when the couple was physically and mentally healthier, and the husband studied Hemingway's work. This literary-inspired visit to their favourite author's house could potentially conjure all that but instead resulted in disappointment.

The following section focuses on an episode of *Paterson*, directed by Jim Jarmusch.

Paterson (2016): Literary Tourism and Literary Landscape

This film, directed by Jim Jarmusch, is about a week in the life of a young poet and bus driver, Paterson (performed by Adam Driver), who lives in Paterson, New Jersey, and keeps his poems in a notebook he only shares with his wife. In Chris Barsanti's (2017) words, "the film is a poignant portrait of the mundane, a singing symphony of the everyday", and, in this chapter, the authors will focus on one of its final sequences in which there is a representation of a different type of literary place, of literary tourist and, hence, of a diverse literary tourism experience, in comparison to *The leisure seeker*.

This 6-minute-long sequence thematises a representation of literary tourism through the presence of a Japanese literary tourist the bus driver meets in a park. The bus driver is sitting on a park bench, gazing at the river waterfall – the Possaic Falls – depicted in William Carlos Williams' *Paterson* (an eighty-five-line long poem set in the town of Paterson, see Conarroe, 1970), when the Japanese tourist arrives, observes the waterfall, asks the bus driver permission to sit, takes a copy of the Japanese translation of William Carlos Williams' poem from his bag and flips through the book. Despite his broken English, the tourist (played by Masatoshi Nagase) quietly starts a conversation with Paterson about poetry and poets, namely, William Carlos Williams and Allen Ginsberg, who was also born in Paterson. On several occasions, they just stay silently gazing at the waterfall.

The waterfall is the literary place, and the city of Paterson is the destination of this tourist's literary-inspired trip. Contrary to the writers' home museums, which are an example of literary places created in a "top-down" movement, the waterfall, as a literary place, results from a "down-top" movement, as it is actively created by the tourist who is knowledgeable about the text and author and, therefore, is capable of identifying the literary "marker" (Müller, 2006). This is also a process of "visual consumption" of places, as John Urry has pointed out, and it implies the recognition, and subsequent decoding, of signs (1992, p.172); a process which turns tourist places and situations into texts and tourists into 'semioticians' (Culler, 1981). Visitors, such as the Japanese tourist character, can produce the literary landscape that derives from the intersection between a location, a writer or text and the tourist's gaze and interpretation of what he observes. Under these circumstances, it is up to the subject to recognise, decode and construct the (literary) landscape, and that process goes beyond perception; it involves active interaction with the space.

Literary landscape, just like any other landscape, refers to a multifaceted product, both an object and a representation, whose production is framed and conditioned by the presence of a subject, the range of his gaze, his knowledge, his emotions and his physicality (Knudsen et al., 2015; Wylie, 2007). In fact, contrary to nature, which exists independently of the human being (von Maltzahn, 1994), the landscape is a product that results from the subject's gaze while interacting as a mobile observer with the physical space. This interaction results in a re-signification process of what is observed. In this regard, the authors recall the metaphor of the landscape as a palimpsest (Ascherson, 2002, p. 193), according to which the landscape corresponds to a succession of meanings attributed to it in terms of human perception and experience. This explains Helena Buescu's statement underlining that landscape is above all a construct that results from a perspective (1990, pp. 190-193), which depends on the hierarchy of the elements the subject chooses to select in a given space. Under such circumstances, the tourist's gaze acts as a factory of places, an attitude that constitutes more than just perception, but action, as asserted before.

In Jim Jarmusch's film sequence, that hierarchy of the elements is summited by the literary motivation and invisible literary baggage of the Japanese tourist, who creates a literary landscape in his interaction with space. In this case, the tourist recognises, decodes and constructs a "down-top" literary place, as mentioned before, meaning that that space has not yet been appropriated or manufactured by the tourism industry, nor has it been pre-interpreted for the visitors. As a literary place, that waterfall is produced by his literary knowledgeable gaze, which allows him to identify literary places, even though they are not signalled with tangible icons.

Considering this tourist's devotion towards the literary, the authors of the chapter believe the Japanese tourist may be the representation of a literary pilgrim: someone "whose eyes are already prepared to see" (Pocock, 1992, p. 240), to read the signs and re-signify the landscape without the intervention of the tourism industry (Santesso, 2004). In David Herbert's view, literary pilgrims are the first category of literary travellers (1996, p. 84), and according to Harald Hendrix (2014, p. 27), they travel with the motivation to "celebrate the creativity of authors in places where their genius is believed to be located, first at their graves, but then also in the places where they were born and lived.". In the character's own words, the only driving force that made him travel to Paterson was the wish to visit the town where William Carlos Williams and Allen Ginsberg were born, and to that end, he takes William Carlos Williams' book, which acted as a map. An observation that recalls Robert Tally Jr.'s statement: "a map can also be made of words" (2013, p. 46).

In *Paterson,* the map that guides this literary pilgrim's gaze, steps and imagination is William Carlos Williams's literary cartography, as well as Allan Ginsberg's biography. An experience that recalls Barbara Kirshenblatt-Gimblett's concept of "ghost landscape" (1998, pp. 169, 171), *i.e.*, a construct based on virtual (literary) associations. This expectation of the experience of virtual places (made possible by the re-signification of the landscape) allows tourism to create an infinite number of virtual places, because as Meinig puts it, "the landscape is composed not only of what we see but also for what is hidden in our minds" ([1976] 2002, p. 35. Our translation). This pilgrim's literary-inspired visit is motivated by his literary knowledge repertoire, devotion and curiosity, all of which enable him to identify and construct the literary place, all in a quiet process, without the "noise" of the tourism industry (with its signs, guides, leaflets and maps filled with points of interest). At a given point of the sequence, the bus driver enquires the Japanese tourist about why he decided to visit Paterson, and the answer is that he wished "to see the city of your interesting poet who lived and wrote his poems here in Paterson, New Jersey."

In this sequence, the spectators may also witness the illustration of an argument presented by the French philosopher Michel Onfray, in *Théorie du voyage: Poétique de la géographie*, in which he has stated that the desire to travel feeds on literary ghosts, and a mere line by an author may play a stronger role in creating the desire to visit a place than photographs or films (2007, p. 26. Our translation). As such, this sequence in *Paterson* represents the driving force of literary art as a road map to travels and a tourist experience motivated and mediated by poems and poets' biography.

FINAL CONSIDERATIONS AND THE FUTURE OF LITERARY TOURISM

Film stories are woven into a space and time, and the nature of film as a representational medium may provide insights for a reflection about literary tourism experiences and products. As stated previously, representation as one the most reliable research approaches for cinema or media studies may offer valuable contributions to understanding realities "outside the text". In a time when the tourist experience is often the experience of a world manufactured, transformed and projected through the travel industry, mass and social media, these two cinematic sequences leave the spectator with two opposed ideas of what literary tourism experiences might be: a shallow, disappointing and inauthentic experience (as showcased in *The leisure seeker*), or a meaningful and authentic event (portrayed in *Paterson*). Although in both tourist representations, the motivation is similar, as all characters share an admiration of the author, the experiences are opposite.

In *The leisure seeker,* the literary tourism experience and product is depicted from the visitor's perspective in a way that reveals disappointment, and that can spark a conversation about the excesses and the over dominating business component in this activity which sometimes overshadows the reason why most of these museums were created in the first place: sites where visitors could feel the presence of the author, its genius, and special aura (Hendrix, 2020, p. 8). As such, in the same way a literary route should be more than a list of stopping points, with chunks of information attached, so writers' houses must be more than a party venue where the loud noise outshines the author's spirit. As mentioned before, cultural tourism and literary tourism derive from human motivations, and in the case of writers' homes, a common motivation is to access information that contributes to the understanding of the work (Santesso, 2004).

In *Paterson*, the literary tourism experience is more authentic because the tourism industry does not mediate it. The whole experience is lived more genuinely, following a solid motivation to be at the places the author (William Carlos Williams) has depicted in his texts and where the authors (William Carlos Williams and Allen Ginsberg) have lived. There is no disappointment but contemplation; there is no loud music but silence.

Regarding the future of literary tourism, the authors share Harrison and Sharpley's belief (2017, p. 6) that it is unlikely that international mass tourism will end. However, shortly, tourism trends might change, and in that context, niche and more sustainable forms of tourism, such as literary tourism, will be more valuable. The authors of this chapter do not have the answer for what will happen in the future. However, they believe literary tourism will continue to expand as there will always be tourists motivated to travel in the footsteps of authors and the maps created by novels (Squire, 1994), as art is a powerful source to make meaning of life and meaningful experiences are the epitome of travelling, in particular, and life, in general. Besides, literary tourism has the potential to develop rural communities (Yiannakis & Davies, 2012); to foster community-based tourism, to preserve the memory of places, to motivate reeditions of literary texts, to diminish the impacts of seasonality (Farmaki et al., 2016, p. 37); to attract tourists who look for more authentic and ethical experience (Richards, 2017, pp. 173-174), and a key role in favouring a "new" literacy (Robinson, 2002/2011, p. 73) in a time when the rise of digital media and overdose of stimuli and the acceleration of life (see Han, 2013; Rennaissance, 2019) may change literary reading habits.

REFERENCES

Agarwal, S., & Shaw, G. (2017). *Heritage, screen and literary tourism*. Channel View Publications.

Ali-Knight, J. (2011). *The role of niche tourism products in destination development* (PhD Thesis). Edinburgh Napier University.

Andsager, J. L., & Drzewiecka, J. A. (2002). Desirability of differences in destinations. *Annals of Tourism Research*, *29*(2), 401–421. doi:10.1016/S0160-7383(01)00064-0

Ascherson, N. (2002). *Stone voices: The search for Scotland*. Hill and Wang.

Baleiro, R., & Quinteiro, S. (2018). *Key concepts in literature and tourism studies*. Faculty of Arts and Humanities, University of Lisbon, Centre for Comparative Studies.

Barsanti, C. (2017). *Film Review: Paterson*. http://fj.webedia.us/reviews/film-review-paterson

Beeton, S. (2005). *Film-induced tourism*. Channel View Publications. doi:10.21832/9781845410162

Bigné, J. E., Sánchez, M. I., & Sánchez, J. (2001). Tourism image, evaluation variables and after purchase behaviour: Inter-relationship. *Tourism Management*, *22*(6), 607–616. doi:10.1016/S0261-5177(01)00035-8

Bonarou, C. (2021). The poetics of travel through unravelling visual representations on postcards: A critical semiotics analysis. *Journal of Tourism. Heritage & Services Marketing*, *7*(1), 44–53. doi:10.5281/zenodo.4519317

Brown, L. (2016). Tourism and pilgrimage: Paying homage to literary heroes. *International Journal of Tourism Research*, *18*(2), 167–175. doi:10.1002/jtr.2043

Buchmann, A., Moore, K., & Fisher, D. (2010). Experiencing film tourism: Authenticity & fellowship. *Annals of Tourism Research*, *37*(1), 229–248. doi:10.1016/j.annals.2009.09.005

Buescu, H. C. (1990). *Incidências do olhar: Percepção e representação* [Directions of the gaze: Perception and representation]. Editorial Caminho.

Busby, G., & Shetliffe, E. (2013). Literary tourism in context: Byron and Newstead Abbey. *European Journal of Tourism. Hospitality and Recreation*, *4*(3), 5–45.

Conarroe, J. (1970). *William Carlos Williams' "Paterson": Language and landscape*. University of Pennsylvania Press. http://www.jstor.org/stable/j.ctv4t80tm

Coshall, J. T. (2000). Measurement of tourists' images: The repertory grid approach. *Journal of Travel Research*, *39*(1), 85–89. doi:10.1177/004728750003900111

Culler, J. (1981). The semiotics of tourism. *American Journal of Semiotics*, *1*(1), 127–140. doi:10.5840/ajs198111/25

Datta, S. (2000). Globalisation and representations of women in Indian cinema. *Social Scientist, 28*(3,4), 71-82. http://www.jstor.org/stable/3518191

del Bosque, I. A. R., Martín, H. S., Collado, J., & de los Salmones, M. M. G. (2009). A framework for tourist expectations. *International Journal of Culture, Tourism and Hospitality Research, 3*(2), 139–147. doi:10.1108/17506180910962140

Derrett, R. (2001). Special interest tourism: Starting with the individual. In N. Douglas & R. Derret (Eds.), *Special interest tourism* (pp. 1–28). Wiley.

Ehrat, J. (2005). *Cinema and semiotic: Peirce and film aesthetics, narration, and representation*. University of Toronto Press.

Farmaki, A., Altinay, L., & Yaşarata, M. (2016). Rhetoric versus the realities of sustainable tourism: The case of Cyprus. In P. Modica & M. Uysal (Eds.), *Sustainable island tourism: Competitiveness, and quality-of-life* (pp. 35–50). CABI. doi:10.1079/9781780645421.0035

Gentile, R., & Brown, L. (2015). A life as a work of art: Literary tourists' motivations and experiences at Il Vittoriale degli Italiani European. *Journal of Tourism, Hospitality and Recreation, 6*(2), 25–47.

Goulding, C. (2000). The commodification of the past, postmodern pastiche, and the search for authentic experiences at contemporary heritage attractions. *European Journal of Marketing, 34*(7), 835–853. doi:10.1108/03090560010331298

Griffin, T. (2002). An optimistic perspective on tourism's sustainability. In R. Harris, T. Griffin, & P. William (Eds.), *Sustainable Tourism* (pp. 24–32). Butterworth Heinmann. doi:10.1016/B978-0-7506-8946-5.50006-X

Grönroos, C. (1990). *Service management and marketing: Managing the moments of truth in service competition*. Lexington Books.

Gupta, S. B., & Gupta, S. (2013). Representation of social issues in cinema with specific reference to Indian cinema: Case study of Slumdog Millionaire. *The Marketing Review, 13*(3), 271–282. doi:10.1362/146934713X13747454353619

Hall, C. M., & Weiler, B. (1992). Introduction: What's so special about Special Interest Tourism? In B. Weiler & M. C. Hall (Eds.), *Special interest tourism* (pp. 1–14). Belhaven Press.

Han, B.-C. (2013). Im Schwarm, Ansichten des Digitalen [In the swarm: Views of the digital]. Matthes & Seitz Berlin Verlag.

Harrison, D., & Sharpley, R. (2017). Introduction. In D. Harrison & R. Sharpley (Eds.), *Mass tourism in a small world* (pp. 1–14). CABI. doi:10.1079/9781780648545.0001

Hendrix, H. (2007). The early-modern invention of literary tourism: Petrarch's houses in France and Italy. In H. Hendrix (Ed.), *Writer's houses and the making of memory* (pp. 15–30). Routledge.

Hendrix, H. (2008). *Writers' houses and the making of memory*. Routledge.

Hendrix, H. (2014), Literature and Tourism: Explorations, Reflections, and Challenges. In S. Quinteiro & R. Baleiro (Orgs.), Lit&Tour: Ensaios sobre literatura e turismo [Lit&Tour: Essays on literature and tourism] (pp. 19-30). Húmus.

Hendrix, H. (2020). The early musealization of writers' and artists' houses through guidebooks. *Nordic Museology, 1*, 8–22. doi:10.5617/nm.7962

Herbert, D. (2001). Literary places, tourism and the heritage experience. *Annals of Tourism Research, 28*(2), 312333. doi:10.1016/S0160-7383(00)00048-7

Herbert, D. T. (1996). Artistic and literary places in France as tourist attractions. *Tourism Management, 17*(2), 77–85. doi:10.1016/0261-5177(95)00110-7

Jain, J., & Rai, S. (2002). *Films and feminism*. Rawat Publications.

Jarmusch, J. (Director). (2016). *Paterson* [Film]. K5 International, Le Pacte, Animal Kingdom, Inkjet Productions.

Jenkins, O. H. (1999). Understanding and measuring tourist destination images. *International Journal of Tourism Research, 1*(1), 1–15. doi:10.1002/(SICI)1522-1970(199901/02)1:1<1::AID-JTR143>3.0.CO;2-L

Keegan, W., Moriarty, S., & Duncan, T. (1992). *Marketing*. Prentice Hall.

Kirshenblatt-Gimblett, B. (1998). *Destination culture: Tourism, museums, and heritage*. University of California Press.

Knudsen, D. (2015). Landscape studies and tourism research. In P. Howard, I. Thompson, & E. Waterton (Eds.), *The Routledge companion to landscape studies* (pp. 286–295). Routledge.

Laws, E., Scott, N., & Parfitt, N. (2002). Synergies in destination image management: A case study and conceptualisation. *International Journal of Tourism Research, 4*(1), 39–55. doi:10.1002/jtr.353

Liu, Z. (2003). Sustainable tourism development: A critique. *Journal of Sustainable Tourism, 11*(6), 459–475. doi:10.1080/09669580308667216

MacCannell, D. (1973). Staged authenticity: Arrangements of social space in tourist settings. *American Journal of Sociology, 79*(3), 589–603. doi:10.1086/225585

Macionis, N. (2004). Understanding the Film-Induced Tourist. In *International Tourism and Media Conference Proceedings* (pp. 24-26). La Trobe University, Monash University and University of Otago.

MacLeod, N. (2020). A faint whiff of cigar': The literary tourists' experience of visiting writers' homes. *Current Issues in Tourism*, 1–16.

MacLeod, N., Shelley, J., & Morrison, A. M. (2018). The touring reader: Understanding the bibliophile's experience of literary tourism. *Tourism Management, 67*, 388–398. doi:10.1016/j.tourman.2018.02.006

McIntosh, R. W., Goeldner, C. R., & Ritchie, J. R. B. (1995). *Tourism: Principles, practices, philosophies* (7th ed.). John Wiley and Sons.

Meinig, D. W. (2002). O olhar que observa: Dez versões da mesma cena [The gaze that observes: Ten versions of the same scene]. *Espaço e cultura, 13*, 35-46.

Morris, C. (2017). *Secular relics: Narrative objects and material biography in the museums of Darwin, Elgar & Holmes* (PhD Thesis), University of the West of England.

Moryson, F. (1617). *An itinerary Written by Fynes Moryson Gent* (Vol. 1). Internet Archive. https://archive.org/details/anitinerarycont17morygoog/page/n422/mode/2up?q=petrarch

Müller, D. K. (2006). Unplanned development of literary tourism in two municipalities in rural Sweden. *Scandinavian Journal of Hospitality and Tourism*, *6*(3), 214–228. doi:10.1080/15022250600667433

Neal, J. D., Sirgy, J. M., & Uysal, M. (1999). The role of satisfaction with leisure travel/tourism services and experience in satisfaction with leisure life and overall life. *Journal of Business Research*, *44*(3), 153–163. doi:10.1016/S0148-2963(97)00197-5

Onfray, M. (2007). *Théorie du voyage: Poétique de la géographie* [Theory of travel: Poetics of Geography]. Le livre de Poche.

Pearce, P., Morrison, A. M., & Rutledge, J. L. (1998). *Tourism: Bridges across continents*. McGraw-Hill.

Pereiro, X. P. (2009). Turismo cultural. Uma visão antropológica [Cultural tourism: An anthropological perspective]. *Pasos (El Sauzal)*.

Pike, S. (2002). Destination image analysis: A review of 142 papers from 1973 to 2000. *Tourism Management*, *23*(5), 541–549. doi:10.1016/S0261-5177(02)00005-5

Pocock, D. (1992). Catherine Cookson Country: Tourist expectation and experience. *Geography (Sheffield, England)*, *77*(3), 236–243. https://www.jstor.org/stable/40572194

Qiao, H., Peterson, A. T., Campbell, L. P., Soberón, J., Ji, L., & Escobar, L. E. (2016). Niche: A Creating virtual species and ecological niches in multivariate environmental scenarios. *Ecography*, *39*(8), 805–813. doi:10.1111/ecog.01961

Rennaissance. (2019). *What kids are reading*. Rennaissance. https://www.renaissance.com/learning-analytics/wkar/

Richards, G. (1996). *Cultural tourism in Europe*. CAB International.

Richards, G. (2001). The market for cultural attractions. In G. Richards (Ed.), *Cultural attractions and European tourism* (pp. 31–54). CAB International. doi:10.1079/9780851994406.0031

Richards, G. (2017). Sharing the new localities of tourism. In D. Dredge & S. Gyimóthy (Eds.), *Collaborative economy and tourism: Perspectives, politics, policies and prospects* (pp. 169–184). Springer. doi:10.1007/978-3-319-51799-5_10

Robinson, M. (2002). Reading between the lines: Literature and the creation of touristic spaces. *Current Writing: Text and Reception in Southern Africa*, *14*(1), 1–28. doi:10.1080/1013929X.2002.9678112

Robinson, M. (2002/2011). Between and beyond the pages: Literature - Tourism Relationships. In M. Robinson & H. C. Andersen (Orgs.), Literature and tourism: Essays in the reading and writing of tourism (pp. 39-79). Thomson Learning.

Robinson, M. & Andersen H. C. (Orgs.). (n.d.). *Literature and tourism: Essays in the reading and writing of tourism*. Thomson Learning.

Robinson, M., & Novelli, M. (2005). Niche tourism: An introduction. In M. Novelli (Ed.), *Niche tourism: Contemporary issues, trends and cases* (pp. 1–11). Elsevier. doi:10.1016/B978-0-7506-6133-1.50007-X

Sampaio, S. (2014). Watching narratives of travel-as-transformation in *The Beach* and *The Motorcycle Diaries*. *Journal of Tourism and Cultural Change*, *12*(2), 184–199. doi:10.1080/14766825.2014.915094

Santesso, A. (2004). The birth of the birthplace: Bread Street and literary tourism before Stratford. *English Literary History*, *71*(2), 377–403. doi:10.1353/elh.2004.0032

Schofield, P. (1996). Cinematographic images of a city: Alternative heritage tourism in Manchester. *Tourism Management*, *17*(5), 333–340. doi:10.1016/0261-5177(96)00033-7

Schofield, P. (2000). Deciphering day trip destination choice using a tourist expectation/satisfaction construct. In A. G. Woodside, G. I. Crouch, J. A. Mazanec, M. Oppermann, & M. Y. Sakai (Eds.), *Consumer Psychology of Travel, Hospitality and Leisure* (pp. 269–293). CAB International.

Sontag, S. (1979). *On photography*. Penguin.

Squire, S. J. (1994). The cultural values of literary tourism. *Annals of Tourism Research*, *21*(1), 103–120. doi:10.1016/0160-7383(94)90007-8

Tally, R. Jr. (2013). *Spatiality*. Routledge.

Taylor, J. E., & Pouliot, A. (2020). Introduction: Placing the author in ecologies of literary tourism. *Nineteenth-Century Contexts*, *42*(4), 381–389. doi:10.1080/08905495.2020.1782014

Thomas, J. (2012). *Shakespeare's shrine: The bard's birthplace and the invention of Stratford-upon-Avon*. University of Pennsylvania Press. doi:10.9783/9780812206623

UNWTO. (2017). *Definitions*. Committee on Tourism and Competitiveness. Available at http://cf.cdn.unwto.org/sites/all/files/docpdf/ctcdefinitionsenweb.pdf

Urry, J. (1992). The tourist gaze "revisited". *The American Behavioral Scientist*, *36*(2), 172–186. doi:10.1177/0002764292036002005

Urry, J., & Larsen, J. (1990/2011). The tourist gaze 3.0. *Sage (Atlanta, Ga.)*.

Virzi, P. (Director). (2017). *The leisure seeker* [Film]. Indiana Production Company.

Von Maltzahn, K. E. (1994). *Nature as landscape: Dwelling and understanding*. McGill-Queen's University Press.

Watson, C., & Saunders, R. (2004). The production of literary landscapes. In M. Robinson, & D. Picard (Eds.), *Conference Proceedings Tourism and Literature: Travel, Imagination and Myth*. Centre for Tourism and Cultural Change.

Watson, N. (Ed.). (2009). Literary tourism and nineteenth century culture. Palgrave.

Watson, N. J. (2006). *The literary tourist: Readers and places in Romantic and Victorian Britain*. Palgrave. doi:10.1057/9780230584563

Weigelt, K., & Camerer, C. (1988). Reputation and corporate strategy: A review of recent theory and applications. *Strategic Management Journal, 9*(5), 443–454. doi:10.1002mj.4250090505

Westover, P. (2012). *Necromanticism: Travelling to meet the dead, 1750-1860*. Palgrave Macmillan. doi:10.1057/9780230369498

Wylie, J. (2007). *Landscape*. Routledge. doi:10.4324/9780203480168

Yiannakis, J. N., & Davies, A. (2012). Diversifying rural economies through literary tourism: A review of literary tourism in Western Australia. *Journal of Heritage Tourism, 7*(1), 33–44. doi:10.1080/1743873X.2011.618538

KEY TERMS AND DEFINITIONS

Literary Place: Literary meaningful portion of space, either a real-life place associated with the life of a writer (e.g., birthplaces, chosen domiciles, gravesides), or an imagined place, associated with written literary works.

Literary Tourism: Literary tourism is a niche of cultural and heritage tourism, and it refers to the act of travelling to visit places and landscapes associated with the presence or memory of literary authors or literary fiction representations.

Literary Tourism Destination: A literary tourism destination is a literary site visited within in the context of a leisure travel, being the main travelling motivation to visit the places renowned for literary depictions and/or connections with literary figures.

Literary Tourist: Individual who is primarily motivated to travel to places associated with literature.

Chapter 2
Authenticity in Literary Tourism Places:
The Case of Asiyan Museum

Samet Çevik
Erdek Vocational School, Bandırma Onyedi Eylül University, Turkey

ABSTRACT

Author-related literary tourism places can create attractive elements for visitors who have an inner interest in the life stories of authors. Literary pilgrims place particular emphasis on authenticity level in literary sites. One of the attractive literary places in terms of authenticity, which is also Turkey's first literary museum, is Asiyan Museum where Tevfik Fikret lived between the years 1906-1915. In 2010, the museum underwent a comprehensive restoration. During this restoration, serious archive research was carried out and the house approached the original condition in the period of the 1910s. This study aims to examine the relationships between the authenticity of the place and the museum management mentality, research, and tourists' expectations by pointing out the prominence of authenticity in literary places. Using the interview technique, data were collected from the museum official in March 2018 on subjects such as the history of the museum, interpretation techniques, restoration works.

INTRODUCTION

A literary place can be regarded as a place to which visitors attach meaning, and it is the value of this meaning to them that draws them there. Meaning can be derived from reading a novel as much as from knowing about the author's life (Herbert, 1996, p. 78). The places where authors spend a part of their lives are considered valuable attractions as a cultural heritage element. These places also contribute to tourism development if their originality is preserved and their sustainability is ensured. Robinson (2002) asserted that authors' houses are arguably the most powerful tourism resource with appeal across a range of markets. These places allow tourists to engage in varied emotional experiences or activities, and they are a source of intimacy, authenticity and insight into the mystery.

DOI: 10.4018/978-1-7998-8262-6.ch002

The most crucial element that literary tourists seek in author-related literary places is authenticity. Suppose a place is a real place where the author lived and reflects the author's period entirely, and the exhibited objects are the items used by the author. In that case, the literary place meets the authenticity expectations of the literary tourists who follow the traces of the author they admire.

However, depending on the message that the literary place wants to give, sometimes tourists looking for the traces of the author may be disappointed in terms of authenticity. For instance, in Sait Faik Abasıyanık Museum, one of the other literary places in Turkey, the management preferred to focus only on the author's life story. The fact that the museum is the house where the author lived, the author's personal belongings are exhibited, and the sharing of anecdotes about essential periods of the author's life through photographs and documents satisfies the authenticity expectations of literary tourists visiting the museum. However, Namık Kemal House and Rıfat Ilgaz Culture and Art House, which are among the other literary places in Turkey, on the one hand, tell the author's life, on the other hand, exhibit some cultural objects from the author's period donated by the local people. Since these places serve as both an ethnography museum and a literary museum, they may fall below the expectations of literary pilgrims in terms of authenticity (Çevik 2018a; Çevik 2018b).

Therefore, authenticity depends on the policies place managers adopt and the interpretation decisions. At this point, doubts about authenticity can also arise. Herbert (2001, p. 317) mentioned that questions about literary places arise commonly about whether these places are genuine and whether they faithfully represent reality. Herbert pointed out that authenticity is a subjective experience, a combination of the developers' intentions, the consumers' interpretation, and their interactions. In another study, Fawcett and Cormack (2001) found out that not all literary site authorities defined the authentic as the real or unreproducible. The case studies they examined indicated that site authorities construct authenticity by setting what they believe is true about the sites the author associated with and the author. Consequently, to ensure the authenticity of the place, it is necessary to investigate the connection of the place with the author, based on accurate data, and present the place in the closest way to the author's period. Explaining this process to visitors and providing evidence of the authenticity of the place or objects is also an essential criterion in meeting the authenticity expectations of tourists.

The present study examines the efforts to ensure the authenticity of the place in line with the policies and interpretation decisions adopted by a literary place. The research is qualitative and descriptive. The case study design, one of the qualitative research designs, was used in the study. Asiyan Museum, Turkey's first literary museum, located in İstanbul city, was chosen as a case study. Asiyan Museum is in an advantageous position in terms of authenticity. It is the house where Tevfik Fikret, the great poet of Turkish literature, lived between 1906-1915 and even the plans, architectural drawings and decorations of the house belong entirely to him. However, as it gradually lost its genuineness due to the restorations made over the years, a comprehensive restoration started in 2010. The primary purpose of the restoration was to transform the house to its original condition in the 1910s. In the study, data were collected on the house's history, the past and present state of the house's sections and the restoration process through the interview technique with the museum official in March 2018 at the Asiyan Museum. The data were analyzed by descriptive analysis.

BACKGROUND

The Concept of Authenticity

Objects are important to the study of human history because they provide a concrete basis for ideas and can validate them. Their preservation demonstrates recognition of the necessity of the past and of the things that tell its story. Preserved objects also validate memories, and the actuality of the object, as opposed to reproduction or surrogate, draws people in and gives them a literal way of touching the past. This, unfortunately, poses a danger as places and things are damaged by the hands of tourists, the light required to display them, and other risks of making an object known and available (UNESCO, 2017). The cultural heritage object in its tangible and intangible form is a meta-cultural production where producers as key stakeholders in the exchange emphasize how it is presented to and understood by tourists. This positions the producer in charge of bringing the cultural heritage object to life. Through the formation and production of the cultural heritage object, it aims to understand what is materially, culturally and historically significant and how this affects the understanding of the object's authenticity (Farrelly et al., 2019, p. 9). The search for originality in cultural heritage properties has led to defining the notion of authenticity as an objective and measurable attribute inherent in the material fabric of historic environments (Lawless and Silva, 2017, p. 148).

The Venice Charter (1964) is the first international document addressing the authenticity of cultural heritage (ICOMOS, 1964):

"The historic monuments of generations of people remain to the present day as living witnesses of their age-old traditions. People are becoming more and more conscious of the unity of human values and regard ancient monuments as a common heritage. The common responsibility to safeguard them for future generations is recognized. It is our duty to hand them on in the full richness of their authenticity."

Trilling (1972, p. 94) points out that authenticity is implicitly a polemical concept (Peterson, 2005, p. 1083). At the Nara Conference on Authenticity, held from 1-6 November 1994, it was noted that there is no word to express precisely the concept of authenticity. There was a consensus among the experts that authenticity is an essential element in defining, assessing, and monitoring cultural heritage. In the conference document in which the relationship between authenticity and values was discussed, it was reported that authenticity appears as the essential qualifying factor concerning values. The understanding of authenticity plays a fundamental role in all scientific studies of the cultural heritage, conservation and restoration planning, and the inscription procedures used for the World Heritage Convention and other cultural heritage inventories (UNESCO, 1994).

Authentication is a process by which something—a role, product, site, object or event—is confirmed as ''original'', ''genuine'', ''real,'' or ''trustworthy'' (Cohen and Cohen, 2012, p. 1296). Authenticity has the abstract origin and roots of the meaning and value of the property, and it is a means of communicating and realizing this concept in time and place in the real world. Authenticity cannot be weakened over time. Although the physical aspect of heritage experiences gradual changes, consciously or unconsciously, over time, in different cultures, the relationship between memory and authenticity continues independently of physical aspects. It does not necessarily require physical continuity (Nezhad et al., 2015, p. 97).

According to Trilling (1972, p.93; quoted in Wang, 1999, p.350), the original usage of the authenticity term was in the museum:

where persons expert in such matters test whether objects of art are what they appear to be or are claimed to be, and therefore worth the price that is asked for them-or, if this has already been paid, worth the admiration they are being given.

Although authenticity is associated with preserving cultural heritage, it also plays a significant role in the tourism experience. MacCannell (1973), who first initiated the relationship of authenticity with tourism experiences, introduced staged authenticity. According to MacCannell, tourists seek authenticity in their experiences, but that is not always possible. MacCannell mentioned environments designed as "front" and "back" regions and argued that the presence of the back region promises more reality to tourists. Still, these regions are actually pre-planned front regions due to commercial concerns. Therefore, tourists who seek authenticity encounter a staged authenticity. Wang (1999) stated that the ambiguity of authenticity gradually increased with the widespread use of the concept and remarked that many tourist motivations and experiences could not be explained with the concept of authenticity introduced by MacCannell. However, authenticity is closely related to many types of tourism, such as ethnic, historical and cultural tourism and their subtypes.

In pursuit of MacCannell's (1973) study, many studies on the relationship between authenticity and tourism have been conducted, and different types of authenticity have been revealed (Andriotis, 2011; Cohen and Cohen, 2012; Fawcett and Cormack, 2001; Gilmore and Pine, 2007; Hughes, 1995; Jia, 2009; Peterson, 2005; Rickly-Boyd, 2012; Waitt, 2000; Wang, 1999; Yi et al., 2017; Yi et al., 2018; Zhu, 2012).

Wang (1999) suggested that due to the complex nature of authenticity in tourism, it should be classified into three different types: objective authenticity, constructive authenticity and existential authenticity. As Trilling (1972) mentions, *objective authenticity* includes a museum-linked usage of the authenticity, and the authenticity of the objects exhibited determines the tourist's experience of authenticity. Hence there is an absolute and objective criterion for measuring authenticity. *Constructive authenticity* refers to the authenticity attributed to objects according to beliefs, opinions, expectations and preferences by tourists, producers or place managers. In constructive authenticity, authentic experiences and the authenticity of toured objects are constitutive of one another. Jia (2009) points out that constructive authenticity is accepting objective authenticity by tourists or producers and its revision through interpretation. Regarding this issue, Hughes (1995) asserted that authenticity in tourism is produced by a wide variety of producers such as entrepreneurs, marketing agencies, tourist guides, animators, and intermediaries rather than being presented as natural. Wang (1999) characterized *existential originality*, or activity-based authenticity, as a potential existential state of being activated by tourist activities. Existential authenticity stresses only tourists' subjective experiences and neglects the authenticity of toured objects (Jia, 2009). It is focused on the idea of individuals feeling free to engage with their true selves (Yi et al., 2017).

Zhu (2012) argued that these three types of authenticity seem to represent the dilemma of objective-subjective orientation in the context of tourism and authenticity and suggested that the search for authenticity cannot be explained solely by objects, socially constructed reality, or the existential state of being. The author focused on performative authenticity and claimed that the ritual performance, integrated with the external world, gives rise to the performative experience of authenticity. It offers a deep understanding of the link between memory, habitus and embodied practice. Distinctly from existential authenticity, performative authenticity emphasizes the dynamism of becoming authentic through the embodied practice. In another similar study, by revisiting Benjamin's (1986) theorizations, Rickly-Boyd (2012, p. 284) remarked that while the authenticity of the object/site is a result of its embodiment in a tradition of which tourism is a ritual, the authenticity of the experience is a part of the engagement with aura. Concerning the tourist experience, being involved in a tradition and its rituals results in an aura experience.

On the contrary of these three types of authenticity, postmodern authenticity, which has received less attention in tourism (Yi et al., 2018), is defined as the deconstruction of authenticity (Wang, 1999). Postmodern authenticity totally deconstructs the conception of authenticity by blurring the boundaries between the original and the duplicated (Eco, 1986; quoted in Yi et al., 2018, p. 413). Yi et al. (2018) pointed out that postmodern authenticity and inauthenticity are different concepts in philosophy. Postmodern tourists do not necessarily consider things that are inauthentic as problematic. According to Wang (1999), postmodern tourists abandon the concept of authenticity since it is increasingly difficult for the authenticity of the original, such as the marginal ethnic culture, to remain immutable with the accelerating globalization. Indeed, authenticity and postmodern authenticity are relative terms that vary depending on the social environment to which tourists belong and their individual perspectives (Yi et al., 2018). Zhu (2012, p. 1497) claimed that the rise of postmodernism is convinced that there is no actual, true, genuine, objective reality, and this notion links to the perspectives of constructive and existential authenticity in tourism research.

Although studies and theories about authenticity were generally carried out in the context of heritage object and tourism experience, Andriotis (2011) made an authenticity assessment for pilgrimage landscapes and religious heritage experience based on the five types of authenticity created by Gilmore and Pine (2007) for consumer studies. These types of authenticity are natural authenticity, original authenticity, exceptional authenticity, referential authenticity and influential authenticity. *Natural authenticity* is the presence of elements that exist in nature and remain untouched by human hands. Therefore, this type of authenticity is also vital for the tourism experience. *Original authenticity* is regarded as the authenticity of elements that are not duplicates or imitations, unique and the first of its kind. *Exceptional authenticity* characterizes the authenticity of activities that are based on the process of human interaction and are performed extraordinarily well, not in an insensitive and insincere way. *Referential authenticity* is the authenticity of the elements inspired by human history, bearing traces of shared memories and emotions. *Influential authenticity* describes the authenticity of the factors that contain meaning and impact and lead people to the better.

Authenticity in Author-Related Literary Tourism Places

Authors' houses, one of the main types of literary tourism, are also a cultural heritage site, as they have a historical value. Janiskee (1996) emphasized that historic houses have intrinsic appeal due to their aesthetic features, educational benefits, novelty, and their ability to stimulate a wide variety of thoughts and feelings associated with the past. In this respect, as Herbert (1996) put it, authors' houses attract people who have an intrinsic interest in the authors' personal life stories. Visits to these houses allow contact with places closely associated with admired authors, allows sight of, and perhaps the chance to touch, artefacts or memorabilia; the setting enhances the experiential quality of these contacts. The visitors have connections not just with the life of the authors but also with the literary works which the authors created. A merging of fact and fiction gives such a special meaning (Herbert, 1996).

Many studies explored various aspects of literary tourism in authors' houses or other actual places related to the author. The authors who are the subject of these studies are; Dylan Thomas (Herbert, 2001), Jane Austen (Herbert, 2001; Orr, 2018), Lucy Maud Montgomery (Fawcett and Cormack, 2001), John Milton (Santesso, 2004), Robert Burns (Bhandari, 2008), Mary Russell Mitford (Booth, 2008),

Honoré de Balzac (Petroman, Petroman, and Brătulescu, 2008), Vladimir Nabokov, Alexander Pushkin (Wallace, 2009), Virginia Woolf (Robertson and Radford, 2009), Lord Byron (Busby and Shetliffe, 2013), Keats-Shelly House (Stiebel, 2013), Anne Frank (Busby and Devereux, 2015; Hartmann, 2013), Gabriele D'Annunzio (Gentile and Brown, 2015), Ismail Kadare (Erkoçi, 2016), Necati Cumalı, Sait Faik Abasıyanık, Namık Kemal (Çevik, 2018a), Rıfat Ilgaz (Çevik, 2018b). In many of these studies, the objective authenticity is the research topic. The studies examined the objects exhibited in author-related houses, whether they were indeed the authors' objects, and investigated the effect of these objects' authenticity on literary tourists' experiences.

One of the most emphasized concepts in literary tourism research is authenticity due to an authentic experience being the most sought thing in a literary site by literary tourists (Çevik, 2020). And authors' houses, as MacLeod (2021) emphasized, offer a full range of authentic experiences as niche attractions. In these places, tourists are seeking the "real thing", in other words, an "authentic experience", but they also want evidence that these things are authentic (Stiebel, 2004). Literary tourists pay special attention to the level of authenticity in literary places. They want to see and touch the real objects the author used, and they want to be able to imagine scenes such as the author walking around, working, reading or eating in the author's own house (Çevik, 2020). These authentic objects play a crucial role in presenting authors' houses, offering the visitor a chance for intimacy (MacLeod, 2021). Therefore the author house with authentic objects that belongs to the author acts as a physical medium through which fans can imaginatively engage in their search for connection (Orr, 2018).

Waitt (2000) argued that although authenticity is used as a promotional device, what is "real'' is open to interpretation. The marketers provide only one possible interpretation of past events, manipulated in the interests of capital or dominant social norms and unquestionably, all representations of the past are selective. Similarly, Fawcett and Cormack (2001) remarked that authenticity is more than a simple idea underlying and animating touristic sites. What is authentic is formed partly by the personal commitments, bureaucratic mandates, and entrepreneurial interests/economic necessities that managers face.

Hence, what is authentic is mainly depends on the elaborative interpretation decisions. According to Freeman Tilden, who made the first definition of heritage interpretation, interpretation is "an educational activity which aims to reveal meanings and relationships through the use of original objects, by firsthand experience, and by illustrative media, rather than simply to communicate factual information" (Tilden, 1977, p. 8). Interpretation of literary places is a crucial determinant of the quality levels of experiences of tourists'. Each literary place makes choices on the extent of interpretation related to the forms taken and the messages selected. Interpretation decisions may have different purposes, or the priority objectives may vary from place to place. For instance, a literary place may adapt economic goals, the need to profit improvement by attracting visitors or cultural purposes, the effort to broaden knowledge, and increasing the recognizability of an author and the author's literary works (Herbert, 1996).

Fawcett and Cormack (2001) argued that authenticity depends on the reality claims adopted by the place managers, focused on three different interpretation forms: modernist, rationalist, and eclectic. The modernist approach asserts elegance and simplicity of style and nostalgia for an authentic or real past. This form of interpretation brings a real touristic interpretation that does not allow uncertainty. The rationalist form denotes a highly conscious plan for separating and channelling certain bureaucratically sanctioned interpretations. It is a preferred form of interpretation in literary places related to fiction, where place managers invite tourists in line with their own imagination.

On the other hand, the eclectic form represents an unstructured, multifaceted form that invites many touristic interpretations. In literary places where this interpretation style is used, fact and fiction are included together. In another study, Santesso (2004) mentioned the "Stratford Model", which is about the interpretation of authenticity of literary sites. This model provides a less intellectual but more immediate resolution to literary-touristic discrepancy through manufacturing a false authenticity designed to match both the tourist-readers' ideal and the argument about the author, which the managers wish to present.

MacLeod (2021) remarked that the notion of the genius loci is a valuable way of conceptualizing the literary visitor's experience of authors' houses and informing their presentation and management as literary tourism attractions. MacLeod identified three components of the notion of genius loci in authors' houses as "domestic sphere"," spirit of the author", and "authors' tools". Literary place managers should emphasize these three components in their interpretation decisions, ensuring access to authentic objects and allowing space and time for visitors' imaginations to roam freely.

THE CASE OF ASIYAN MUSEUM

Tevfik Fikret

Tevfik Fikret (Mehmed Tevfik), one of the most important figures of Turkish poetry, was born on December 24, 1867, in Istanbul. His grandfather and grandmother were two Greek children relocated to Turkish families with the devshirme method due to the suppression of the Greek uprising in Chios in 1821 by the Ottoman army. Hüseyin Bey, who was the Ottoman constabulary-official for public order, took these children for adoption and named them Hüsrev and Sıdıka. Hüseyin Bey, who adopted the children as a father, provided them with the best education and upbringing, and then married these two children to each other. They had five children. Tevfik Fikret's mother, Hatice Refia Hanım, is one of these five children. Hatice Refia Hanım married Hüseyin Efendi, who was known to be the son of Ahmet Ağa from Çankırı Çerkeş and they had three children: Şevki, Mehmed Tevfik and Sıdıka. In 1879, when Fikret was 12 years old, his mother, Hatice Refia, and his uncle died due to a cholera epidemic during the pilgrimage. His mother's death had a significant effect on him. Tevfik Fikret's father was denounced to Sultan Abdulhamid II due to he visited the Shaykh al-Islām twice to consult while working at the Pertevniyal Valide Sultan Foundation. Abdulhamid, who saw the Shaykh al-Islam as a potential threat to his reign, exiled Hüseyin Efendi in 1886, first to Hama, then to Nablus, Akka, Urfa and Aleppo. Hüseyin Efendi spent 19 years apart from his children and passed away in 1905 in Antep. His sister Sıdıka died at a young age in 1902 due to his unhappy marriage (Ayvazoğlu, 2019; Kuran N., 2016).

When Tevfik Fikret reached school age, he was enrolled in the Mahmudiye Secondary School, built by Pertevniyal Valide Sultan next to the Pertevniyal Valide Sultan Mosque in Aksaray. As a result of the closure of this school at the end of the Russo-Turkish War of 1877-1878, he enrolled in Galatasaray High School the year he lost his mother. He graduated with a first from this school in 1888. After graduation, he became a clerk at the Consultation Office of the Ministry of Foreign Affairs. However, he resigned after a short time as that job went against his artistic spirit. In 1889, he was promoted and reassigned to the Consultation Office of the Ministry of Foreign Affairs. In 1890, he took up to teach Turkish, calligraphy and French at Gedikpaşa Commercial School. In the same year, he married his uncle's daughter, Nazime Hanım, and they had a son named Haluk. Afterwards, he took up a position as a literature teacher at Galatasaray High School. He resigned from his duty here in 1897 and began to lecture at Robert College upon the proposal. He was the director of Galatasary High School between December 28 1908, and April 9 1910. In this short period, he performed very successfully and became a symbolic name of the institution. He died on August 19, 1915, at the age of 48 (Alper, 2006; Ayvazoğlu, 2019; Engin; 2006; Kaplan, 1987; Kuran N., 2016).

He began poetry at Galatasaray High School at 15-16 (Kaplan, 1987). In the 1890s, his first poems began to be published in Mirsad, Ma'lûmat, Maarif and Mekteb. There was no innovation in terms of form and content in these poems yet. After getting to know Charles Baudelaire, Sully Prudhomme and especially François Coppée, he moved away from imitation and found his true personality as a poet and determined his actual line in poetry. In 1896, he became the editor of the *Servet-i Fünûn Journal* and wrote poems emphasizing individuality with the understanding of "art for art". He brought together the poems published in this journal between 1896-1900 in his poetry book *Rübâb-ı Şikeste*. His other works are *Tarih-i Kadîm, Haluk'un Defteri, Rübab-ın Cevabı* and *Şermin* (Uçman, 2016).

Although Tevfik Fikret is one of Turkish literature's most important poets, he had a versatile artist personality. He was well-versed in different branches of art such as painting, calligraphy and architecture. Before poetry, his interest in painting appeared. He was always involved in painting in his childhood and called himself "Artist Tevfik". He took painting lessons at Galatasaray High School, and his teachers appreciated him (Kaplan, 1987). He ran painting and poetry together throughout his life. Today, there are more than twenty paintings in the Asiyan Museum. These paintings consist of three groups as still life, landscape and portrait (İnankur, 2016).

Kaplan (1987) mentioned three prominent features of Tevfik Fikret's character: hypersensitivity, introversion, and formalism. Undoubtedly, all these features of his character influenced his art. Tevfik Fikret's formalism influenced his poetry, calligraphy and painting, and domestic life, namely his understanding of architecture. As will be discussed in more detail below, Asiyan is Fikret's own work, from outside to inside, down to the last detail.

The Story of Asiyan Museum

Asiyan is a Persian word and is used in the sense of "bird's nest", "house", "dwelling". According to Kaplan (1987), Fikret had a desire to establish Asiyan at an early age. Fikret had a small hut in the garden of their home in Aksaray. He gained all the features that made him Fikret in this tiny hut, namely in the first Asiyan (Kaplan, 1987). He personally asked for this hut under the fig tree in the mansion's garden and had it built. He decorated the hut himself with paintings, books, small ornaments, flower pots, vases, and mini-sculptures and spent most of his time there. Years later, about this hut in his childhood, Fikret said: "*My room had a window. I used to sit there, write, paint. I wrote Artist Tevfik in a great script and hung it on the window from the inside. When I looked from the outside, I saw the text upside down, and I was bored*" (Ayvazoğlu, 2019, p. 54-55).

Yersu (2016) mentioned that Asiyan is an embodied part of Fikret's childhood dreams. He remarked that the loss of his mother at a young age and his father's exile made him yearn for a home where his family lived together, like every child. After he began to lecture at Robert College, the road he used to go to the college was on Asiyan's land, which made it a place where he came with his wife to dream, write poems and paint pictures.

While getting about to the college, Fikret was admiring the Rumelihisarı ridges every day. He was thinking of building a house there, and after the death of his father, he took more kindly upon this project. Instead of living in a mansion in Rumelihisarı, it was more attractive to move to a house on the ridges and lead a life away from society. When he sold the mansion in Aksaray in 1905, he immediately embarked on the Asiyan project. He bought the land by adding some borrowed money to the money he received from the sale of the mansion in Aksaray. As Asiyan's architect, he first drew the project of the building. He thought about where to put furniture, what to hang on the walls, and placed them in the project (Topuz, 2015). There are plans and drafts prepared as the first examples of this project in the collection of the Asiyan Museum, and Yersu (2016) mentioned that the plans of Asiyan in the artist's imagination could be seen from different angles in these drawings which he drew and coloured up to the garden landscape. The construction of Asiyan was started in 1905 and was completed in 1906.

Figure 1. The view of Aşiyan in the early 20th century[1]

Asiyan is a house consisting of a basement, ground floor and first floor, located on the hillside in front of today's Boğaziçi University on the Rumeli Fortress cemetery, overlooking the Bosphorus (Koran Ö., 2016). Asiyan is Fikret's own work, from outside to inside, down to the last detail (Kaplan, 1987). Everything from the furniture he used here to the ceiling decorations was his own design, and everything from wallpapers to floor coverings reflected his refined taste (Kuran Ö., 2016). Asiyan that fully reflects Fikret's aestheticism and sense of morality, was described in detail by Ruşen Eşref in his 1919 work (Ayvazoglu, 2019). Ruşen Eşref said, *"Fikret, the architect of Asiyan, Fikret, the decorator of its rooms, Fikret, the artist of the paintings on the walls, Fikret who chooses its furniture and landscapes its garden, its soul is Fikret"* (Ruşen Eşref, 1919; quoted in Kaplan, 1987, p.68). Ruşen Eşref portrayed in detail the living room of the house, every object and furniture in the living room, the room next to the living room, the library and the dining room. In addition, some other literary friends of Fikret who visited the house also presented some anecdotes about the place in their writings, although not as detailed as the narrative of Ruşen Eşref (Ayvazoğlu, 2019).

Tevfik Fikret, who lived in Asiyan for a short period of 9 years, died on August 19, 1915, at 48. In his will, he wanted to be buried in Asiyan, but family members did not agree to this, thinking that Asiyan could be taken possession of other people in the future. Fikret's body was buried in the family cemetery in Eyüp. As a result of the work of Tevfik Fikret Association, Fikret's body was taken from Eyüp on December 24, 1961, and moved to Asiyan's garden (Yersu, 2016).

After Tevfik Fikret's death, his wife Nazime Hanım continued to live in Asiyan until 1940. She had various arrangements and renovation works done in Asiyan. Three rooms on the upper floor of the building facing the college were added during her period, and she used this section. She rented other rooms as a hostel to college teachers and students. Nazime Hanım put Asiyan for sale in 1940 because of the maintenance costs of the building. Istanbul Municipality bought the building and the Asiyan land on behalf of the state for 10000 Liras, and then the works to establish a museum began. On August 19, 1945, on the 30th anniversary of Tevfik Fikret's death, the first Turkish literature museum, "Edebiyat-ı Cedide Museum", was opened with a ceremony (Yersu, 2016). The great hall on the ground floor of the Edebiyat-ı Cedide Museum was designed as the Abdülhak Hamit Hall. The glazed room on the ground floor overlooking the sea was arranged as the Edebiyat-ı Cedide Room. In the hall, there were pictures of Recaizade Mahmut Ekrem and other Servet-i Fünun authors, their works, writing sets of Ali Ekrem Bolayır and Recaizade. The small room at the back of the ground floor was arranged as the Şair Nigâr Hall (Pamukçu, 1980; quoted in Törenek, 2011). Two rooms on the upper floor of the Edebiyat-ı Cedide Museum were reserved for the belongings of Tevfik Fikret. The museum has been named "Asiyan Museum" since Fikret's body was moved to Asiyan.

Asiyan Museum and Its Authenticity as a Heritage

Asiyan Museum consists of three floors. There was a dining room, kitchen, laundry, and cellar in the basement in Fikret's period. The large room on the ground floor was used as a guest lounge, while the small glassed room was used as a summer living room. On the first floor, there were the study room and bedroom.

The museum official remarked that the museum had undergone restoration many times since its opening. Before the museum was opened in 1945, a restoration expenditure was made. Since the building was dilapidated, necessary renovations were made inside the building; outside the building, garden arrangements and road-stair construction were carried out. Similar maintenance was performed in subsequent restorations in the museum building. However, the museum technique in 1945 was not changed. With the decision taken in 2010, the museum was undergone a comprehensive restoration with a very different mentality from previous restorations. The main aim of this restoration was determined to make the museum as similar as possible to the period between 1906 and 1915 when Tevfik Fikret lived. The museum official indicated that the place began to move away from its original condition for two reasons. First, the museum opened in 1945 as a literary museum encompassing many other authors, and second, the previous restorations in the building. During past maintenances, many arrangements were carried out that led to the loss of the house's authenticity. Among them remaining the hand-drawn paintings under the plaster, losing the originality of the wall colours, being neglected of the rooms, modifying the bathroom-kitchen-laundry room, raising the floor, and laying wood on the original stones can be given as examples. After two years of intensive and detailed work, Asiyan was put into service again in 2012, in line with the period of Tevfik Fikret.

Extensive archive research was conducted to approach the Tevfik Fikret period of the house. At this stage, they used people's memories of the place, books portraying that period, photographs reflecting the original condition of the house as sources. The goal was to take visitors to a century ago and ensure that visitors feel as if Tevfik Fikret is living there while they are touring. The museum official who highlighted that archive research is a vital element in restoring cultural heritage sites explained that they tried to seek answers to many details such as what the sections were used for, which colours, patterns or curtains were utilized through the sources mentioned above. He remarked that they obtained around 20 photos of each part of the house from an auction in 2005. These photos were beneficial at this stage. Tevfik Fikret took these photos in person. Fikret, who also designed the interior of his house, photographed this work under the conditions of that period. Moreover, Fikret directed the photographs for only the creation made the details more obvious.

Alongside the photographs, another very beneficial resource at this stage is Ruşen Eşref's 1919 book Tevfik Fikret, as mentioned above. In this book, the comprehensive description of Tevfik Fikret's house, the representation of each part and every object, from coffee tables to vases, from carpets to paintings, in detail, provided a substantial advantage in the success of the restoration process. The museum official, who stated that they progressed step by step by comparing the photographs with those characterized in the sourcebooks, pointed out that they elaborated to put the original objects belonging to Tevfik Fikret in their actual places. In this context, the management evicted all the personnel from the place they occupied in the house. All parts in Fikret's period opened to exhibit within the museum's scope.

Some original items in the house at that time could not be found. After the death of Tevfik Fikret, during the period when Nazime Hanım lived at home, many items were sold or distributed to other people in the process of renting some rooms as a hostel. Although, before 1945, the municipality tried to recover these items during the museum's preparation stage, some of them could not be found. The museum official specified that a few such things were made as replicas and underlined that they gave close attention to replicas and acted in a way that would make them look exactly like their original condition.

Figure 2. The hall in the early 20th century and today

In which Tevfik Fikret hosted his guests, the guest hall was designed as Abdülhak Hamit Hall during the Edebiyat-ı Cedide Museum period in 1945. In the last restoration, the hall was decorated with the objects used by Fikret at that time. Since the main hall is the most spectacular part of the museum, the official expressed that they paid particular attention to this part. Main components such as a nacreous sofa and armchairs in the hall were made as replicas. The wall colours are original and are the colour in the sourcebooks and under plaster. Chandeliers, carpets, curtains were specially made in accordance with the photographs of that period.

On the left side of the hall, there are two armchairs and a waxwork of the poet sitting on one of these armchairs in front of the window. The museum official emphasized that the waxwork was made to impress the visitors as if the poet greets the visitors. The fact that the poet's face in the waxwork is already original increases this effect. It was made from the first death mask in Turkey, taken from Fikret right after his death by artist Mihri Müşfik. Since it is a real face, visitors encounter an imposing facial structure when they look at it. In this part, "The Mist Painting" was painted by the last caliph Abdülmecit, who was impressed by Tevfik Fikret's The Mist poem and gave this painting to Fikret as a gift is displayed. There is an Istanbul silhouette in the picture. When viewed from a little further away, minarets, domes, Hagia Sophia and Galata Bridge continue under the sun. In this sense, the official stated that the painting coincides with the poem and underlined that the painting is an exceptional painting regarding having Abdülmecit's tughra, his signature, and exhibiting in its original place.

In the other part of the hall, a nacreous sofa where the poet rested and wrote his poems is another piece made as a replica. There are paintings on both sides of the sofa titled "Vase of Chrysanthemums" and "Nazime Hanım on Ridges of Bebek". These paintings are Tevfik Fikret's works, and they were restored. The tapestry behind the sofa was restored and put in its original place. The toss pillows on the sofa are similar to those that evoke the atmosphere of that period. Besides this sofa, there is a fireplace around which some remembrances of the poet's father. In this section, located opposite the main entrance, charcoal paintings of the poet's father Hüseyin Efendi, made by Tevfik Fikret, and photographs of Hüseyin Efendi are exhibited on the wall. The clock on the fireplace was stopped to show the poet's death time. There are Chinese porcelain vases and various decorations on both sides of this clock. A wooden bookcase and a rocking chair are also located here. The sofa set and copper barbecue in the bay window of the hall are original pieces.

Figure 3. The other part of the hall in the early 20th century and today

With the last restoration, Abdülhak Hamit Tarhan's property, which was previously displayed in the main hall, were moved to two rooms in the background of the house. The official asserted that these rooms were probably used as his son Haluk's room or as a guest room during the Tevfik Fikret period. After the death of the poet Abdülhak Hamit Tarhan, his wife Lüsyen Hanım donated the poet's property to the Istanbul Municipality. The municipality included his property in the collection when transforming Asiyan into the Edebiyat-ı Cedide Museum. Therefore, the objects displayed in these rooms are also original. The poet's desk, armchair, clothes, books and photographs of his family are exhibited in the first room. There is also the portrait of Abdülhak Hamit, another painting signed by the caliph Abdülmecit. In the second room, there are personal objects of the poet, his badges and portraits of family elders.

The summer room next to the guest hall was Tevfik Fikret's restroom. This room is used as the Edebiyat-ı Cedide Room today to highlight the house's Tevfik Fikret period. Although they understood from the old photographs of this room and Ruşen Eşref's book that the walls were covered with dark brown wallpaper, the museum official shared the information that there was no wallpaper in this room after the previous restorations and the walls were white. In the last restoration, the wallpaper was enlarged proportionally to the object next to it from the photograph in the AutoCAD program. It was redrawn and mounted on the wall with a unique print. The portrait of Recaizade Mahmut Ekrem hanging on the room wall is another painting signed by the caliph Abdülmecit. On both sides of the picture, there are photographs of the representatives of the literary movement Edebiyat-ı Cedide.

Figure 4. Restroom in the early 20th century and today

Tevfik Fikret's study room, bedroom and bathroom are located on the first floor of the museum. The rooms on this floor are the most remarkable part of the house regarding literary tourism, as Tevfik Fikret's personal objects are exhibited. The rooms on this floor reflect the daily life of Tevfik Fikret entirely. In front of the window overlooking the Bosphorus and the Asian continent on the east side of the study room, there is the original study desk and chair designed by the poet himself, and Asiyan's first designs, his writing sets and calligraphy plates are displayed on the desk. In front of the desk, there are two original armchairs and a lampshade right next to it. There is the poet's library on the other side of the room, where his books, magazines, and various publications of that period. In this section, it is also possible to see the poet's paintings. Original paintings such as his self-portrait, a charcoal humour work in which he portrays himself while looking in the mirror and the portrait of Darwin show the poet's artistic side. Fikret's birth certificate, recorded as "December 24, 1867, Tuesday night at 5 o'clock 26 minutes", is also exhibited here. This room also has a door to the drawbridge that connects to Robert College. The official, who claimed that the poet found such practical solutions in architecture, mentioned that the poet could reach the institution he lectured quickly.

Figure 5. Study room in the early 20th century and today

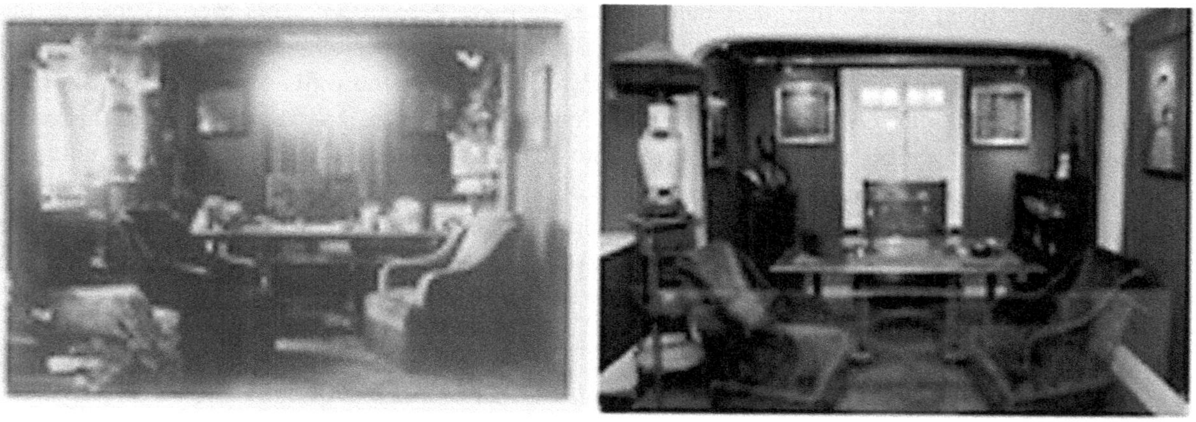

The room next to the study room is the bedroom where the poet died on August 19, 1915, at 48. The poet's bed is not original and was brought to the museum in 1945. There is an original nightstand next to the bed. There is also the face mask of Tevfik Fikret on the nightstand. This mask is the first example of death mask tradition in Turkey, and it was taken from the poet right after his death by the first female artist Mihri Müşfik. Besides, the bedroom has a magnificent Bosphorus view. The museum official mentioned that the window here is described expressions such as "the most beautiful bedroom window overlooking the Bosphorus in the world" and "the window opening to heaven".

Figure 6. Bedroom in the early 20th century and today

On the basement floor of the house, there are the dining room, kitchen, laundry and Şair Nigâr sections. The door of the dining room is decorated with arched, coloured glass. There are still lifes by Tevfik Fikret on the walls of the dining room. The dining table was belonged to Şair Nigâr, not Tevfik Fikret's family. However, it reflects the same period. The dinnerware on the table is the original ones used by Tevfik Fikret and his family. The official who explained that the stones on the dining room floor were covered with wood in previous restorations highlighted that they reached the original stones after a meticulous study after discovering that the original floor was stone. The small room in front of the dining room belongs to Şair Nigâr. Şair Nigâr was one of the first female poets after the Tanzimat Reform Era and was also the first female playwright in the Ottoman Empire. In this room, Şair Nigar's personal objects, books, and photographs donated to the museum by her family in 1959 are displayed.

Next to the dining room is the kitchen. This section was for museum personnel after 1945. Personnel was evicted from here in the last restoration. The official stated that they reached the original kitchen bench used under the old kitchen bench during the restoration. The kitchen also includes a cooking stove and a cellar with a hatch under the floor. The window in the kitchen has a sea view. Tevfik Fikret named this window "Socrates' Window" because of his admiration for Socrates. There is a laundry next to the kitchen. Here, there is a system established under the conditions of that period for the servants to wash the clothes. The laundry reveals the features of the 20th century with its washpot, stove, hood, British-made laundry tub.

DISCUSSION

It has been revealed in the literature that authenticity is among the most significant expectations of tourists in literary tourism places (Bhandari, 2008; Booth, 2008; Çevik, 2018a; Çevik, 2020; Fawcett & Cormack, 2001; Gentile & Brown, 2015; Hazard, 2016; Herbert, 2001; Jia, 2009; MacLeod, 2021; Orr, 2018; Petroman, Petroman, & Brătulescu, 2008; Santesso, 2004; Stiebel, 2004; Stiebel, 2013; Wallace, 2009). The fact that the literary place also has a cultural heritage value increases the importance of authenticity in literary places within the relationship between cultural heritage and authenticity.

The data obtained show that the museum carried out meticulous work to transform this cultural heritage value to its condition approximately 115 years ago. The museum exceeds the authentic expectations of literary tourists. Because it was the house where Tevfik Fikret and his family lived, and even many aspects such as the house's architecture, decoration, and landscaping of the garden that he undertook. Most of the objects exhibited in the place are the belongings of the poet and his family. Regarding items that could not be reached for various reasons, the museum has adopted a research-based approach and made identical replicas using resources such as photographs and books. At this point, instead of adopting the Stratford model, as Santesso (2004) suggests, and choosing to exhibit a few items that are thought to reflect the characteristics of that period, on the contrary, it adopted a policy of interpretation that will overlap the original condition of the house as closely as possible. The museum applied this interpretation policy to the items on display and the house itself as a whole. Walls, ceilings, floors, etc., were restored to reveal their original condition in the sources. Within this context, one could argue that the museum has adopted the modernist approach from the three interpretation types mentioned by Fawcett and Cormack (2001). In the modernist approach, a real past and nostalgia are at the forefront. When touring the Asiyan Museum, the questions that require different interpretations do not arise in the minds of the tourists.

It could be considered that changes were made in the previous restorations of the museum in line with the constructive authenticity approach determined by Wang (1999). However, with the last restoration, it was understood how essential objective authenticity is regarding tourists' expectations, and motivations and efforts were performed in this direction. The objective authenticity is prominent not only for the motivations and expectations of literary tourists but also an influential determinant of their ultimate experience. For literary pilgrims who follow the traces of the author, imagining how the author lived in the house, seeing and touching the real objects the author used, and establishing an emotional connection with the author through these objects are the most strong motivation sources.

MacLeod's (2021) study indicated that the notion of genius loci is very useful in the management and interpretation decisions of literary places. It would not be wrong to say that the Asiyan Museum acted with the awareness of this notion. MacLeod identified three components of the notion of genius loci in authors' houses as "domestic sphere"," spirit of the author", and "authors' tools". The museum management tried to establish the setting in the author's own house as it was 115 years ago. In such a way, literary tourists can feel as if the author is living there now, and they can witness that period. The exhibition of the author's works and personal belongings in the house also strengthens this feeling.

CONCLUSION

The most crucial element that literary tourists seek in author-related literary tourism places is authenticity, which is also a principle of heritage interpretation. Literary tourists, who are enthusiasts of the author, want to see the real objects used by the author in a place where the author lived and want the evidence that these are real. Consequently, literary places that present authentic experiences tourists seek with objects and documents meet the expectations of literary tourists. The Asiyan museum gives literary tourists the experience of the Tevfik Fikret period and communicates this period with its visitors both with narratives in guided tours and with visual documents.

Apart from its functions, such as authenticity, nostalgia, and emotional connection with the author, another particular role of the Asiyan Museum is to shed light on a highly significant Turkish literature period through literary tourism to introduce this period and increase awareness. The fact that students are one of the most prominent target markets of the Asiyan Museum, which also performs its educational function that one of the crucial functions of cultural heritage sites and museums, confirms this argument.

Asiyan Museum, a cultural heritage value, took a considerable step in transferring this value to future generations in its original form, ensuring its sustainability, in line with its adopted policies. It has to be remembered that authors' houses are a literary tourism place and a heritage value. The authorities should consider the principle of sustainability of cultural heritage in the policies and the decisions. Consequently, authenticity in authors' houses is a vital element in meeting the expectations of literary tourists and having quality experiences, as well as preserving the cultural heritage and keeping it alive in its original form.

This study aims to evaluate the efforts to ensure authenticity in the Asiyan Museum, Turkey's first literary museum. It demonstrates how essential the museum management's mentality and its perspective on literary heritage ensure authenticity in literary places. Hence, literary place authorities should act based on accurate data in arrangements or renovations of the site. They should conduct comprehensive research through experts or secondary sources. The principles of museology and cultural heritage protection should be guiding in the efforts. As mentioned above, literary places have a significant social role in keeping the memory of the author alive and the sustainability of the cultural heritage. Therefore, literary place managers should act with the awareness of this responsibility.

In future research, quantitative or qualitative research can be carried out to determine the effect of authenticity presented to tourists in literary heritage places regarding different aspects such as how it affects tourists' experiences and satisfaction levels and how much it meets their expectations. In addition, the relationship between different interpretation strategies in literary places and the authenticity of the place can be the subject of research. Research can be conducted on the reflection of this relationship on literary tourists from various aspects.

REFERENCES

Alper, S. (2006). Tevfik Fikret'in kırk sekiz yılı. In A. Yaraman (Ed.), *Biyografya 7: Tevfik Fikret* (pp. 11–16). Bağlam Publishing.

Andriotis, K. (2011). Genres of heritage authenticity: Denotations from a pilgrimage landscape. *Annals of Tourism Research, 38*(4), 1613–1633. doi:10.1016/j.annals.2011.03.001

Ayvazoğlu, B. (2019). *Fikret: Kendi cevvim, kendi eflâkimde kendim tâirim.* Everest Publishing.

Bhandari, K. (2008). Touristification of cultural resources: A case study of Robert Burns. *TOURISM Preliminary Communication, 56*(3), 283–293.

Booth, A. (2008). Revisiting the homes and haunts of Mary Russell Mitford. *Nineteenth-Century Contexts, 30*(1), 39–65. doi:10.1080/08905490801945538

Busby, G., & Devereux, H. (2015). Dark tourism in context: The diary of Anne Frank. *European Journal of Tourism. Hospitality and Recreation, 6*(1), 27–38.

Busby, G., & Shetliffe, E. (2013). Literary tourism in context: Byron and Newstead Abbey. *European Journal of Tourism. Hospitality and Recreation, 4*(3), 5–45.

Çevik, S. (2018a). Interpretation of author-related literary tourism sites: Evidence from three sites in Turkey. *Journal of Tourism Research, 21*, 118–136.

Çevik, S. (2018b). Sarı yazma diyarında bir edebiyat turizmi mekanı: Rıfat Ilgaz Kültür ve Sanat Evi. In *II. International Congress of Tourism and Cultural Heritage* (pp. 166-178). Şanlıurfa.

Çevik, S. (2020). Literary tourism as a field of research over the period 1997-2016. *European Journal of Tourism Research, 24*, 2407.

Cohen, E., & Cohen, S. A. (2012). Authentication: Hot and cool. *Annals of Tourism Research, 39*(3), 1295–1314. doi:10.1016/j.annals.2012.03.004

Engin, V. (2006). Tevfik Fikret'in Mekteb-i Sultani müdürlüğü ve eğitimci kişiliği. In A. Yaraman (Ed.), *Biyografya 7: Tevfik Fikret* (pp. 237–252). Bağlam Publishing.

Erkoçi, I. (2016). Ismail Kadare as tourist attraction. *Dos Algarves: A Multidisciplinary e-Journal, 28*, 1-12.

Farrelly, F., Kock, F., & Josiassen, A. (2019). Cultural heritage authenticity: A producer view. *Annals of Tourism Research*, *79*, 102770. doi:10.1016/j.annals.2019.102770

Fawcett, C., & Cormack, P. (2001). Guarding authenticity at literary tourism sites. *Annals of Tourism Research*, *28*(3), 686–704. doi:10.1016/S0160-7383(00)00062-1

Gentile, R., & Brown, L. (2015). A life as a work of art: Literary tourists' motivations and experiences at Il Vittoriale Degli Italiani. *European Journal of Tourism, Hospitality and Recreation*, *6*(2), 25–47.

Gilmore, J. H., & Pine, B. J. Jr. (2007). *Authenticity: What consumers really want?* Harvard Business School Press.

Hartmann, R. (2013). The Anne Frank House in Amsterdam: A museum and literary landscape goes virtual reality. *Journalism and Mass Communication*, *3*(10), 625–644.

Herbert, D. (2001). Literary places, tourism and the heritage experience. *Annals of Tourism Research*, *28*(2), 312–333. doi:10.1016/S0160-7383(00)00048-7

Herbert, D. T. (1996). Artistic and literary places in France as tourist attractions. *Tourism Management*, *17*(2), 75–85. doi:10.1016/0261-5177(95)00110-7

Hughes, G. (1995). Authenticity in tourism. *Annals of Tourism Research*, *22*(4), 781–803. doi:10.1016/0160-7383(95)00020-X

ICOMOS. (1964). *The Venice Charter 1964*. https://www.icomos.org/charters/venice_e.pdf

Inankur, Z. (2016). Ressam Tevfik Fikret. In A. Yersu (Ed.), *Ölümünün 100. Yıldönümünde Tevfik Fikret Sempozyumu* (pp. 33–48). Istanbul Metropolitan Municipality Directorate of Libraries and Museums.

Janiskee, R. L. (1996). Historic houses and special events. *Annals of Tourism Research*, *23*(2), 398–414. doi:10.1016/0160-7383(95)00069-0

Jia, H. (2009). The construction of literary tourism site. *Tourism (Zagreb)*, *57*(1), 69–83.

Kaplan, M. (1987). *Tevfik Fikret: Devir-Şahsiyet-Eser* (2nd ed.). Dergâh Publishing.

Kuran, N. (2016). Tevfik Fikret'in aile hayatı ve hayatında sanatın önemi. In A. Yersu (Ed.), *Ölümünün 100. Yıldönümünde Tevfik Fikret Sempozyumu* (pp. 7–22). Istanbul Metropolitan Municipality Directorate of Libraries and Museums.

Kuran, Ö. (2016). Mimar Tevfik Fikret. In A. Yersu (Ed.), *Ölümünün 100. Yıldönümünde Tevfik Fikret Sempozyumu* (pp. 77–90). Istanbul Metropolitan Municipality Directorate of Libraries and Museums.

Lawless, J. W., & Silva, K. D. (2017). Towards an integrative understanding of 'authenticity' of cultural heritage: An analysis of World Heritage Site Designations in the Asian context. *Journal of Heritage Management*, *1*(2), 148–159. doi:10.1177/2455929616684450

MacCannell, D. (1973). Staged authenticity: Arrangements of social space in tourist settings. *American Journal of Sociology*, *79*(3), 589–603. doi:10.1086/225585

MacLeod, N. (2021). 'A faint whiff of cigar': The literary tourist's experience of visiting writers' homes. *Current Issues in Tourism*, *24*(9), 1211–1226. doi:10.1080/13683500.2020.1765996

Nezhad, S. F., Eshrati, P., & Eshrati, D. (2015). A definition of authenticity concept in conservation of cultural landscapes. *Archnet-IJAR*, *9*(1), 93–107. doi:10.26687/archnet-ijar.v9i1.473

Orr, A. (2018). Plotting Jane Austen: Heritage sites as fictional worlds in the literary tourist's imagination. *International Journal of Heritage Studies*, *24*(3), 243–255. doi:10.1080/13527258.2017.1378911

Peterson, R. A. (2005). In search of authenticity. *Journal of Management Studies*, *42*(5), 1083–1098. doi:10.1111/j.1467-6486.2005.00533.x

Petroman, I., Petroman, C., & Brătulescu, M. (2008). Virtual tourism: Balzac's house in Paris (France). *Research Journal of Agricultural Science*, *40*(3), 303–306.

Rickly-Boyd, J. M. (2012). Authenticity & aura: A Benjaminian approach to tourism. *Annals of Tourism Research*, *39*(1), 269–289. doi:10.1016/j.annals.2011.05.003

Robertson, J. P., & Radford, L. A. (2009). The private uses of quiet grandeur: A mediation on literary pilgrimage. *Changing English*, *16*(2), 203–209. doi:10.1080/13586840902863186

Robinson, M. (2002). Reading between the lines: Literature and the creation of touristic spaces. *Current Writing: Text and Reception in Southern Africa*, *14*(1), 1–28. doi:10.1080/1013929X.2002.9678112

Santesso, A. (2004). The birth of the birthplace: Bread Street and literary tourism before Stratford. *English Literary History*, *71*(2), 377–403. doi:10.1353/elh.2004.0032

Stiebel, L. (2004). Hitting the hot spots: Literary tourism as a research field with particular reference to KwaZulu-Natal South Africa. *Critical Arts: South-North Cultural and Media Studies*, *18*(2), 31–44. doi:10.1080/02560240485310151

Stiebel, L. (2013). 'When in Rome…?': Literary tourism in Rome from a South African perspective. *Journal of Human Ecology (Delhi, India)*, *43*(1), 93–99. doi:10.1080/09709274.2013.11906615

Tilden, F. (1977). *Interpreting our heritage* (3rd ed.). The University of North Carolina Press.

Topuz, H. (2012). *Elbet Sabah Olacaktır* (10th ed.). Remzi Publishing.

Törenek, M. (2011). Tevfik Fikret ve Aşiyan. *Yeni Türk Edebiyatı Araştırmaları*, *3*(5), 7–27.

Uçman, A. (2016). Bir Servet-i Fünun şairi olarak Tevfik Fikret. In A. Yersu (Ed.), *Ölümünün 100. Yıldönümünde Tevfik Fikret Sempozyumu* (pp. 23–32). Istanbul Metropolitan Municipality Directorate of Libraries and Museums.

UNESCO. (1994). *Nara Document on Authenticity*. https://whc.unesco.org/archive/nara94.htm

UNESCO. (2017). *Tangible Cultural Heritage*. https://webarchive.unesco.org/20170130173132/http://www.unesco.org/new/en/cairo/culture/tangible-cultural-heritage/

Waitt, G. (2000). Consuming heritage: Perceived historical authenticity. *Annals of Tourism Research*, *27*(4), 835–862. doi:10.1016/S0160-7383(99)00115-2

Wallace, C. (2009). Yeats's Country and 'Yeats Country': Conceptualizing literary spaces. *Journal of Tourism and Cultural Change*, *7*(1), 48–60. doi:10.1080/14766820902803424

Wang, N. (1999). Rethinking authenticity in tourism experience. *Annals of Tourism Research*, *26*(2), 349–370. doi:10.1016/S0160-7383(98)00103-0

Yersu, A. (2016). Tevfik Fikret'in Aşiyanı ve yüzyıllık serüveni. In A. Yersu (Ed.), *Ölümünün 100. Yıldönümünde Tevfik Fikret Sempozyumu* (pp. 131–153). Istanbul Metropolitan Municipality Directorate of Libraries and Museums.

Yi, X., Fu, X., Yu, L., & Jiang, L. (2018). Authenticity and loyalty at heritage sites: The moderation effect of postmodern authenticity. *Tourism Management*, *67*, 411–424. doi:10.1016/j.tourman.2018.01.013

Yi, X., Lin, V. S., Jin, W., & Luo, Q. (2017). The authenticity of heritage sites, tourists' quest for existential authenticity, and destination loyalty. *Journal of Travel Research*, *56*(8), 1032–1048. doi:10.1177/0047287516675061

Zhu, Y. (2012). Performing heritage: Rethinking authenticity in tourism. *Annals of Tourism Research*, *39*(3), 1495–1513. doi:10.1016/j.annals.2012.04.003

ENDNOTE

1 All photographs used in this study are copyrighted by the Istanbul Metropolitan Municipality Directorate of Libraries and Museums.

Chapter 3
Literary Tourism:
A Perspective of the Visitor Experience in the Museum of Innocence

Ozan Atsız
https://orcid.org/0000-0003-2962-1903
Yozgat Bozok University, Turkey

Selman Temiz
https://orcid.org/0000-0001-7899-9625
Yozgat Bozok University, Turkey

ABSTRACT

This chapter aims at understanding travelers' literary experience in "The Museum of Innocence," which is known as a museum based on a book that was written by Orhan Pamuk. To reach the purpose, a qualitative research approach was adopted in this study, and online reviews posted by visitors of the museum were used to explore the main components of their experience in a literary context. The collected data was examined through content analysis. As a result of analysis, visualization, a sense of nostalgia, awe, and memorable components were revealed. These components were interpreted in different ways by visitors. Theoretical and practical implications were provided as well as limitations for future research lines.

INTRODUCTION

People are motivated to visit some destinations for a wide range of purposes such as leisure, seeking for a novelty, experiencing literary objects and filmed areas within the different types of tourism, including film tourism and literary tourism (Hoppen et al., 2014; Kassean & Gassita, 2013). Among these, literary tourism has become a popular market segmentation over the past decades and has gained significant attention from tourism scholars (Çevik, 2020; Quinteiro et al., 2020). According to Quinteiro et al. (2020, p. 361; as cited in Butler's (2000) study), it is described as "the form of tourism in which the primary motivation for visiting specific locations is related to interest in literature." In sum, it has been defined

DOI: 10.4018/978-1-7998-8262-6.ch003

as the act of visiting literary places related to writers, authors' biographies and their texts. In the literary destinations, understanding visitor experience is crucial for destination practitioners and extant literature.

As an important literary attraction, Museum of Innocence located in Istanbul (Turkey) plays a major role in the memorable tourist experience and is considered as a literary attraction. The Museum of Innocence, created by Orhan Pamuk, is a small Istanbul Museum consisting of installations that show the memories and meanings of daily life items described in the author's novel with the same name. Orhan Pamuk is a recipient of the 2006 Nobel Prize in Literature. After this award, his recognition increased. People who read his novel (Museum of Innocence) wondered his museum. Despite tourists' visiting to this museum, very limited research in the context of literary tourism was conducted on it (Çevik, 2018). In particular, to the best of authors' knowledge, there is no research that examines visitor experiences at this museum.

In order to fill this research gap, this book chapter will attempt to investigate visitor experiences of Museum of Innocence using User Generated Content within the context of literary tourism. To attain our research gap, reviews posted by international travelers on TripAdvisor will be examined by descriptive analysis. In accordance with this book chapter purpose, the content of the present research is configured as follows: in the first section, a literature review about the literary tourism, the role of fictional museums in the visitor experience and User Generated Content and visitor experience will be presented. The research method and findings will be presented in the next section. Finally, a clear conclusion will be offered.

Literature Review

Literary Tourism Experiences

Literary tourism is defined as the form of tourism where the primary motivation to visit certain places is related to literature (Baleiro & Quinteiro, 2018). In other words, literary tourism means traveling to literary places such as authors' biographies and works (Quinteiro et al., 2020, p. 361). According to Topler (2016, p.129), a literary visitor can better understand the author's life and works. Fawcett and Cormack (2001, p.987) stated that literary tourism is shaped by the interaction of author (biographical facts and real places about the author) and fiction (environments and characters). Stating that literary tourism has a multidisciplinary nature, Jiang and Yu (2019, p.128) define literature tourism as visiting places in famous literary works and the place of birth and dwellings of famous writers and poets. To sum up, literary tourism, which has attracted the attention of researchers in the last decades and has various definitions in tourism studies, is a type of tourism where the main motivation of the visitor is to visit places related to the author or the work (Baleiro & Quinteiro, 2018, p.32; Hoppen et al., 2014, p.39). Literature tourists, also called literary pilgrims, have a sense of admiration for the places associated with the writers. Literary pilgrims want to visit the place of birth of the author and imagine the environment in which his/her childhood passed, to see places that inspired the work or to respect to the author's heritage (Herbert, 2001, p.312).

There may be several reasons for people (literary pilgrims) to visit literary places. Herbert (2001) classified these reasons under three main topics. Firstly, literary pilgrims may want to see places the lives of writers because places where the writer lived and worked can create a sense of nostalgia, awe or respect in the literary pilgrim. When literary pilgrims visit such places, they may feel as if the writer is leaving the house and as if the writer is breathing with the visitor. Secondly, literary pilgrims may want to see the fictional place in the book. These kinds of places are places where reality and imagine intersect. In the mind of the literary pilgrim, there is such a powerful picture of the place and the fictional character that it is as if he/she encounters the fictional character through the visit to the literary place or feels more deeply what the fictional character experiences in the visited place. The last reason is related to willingness of literary pilgrims to see literary locations for broader and deeper feelings from a specific author or literary work. In addition to these three reasons, there are more general qualifications that can be used to increase the attractiveness of the literary space. For example, literary places can become a well-known tourism route. Further to this, considering the importance of ambiance, natural settings, landscapes, and facilities ranging from afternoon teas to souvenirs can be considered part of the experience, leading to a visit to the literary destination (pp.314-315).

Herbert (1996; 2001) emphasized that tourists visiting literary places do not always need to know about the author or read literary works. In his research on tourists visiting the literary tourism place of Cabourg, Pont-Aven and Auvers in Northern France, Herbert concluded that some tourists did not know the author, did not know about the author's works, and most tourists were not even interested in literature. Participants of the study stated that they enjoyed their visit because they had the opportunity to enjoy the ambience of a place, relax, have a nostalgic experience and watch beautiful views. In the study conducted on visitors to Hill Top Farm, which is mentioned in the stories of Beatrix Potter, it was determined that the reasons for visiting the visitors were the childhood memories inspired by the stories they read, and the desire to experience nostalgia with their family and friends (Wang & Zhang, 2017, p.235).

Determining that the participants who visited literary places have different motivations revealed that there are different types of literary tourism. Therefore, the need to classify the activities within the scope of literary tourism has emerged (Busby & Klug, 2001; Hoppen et al., 2014, p.40). Six literary tourism types have been identified:

- Travels to real places where the author was born, lived, and wrote in order to learn more about the work and the author, and to better understand the background from which the work was produced. Charles Dickens in England, Kafka in Czechia and Tolstoy tours in Russia are examples of these travels. Besides, visits to bars in Dublin where literary figures such as James Joyce, Samuel Beckett, and Oscar Wilde were regulars, where they met and socialized with their friends and even wrote most of their works under candlelight or kerosene lamps can be considered within this scope.
- Travels aiming to see the fictional place where the event described in the work takes place. Wuthering Hills Tour in England, Crime and Punishment walking in the footsteps of Raskolnikov in St. Petersburg or Dan Brown's work of The da Vinci's Code in the Secret Halls of Istanbul were some of the examples.
- Travels by private or public sector with the aim of promoting the destination, obtaining economic benefits and seeing places that add charm to literary figures.
- Travels to real life locations mentioned in the literary work.

The following literary tourism types have been determined by Busby and Klug (2001);

- Travel writing as a type of literary tourism refers to travels where places and people are reinterpreted and communicated to a wider audience. Bill Bryson's humorous travel writings "Notes from a Small Island", the book that best represents England, can be an example of this type (Hoppen et al., 2014, p. 41).

- Film-based literary tourism as a result of an interest in a destination by reading the work after watching the film. Visiting New Zealand after watching The Lord of the Rings movie is a good example for this group of literary tourism.

In the trend of the commercialization of art, people's interest in literary tourism and consumption society has been effective. Accordingly, as a product of the popularization of literature and art, literary tourism destinations are considered as a new way of reading and understanding literary works in the age of popular culture and art. Indeed, the purpose of better understanding of the literary work by visiting the literary place is one of the main motivations that drive tourists to a destination. Literary tourists are impressed by the literary works they read, and they want to trace the story by experiencing the characters, environment and story scenes in the literary tourism destination. In line with this purpose, emotional and imaginary power is put into the fictional literary tourism destinations in order to make real the expectations of tourists, revive their desires and enable them to be experienced in every aspect of the literary place (Jiang & Yu, 2019, pp.144-146).

Today, it is possible to visit many different literary tourism destinations from the places where the most favorite writers were born, grew up, lived or died to the places where the favorite books were written (Hoppen et al., 2014, p. 37). For example, Lucy Maud Montgomery (1874–1942) is one of Canada's best known and translated authors. Montgomery published 530 short stories, more than 500 poems and 20 novels during her long and prolific career. His works often depict the idyllic and fascinating Prince Edward Island. The themes in Montgomery's works about the island make a serious contribution to the tourism industry. In 1998, about 1.24 million tourists visited Prince Edward Island, of whom about 210,755 stated that they traveled to experience the island's history and culture. In addition, more than half of the visitor group reported that they wanted to visit the Montgomery / Anne attraction centre (Fawcett and Cormack, 2001, pp.686-687). Moreover, Margaret Mitchell House and Museum, Herman Melville's Arrowhead and Pearl S. Buck International are other major literary attractions that can be visited by international travelers (Neace, 2015).

In Europe, which is seen as the center of literary tourism, especially Ireland and England have become a literary tourism phenomenon with their literary tourism resources and their tourism policy (Hoppen et al., 2014, p.37). Both the UK Tourism Authority and regional tourism boards have produced maps, brochures and other promotional materials to inform visitors about local literary tourism destinations. The literary centers include many famous venues such as Shakespeare's Stratford, Brontë's Yorkshire, Hardy Wessex, Wordsworth's Lake District and Dickens's London (Squire, 1996, p. 120). Literary tourism is so naturalized as a cultural phenomenon in the British Isles that tourists visiting the area can see literary places detailed in guides and marked on the roadmap. There are many souvenir shops selling soap, t-shirts, handbags, postcards and bookmarks, especially museums related to literary tourism (Hoppen et al., 2014, p. 38). As a result of the increasing interest in literary tourism, the increase in the number of literary destinations has made it necessary to set a standard in this regard. With the aim of resolving this situation, UNESCO first registered the city of Edinburgh as the city of literature in the program initiated in 2004. Today, there are a total of 39 literary creative cities, covering 28 countries and six continents, registered by UNESCO.

However, Quinteiro et al. (2020, p.363) stated that literary tourism should not be seen as a strategy for the development of the destination as it is in many places. Instead, the authors state that literary tourism should be considered as a differentiated cultural tourism product that can distract tourists and visitors from overcrowded areas. Notre Dame, for example, receives millions of visitors each year, but for the vast majority, this is only part of the trip to Paris. In this respect, literary tourism should be considered as a holistic approach with other tourism types in the destination (Herbert, 1996, p.78). What is driving the literary pilgrim or other tourist away from their homes, from everyday life? What do tourists hope to find that they cannot find at their home? Typically, except from recreation and leisure, the academic answer to this question is that modern tourists seek "authenticity" in their travel, something that is not sufficiently provided in their everyday life experiences (Fawcett and Cormack, 2001, p. 687). At this point, the way to create authenticity is the synergy to be created between literary tourism and other tourism types. These creative synergies have been used by many destinations in recent years, especially through products such as film-based tourism, literary tourism, architectural tourism and music tourism (UNWTO, 2018, p.82). In other words, literary destinations can be a tourist attraction in their own right, or they can be a component of a more broadly defined tourist destination. Thus, the authenticity, representation and experience will contribute to the increase in attractiveness (Herbert, 1996, p.79). Within the scope of creating authentic experience, there are many literary tourism activities that can be done (Baleiro & Quinteri, 2018). These activities are presented in the Table 1.

Table 1. Literary tourism products and experiences

Literary Tourism Products and Experiences	Description
Visits to authors' houses	Travel to visit the houses where authors were born, lived or died, the places where they wrote, the schools where they studied, the cafés, restaurants and hotels they went to, the churches where they were baptized, married or had their funerary ceremonies, and their graves.
Visits to the places of the literary text	Travel to visit the places where the action unfolds, see the statues of the characters, the places where the work was written or places where texts were staged.
Literary walks	Tours through the authors' and/or the characters' places/routes (*e.g.*, places that were once the site of writers' inspiration and/or the route taken by a character: Leopold Bloom's promenades around Dublin or Mrs. Dalloway's journeys across London. These tours can either have a guide or be self-guided. In these situations, tourists have a map-brochure that helps them follow the right route (sometimes there may be signs at each stopping point on the literary walk).
Literary pub crawls	Tours taken in several historical pubs, with professional actors performing the works of famous writers (*e.g.*, Dublin Literary Pub Crawl and Edinburgh Literary Pub Crawl).
Visits to literary festivals	Travel to participate in literary festival activities (lectures, dinners, walks, and many other creative activities with writers and performers).
Visits to libraries	Tours to libraries to visit the library building for its beauty and architecture and/or to see a rare first edition or a historic manuscript.
Visits to bookshops	Tours to bookshops because of their beauty, original concept or because the places are associated to a text or author.
Visits to literary parks	Visits to parks, whose design was inspired by literary texts and/or their authors. Most of these parks are associated with children's literature, although parks for adults are increasingly common.
Stays in literary hotels	Stays at hotels that are somehow related to literature, an author, a work or a character.
Stays in library-hotels	Stays in hotels that have a strong literary connection, either because they place a vast book collection at the disposal of their guests or because they were once the house of writers and poets (*e.g.*, John Fowles's house, the author of *The French Lieutenant's Woman*, is now a literary lodging). These spaces are meant to be privileged reading places, where guests can go on a literary retreat.
Visits to literary cities/villages	Visits to the United Nations Educational, Scientific and Cultural Organization (UNESCO) literary cities as well as others that due to their strong association with literature can be classified as literary cities/villages/towns.
Participation in literary competitions	Participation in literature inspired competitions, such as literary paperchases, literary quizzes and detective games inspired by police narratives.
Participation in literary dinners	Participation in dinners where an author, work or character is celebrated. The dishes that are served, the guests' outfits and the readings are all literary inspired.
Participation in lectures and reading sessions	Participation in reading sessions or lectures where an author, a text or a character is discussed and celebrated. In these sessions, the tourist can actively participate or just attend.
Attendance of literary text staging/ performances	The staging of texts in places associated with the text and/or the author. In these sessions, the tourist can actively participate or just attend.

Source: (Baleiro & Quinteiro, 2018, pp.38-40).

When the activities that can be carried out within the scope of literary tourism are considered, it is seen that these activities have an emotional, artistic, cognitive or social participatory structure. Thanks to this participatory structure, it is aimed to make experience authentic by providing situations such as the formation of associations about the work or visitor's own past, the revival of past memories or the formation of new memories and the intensity of the emotions felt. Thus, the distinctive feature and authenticity of the literary destination will be strengthened thanks to memorable and enriched experiences.

The Case of the Museum of Innocence

Literary tourism includes not only visiting homes, monuments, tombs or literary festivals belonging to the authors of literary works, but also the environment in which the fictional hero in the literary work lives. Until the early twentieth century, literary tourism was translated into a devotional pilgrimage to the author figure. However, in the following decades the places represented in literary works also became a center of attraction. Instead of focusing solely on the author, tourists have sought to trace literary places or fictional characters (Baleiro & Quinteiro, 2018, p.32-36). Accordingly, the Museum of Innocence, which was constructed based on the novel "Museum of Innocence" written by the Nobel Prize Winner Turkish writer Orhan Pamuk, has become one of the places visited by literary pilgrims.

Orhan Pamuk, born in 1952, is one of the youngest names to be awarded the Nobel Prize for Literature at the age of 54. Orhan Pamuk's books have been translated into 63 languages and reached approximately 13 million prints in more than 100 countries. His work named "Museum of Innocence" is one of these works and was published in 2008. The plot of the novel is a love story based on a social class distinction. The novel was written after ten years of work. The book was first published in Germany after Turkey. The Museum of Innocence was included in the "best books of 2009" list by the New York Times in 2009 and it has been translated into over 30 languages. In fact, when Orhan Pamuk started writing his book, The Museum of Innocence, he planned to open the Museum of Innocence, bearing the same name as his book. In other words, while the book is finished and put on the market, the museum bearing the same name as the book will also be opened. For this purpose, from the 1990s on, he started collecting old items that he thought would be appropriate to be in the museum, while carrying out studies on the content and writing of the book. In the meantime, the author, who was looking for a building suitable for the museum, bought a historical three-storey house built in 1897 in Çukurcuma/Beyoğlu of Istanbul. The house purchased is restored by famous Turkish and German architects.[1]

In order to understand better how the storyline of the novel intersects with the idea of a museum, it would be beneficial to mention briefly the novel. The novel takes place in Istanbul in the second half of the 20th century. Kemal, the son of a wealthy family, goes to a shop to buy a gift for his fiancée, Sibel. In the shop, he encounters Füsun, who is a distant relative and has not seen for years. She works in the shop. After this encounter, Füsun who is preparing for the university exam, starts to come to Kemal's house to prepare the exam. Over time, this relationship turns into a strong love. However, Kemal has to be engaged to Sibel and after this, he cannot meet or hear from Füsun again. Kemal, who could not hear from Füsun despite all his efforts, is so unhappy and begins to spend most of his time in his house with the belongings of Füsun. Realizing that he cannot forget Füsun, Kemal breaks off his engagement with Sibel. Meanwhile, Kemal's father passed away, and for this reason, a letter of condolence comes from Füsun's family. Thanks to the address on the letter, Kemal goes to Füsun's house for dinner one evening to propose marriage to Füsun. However, he learns that Füsun is married to a young filmmaker named Feridun. Feridun willing to make a movie, but he does not have enough finance for this. Kemal says that he can offer required finance for the film in order to be closer with Füsun and offers to be a producer for the film. This is accepted by Feridun and then Kemal is able to meet with Füsun more often. These meetings last 7 years and 10 months. In these meetings, Kemal takes Füsun's belongings to his home as a memory and begins to create a collection with the fear of losing Füsun. Kemal and Füsun, who are about to come together in the later parts of the novel, have a crash and Füsun dies. Upon this, Kemal wants every item he has collected until that time and the house where Füsun lived to be a museum. In addition, Kemal agrees with Orhan Pamuk to write a novel about his love with Füsun. Kemal meets with Orhan Pamuk many times at Füsun's house, which will turn into a museum, and tells Orhan Pamuk his relationship with Füsun from beginning to end. The novel is written by Orhan Pamuk and ends with the death of Kemal.

When the storyline of the novel is considered; It is seen that the fiction of the novel and the idea of a real museum based on the fictional world quite overlap. While writing his novel, Orhan Pamuk put some of the old items he collected for the museum on his worktable and tried to ensure the amalgamating of the items and the novel. The novel The Museum of Innocence, which was shaped in this way, was published in 2008. However, due to some problems, the museum was opened in 2012. Those who buy the book also get the entrance ticket to the museum and a map of the place where the museum is located on the last page of the book.

Istanbul, where the Museum of Innocence, bearing the traces of almost every detail of Kemal and Füsun's life, is a world-renowned tourism destination. The history of Istanbul dates back to 8500 years ago as a settlement and 3000 years ago as an urban area. Besides, Istanbul as a capital city in which many civilizations lived, dates to about 2000 years ago. The city was the capital of the Roman, Byzantine and Ottoman Empires, respectively. Istanbul was declared as the "European Capital of Culture" by the European Union in 2010. It has been on the UNESCO Cultural Heritage List since 1985 with its Historical Peninsula, where many historical monuments such as Istanbul Topkapı Palace, Hagia Sophia and Sultan Ahmet Mosque are located. Istanbul was determined as the fastest growing tourism city in 2018 by the World Travel and Tourism Organization (WTTC) (WTTC, 2020, p.10). Before the COVID-19 pandemic, approximately 15 million foreign tourists visited Istanbul in 2019.[2]

Beyoğlu where the Museum of Innocence is located is one of the most historical, popular and dynamic areas of Istanbul with its districts each of which has their own characteristics and streets that never sleep. There are many places of interest in Beyoğlu including Istiklal Street, Taksim Square, Galata Tower, Galatasaray High School (Mekteb-i Sultani), and Sent Antuan Church. In addition, Beyoğlu reflects the harmony of eastern and western civilization with its famous art galleries Pera Museum and other art galleries, Tophane Fountain, French Passage, many historical passages and unique historical texture. Çukurcuma District, where the Museum of Innocence is located, is relatively quiet, famous for its antique shops, and has many cafe, restaurants and quality boutiques. In addition to the historical and cultural texture of Istanbul, the mentioned features of Beyoğlu / Çukurcuma, which hosts the Museum of Innocence, make the region one of the most preferred places by tourists.

The Museum of Innocence, as one of the best examples of literary tourism for fictional places, constitutes an authentic part of the cultural richness of the region. The Museum of Innocence offers an experience that emerges from the intersection between reality, fiction, and the reader's imagination, and intertwines fiction with the real world. Today, the Museum of Innocence is a literary tourism place that attracts all kinds of local and foreign tourists and is included as a private museum in the Istanbul tourism travel guide. According to Orhan Pamuk, it is not necessary to read the novel to visit the museum. However, visiting the museum after reading the novel may allow a better embrace of the different meanings of the museum and many things that are not discovered while reading the novel.

Research Methodology

The purpose of this book chapter is to understand which components are given importance by international travelers who visited the Museum of Innocence in a literary tourism context. For this purpose, a qualitative research design was adopted for this study since limited knowledge exist in the literary tourism field and no research cover the international travelers' experiences in the literary context. To address this knowledge gap, the chapter examined the travelers' reviews toward the Museum of Innocence posted on TripAdvisor (URL: https://www.tripadvisor.com/). This research design is seen as proper approach to understand travelers' opinions, feelings, behaviors, and experiences. Furthermore, previous studies have used online reviews to understand main dimension or components of travelers' experience. Thus, this chapter will be extremely helpful for researchers who desire to know tourist experience.

Online data collection area was chosen as TripAdvisor and its website mentions that "With more than 884 million reviews and opinions of nearly 8 million businesses, travelers turn to TripAdvisor to find deals on accommodations, book experiences, reserve tables at delicious restaurants and discover great places nearby. As a travel guidance company available in 49 markets and 28 languages, TripAdvisor makes planning easy no matter the trip type." (Tripadvisor, 2021). This platform is intensely utilized by international travelers and post numerous comments about their experiences. Due to that, we selected this travel platform as an online community area.

The data collection period started between 15-20 February 2021. When the authors decided to collect data, some criteria were put forward. First, all reviews will be collected. Secondly, there were some reviews posted in different languages such as Chinese. Therefore, the authors disregarded these reviews since we do not know different language. After we determined these criteria, we selected carefully all reviews and transferred these into the Microsoft Word to be analyzed. A total of 341 reviews were included in this study.

The collected data was examined through a content analysis which is defined as "a research method which allows the qualitative data collected in research to be analyzed systematically and reliably so that generalizations can be made from them in relation to the categories of interest to the researcher" (Haggarty, 1996, p.99). By adopting this technique, the main components occurred through a data reduction. As a data reduction strategy, we used a coding process in our data analysis. Two independent coders involved in the data analysis. At first, the coders independently examined the raw data. Then, they revealed some themes that can identify the literary tourism experience. After this process, the coders compared the main themes with each other. Till a consensus, a debate on these themes was performed. After the coders determined the main themes, they consulted an exterior expert for a final decision. So, a reliability and validity has been ensured. As a result of this stage, the key components that identify the literary tourism experience were revealed.

Findings and Discussion

Some travelers who visit the museum reviewed about how to arrive the location. They tried to give some helpful suggestions for potential visitors. This can be a major aim of the User Generated Content (UGC) travelling areas because this aims to give important advice for travelers. Further to this, few travelers highlighted the importance of price in their experience. Some of travelers commented that it is worth to visit for entrance fee whereas some of them reviewed that the entrance fee is high for visit. For example, one traveler said following: "*Paying 30 lira for admission (without the guide) might not be worth if you haven't read the book.*" Thus, the utilitarian function of experience depends on reading the book at first. The majority of travelers commented about the content of the book in their reviews. Generally, the common characters, the objects that was read in the book, and the environment that the Kemal and Füsun visited was the frequently mentioned topics about the book.

As a result of the analysis, this study occurred four main themes that constitutes the literary tourism experiences. These were listed as visualization, a sense of nostalgia, awe, and memorable. All components including indicators are presented. Then, we illustrated some direct quotations from the travelers. Lastly, we discussed each component by highlighting the role of it in previous research.

Figure 1. Components of travelers' literary tourism experience

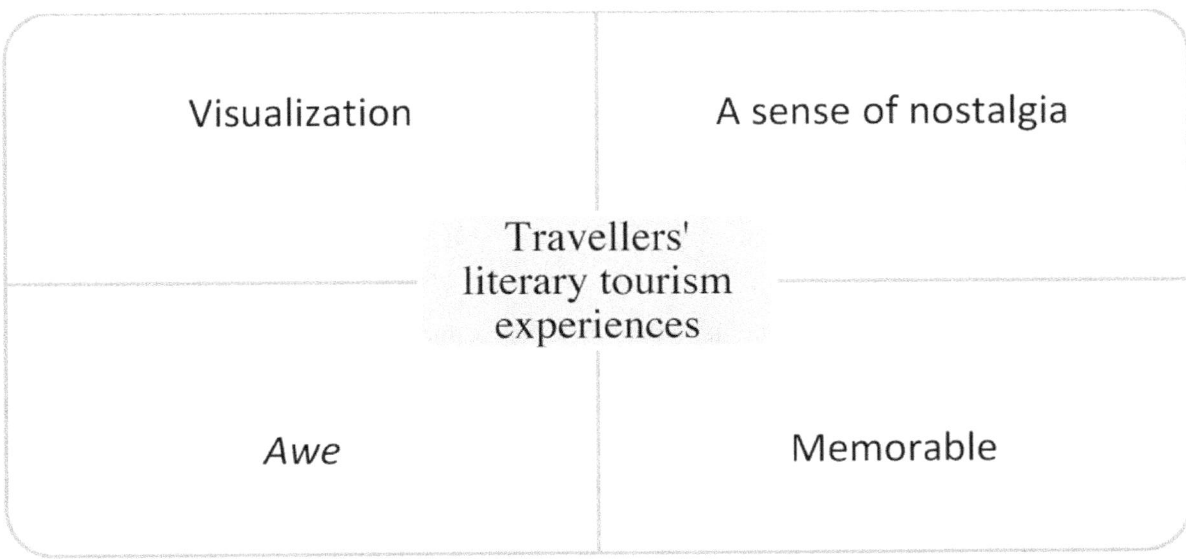

Visualization

The majority of travelers emphasized the importance of the characters and events' visualization. This situation has been led to make a real connection between museum and book for travelers. The most mentioned indicators for visualization were picturing shops, and market stalls that were visited by the novel character, some scenes of the novel, objects that travelers read about it in the book, the stuff remembering Kemal and Fusun, and physical manifestation of the book. Three travelers emphasized these indicators as below.

- A special type of museum. It visualizes the charters and the scenes of the novel. So creative!
- You must check it! I loved it so much! The atmosphere, the small displays showing all the stuff remembering Kemal and Fusun.

- It visualizes and confirms the atmosphere you already had in your mind while reading the book (at least for me).

All these indicators evoked a sense of reality feelings for travelers. For example, two travelers reviewed the following statements.

- The museum is a fascinating approach to blurring fact and reality through actual objects belonging to a particular period, in the context of fictional lives.
- The items in the boxes are so real, again, it is such an effective illusion you believe these people actually existed and lived in this house.

In particular, one of the travelers commented that this museum gives a sense of visiting the novel's page by going to seeing every object. This was highlighted as such:

- It is like you are walking through the chapters of the novel recalling the stories by looking at the little objects that gathered in).

Travelers who are sought to experience a place, object, or attraction desire to visualize in their mind. Thus, this is particularly essential for travelers desiring to see literary attractions in the destination. As abovementioned, these indicators will serve the literary tourism experience literature.

A Sense of Nostalgia

Travelers emphasized the role of nostalgia in their experience, and they interpreted their experience in a different way. According to them, all objects that evoke novel and its fiction have encouraged them to feel a sense of nostalgia. Furthermore, the fact that they see these objects brought them to past livings of Kemal and Füsun. Moreover, surroundings of museum were giving them a sense of being lived. In particular, a reviewer emphasized the following: Even for those who have not (yet) read the book, it's totally worth of visiting. It has a great collection of items on display, which together with a beautiful building, in which the museum is located, create a nice nostalgic atmosphere.

Another important indicator that addresses to a sense of nostalgia was related to past livings of Istanbul. Longing for the past was the major component of a sense of nostalgia for travelers. The majority of travelers who commented about a sense of nostalgia highlighted that the book already explains the past of Istanbul (i.e., 1970's or 1950's) and insider of the museum gives this sense about life of Istanbul, middle classes, and cultures. To address this, some reviewers directly commented about nostalgia of Istanbul's life. For example, four travelers visiting the Museum Innocence highlighted their experience as below.

- A fascinating idea and he really pulls it off. It evokes a whole era in the 1970s when Istanbul was becoming more westernized. It cleverly narrates a summary of the book through the pieces of 1970s Istanbul that were collected.
- Museum is located in beautiful Cukurcuma neighborhood and shows everyday life and culture of Istanbul during the period in which the novel is set.
- Pamuk's fantastic concept of a novel + museum is fully realized with a fascinating collection of everyday objects offering a detailed insight into middle-class life in Istanbul from the 50s onwards.

- This museum brings you to a new world, the magical world from the book, a different time, different culture.

Many researchers try to find out the role of nostalgia and longing for the past in tourism experience (Christou, 2020; Suntikul, 2017). Nostalgia is known as "a self-conscious, bittersweet but predominantly positive and fundamentally social emotion" (Sedikides, et al., 2015, p.190). In tourism context, travelers frequently desire to visit some destination or see various attractions in order to meet their nostalgia wants (Shi et al., 2021). It is considered as a major travel driver for potential travelers (Sellick, 2004). Nostalgia emotions are not only positive but also can be negative (Holak, & Havlena, 1998); however, in our study, it was determined that nostalgia experience in literary places are positive emotions. In general, nostalgia gives individuals a sense of lived old times. Our research found out that travelers lived the places and times in the novel. Moreover, it is emphasized that museums can be a good example of nostalgia attractions for travelers (Shi et al., 2021).

Awe

A sense of awe during travelers' visitation to museum was the frequently mentioned component of literary tourism experience. This sense was generally related to their experience in the museum. Furthermore, they saw the objects and the idea of creating a museum through objects that was mentioned in the novel as awe in their reviews. The frequently mentioned indicators for this component were listed as: amazing, wonderful, great, perfect, excellent, unique, incredible, spectacular, pleasant, fantastic, fabulous, extraordinary, interesting, fascinating, and outstanding. This was commented by the following four travelers.

- Tricky to find but well worth the visit and quite incredible to see the collection of objects mentioned in the novel.
- This museum is truly an interesting, eye-opening exhibit. it is directly related to the book of the same name, by Orhan Pamuk, but having not read the book I still found this museum quite extraordinary. it tells a story and showcases different aspects of the city in a unique and fascinating way and would definitely recommend a visit to anyone looking for something to do in Istanbul. not far off Istiklal.
- I would love to meet the author as he must have an amazing mind to have conjured up this unique experience for us to share.
- Perfect viewing of all the details of the novel...brought back memories of the read but also an interesting visual insight into life in Istanbul in 1970's and 1980's.

Studies investigating a sense of awe are limited examined in the tourism literature. This component is the highly desired sense by travelers and tourism service providers because it shows that a praise occurs (Coghlan, Buckley, &Weaver, 2012). According to Coghlan, Buckley, and Weaver (2012), three main components of awe exist. They are listed as: 'Physiological response,' 'Comparative uniqueness' and 'Schema-changing.' Physiological response consists of responses including stunning, breathtaking, overwhelming and so on whereas comparative uniqueness consists of responses including amazing, surprising, wonderful, unique and so on. The last one includes responses such as exiting, fascinating, motivating and so on. The sense of awe has a determining role on tourist satisfaction and tourists' behavioral consequences (Faullant, Matzler, & Mooradian, 2011). This empirical finding was confirmed by Tian et al. (2015).

Memorable

For the majority of travelers, experiencing this museum and reading the book in previous makes their moment as memorable post-trip. When considering the abovementioned components of literary tourism experience, all of these have created a memorable experience for travelers. When travelers described their memorable experience, they commonly mentioned some indicators such as unique experience, memorable, different experience, one of the most important highlights of their trip, an experience to remember, and memorable adventure. The memorable aspect of travelers' experience has been expressed differently. Four travelers emphasized the role of memorable literary experience as following.

- It generates discussion with others, and it is a memorable experience in this wonderful city.
- Unique museum, very detailed and lovingly arranged. It is about the sad and very romantic love story of Füsun and Kemal.
- It was certainly one of the highlights of our trip. Not easy to get to or find but well worth it.
- You'll remember details to appreciate the exhibits and will get in free if you bring book!

All reviews posted about memorable were positive experiences. So, a positive memorable tourism experience is defined as "a tourism experience positively remembered and recalled after the event has occurred" (Kim, Ritchie, & McCornick, 2012, p.13). This positive experience enables for destinations to create loyal visitors (Coudounaris & Sthapit, 2017). Moreover, it is a major antecedent of tourist' satisfaction and affective commitment (Zhong, Busser, & Baloglu, 2017). Further to this, tourists desire to experience memorable moments in the services that they received from the destination or businesses (Tung & Ritchie, 20211). Kim, Ritchie, and McCornick (2012) illuminated the role of memorable tourism experience and underlined that tourist who experience memorable moments in the destination have some aspects such as hedonism, novelty, local culture, refreshment, meaningfulness, involvement and knowledge. As an important attraction, museums have a vital role in creating a memorable tourist experience (Seyfi, Hall, & Rasoolimanesh, 2020). In addition, when considering that memorable tourism experience is impacted by destination and country image (Zhang, Wu & Buhalis, 2018), the memorable part of literary tourism is experience is extremely important for destination.

CONCLUSION AND IMPLICATIONS

Contribution

The purpose of this research is to explore components of international travelers in a literary place, namely the Museum of Innocence. This study adopted a qualitative research approach to reach our purpose. For this purpose, we examined international travelers' reviews posted on a popular travel website (i.e., TripAdvisor). As a result of data analysis, four main components revealed: visualization, a sense of nostalgia, awe, and memorable. These components are extremely useful for relevant literature to capture the main picture of literary tourism experience and we believe that these findings would contribute to field of tourism experience in a literary tourism context.

Theoretical Implications

This study has considerable theoretical implications for relevant literature. The role of awe has been implied in the literary tourism by Herbert (2001) by grouping the tourists into some classifications. The importance of this component was highlighted by previous tourism studies (Coghlan, Buckley, &Weaver, 2012; Faullant, Matzler, & Mooradian, 2011; Tian et al., 2015). Furthermore, Atsız, Cifci, & Rasoolimanesh (2021) explored as a major dimension of food experience in their qualitative research. Although some studies cover the awe and explored it as a dimension in the prior research, it has not yet received vital attention from tourism scholars. Somewhat surprisingly, our study ascertained the awe as a major part of literary tourism experience. This will be extremely useful for literary tourism experience literature in terms of completing the main dimensions of this type of experience. Further, we believe that this component will help to tourism experience literature as well. Thus, it can be regarded that understanding the awe component in this research will leverage the tourism experience and literature tourism literature.

The pertaining literature concluded that memorable experiences of tourists have been more important than standard experiences of tourists in the destination (Kim et al., 2012). Memorable tourism experience has been investigated in various types of tourism such as food tourism (Sthapit, 2017), adventure tourism (Liew et al., 2020), cultural tourism (Seyfi et al., 2020), etc. In addition, many studies have examined the memorable tourism experience in terms of conceptualizing and examining memorable experiences of tourists (Seyfi et al., 2020). To the best of our knowledge, there is no research the examines the role of memorability in the literary tourism experience. Further to this, all findings that was put forward by different scholars are dissimilar and all conceptual frameworks that was recommended are repeatedly subjective, thereby lacking a coordinated direction (Chen & Rahman, 2018). This research, in a literary tourism context, revealed the importance of memorable experiences. Thus, it is concluded that memorable aspect is so crucial in terms of tourist experience. This component will be particularly helpful for tourism scholars who investigate the literary tourism experience in the destination or specific literary attractions.

Finally, other components that revealed visualization and a sense of nostalgia as a components of literary tourism experience are vital part of travelers' experiences. These components are actually well explained in the literature. Thereby, it is not surprising to explore these components as the major themes of literary tourism experience. However, we believe that these components are confirmed the previous scholars' statements and research that explores the literary tourism experience.

Managerial Implications

Our study has important implications that should be taken into consideration by managers and practitioners. First, it was seen that literary attractions have a major role in travelers' experience. Therefore, destination planners can investigate which books promote Istanbul and can build some literary buildings or items for travelers who desire to experience literary attractions because these create memorability and a sense of awe for travelers. Moreover, destination managers can promote their attractions via literary tourism context. For instance, in a sense of nostalgia component, travelers highlighted that the museum gives them a sense of past livings of Istanbul. Thus, promoting their destination via literary attractions can be extremely beneficial and motivate other potential travelers who want to seek literary experience.

LIMITATIONS AND FUTURE RESEARCH LINES

This research is not free from limitations that should be benefited by future studies. This study examined online reviews posted by international travelers. Thus, future research can reach to travelers visited this place and make an interview for a better understanding of literary tourism experiences through various techniques. Moreover, this study adopted a qualitative research approach in the research design. Future research can overcome this issue by adopting quantitative research approach. By doing so, future studies can examine the literary tourism experience through a questionnaire. Finally, other variables that play a vital role in tourist experience such as tourist satisfaction and behavioral consequences of tourists can be examined within the scope of literary tourism experience.

REFERENCES

Atsız, O., Cifci, I., & Rasoolimanesh, S. M. (2021). Exploring the components of meal-sharing experiences with local foods: A netnography approach. *Current Issues in Tourism*, 1–18.

Baleiro, R., & Quinteiro, S. (2018). *Key concepts in literature and tourism studies*. Academic Press.

Busby, G., & Klug, J. (2001). Movie-induced tourism: The challenge of measurement and other issues. *Journal of Vacation Marketing*, *7*(4), 316–332. doi:10.1177/135676670100700403

Çevik, S. (2018). Kurgusal Mekânlarda Edebiyat Turizmi Deneyimi: "Masumiyet Müzesi" Örneği. *Art-e Sanat Dergisi*, *11*(22), 281–311. doi:10.21602duarte.466075

Çevik, S. (2020). Literary tourism as a field of research over the period 1997-2016. *European Journal of Tourism Research*, 24.

Chen, H., & Rahman, I. (2018). Cultural tourism: An analysis of engagement, cultural contact, memorable tourism experience and destination loyalty. *Tourism Management Perspectives*, *26*, 153–163. doi:10.1016/j.tmp.2017.10.006

Christou, P. A. (2020). Tourism experiences as the remedy to nostalgia: Conceptualizing the nostalgia and tourism nexus. *Current Issues in Tourism*, *23*(5), 612–625. doi:10.1080/13683500.2018.1548582

Coghlan, A., Buckley, R., & Weaver, D. (2012). A framework for analysing awe in tourism experiences. *Annals of Tourism Research*, *39*(3), 1710–1714. doi:10.1016/j.annals.2012.03.007

Coudounaris, D. N., & Sthapit, E. (2017). Antecedents of memorable tourism experience related to behavioral intentions. *Psychology and Marketing*, *34*(12), 1084–1093.

Faullant, R., Matzler, K., & Mooradian, T. A. (2011). Personality, basic emotions, and satisfaction: Primary emotions in the mountaineering experience. *Tourism Management*, *32*(6), 1423–1430. doi:10.1016/j.tourman.2011.01.004

Fawcett, C., & Cormack, P. (2001). Guarding authenticity at literary tourism sites. *Annals of Tourism Research*, *28*(3), 686–704. doi:10.1016/S0160-7383(00)00062-1

Haggarty, L. (1996). What is content analysis? *Medical Teacher*, *18*(2), 99–101. doi:10.3109/01421599609034141

Herbert, D. (2001). Literary places, tourism and the heritage experience. *Annals of Tourism Research*, *28*(2), 312–333. doi:10.1016/S0160-7383(00)00048-7

Herbert, D. T. (1996). Artistic and literary places in France as tourist attractions. *Tourism Management*, *17*(2), 77–85. doi:10.1016/0261-5177(95)00110-7

Holak, S. L., & Havlena, W. J. (1998). Feelings, fantasies, and memories: An examination of the emotional components of nostalgia. *Journal of Business Research*, *42*(3), 217–226. doi:10.1016/S0148-2963(97)00119-7

Hoppen, A., Brown, L., & Fyall, A. (2014). Literary tourism: Opportunities and challenges for the marketing and branding of destinations? *Journal of Destination Marketing & Management*, *3*(1), 37–47. doi:10.1016/j.jdmm.2013.12.009

Jiang, L., & Yu, L. (2019). Consumption of a literary tourism place: A perspective of embodiment. *Tourism Geographies*, *22*(1), 127–150. doi:10.1080/14616688.2019.1586985

Kassean, H., & Gassita, R. (2013). Exploring tourists push and pull motivations to visit Mauritius as a tourist destination. *African Journal of Hospitality, Tourism and Leisure*, *2*(3), 1–13.

Kim, J. H., Ritchie, J. B., & McCormick, B. (2012). Development of a scale to measure memorable tourism experiences. *Journal of Travel Research*, *51*(1), 12–25. doi:10.1177/0047287510385467

Liew, J. F., Sulaiman, Z., Zakwan, N., & Awang, S. R. (2020). Memorable adventure tourism experience: Constructs and consequences. *Asian Journal of Arts. Culture and Tourism*, *2*(3), 14–22.

Neace, C. (2015). *Literary Tourism: American Author Museums*. Available at: https://bookriot.com/literary-tourism-american-author-museums/

Quinteiro, S., Carreira, V., & Rodrigues Gonçalves, A. (2020). Coimbra as a literary tourism destination: Landscapes of literature. *International Journal of Culture, Tourism and Hospitality Research*, *14*(3), 361–372. doi:10.1108/IJCTHR-10-2019-0176

Robinson, M., & Novelli, M. (2005). Niche tourism: An introduction. *Niche tourism: Contemporary issues, trends and cases,* 1-11.

Sedikides, C., Wildschut, T., Routledge, C., Arndt, J., Hepper, E. G., & Zhou, X. (2015). To nostalgize: Mixing memory with affect and desire. *Advances in Experimental Social Psychology*, *51*, 189–273. doi:10.1016/bs.aesp.2014.10.001

Sellick, M. C. (2004). Discovery, connection, nostalgia: Key travel motives within the senior market. *Journal of Travel & Tourism Marketing*, *17*(1), 55–71. doi:10.1300/J073v17n01_04

Seyfi, S., Hall, C. M., & Rasoolimanesh, S. M. (2020). Exploring memorable cultural tourism experiences. *Journal of Heritage Tourism*, *15*(3), 341–357. doi:10.1080/1743873X.2019.1639717

Shi, Y., Bettache, K., Zhang, N., & Xue, L. (2021). Constructing nostalgia in tourism: A comparison analysis of genuine and artificial approaches. *Journal of Destination Marketing & Management, 19*, 100488. doi:10.1016/j.jdmm.2020.100488

Squire, S. J. (1996). Literary tourism and sustainable tourism: promoting'Anne of Green Gables' in Prince Edward Island. *Journal of Sustainable Tourism*, *4*(3), 119–134. doi:10.1080/09669589608667263

Sthapit, E. (2017). Exploring tourists' memorable food experiences: A study of visitors to Santa's official hometown. *Anatolia*, *28*(3), 404–421. doi:10.1080/13032917.2017.1328607

Suntikul, W. (2017). Nostalgia-motivated Thai domestic tourism at Amphawa, Thailand. *Asia Pacific Journal of Tourism Research*, *22*(10), 1038–1048. doi:10.1080/10941665.2017.1363064

Tian, Y., Lu, D., & Wu, T. (2015). Impacts of awe emotion and perceived value on tourist satisfaction and loyalty—The case of Tibet. *East China Economic Management*, *29*(10), 79–85.

TripAdvisor. (2021). *About us*. Available at: https://tripadvisor.mediaroom.com/us-about-us

Tung, V., & Ritchie, J. (2011). Exploring the essence of memorable tourism experiences. *Annals of Tourism Research*, *38*(4), 1367–1386. doi:10.1016/j.annals.2011.03.009

UNWTO. (2018). *Tourism and culture synergies*. UNWTO.

Wang, H. J., & Zhang, D. (2017). Comparing literary tourism in mainland China and Taiwan: The Lu Xun native place and the Lin Yutang house. *Tourism Management*, *59*, 234–253. doi:10.1016/j.tourman.2016.08.008

World Travel and Tourism Council. (2020). *Travel and tourism economic impact and trends 2020*. Author.

Zhang, H., Wu, Y., & Buhalis, D. (2018). A model of perceived image, memorable tourism experiences and revisit intention. *Journal of Destination Marketing & Management*, *8*, 326–336. doi:10.1016/j.jdmm.2017.06.004

Zhong, Y. Y. S., Busser, J., & Baloglu, S. (2017). A model of memorable tourism experience: The effects on satisfaction, affective commitment, and storytelling. *Tourism Analysis*, *22*(2), 201–217. doi:10.3727/108354217X14888192562366

ENDNOTES

1 https://tr.masumiyetmuzesi.org can be visited for more information.

2 https://yigm.ktb.gov.tr can be visited for more information.

Chapter 4
The Jerilderie Letter and Others:
Using Personal Letters to Enrich Heritage-Based Tourism

Michael Thomas Fagence
University of Queensland, Australia

ABSTRACT

It is the purpose of this chapter to investigate the opportunities that lie with drawing on the personal letters of an iconic historical figure so as to implant into the process of telling his story through the medium of tourism a sensitive interpretation and meaning of historical events in which he was involved. The claim will be that input from the letters could be used to conjure a particular 'spirit' of and 'personality' for any place at which significant events occurred, and in so doing give to that place a special identity, its genius loci. It is a process of enrichment and value-adding. Using a story embedded in the history of 19th-century Australia, the study uses a custom-built form of narrative analysis to disaggregate six selected personal letters authored by the selected historical figure and to interpret their potential for enriching the telling of his story through tourism.

INTRODUCTION

With this letter [the Jerilderie Letter] Kelly inserts himself into history, on his own terms, with his own voice ... We hear the living speaker in a way that no other document in our history achieves ... the logic is associational rather than linear ... Kelly talks as though his listeners already know the details, but have failed to understand them. McDermott (2001, pp. xxviii, xxix)

One of Germany's greatest literary figures of the modern era, Johan Wolfgang von Goethe (1749-1832), is credited with the assertion that 'the written word has this advantage, that it lasts and can await the time when it is allowed to take effect'. That assertion is particularly important for the study in this chapter because it points to the almost limitless timescale through which the significance of what is written can be effective; and that is crucial for this study as its focus is on what was written about a number of inci-

DOI: 10.4018/978-1-7998-8262-6.ch004

dents that took place in the history of Australia in the nineteenth century, and the subsequent expression of them through tourism in the twenty-first century. It is also important because, in a complementing comment, Goethe claimed that 'letters are among the most significant memorial a person can leave behind them'; and that observation is especially significant for this study.

The personal letter is a particularized form of nonfiction with bedfellows such as autobiographies, diaries, essays, and memoires, and although there is evidence that these others have been drawn upon to tell stories from history there is scarce evidence that the personal letter has been as well used as those others; that is one of the challenges that is faced by this study. It is difficult to be certain about the use being made of personal letters as a resource category from the recent competently authored reviews of the interface of literature and tourism by Mura and Sharif (2017) and by Cevik (2020); but, from accumulations of past published experience, the evidence seems to show that historians and others have focused much more attention on what has been written in autobiographies, diaries, or memoires, than in personal letters.

However, for the study in this chapter it is the personal letter form of literature that has been positioned front and centre both as the subject and as the primary resource for the telling of a story that has become an important marker of significant events in the history of Australia.

One of the basic inspirations for this study has been Herbert's (2001) landmark paper in which he addressed literary places and 'the heritage experience'. In that paper, in refraining from actually denying the significance of what he referred to as literature-linked places as 'accidents of history' – such as where an author was born/lived/died, or authored a particular novel and so on – he offered the opinion that places could become the target of tourist interest because of their associations with events and spectacles, with ideas, with issues of identity, and with nostalgia. That has become an underlying theme for this study as it considers a particular event in Australian history as it is recorded in a number of personal letters – all from the same author. Between them the letters not only give attention to the facts of the situation but also, by the tone of the text and by the contextual issues that are drawn into it, generate, for the site of that particular event, what has been referred to in a number of studies as a 'spiritual loci' (Ramsay, 2008), or a *genius loci* (Lappin, 2015), with their distinctive complements of 'atmospherics' (Volgger, 2020). As Park and Stephenson (2007) have suggested, it is 'the meanings and symbols attached to [the heritage resources] that are of greater significance in shaping an essence of identity'. For this study, it is just that: it is the meanings, feelings and sentiments that are generated from the story, and the symbols that create a particular atmosphere that enfolds any historical action that has taken place at a site. After all, the letters, if from an eyewitness, offer a first-hand and sensitive account; and, as Landsberg (2004) has suggested it is at the 'experiential site … [that] an experience occurs through which the person sutures himself or herself into a larger history … the person does not simply apprehend a historical narrative but takes on a more personal, deeply felt memory of a past event through which he or she did not live' (p.2). For Landsberg, what are carried away from those sites by the tourist are 'prosthetic memories'.

What this study does is take the first-hand accounts, perceptions, and impressions from an eyewitness as they have been recorded in a number of his letters and draws from them a narrative that is respectful of the ambient circumstances of an historical event in which he participated. The pathway from this introduction to the case study gains enlightenment from brief commentaries about the position of letters within the scope of literature, the mysteries of the creation of *genius loci*, and the story-drawing power of major historical figures. A case study follows and adopts the same sequence; after the case study a concluding section offers an estimation of the usefulness of the letter form for contributing to the telling of stories from history through tourism.

BACKGROUND

Underpinning the investigations and interpretations in this study are expectations that the letter form is a *bona fide* category of literature, that in addition to expressing the facts of a situation letters have the capacity to generate a sense of the 'spirit of place' or *genius loci*, and that the degree of success in communicating the story widely is the iconic status of the writer. These are the matters considered in this section; and the sequence is repeated in the Case Study.

Literature; Letters

For a study such as this it is probably only necessary to acknowledge that literature forms 'a unique canon' of discourse – a tripartite of poetry, prose, and drama – and that it is contributed to in a diversity of 'writing' styles that are sometimes custom designed to suit the subject being written about and the purpose of the writing (Culler, 1997; Fowler, 1982). It is evident that the writing of letters seldom attracts a high level of commentary in most overviews of styles of literature; but, despite this general lack of attention, useful backgrounding commentaries can be found in Najemy's (1993) interpretation of three 'traditions' of letter writing, in Stanley's (2015) speculations about epistolary writing, and in Borich's (2020) commentary about 'creative nonfiction'.

Commentaries about this epistolary form – that is, letter-writing (see, for example King, 2018; Najemy, 1993; Spengemann, 1980) – have referred to the special contribution that it can make to the telling stories from history. They point out that letter writing is not typically bound by the conventions of the accounts recorded in such forms as the biography, and it will most likely not have been subjected to the rigours of censorship on the basis of accuracy and authenticity, of political correctness and sensitivity. Authors of letters have an unfettered opportunity to list information, and to write with emotion, passion, frankness, and with sentiment, because they suspect that their letters will not be disclosed widely, and that they might not even elicit a response.

It is usually claimed that Alexander Pope (1688-1744) was the first writer to have his correspondence published, and that this set a precedent that has been followed since the eighteenth century by many of the so-called educated middle class of professionals, politicians, novelists, and artists; there is no easily discoverable evidence that bandits, criminals, and outlaws were among these correspondents. In a quite separate strand of development some novelists have turned to the letter form for some of their texts. Examples of this would include Bram Stoker's novel *Dracula* (1897) that incorporates letters, diary entries and excerpts from newspapers, and C.S. Lewis' *The Screwtape Letters* (1942) that was based around communications exchanged between a senior demon and his nephew. More recently, and directly related to the subject matter of this study is Peter Carey's award-winning novel, *True history of the Kelly Gang* (2000). It was written in the language style use by Ned Kelly in the most famous of his letters, the Jerilderie Letter (1879).

Whilst many commentaries and critiques of the letter form point to the advantage that it can be written in a style that might readily connect with the reader, binding the reader to the story being told, and informing the reader of some of the situational cultural, environmental, and social nuances of the story; the same commentaries also point to the complications that may flow from the single viewpoint of the author. These are among the challenging issues faced by the commentary in the case study in this chapter.

'Spirit of Place'; *Genius Loci*

There is no shortage of published literature about *genius loci*, what it means and how it may be interpreted, and there are concentrations of sources that have close associations with culture, design, environment, geography, and psychology. What is being referred to by the term is a distinctive, but perhaps loose set of characteristics of uniqueness, of distinctiveness, of separateness, that can be associated with a particular place – that is, they are 'special' to that place. So fascinating is the subject that there has emerged a profusion of similar terms that includes, for example, 'spirit of place, 'sense of place', 'symbolic landscape', 'emblematic landscape' among others.

Urry (1992), without specifically using the term *genius loci* is, nevertheless, referring to something like that with his commentaries about the significance of a place as one of the factors of attraction for tourism. There are examples of such 'special places' in Ramsay's (2008) study for ICOMOS, and Lappin (2015), with her handbook on creative writing has perhaps unwittingly – but certainly usefully – set up a bridge between writers, their literary works, and touristic attraction by expressing the opinion that it is a writer's responsibility to tease out the wealth of memories, emotions, stories, voices, imagery, and inspiration that are embedded in a place that is central to any story being told. Her reference is to being 'touched by the power of place', to being aware of 'the genius abiding in a place', and to the 'guardian spirits' of the personality of places. For her, it is for writers, poets, and artists 'to interpret and recreate reality … to capture the qualities or mood of a location … to show how places can shape our personality, history and even our fate' (p.10). The general thrust of the commentaries about 'place' from these various authors is that *genius loci* or whatever similar-meaning term is used, can be fashioned into a core issue across the spectrum of literary works, and can become an important element of explanation in all forms of communication. Conceptualizations of 'place' are pitched at a number of scales. For example, at the broadest of scales Rodaway (1994) has conceptualized theme-based spaces as 'themescapes', while at the miniaturized and disaggregated scale, Schmitz (2016) has made references to 'atmospheric spaces'. As was mentioned in a previous paragraph, Park and Stephenson (2007) have participated in this exercise with their interpretations about the significance of meanings and symbols. These, and similar commentaries by Edensor and Sumartojo (2015) and Volgger (2020), position this discussion in what has been referred to as 'the atmospheric turn in culture and tourism', acknowledging that the 'atmospherics' of 'place' are, in fact, a dynamic element of constant transformation, communicating different stories to different interpreters at different times, and generating responses across a range of powerful emotions. This is the capacity of the resource to transition stories to tourism attractions; it is this that is captured in the quotations attributed to Goethe that were cited in the opening paragraph.

Iconic Historical Figures

The reasons for marshalling stories about iconic historical figures include the significance that it has for developing cultural self-improvement. For example, Jansen-Verbeke (2016) has referred to the need to be aware of 'icons and local stories to identify the places we live in and visit … [to] tell the story of a place … and [to] construct our identities' (p.279), while Herbert (2001) has highlighted the confluence of interest in people and their contributions to history with evidence of the places associated with them, and Pearce and his colleagues (2003), have taken that further by advocating the use of such icons (and particularly people as icons) as focal points for planning, development and marketing in tourism. The literary scope of biographies, diaries and memoires is dominated largely by the intimate secrets of historically important and iconic architects, artists, educationalists, entrepreneurs, judges, military leaders, musicians, philosophers, politicians, scientists, and sportspeople, while there are very few examples of similar levels of interest being shown in the lives and times of convicted felons.

Embedded within the enormous repertoire of historically famous and important people that have been exposed to public gaze through forms of biographical literature are subgroups, many of which have attracted special attention. One such subgroup is composed of folk heroes, legends and magical heroes. In his study of this group, Seal (2001) generated thirty-six hero types, each of which could be linked to a celebrity, many of which have attracted an almost cult-like following. The prospect of being faced with such an amorphous group caused the historian Max Jones (2007) to ask, 'what should historians do with heroes ?' To this group, collectively but especially particular 'members' of it, is credited a degree of importance that transcends its many constituents. For example, many commentators that write about folk heroes of one kind or another suggest that what is most important is not necessarily what the hero did (or did not do); rather it is the meanings and messages that were encoded in their actions that revealed the social and other prevailing circumstances that would be Important in the broad sweep of history. The commentary of Coffin and Cohen (1978) captures this sentiment appropriately: 'it doesn't matter … that their lives are beclouded with legend … it is not what he was but what people need him to be and make him that is important' (p. xxxiv). This is especially the case for the transitioning of an iconic figure's story into tourism, and this is demonstrated in thee case study in a later section.

RESEARCH PERSPECTIVES

Rather than working to a pre-existing template or using a particular study as a guide, this study works within the freedom advocated by Moustakas (1990) that 'each research process unfolds in its own way … any course that a researcher's ingenuity is capable of suggesting is an appropriate method for scientific investigation' (p. 43). However, by taking that route it is not a wholly one-off process that has been devised for this study. It has been structured by responding to, for example, Hsieh and Shannon's (2005) encouragement to engage in 'the subjective interpretation of the content of text data through the systematic process of coding and identifying themes or patterns' (p.1278). In addition, the choice of approach being used here has been given its shape by drawing from what has been learned from such commentaries as Mura and Sharif (2016), Riesman (2008), and from Zhang and Wildemuth (2018) on narrative analysis, and from Anderson (2007) and Braun and Clarke (2006) on thematic analysis. By responding to advocacies such as these, and with the outcomes that have emerged from the ensuing study, this study can be fitted as one more niche investigation into a range of probes into various elements of the Ned Kelly story that have implications and opportunities for expression through tourism (Fagence, 2019).

The exploration of the usefulness of Ned Kelly's personal letters offers a new and a different challenge, a niche of its own because of the nature of the letters as a resource, and because of the investigatory and interpretive perspectives and processes that are drawn into the study. It will be mentioned in the case study section that it is no longer possible to access the letters in their original form; firstly, they are in a delicate physical state, and secondly, the calligraphy of the original hand-written versions present particular challenges. So, in order to achieve access to versions that share a uniform standard of presentation, advantage was taken of access to the digital transcripts that are readily available on the internet site https://ironoutlaw.com.

Study Process

Figure 1 summarizes, in diagrammatic form, the approach taken to construct an investigatory and interpretive coding process that drew the raw information from the six selected letters authored by Ned Kelly through a process of text disaggregation, using the advocacies of the previously mentioned commentaries on narrative and thematic analysis. Although expressed in Figure 1 as a linear process, in operation, a strictly linear process that was not followed. After working through an initial information gathering, sorting, and coding phase it was realized that the conventional approach to narrative analysis was not working well. This prompted three decisions: one of these was to select as the unit of 'measurement' lines of text rather than words; the second was to confine the scope of the storytelling to a single episode; and the third was to focus the search on those matters that 'spoke' to the matter of *genius loci* and of the distinctive 'spirit' of a place about which there was an aura and a sense that something important to Australian history had happened there.

Figure 1.

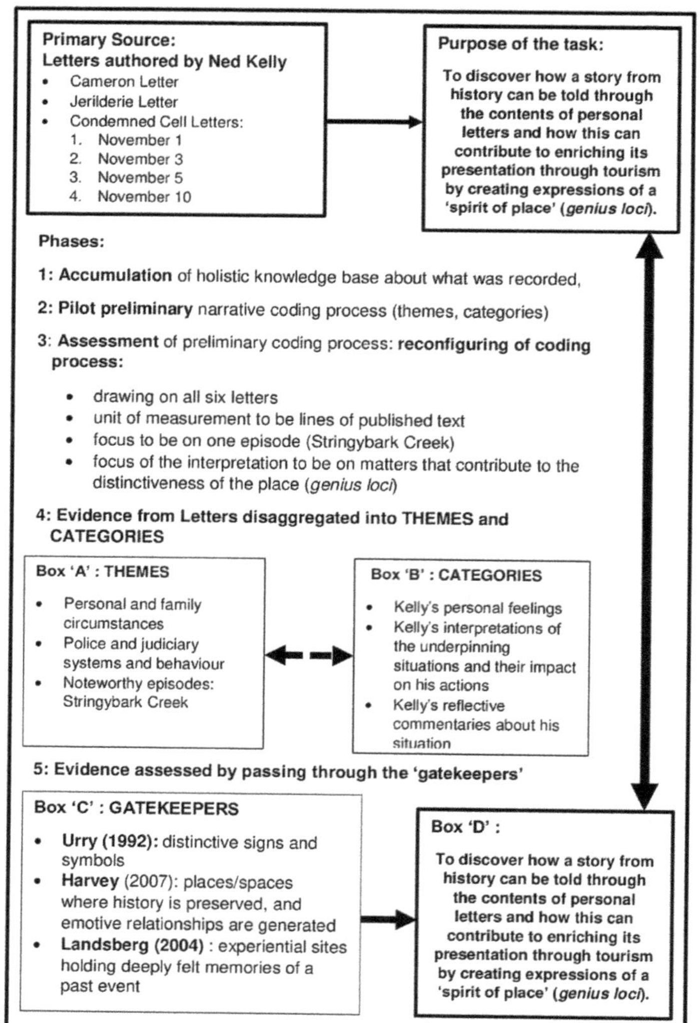

Figure 1: Study Process

The texts of each of the letters was read so as to conjure a holistic interpretation of what was reported, and the perspectives taken by the author on each of the matters he raised. Each was then re-read on at least two further occasions, and from this process a number of ideas emerged about a suitable framework that could engage with what was emerging as the most significant issues and elements. By a process of topic definition, differentiation, and then aggregation the raw data could be segregated into three themes (Figure 1: Box 'A': personal and family circumstances, police and judiciary systems, and then three noteworthy episodes). This bank of information was then reinterpreted as categories (Figure 1: Box 'B': Kelly's personal feelings, references to the underpinning situation, and then Kelly's reflective commentaries). No attempt was made to integrate the two 'boxes': rather, each of the items in the two 'boxes' were separately interrogated through the particular persuasions of the three 'gatekeepers' (Figure 1: Box 'C': Urry, 1992; Harvey, 2007: Landsberg, 2004). This had the benefit of applying each of the gatekeeper tests to every listed item, of at least minimising the possibility that important items would be overlooked, and of steering the interpretation of the data from the letters towards Box 'D' that was focused squarely on identifying story-relevant 'spirit of place' matters that could contribute to enriching the telling of the story. The target was to meet the challenge encapsulated in Harvey's (2007) comment that 'heritage sites are not simply spaces where history is preserved but by spaces where emotive relationships are generated' (p. 65)

CASE STUDY

Although Ned Kelly's most dramatic episodes occurred over only a three-year period 1878 to 1880 – often referred to as the Kelly Outbreak (McQuilton, 1979) – they have managed to capture such levels of interest, interpretation, and imagination that they have been responsible for the generation of an extensive catalogue of artworks of various kinds, dramas and music, documentaries, films, literary works (both fiction and nonfiction), and TV mini-series.

Background to the Story of Ned Kelly

The Kelly era was set in the midst of an awkward confluence of past and contemporary issues. Among the influences was a period of convict immigration from Britain, and especially from Ireland. Some Irish immigrants took advantage of recently enacted land laws, and from gold mining opportunities which they shared with many other national groups, and many Irish seemed to gravitate to careers in the police and the judiciary. Although both of Kelly's parents were of Irish decent, he was born in Australia. Pockets of tension were emerging through the nineteenth century with British Colonial rule being the principal target, with such noteworthy outbreaks of rebellion at Castle Hill (in Sydney, March, 1804), at Bathurst (September, 1830), at Forest Creek (December, 1851), at Eureka (in Ballarat, Victoria, December, 1854), and the more diverse disturbances during the Kelly Outbreak (in north east Victoria, 1878-1880). Within that general trend of disturbance there was simmering tension between urban and rural communities, between those who owned property and those who did not, and between members of the rural communities and the often intimidatory police forces and judiciary.

This was a situation that, in fiction, would have needed a 'hero' to rescue the impoverished and the intimidated from their fate; in the real circumstances of nineteenth century Australia, this situation was filled in part by the fraternity of bushrangers, of which Ned Kelly has been one that has attracted a high profile. These various circumstances are comprehensively documented by, for example, Ian Jones (2008), McQuilton (1979) and Molony (1980). It is the personal letters authored by Ned Kelly, however, that offer more specific insight into the prevailing circumstances from someone who was there to experience them.

The Letters of Ned Kelly

What are the Kelly letters, what is significant about them, and what do they contribute to the telling of his story through tourism ?

Ned Kelly was not a prolific writer of anything. No diary has been found, and the balance of probabilities is that he did not commit to any form of autobiography, diary, or memoires. It therefore falls to an examination of what may be suggested to be his six most important letters to provide some insight into his perspectives on various community and other happenings. In Table 1 there is a summary of important matters embedded in each of the six letters used for this case study. The six letters are listed in chronological order, and aggregate to 15,739 words. They offer Kelly's interpretation of events and circumstances that have been reported elsewhere in more substantial publications (including those listed previously), but the letters include insights, interpretations and observations that would be possible only from the position of a participant. As a 'setting the record straight' exercise the six letters contribute more to 'setting the scene' and to conjuring the sensations and emotions of the various incidents and episodes, than to the quantum of 'facts'.

To achieve an assessment and commentary from a consistent baseline of substance and presentation format, the sources used for this case study have been drawn from the internet site https://ironoutlaw.com where the original hand-written versions have been converted into conventional print versions. Using the approach to narrative analysis referred to in the Research Perspectives section of this chapter, the text of each of the letters was disaggregated into three broad themes (see Box 'A' in Figure 1), then into three broad categories (Box 'B'), and then subjected to interpretation using three foundation or 'gatekeeper' elements (Box 'C') derived from Urry's (1992) references to distinctiveness and symbolism, Harvey's (2007) references to places where history is preserved and emotive relationships generated, and Landsberg's (2004) references to experiential sites holding memories of past events. Although analysis reveals that there is a consistency across all six letters of criticism of the police, the judiciary and the general economic, political and social circumstances in the State of Victoria in the second half of the nineteenth century, and the impact of them on him, his family and his friends and sympathisers in particular, Kelly seems to have been particularly anxious to place his interpretation of the Stringybark Creek episode on record. It seldom occupies less than one-quarter of the content of the two letters written before his capture; however, only in one of the Condemned Cell Letters – albeit, the longest – does he refer to it again.

An early conclusion that can be drawn from the letters is encapsulated in the quotation from McDermott (2001) that precedes the Introduction to this chapter. McDermott's claim is that with the Jerilderie Letter (in particular) Kelly was 'inserting himself into history, on his own terms, with his own voice … Kelly talks as though his listeners already know the details but have failed to understand them' (2001, pp. xxviii, xxix). A second matter of importance was that his letters were targeted to particular addressees (firstly to Donald Cameron, a member of the Victorian Parliament; and secondly to the Governor of the State of Victoria). As far as can be established, Kelly had not intended that they would gain widespread dissemination. It is only in recent years that the national significance of them has emerged, that they would achieve public display and attention, and that they would make a serious contribution to framing and shaping the telling of the Kelly story through tourism.

Table 1.

Letter	Date	Words	Comments
Cameron	December 14, 1878	3,703	• Sent to Donald Cameron, in the belief that he would raise the issues of the behaviour of the police in the Victorian parliament; • Includes specific comments about the fairness of court proceedings, and about the Fitzpatrick incident at the Kelly homestead at Eleven Mile Creek; • Includes a version of the encounter of the police with the Kelly Gang at Stringybark Creek.
Jerilderie	February 10, 1879	7,386	• Includes a chronological account of the charges levelled at the Kelly family members, the court proceedings, the sentences imposed, and the occasions of false identification; • Includes a version of the Stringybark Creek incident, with some repetition from the Cameron Letter; • One of the early references to the creation of a break-away state.
Condemned Cell Letter 1	November 1, 1880	330	• Begins 'I do not pretend that I have led a blameless life'; • Includes a comment that people that live in large towns 'have no idea of the tyrannical conduct of the police in their country places', setting the scene for the persistent criticism of the police that was first raised in the Cameron Letter.
Condemned Cell Letter 2	November 3, 1880	1,609	• Includes a commentary on the failings of the court system, the issue of warrants; • Includes further commentary on the Fitzpatrick incident and the Stringybark Creek incident; • This is Kelly's last attempt to have an investigation into the court proceedings.
Condemned Cell Letter 3	November 5, 1880	1.609	• This is the only letter in which reference is made to the siege at Glenrowan, and how Kelly's plans unravelled because the police strategy was not what he expected.
Condemned Cell Letter 4	November 10, 1880	1,102	• This is a letter of a man now accepting the consequences of the court's decision; includes the statement that he 'tried every legal means for justice … it never crossed my mind for revenge'; • His final claim was that his body would be buried in consecrated ground, and that his mother would be released from Gaol before he was executed.

Notes:

1. The texts used for this tabulation can be found at https://www.ironoutlaw.com.
2. All except the first letter (to Donald Cameron, MLA) were addressed to the Governor of Victoria.

Taking McDermott's claim into account, the approach that was taken by Ned Kelly in these letters is encapsulated in the opening sequences of sentences to a number of the letters. For example,

- in the Cameron Letter he wrote 'It seems impossible to get any justice without I make a statement to someone who will take notice of it';

- in the Jerilderie Letter he wrote 'I wish to acquaint you with some of the occurrences';
- and later, in Condemned Cell Letter 3, he confirmed that 'I tried every legal means for justice … it never crossed my mind for revenge'

Throughout all of the letters Kelly sets out his account of events, and to correct what he claimed were the often-contradictory accounts from the police. It may be considered as unfortunate that his letters sometimes lapsed into colourful and often disparaging language, especially about the police and their actions, that the letters were composed with paragraph-long sentences that tended to meander across a range of subjects, and that statements were repeated (almost verbatim) across and between letters. Whilst the style and content of the letters may have appeal for film and drama scriptwriters (there have been more than ten films on various aspects of the Kelly story), they demonstrate little regard for the historian's craft of reporting. What they achieve, though, is evidence of his passion for his cause, and there is some agreement amongst commentators that it is this passion that would be likely to lead to a confrontation with the police – as it did at Stringybark Creek and Glenrowan – and that the outcome would be a site (or sites) imbued with the 'atmospherics' of anger, antagonism, danger, violence, and death, but also of bravery, courage, loyalty and survival. Drawing on Rodaway's (1994) 'themescape' terminology, the outcome would be most like a 'traumascape'. Three of the letters in particular – the Cameron, Jerilderie, and Condemned Cell Letter 2 – give special attention to the Stringybark Creek episode, the happenings at which culminated in a gunfight between a police patrol and Ned Kelly, his brother Dan and two friends Steve Hart and Joe Byrne. This episode is considered further in the next section.

The letters contribute to tourism in two ways. One of these is the very existence of the texts, facsimiles of which can be viewed in a permanent exhibition in the State Library of Victoria (Melbourne), while the original letters are on deposit in the Kelly Historical Collection of the Public Records Office (also in Melbourne, Victoria). It is for the second of the two purposes that the letters hold particular significance for tourism. The texts progress beyond the bald statement of facts to become foundational elements for generating sympathy for, empathy with and even admiration for both sets of protagonists in each of the episodes, and perhaps most particularly for that at Stringybark Creek. Using information drawn from the letters it is possible to conjure for any of the sites involved in the dramatic events, a particular 'spirit' and *genius loci,* that converts to 'an experience'. The episode at Stringybark Creek is used here to demonstrate that.

Genius loci: Stringybark Creek

The particular study site of Stringybark Creek lies in the southern quadrant of what may, for convience, be referred to as the Kelly 'footprint'. Figure 2 (Location map: Ned Kelly 'footprint') summarizes the evidence of the locations of important places in the unfolding story, and includes brief place-specific statements about their significance to the telling of the story. Stringybark Creek's remoteness was of advantage to the Kelly Gang with reasonable cross-country access to most of the family's homesteads, to the adjoining hill country, and to many of the small townships that litter this region of Victoria; but that same remoteness is a locational impediment for tourism in the modern peirod as it is more than sixty kms from the major highway that links to the two most likely sources of tourists and travellers. From Sydney, Stringybark Creek would bemore than a seven hour drive (say 700 kms) and from Melbourne it would be more than a three hour drive (say 285 kms). The site is remote, but that doesn't diminish its historic significance – just its accessibility; there is an important story from Australian history to be told and an experience to be had from an encounter with the site, no-matter how remote it is.

Figure 2.

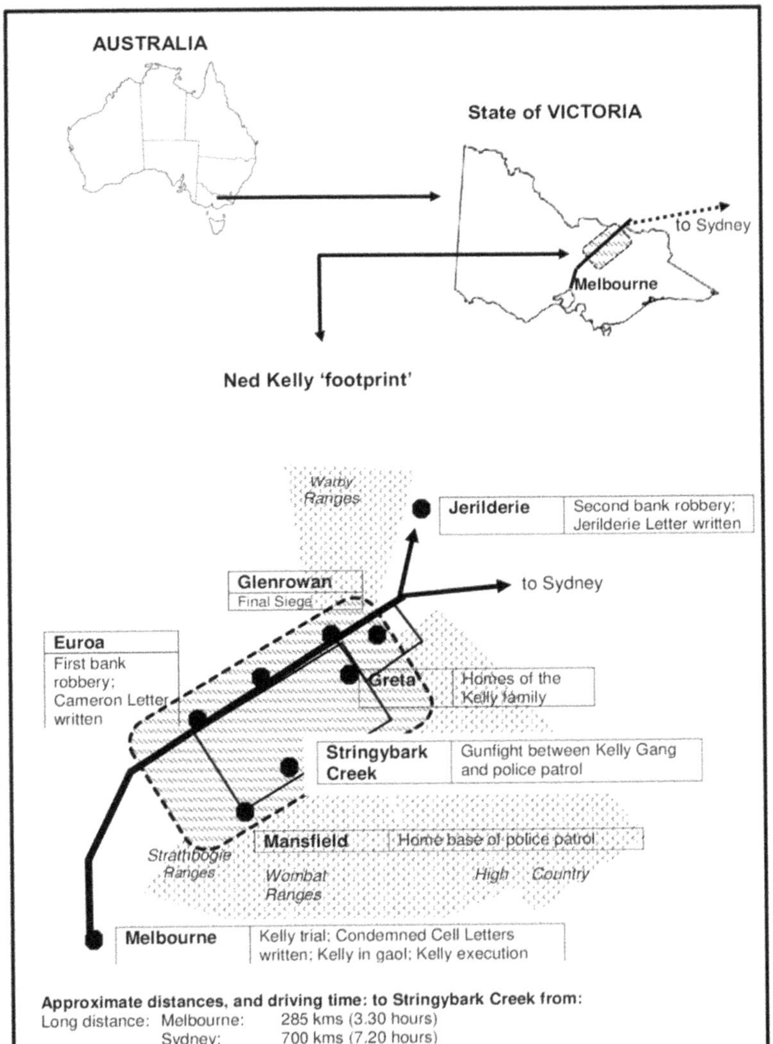

Figure 2: Location Map:

Ned Kelly 'footprint'

So, what do the Kelly letters report about the Stringybark Creek episode, and what is that episode's lasting impact, especially for tourism ?

The animus between the Kelly family and its sympathizers and the police and judiciary in north east Victoria was probably sufficient to ensure that any encounter in remote bushland would become an 'us-versus-them' confrontation; the rough bushland setting of Stringybark Creek provided almost the perfect setting with a confrontation that was not only between the 'hunters' (police) and the 'hunted' (Kelly family and sympathisers) but also between two competing ideologies. At the time of the engagement between the two forces there would have been four significant material components. The fundamental component would have been the natural bush setting, with a loose pattern of swampy sites, a number of interlocking streams and tributaries, a ground cover of sassafras scrub and speargrass, a generally intensive and extensive cover of gum and blackwood trees, and a few trails. A second component would have been another loose pattern, this time of remnant and largely open shafts from former gold ming ventures. The third would have been miner's huts and camp sites, and it is from these there is ongoing speculation that there are objects and remnants of the ballistics related to the gun-fight that are still to be found. The heritage significance and importance of the site for Victoria's cultural history relies heavily on the material evidence of these three components. As the briefing notes for an interpretive strategy of this site have suggested, 'it is the place itself that underpins its heritage'. But there is a fourth component, and this has casued disruption to the site since the time of the gunfight (1878) with bushfires, commercial logging, and clearance for use as horse paddocks. It has been this disruptive action that has obliterated many of the key physical markers on the site, especially those that were mentioned in Kelly's letters, and this has created doubt about the accuracy of the locations of markers such as the former miner's huts, the camping sites of the police, the Kelly camp site, and two prominent trees that have often been referred to as the Kelly 'target' tree, and the Kennedy tree (where the third of the policemen was killed). Melding tourism responses to these various problem areas at the site could be dealt with by summoning the structure used by Kerstetter, Confer and Graefe (2001); in their case study, they considered touristic interest could be differentiated along a spectrum that extends from interest to a 'full-fledged history buff' to an opposite pole where the level of interest would appeal to a 'history greenhorn'. This would meet the challenges posed by the three 'gatekeepers' (Urry, Harvey and Landsberg) used in this study.

Iconic Historical Figure: Ned Kelly

Finally, what do the letters tell about the writer ? There are, in fact, two questions here, with one asking 'what the letters tell' about the Ned Kelly story, and the other asking what the auther tells about himself, either directly, or by inference in the texts of the letters.

In response to the first question, Ned Kelly incorporates in the text of many of his letters commentaries about the economic, political, and social problems that he and his family and sympathisers were experiencing. In miniature it is a history lesson about conditions in nineteenth century Australia. But there are commentators that insist that the Kelly story is less than typical. MacFarlane (2012) would be one such author; he has suggested that other families that were experiencing the same conditions sought to respond to them less aggressively, thereby causing less distress and disruption to themselves and to the community at large. Molony's (1980) interpretation of the situation is that the Kelly Outbreak was almost a typical Irish response; his view was that 'The Irish have always been singularly unfortunate in the manner in which they have conducted their uprisings and in their outcome' (p.168). Ned Kelly's ancestors on both sides of the family migrated to Australia from Ireland, although he was born in Australia (near Wallan, north of Melbourne).

The pockets of community tension that emerged in north east Victoria throughout the nineteenth century has been designated as one of the provocations for the growth of the bushranging fraternity (Cashman, 2000; Seal, 1996, 2011); in other national contexts fraternities such as this have been referred to as social bandits, highwaymen, outlaws, and others. Ned Kelly's story and his experiences sit well with the story of the fraternity of bushrangers around which so many legends of the period have been formed. Gammage (1998), with his entry about Ned Kelly in *The Oxford Companion to Australian History* suggests that any legends constructed around the Kelly story have been energised 'because he offered more than bushranging .... his courage and strength spoke for the frontier ... in his daring and fatalism he chose both a short and merry life ... resistance more than crime .... protests at injustice and hints at a republic in north-east Victoria' (p.362).

As for the second question – what do the letters tell about the author ? – and although in the first of the Condemned Cell letters, Kelly claims that 'I do not pretend that I have led a blameless life', it can be suggested that through most of his letters he is attempting to justify his actions, and to offer rationales why the perpetrator of the alleged crimes could not have been him. He even claims (in the Cameron Letter) that 'I need no lead or powder to revenge my cause, and if words be louder, I will oppose your laws with no offence', and later in his fourth Condemned Cell Letter his claim was that 'I tried every legal mean for justice ... it never crossed my mind for revenge'. Even so, his prowess with armed weapons clearly exceeded that of his pen. For his letters he could draw only on the experience of a primary school education, and of a legthy period of frustration with the agents of authority (the police, judiciary). He was persistent with declarations of support and passion for his family and his many sympathisers, and he tucked away in his letters suggestions that he might create a new territory or state with an adminstration that would be more synmpathetic to the causes he espoused (a gesture of this kind was made in the Jerilderie Letter). This mixture of abilities and emotions are embedded in the texts of the letters where, it may be claimed, he adopted a significant degree of literary license with his flamboyant descriptions of the police service in Victoria, of some individuals in particular, and of the playing out of the judicial system. His letters betray no special negotiating and diplomatic skills, and his literary skills, whilst pitched at a reasonable level in the Cameron Letter, declined by the Jerilderie Letter to describe the local police as 'the ignorant unicorns' and 'a parcel of big, ugly, fat-necked, wombat-headed, big-bellied, magpie-legged, narrow-hipped, splay-footed sons of irish bailiffs'.

By the fourth of the Condemnd Cell letters he had reconciled himself with his fate, and asked the Governor to release his mother from gaol before the execution, and for his body to be buried in consecrated ground. Neither of these requests was granted; however, his body was recoverd from its interment and he was reinterred with his mother and brother in Greta cemetery in January 2013.

Kelly's life and profile, as it is recorded in conventional sources, might not achieve the status often reserved for more typical heroes, but, nevertheless, it has been more than sufficient to achieve for him a status that has attracted communication through many forms of art and literature, and his writing style of the Jerilderie Letter was imitated in Peter Carey's (2000) Booker Award prize-winning novel *True History of the Kelly Gang.*

CONCLUSION; FURTHER RESEARCH

This study has been, in some ways, a step into the 'great unknown'. By venturing into the mysteries of letter-writing as a form of literature and then drawing on personal letters as a contribution to telling a story from history through the medium of tourism, this study has had to nudge the parameters not only of literature studies, but also of narrative forms of analysis, and to a degree, visitor sensitivity to stories, so that those mysteries may be unlocked for the benefit of tourism. As the study has progressed the focus has sharpened to interpreting how letters can be used to add value to the telling of the story by conjuring a sense of the 'spirit' of what is told through the letters, and, particularly, to create a 'spirit of place', a distinctive *genius loci,* into which the sensitivities and emotions of the story are bound.

With that as the task, this chapter first considered three important backgrounding issues – the positioning of personal letters in the prospectus of literature forms, the construction of a 'spirit' of place' from meanings and symbols in and of the story being told through the letters, and the focusing of attention on letters from iconic historical figures. These three issues were carried into the Case Study by subjecting six letters authored by the iconic historical figure Ned Kelly from the history of Australia, and to assess them using a custom-built process of investigation, interpretation and coding derived from narrative analysis. Each of the selected letters was subjected to a series of disaggregations so as to determine what was being written in them, what was the intended purpose of them, how the message was expressed, what they communicated about the author, and how they may be used to enrich the telling of the story through tourism. Figure 1 reveals the process, and Table 1 summarizes the logistics and the content of the letters. As the trend of the considerations was in need of being refined, it was decided to focus important elements of the study on the specific story episode of Stringybark Creek, and this led to a conclusion that one of the most important contributions of personal letters to storytelling about events in history was their capacity to communicate (from an eye-witness, personal and first-hand account) attitudes, feelings, and sentiments that were the special ingredients of a place's 'spirit', and its *genius loci.*

Further Research

Most sections within the Case Study section were primed to address a variant of the question 'what do the letters tell us about …?'. This was the key to the conclusion mentioned in the last sentence of the previous paragraph; and it is the same key that could unlock the future for further research into the contribution that letters could be expected to make to telling stories from history through the medium of tourism. The key, whilst embracing the capacity to influence the factual content of the storytelling, is to draw out from the letters almost anything that will help to establish the sensitivities of the situation; and that includes the especially important matter of the *genius loci,* the 'spirit' of any location that is involved in the action. This is particularly so for tourism because one of its core attractions is that it takes place where something happens, where something can be seen, where something can be experienced, and so on. Each of the three 'gatekeepers' used for the assessment phase of this study referred to this as the 'distinctive signs and symbols' (Urry, 1992), as 'where history is preserved and emotive relationships established' (Harvey, 2007), and where 'deeply felt memories of a past event' are forged (Landsberg, 2004). This will be a persistent challenge for expressing history through tourism – it is not only about what happened, but why ?, why there ?, how and with what outcomes ?, who was impacted, why and how ?, and importantly, what is the legacy ? All of these issues have a need of facts that can be gathered from conventional sources; but they also have an emotions-based need to be met on-site, where the imagination, whilst at the site, can be enriched beyond the scope of the facts and can add value to them. Although the Canary and Kozicki (1978) edited volume was designed to address the writing of history as a literary form, it has messages for the research conducted here. For the transformation from literary sources to a tourism expression those sources need to be a 'cognitive instrument' that offers challenge, comfort, explanation, identity, knowledge, stimulation and so on; as Goethe has claimed 'the written word … lasts and can await the time when it is allowed to take effect' and letters that can be read and re-read 'are among the most significant memorial a person can leave'. The treatment here of the Kelly letters has been a sample of the evidence.

REFERENCES

Anderson, R. (2007). *Thematic Content Analysis (TCA)*. Retrieved from www.rosemarieanderson.com/wp-content/uploads/2014/08/ThematicContentAnalysis.pdf

Borrich, B. (2020). Creative Nonfictions. In S. Belasco, T. Gaul, L. Johnson, & M. Soto (Eds.), A Companion to American Literature (Vol 3, pp. 448-464). John Wiley and Sons. doi:10.1002/9781119056157.ch90

Braun, V., & Clarke, V. (2006). Using thematic analysis in psychology. *Qualitative Research in Psychology, 3*(2), 77–101. doi:10.1191/1478088706qp063oa

Canary, R., & Kozicki, H. (Eds.). (1978). *The Writing of History: Literary Form and Historical Understanding*. The University of Wisconsin Press.

Carey, P. (2000). *True History of the Kelly Gang*. University of Queensland Press.

Cashman, R. (2000). The heroic outlaw in Irish Folklore and Popular Culture. *Folklore, 111*(2), 191–205. doi:10.1080/00155870020004602

Cevik, S. (2020). Literary Tourism as a field of research over the period 1997-2016. *European Journal of Tourism Research, 24*, 2407. https://ejtr.vumk.edu/index.php/about/article/view/409

Coffin, T., & Cohen, H. (1978). *The Parade of Heroes*. Doubleday.

Culler, J. (1997). *Literary Theory: A Very Short Introduction*. Oxford University Press.

Edensor, T., & Sumartojo, S. (2015). Designing atmospheres: Introduction to Special Issue. *Visual Communication, 14*(3), 251–265. doi:10.1177/1470357215582305

Fagence, M. (2019). Creating heritage for tourism: 'consuming history', 'prosthetic memories' and the popularization of a folk hero's story. In C. Palmer & J. Tivers (Eds.), *Creating Heritage for Tourism* (pp. 101–114). Routledge.

Fowler, A. (1982). *Kinds of Literature*. Clarendon Press.

Gammage, B. (1998). Kelly, Edward 'Ned. In G. Davison, J. Hirst, & S. Macintyre (Eds.), *The Oxford Companion to Australian History* (pp. 362–363). Oxford University Press.

Harvey, D. (2007). (Re)creating culture through tourism: Black heritage sites in New Jersey. In M. Smith (Ed.), *Tourism, culture, and regeneration* (pp. 59–68). CABI.

Herbert, D. (2001). Literary Places, Tourism, and the Heritage Experience. *Annals of Tourism Research, 28*(2), 312–333. doi:10.1016/S0160-7383(00)00048-7

Hsieh, H.-K., & Shannon, S. (2005). Three approaches to qualitative analysis. *Qualitative Health Research, 15*(9), 1277–1288. doi:10.1177/1049732305276687 PMID:16204405

Jansen-Verbeke, M. (2016). Tourismification of cultural landscapes: synergies between tangible and intangible heritage resources. *Proceedings of TCL2016 Conference*, 276-281.

Jones, I. (2008). *Ned Kelly: a short life*. Lothian Books.

Jones, M. (2007). What should historians do with heroes? Reflections on nineteenth- and twentieth-century Britain. *History Compass*, *5*(2), 439–454. doi:10.1111/j.1478-0542.2007.00390.x

Kerstetter, D., Confer, J., & Graefe, A. (2001). An exploration of the specialization concept within the context of heritage tourism. *Journal of Travel Research*, *39*(3), 267–274. doi:10.1177/004728750103900304

King, R. S. (2018). *Writing to the World*. Johns Hopkins University Press.

Landsberg, A. (2004). *Prosthetic Memory: The transformation of American remembrance in the age of mass culture*. Columbia University Press.

Lappin, L. (2015). *The Soul of Place*. Travellers' Tales.

MacFarlane, A. (2012). *The Kelly Gang Unmasked*. Oxford University Press.

McDermott, A. (2001). *The Jerilderie Letter*. Faber and Faber.

McQuilton, J. (1979). *The Kelly Outbreak: The geographical dimensions of social banditry*. Melbourne University Press.

Molony, J. (1980). *Ned Kelly*. Melbourne University Press.

Moustakas, C. (1990). Heuristic Research: Design, methodology, and applications. *Sage (Atlanta, Ga.)*.

Mura, P., & Sharif, S. (2017). Narrative Analysis in Tourism: A critical review. *Scandinavian Journal of Hospitality and Tourism*, *17*(2), 194–207. doi:10.1080/15022250.2016.1227276

Najemy, J. (1993). *Between Friends*. Princeton University Press.

Park, H., & Stephenson, M. (2007). *A Critical Analysis of the Symbolic Significance of Heritage Tourism*. Retrieved from https://repository.usp.ac.fj/8692/1/Article_49_HBMSU.pdf

Pearce, P., Morrison, A., & Moscardo, G. (2003). Individuals as tourist icons: A developmental and marketing analysis. *Journal of Hospitality & Leisure Marketing*, *10*(1-2), 63–85. doi:10.1300/J150v10n01_05

Ramsay, J. (2008). *Cultivating Landscape spirits to create an identity*. Retrieved from https://openarchive.icomos.org/id/eprint/126 on 31/01/2021

Riesman, C. (2008). Narrative Methods for the Human Sciences. *Sage (Atlanta, Ga.)*.

Rodaway, P. (1994). *Sensuous Geographies*. Routledge. doi:10.4324/9780203306864

Schmitz, H. (2016). Atmospheric Spaces. *Ambiances*. https://journals.openedition.org/ambiances/711

Seal, G. (1996). *The Outlaw Legend*. Cambridge University Press.

Seal, G. (2001). *Encyclopaedia of Folk Heroes*. ABC-CLIO.

Seal, G. (2011). *Outlaw Heroes in Myth and History*. Anthem Press.

Spengemann, W. (1980). *The Forms of Autobiography*. Yale University Press.

Stanley, L. (2015). The Death of the Letter? Epistolary Intent, Letterness and the Many Ends of Letter-writing. *Cultural Sociology*, *9*(2), 240–255. doi:10.1177/1749975515573267

Urry, J. (1992). The Tourist Gaze Revisited. *The American Behavioral Scientist*, *36*(2), 172–186. doi:10.1177/0002764292036002005

Volgger, M. (2020). Staging *Genius Loci*: Atmospheric Interventions in Tourism Destinations. In M. Volgger & D. Pfister (Eds.), *Atmospheric Turn in Culture and Tourism* (pp. 139–151). Emerald Publishing.

Wildemuth, B. (Ed.). (2018). *Applications of Social Research Methods* (2nd ed.). ABC-CLIO.

Zhang, Y., & Wildemuth, B. (2018). Qualitative analysis of Content. In B. Wildemuth (Ed.), *Applications of Social Research Methods* (2nd ed., pp. 318–329). ABC-CLIO.

Chapter 5
Death and Resurrection on the If Island:
Revisiting Episodes of The Count of Monte Cristo

Adriana Coelho Florent
Aix-Marseille University, France

ABSTRACT

The castle of If is one of the most popular tourist attractions in Marseille. This 16th-century fortress was a prison from 1540 until the First World War. Tourists from all over the world are drawn by the presence of a prisoner who has never been there because he never existed: Edmond Dantès, the Count of Monte-Cristo. This chapter analyses how this castle became a confluence space between fiction and reality and a venue for various film adaptations of Dumas' novel. It also shows how this romantic pamphleteer managed to overcome the paradox of giving authenticity to a tourist spot without paying any attention to historical facts or verisimilitude. The chapter draws on the concepts of literary and dark tourism, the concept of aura linked to authenticity in the work of art, and the concept of literary tourist.

INTRODUCTION

The Château d'If, situated on the island of If, is twenty minutes away by ferry from Marseille's Vieux Port. It is one of the most visited locations of France's second biggest city (100,000 visitors a year). This small 800-square-meter fortress was built in the sixteenth century to defend the city's sea entrance. As early as 1540 and until World War I, it was used as a political prison. Although the Château d'If has a limited number of cells, its major influence inside the dark tourism movement is undeniable. This is partly due to its attractive historical background linked with the specificities of the prison islands: the impossibility to escape, the presence of dissident groups throughout French history - Huguenots, Napoleon's defenders, rebels of 1848… It is also known for its stories of quaint and famous characters,

DOI: 10.4018/978-1-7998-8262-6.ch005

like the rhinoceros sent by the king of Portugal to the Pope or also the one about the wizard that could bring fish to life.

However, neither the natural beauty of the island nor the historical interest of the fortress could explain how delightful the walk around the Château d'If really is. Moreover, what attracts the tourists visiting from all over the country, and even from all over the world, is the presence of a prisoner that has never set foot on the island, for the simple reason that he did not exist: Edmond Dantès, best known as the Count of Monte-Cristo. The aim of this article is to show in what ways the Château d'If has become a turning point between fiction and reality through literary tourism, but also through its use as a setting for numerous adaptations of the famous novel written by Alexandre Dumas. Furthermore, it is important to note that this romantic novelist, himself fond of cultural tourism, has given authenticity to a touristic location, beyond all historical accuracy and likelihood.

After having introduced the island of If as a geographical and historical space, this chapter will look at Dumas' novel and its protagonist, a romantic hero whose fame turned the island of If into an international tourist attraction. Paradoxically, the "willing suspension of disbelief" has been reinforced by the multiple movie adaptations of *The Count of Monte-Cristo*. The example of the celebrity of the Château d'If through a literary character raises the delicate question of authenticity in relation to literary tourism.

THE REFERENTIAL LOCATION: AN ISLET FULL OF HISTORY

The island of If is located on the archipelago of Frioul, situated in the bay of Marseille, approximately 4 km from the coast. the origin of its name is mysterious, but it is well known that it does not refer to the tree of the same name in French, "If", the yee tree, since this tree is not native to the Mediterranean (Bouvier, 2014, p.8). It is the closest and the smallest of the four islands that constitute the archipelago. It is filled with limestone reef with uneven rocks. Because of its position, the archipelago, which controls the maritime access to the bay of Marseille, was occupied during the Antiquity. However, If, which was 290 meters long and 168 meters wide, was first a refuge for fishermen and hunters. In 1516, King Francois the 1st, travelling through Marseille after the battle of Marignan, notices the strategical location of the islet and starts the construction of the Château d'If. Legend has it that the king's visit was due to the presence of a peculiar guest that came for a short stay on the island: the famous Indian rhinoceros, sent by D. Manuel, king of Portugal, to the Pope Léon X, the first animal of its kind to come to Europe. Its portrait was painted by Albrecht Dürer in 1515 following anonymous descriptions.

The fortress would allow to counter any attempts of invasion from the foreign boats. It would also prevent the eventual rebellions of the people of Marseille, which only became part of France's kingdom in 1481. The structure, built from 1529 to 1533, had first a great round tower, then a dungeon flanked by two short towers. It also had outbuildings, three wells, one windmill and a chapel.

Although well-built and well-armed, the fortress has rarely been used to defend the city. However, its configuration was adequate for another function: being a prison. The idea was not to make it a penitentiary but rather use the few dungeons carved into the rocks to imprison political dissidents that were suspected to escape with allies and even with the help of the Marseille population. One of the first prisoners was the brother of Valère des Champs, who, during the plague of 1580, had acquired the reputation of a holy healer and years later was accused of witchcraft. He was burnt alive in Aix-en-Provence, in 1588, on the 23rd December.

The rooms on the first floor, which could be rented for the richest visitors, welcomed the Polish prince Jean Casimir in 1638 and the soon to be Count Mirabeau in 1774. Finally, General Kléber, or rather its corpse, brought from Cairo in 1801 by Napoleon's troops back from Egypt, remained in a lead casket under the chapel until 1818.

From 1681 onward, the dungeons, on the ground floors, were used to imprison Protestants, due to the revocation of the edict of Nantes. The prisoners were also rioters from the 1848 revolution and Republicans from Napoleon the 3rd *coup d'état* in 1851. Finally, German and Alsatians prisoners were transferred to the Château d'If during the arrival of the French army in Alsace at the beginning of World War I, but since March 1880, the castle has not been a prison anymore (Laget, 1954).

The fortress was nearly abandoned, until 1926, when the castle was classified as a historical monument. However, the castle was unofficially open to visitors at the beginning of the nineteenth century. Shortly after, a serial novel, corresponding fully to the second category of tourist literature established by Harald Hendrix, gives the site an unexpected fame. It has now become one of the most visited locations in France. *The Count of Monte-Cristo* is indeed a text " that offers a literary representation of space which adds to its value as a tourist attraction. " (Hendrix, 2014, p .23)

EDMOND DANTÈS, ROMANTIC HERO, TRUER THAN NATURE

When the romantic writer Alexandre Dumas starts publishing the first chapter of his novel *The Count of Monte-Cristo* in *Le journal des Débats*, on the 28th August 1844, his celebrity is already well assured. It is mainly due to plays such as *Henry III et sa cour* (1829) or *Christine* (1830): controversial pieces where all the rules of classical drama are broken and where Dumas' taste for History is appreciated by the spectators. However, in 1840, his meeting with Auguste Maquet, a playwright just like him, will draw him to another genre, the serial novel. Without being the first, he becomes the most famous among the writers of the same genre, publishing eleven serial novels in seven years, always in collaboration with Maquet. Among those eleven novels, two were destined to become international best-sellers: *The Three Musketeers* (1844) and *The Count of Monte-Cristo* (1844-1846). The sources of this novel were indicated by Dumas himself, in one of his chronicles called "causeries" entitled *L'état civil du Comte de Monte-Cristo* (Dumas, 1867/1981).

It tells the journey of an unexpected trip in 1842 on the island of Monte-Cristo, located between the Island of Elba and the island of Pianosa, accompanied by Jérôme Bonaparte's son, brother of Napoleon the 1st. If the islet, populated only by many goats and a few smugglers, was to become the title of his next novel, the plot at the heart of the story comes from a true event, described in the memoirs of Jacques Peuchet, a journalist and member of the police ministry, published in 1838. Dumas and Maquet stayed faithful to the bleak incident about crime, betrayal and revenge, while developing the causes and consequences of the reported events.

Dumas and Maquet bring many changes to the spectacular, but rather simple, true story told by Peuchet. The main ingenious idea of Dumas was to project all his ardour and thirst for justice and social revenge inside the character of Edmond Dantès. Indeed, because of his fourteen apprenticeships with Abbot Faria, the young and naive sailor becomes an incredible man, capable of controlling time and space, and through this power, manipulating the heart of men. The endless fortunes of the Island of Monte-Cristo's treasure allow him access to the beginning of the capitalistic society. The corruption, the blackmail and the credit allow him to have servants, blindly obedient, and all the luxuries of modern civilisation: faster transportation, efficient communication, and changeable setting depending on the masks he chooses to wear. Nevertheless, the means will not suffice without the amount of knowledge accumulated in his dungeon: an accomplished polyglot, an expert physicist and chemist, a skilled financier, a geographer, a historian and a philosopher, he combined the encyclopaedical knowledge of his time. In the last chapter of the book, his young protégé Maximilian Morrel summarises the Count's hold over him:

'Count,' said Morrel, 'you are the epitome of all human knowledge, and you seem to me a being descended from a wiser and a more advanced world than ours.'

'There is something true in what you say,' said the Count, with that smile which made him so handsome, 'I have descended from a planet called suffering.'. (Dumas, 1846/2008, p. 1075).

Furthermore, as the omniscient narrator highlights it in an unusual way, the Count of Monte-Cristo is the masculine incarnation of the romantic idea of beauty. Even his deadly pale skin, due to his dark stay in the dungeon, adds to his charm:

Although of a paleness that was almost livid, this man had a remarkably handsome face; his eyes were piercing and sparkling; a nose, straight and projecting direct from the brow, gave out the Greek type in all its purity, whilst his teeth, as white as pearls, were set off to admiration by the black moustache that encircled them. (Dumas, 1846/2008, p. 278)

In chapter 34, Dantès is even compared to a vampire figure, fashionable since the 1819 publication of *The Vampire*, then attributed to Lord Byron. Franz d'Epinay, friend of Albert de Morcerf, son of Mércèdes, the previous fiancée of Edmond Dantès and of his enemy Fernand Mondego, is in the Argentina theatre, in Rome. After having seen the Count of Monte Cristo inside the theatre, he talks about it with a Venetian countess that has also noticed him:

'Well,' inquired Franz, after the countess had a second time directed her lorgnette at the loge of their mysterious vis-à-vis, 'what do you think of our neighbour opposite?'

'Why, that he is no other than Lord Ruthven himself in the flesh.' This fresh allusion to Byron drew a smile to Franz's face; although he conceded that if anything was likely to induce belief in the existence of vampires, it would be the presence of such a man as the mysterious personage before him.

'I must positively find out who and what he is,' said Franz, rising from his seat.

'No, no!' cried the countess, 'you must not leave me. I depend upon you to escort me home. I cannot permit you to go.'

'Is it possible,' whispered Franz, 'that you are afraid?'

'I'll tell you,' answered the countess. 'Byron believed in the existence of vampires, and even assured me he had seen some. The description he gave me corresponds with the features and character of the man before us. Oh! it is the exact personification of what I have been led to expect. The coal-black hair, large bright, glittering eyes, in which a wild, unearthly fire seems to burn—the same ghastly paleness. […] (Dumas, 1846/2008, p. 330).

Notice the wink from the novelist: the countess who identifies with Lord Ruthven, name of a Byron's character, is nothing else but his actual mistress, the countess Teresa Giuccioli, discretely named "Countess G…" by Dumas. The construction of Edmond Dantès is not only about melancholia and macabre features, but also about the bewitching gaze of the vampire who controls the will of his victims.

Indeed, even though this part of his apprenticeship is not explicitly present during his stay at the Château d'If, Edmond has Abbot Faria for a mentor. This name has not been chosen randomly. As always, mixing fiction and reality, the novelist gives the Italian abbot the name of another abbot, also part of French Revolution, but from Portuguese origins. José Custódio de Faria, born in Goa in 1756, owed his fame to his position as one of the first scientists to study and practice hypnosis when he was living in Paris. He was more famous than his enemy, Franz Anton Mesmer, inventor of animal magnetism. Faria called it "lucid sleep". Moreover, in all the scenes where Edmond Dantès, with the features of an Italian priest or sometimes with the ones of a British clerk, or under the orientalist pseudonym of Sindbad the Sailor, or even under his new title of Count, makes his enemies confess or gives orders to his faithful servants, it is indeed with his fierce look that he manages to impose his will.

But the great qualities of Dantès would be meaningless if they were not aimed at one goal: the revenge of an innocent, victim of the paltry interests of jealous foes. In his essay "From Vengeance", about *The Count of Monte-Cristo*, the literary critic Antonio Candido sees in Edmond Dantès the exact type of the romantic rebellious and powerful hero, who he describes as follows:

In order to elevate himself to such a height, from which it is possible to manoeuvre feelings and destinies freely, Monte Cristo starts from a psychological simplification and a sociological verification. Psychologically, his strength comes from the circumstance that he divides men, without nuance or refinement, into good and bad, without appeal, having as a point of reference the role of those who caused his disgrace in his youth: Danglars, Morcef, and Villefort. Sociologically, his strength comes from the knowledge that the society of his time, in the midst of structural reconstruction, was composed in good part of adventurers and arrivistes who have climbed, as he himself has, and that therefore the links of the highest spheres with the dregs of misery or crime were still fresh, palpable and manipulatable. (Candido, 1976/1995, pp. 17-18)

The strength and success of *The Count of Monte-Cristo* can then be attributed to his superhuman hero, undertaking the role of victim and executioner. Subordinated to his destiny until chapter 23, Edmond Dantès takes the lead of the plot to the point where he plays the role of Providence itself for most characters. The process of identification for the reader becomes satisfying in two ways: he can show his compassion for the young Dantès, while enjoying his accomplishment of the unstoppable system of Vengeance, carefully plotted by the Count of Monte-Cristo.

CHÂTEAU D'IF, FROM SERIAL NOVEL TO FILM SERIES

In his writings on *Literature and The National Life*, the Italian philosopher Antonio Gramsci, while reflecting on the sociological importance of the "popular and commercial" literature, analyses the success of *The Count of Monte-Cristo* as such:

This literature is a 'narcotic' of the people, an 'opium'. (A. Dumas's Count of Monte-Cristo, which is perhaps the biggest opiate of the popular novels, be analysed from this perspective. What man in the street does not believe he has been treated unjustly by the powerful and does not fantasise about the " punishment " to inflict upon them? Edmond Dantès offers him the model, " intoxicates " him with excitement, replaces the belief in a transcendental justice in which he no longer " systematically " believes). (Gramsci, 1950/2011, p. 313)

Indeed, almost a century after the publication of Alexandre Dumas' novel, the "intoxication" provoked among the popular readers, instead of slowing down, takes a new dimension, thanks to the cinematographic adaptations, which can be dated back to 1908. Incidentally, as Joël Augros and Kira Kitsopanidou emphasise in their book about the economics of American cinema, this adaptation, directed by Francis Boggs and Thomas Persons, and produced by William Selig, was the first movie shot in Hollywood, which was, until then, a quiet village close to Los Angeles (Augros & Kitsopanidou, 2009, p. 20).

From the silent film era to the new adaptation of Alexandre de la Patellière and Mathieu Delaporte, which will be released in 2022, the number of adaptations of Dumas' novel is now well over fifty. It is to be expected that an international best-seller such as *The Count of Monte-Cristo* would lend itself so well to film adaptation, especially considering that, at its origins, popular cinema itself had also borrowed from literature the serial format published in the press. Moreover, as Gérad Le Blanc, researcher at the I.R.C.A.V. (Institute for cinema and audio-visual research), highlights it, the work of Alexandre Dumas is fit to the cinematic language:

Alexandre Dumas a influencé un cinéma qui tendait à rendre l'écriture transparente à la représentation. [...] Celle qui permettait au lecteur de visualiser ce qui était écrit sans s'arrêter sur les mots. [...]. L'écriture pré-cinématographique à la Dumas, une écriture scénaristique avant la lettre du cinéma, génère des images dans le cerveau du lecteur et on les dirait aussi précises et vivantes que si elles eussent été cinématographiées. Il s'agit d'images concrètes, presque matérielles. Lèverait-on la main qu'on aurait l'impression de pouvoir les toucher du doigt. (Le Blanc, 2013, p. 2)

The detailed analysis of *Count of Monte-Cristo*'s adaptation would be difficult to achieve in one chapter; therefore, the examples given will mostly be extracted from *Edmond Dantès,* first episode of *The Count of Monte-Cristo* directed by Henri Pouctal. This "cinema novel" was shot in six parts, with a total of fifteen episodes, from 1915 to 1917. Distributed by *Pathé frères* in 1918, the film was so successful that it was distributed in Fortaleza, Brazil, from 1919. It corresponds to one of the most faithful and complete adaptations of the novel.

The scenes about the imprisonment of Edmond Dantès were partly shot inside the Château d'If, which is easily recognisable. The first scene corresponds to the arrest of Edmond Dantès, and his arrival by canoe to the pier of If islet:

At this moment, a violent shock made the bark tremble. One of the sailors leaped on shore, a rope creaked as it ran through a pulley, and Dantès guessed they were at the end of the voyage.

His guards, taking hold of his arms, forced him to rise, and dragged him towards the steps that lead to the gate of the fortress, while the exempt followed, armed with a carbine and bayonet. [...] They halted for a minute, during which he strove to collect his thoughts; he looked around; he was in a court surrounded by high walls; he heard the measured tread of sentries, and as they passed before the light he saw the barrels of their muskets gleam. (Dumas, 1846/2008, pp. 66 - 67).

In Henri Pouctal's version, the actor Léon Mathot seems to follow his literary counterpart almost line by line (Delac & Pouctal, 1918, 0:24:42). Moreover, "the steps that lead to the gate of the fortress" and the "court surrounded by high walls" are easily recognisable for the tourists that have visited the place before. In one of the set pictures sold as postcards when the film aired, it is possible to read the inscription "hostel of the sovereign people" on one of the arches (Figure 1).

Figure 1. Edmond Dantès' arrival at "hostel of the sovereign people" (Société des Amis d'Alexandre Dumas)
Source: https://www.dumaspere.com/pages/phototheque/cartesMonte-Cristo.html

This ironic inscription was made by Marseille rebels in 1848; they were building workers that were locked inside the Château d'If for several months, after a labour mobilisation for an eight-hour working day.

Of course, for technical reasons, the episodes set in the dungeons had to be filmed in the studios. However, Henri Pouctal remains, once again, faithful to the writings of Dumas, making the suspense last longer, for example in the first meeting between Edmond and Abbot Faria (Delac, & Pouctal, 1918, 00:32:12). Edmond first hears knocks on the wall, which indicates the presence of another prisoner, but then no sign of life is heard for days on end. Edmond spends his days waiting for other signs of life. He starts to dig a hole, until he hears the "sepulchral accent" of the Abbot. The introductions are again made through the walls, and it is only after twenty-four hours that Faria finally manages to enter Edmond's cell:

In an instant the portion of the floor on which Dantès (half buried in the opening) was leaning his two hands, gave way; he cast himself back, whilst a mass of stones and earth disappeared in a hole that opened beneath the aperture he himself had formed. Then from the bottom of this passage, the depth of which it was impossible to measure, he saw appear, first, the head, then the shoulders, and lastly the body of a man, who sprang nimbly into his cell. (Dumas, 1846/2008, p. 120)

The actors' theatrical expressiveness characteristic of silent film aesthetics corresponds perfectly with Dumas' style. It combines the intensity of emotions and the fast-paced actions that unfold just before the final meeting. Marx Charlier, who plays the role of Faria, seems more solemn than Léon Mathot. However, his grand hand gestures are in perfect cohesion with Dumas' idea of the character.

Moreover, the strangely agile old man comes out of the walls as if he was reborn from the darkness, stepping out, headfirst, then with his shoulder and his whole body. The metaphor is only suggested in the novel, but this scene, which already carries a cinematic quality, is represented in most movie adaptations. This rebirth of hope among darkness constitutes one of the strong images of the story.

Nevertheless, in a paradoxical way, it is only after the death of Abbot Faria, caused by a final attack of the sick old abbot after many years of apprenticeship, that the hero can come back to life. Indeed, Dantès carries the corpse of his friend in his own dungeon, and hides in a linen bag, which is meant to be put outside the dungeon. This scene is another important moment of the novel. Indeed, Edmond, and the reader, awaits one last trial to achieve freedom:

They deposited the supposed corpse on the stretcher. Edmond stiffened himself in order to play the part of a dead man, and then the party, lighted by the man with the torch who went first, ascended the stairs.

Suddenly he felt the fresh and sharp night air, and Dantès recognised the mistral. It was a sudden sensation, both wonderful and frightening. [...]

The [bearer] came towards Edmond, who heard a heavy and sounding object laid down beside him, and the same moment a cord was fastened round his feet with sudden and painful violence. [...]

They ascended five or six more steps, and then Dantès felt that they took him, one by the head and the other by the heels and swung him to and fro.

'One!' said the gravediggers. 'Two! Three, and away!'

And at the same instant Dantès felt himself flung into the air like a wounded bird, falling, falling with a rapidity that made his blood curdle. Although drawn downwards by the same heavy weight which hastened his rapid descent, it seemed to him as if the time were a century. At last, with a terrific splash, he entered the ice-cold water, and as he did so he uttered a shrill cry, stifled in a moment by his immersion beneath the waves.

Dantès had been flung into the sea, into whose depths he was dragged by a thirty-six-pound shot tied to his feet.

The sea is the cemetery of the Château d'If. (Dumas, 1846/2008, pp. 173 - 174)

Since in cinema, it is hard to adopt the internal point of view of the character, as is possible in literature, Henri Pouctal makes the scene where the gravedigger carries Dantès last longer, therefore creating a Hitchcock-like suspense. Indeed, the spectator is the only one to know that the hero is hidden in the bag. By using a plot device that is familiar to Dumas, the director multiplies each threatening action that leads to the final throwing of the bag into the sea. The sequence is then divided into many shots before the final reveal: the moving of the bag, the difficulty of the gravediggers because of Dantès' weight, the bag to which weights are tied, then the soldier holding their guns as the gravediggers throw over what they believe to be a dead body. We then see a medium shot of the white walls on the cliff, with the dark shadows of the soldiers and gravediggers. We witness the bag falling, still from a distance. Finally, as in the novel, we see Dantès easily breaking free from his ropes, and swimming to the closest island of Tiboulen (Delac & Pouctal, 1918, 0:44: 50).

The rescue of Edmond Dantès, who not only manages to set himself free but also to swim to the shore, seven kilometres away, could be a source of disbelief from the reader. But the Marseille population has a different opinion: in order to save the reputation of their hero, they have organised, since 1999, the " Monte-Cristo challenge ". It is now the first swimming challenge in Europe, with 10,000 competitors planned for the 2021 session (retrieved https://www.defimonte-cristo.com/fr).

Edmond Dantès then becomes the only prisoner that has managed to escape from the Château d'If: a way to perpetuate his legend. The question remains to determine his role in the prison's transformation into a major location for French tourism.

Between literary studies and tourism studies, there is an important place given to the notion of space. In the specific case of the Château d'If, it corresponds to the definition of literary location, as Rita Baleiro and Sílvia Quinteiro have established it:

[W]e define literary place as a segment of space whose significance emerges from the individual's knowledge of the literary text and the author's biography. This knowledge will act as a reading protocol for the space and as a stimulus for the identification or even creation of literary places. (Baleiro & Quinteiro, 2018, p. 48)

The authors insist on the key role played by the tourist-reader, who gives to the location its status of literary place through the memory of the original book (Baleiro & Quinteiro, 2018, p. 49). For the Château d'If, the images merge with a multitude of references that gives the rustic and simple fortress an attraction constantly renewed.

A TOUR OF YEW CASTLE, REMARKS AND SUGGESTIONS

On the Center of National Monuments' website, dedicated to the *Yew Castle* (as Château d'If is named there), the audience can visit the fortress and its surroundings both on location and online. (Retrieved http://www.chateau-if.fr/Explorer/Video-visite-guidee). On the main page, the authors of its touristic enhancement describe it first as a crucial fortress, then as a state prison, to then establish it as "the stuff of legend", renowned worldwide. Thus, historical data gives greater credibility to the heritage site and the legend created by the literary character contributes to the never-ending enchantment process of the place.

In order to maintain its delicate equilibrium, between the historical truth and the literary likelihood, the staging of Edmond Dantès and Abbot Faria's dungeons is minimalist: no reconstitution of the furniture nor mundane objects described in the novel (the bed, the pitcher, the pot), nor even the drawings made by Abbot Faria during his lessons to Dantès. However, a small screen cast on the back wall of the dungeon introduces silent extracts from several cinematographic adaptations of Dumas' novel (Figure 2).

Figure 2. Edmond Dantès' dungeon with its audiovisual device (2014)
Source: http://anemomili.canalblog.com/archives/2014/06/24/30133173.html

Without compromising this laudable choice for sobriety, the recorded reading of extracts from the novel could go along the pictures from the films (particularly in the silent version of Henri Ponctal). The device could be accessible through a code on mobile phones.

Furthermore, the If Castle is also introduced as a shooting location, particularly for the French adaptations of *The Count of Monte-Cristo*, through the use of several panels, in the aisle dedicated to temporary exhibits. Pictures of the movies, posters and postcards are all displayed with a fairly documented context. In 2022, a new adaptation of Dumas' novel by Alexandre de la Patellière and Matthieu Delaporte is planned, coincidentally produced by Pathé, like the first adaptation of 1908. The release of the film will probably attract renewed public attention to the Château d'If.

Finally, it would be interesting to map out a broader itinerary based on Alexandre Dumas' novel and its cinematographic adaptations. The journey could follow the one suggested by the novelist himself in chapter 113 of the British edition, "The house in the Allées de Meillan". Starting from the house where his father lived, as indicated in the chapter title, he takes the same path that he once crossed during his arrest, on foot, then on boat.

[M]onte Cristo walked down the rue de la Caisserie. This was the self-same road along which he had once been conducted at night by a silent guard. [...]

He walked down to the port along the rue Saint-Laurent and marched to the Custom- House: it was from this exact spot on the wharf that he had been embarked. A pleasure boat passed by with its canopy of cotton fabric. Monte Cristo hailed the captain who swam directly to him with all the eagerness of boatmen who have a smell of profitable business in their nostrils. [...]

For all that the sky was blue, that the shapes of these vessels was pleasing to the eye, that the light which flooded sea and land was golden, the Count, wrapped close in his cloak, remembered, each in turn, the details of that terrible voyage: the light incomparable and remote which had caught the village of the Catalans in a blazing noose, his first sight of the Château d'If which informed him where he was being taken, his struggle with the gendarmes when he tried to throw himself into the sea, his despair when he realised he was beaten, and the cold sensation of the end of the carbine barrel against his temple, like a ring of ice. (Dumas, 1846/2008, pp.1043-1044).

The novelist himself encourages the reader to step into the position of the character and to follow Edmond Dantès, as an irresistible way to practice literary tourism. The touristic aspect of the Count of Monte-Cristo's dungeon dates back to the moment of its publication, as Dumas tells himself the story. This anecdote paves the way for a reflection on the controversial question of authenticity in literary tourism.

THE QUESTION OF AUTHENTICITY IN TOURISM: THE CASE OF CHÂTEAU D'IF

In a chapter dedicated to Marseille, during his first stay in the city in 1834, Alexandre Dumas visits the cell of Count Mirabeau, a famous figure of the French Revolution, whose "Essai sur le despotisme" was written in this very castle (Laget, 1954, pp. 34, 35). The choice of this location as one of the main settings for *The Count of Monte-Cristo* was not random: it is a way to protest against the despotic power that it embodies, throughout the black legend that occurred before the invention of Edmond Dantès (Lastécouères, 2014).

Nonetheless, The Count of Monte-Cristo has a significant success that has given the location the "aura" of imagination much more than the actual historical background ever could. As the sociologist of contemporary art Nathalie Heinich points out, Walter Benjamin's concept of aura has a complex relationship with authenticity:

Ce qui donc, en résumé, se perd pour Walter Benjamin avec les techniques de reproduction, c'est ce qu'il appelle "l'aura" de l'œuvre d'art: son unicité, son authenticité, sa présence "ici et maintenant" (ou, en d'autres termes, son caractère sacré). [...] Mais ce que l'argumentation de Walter Benjamin dissimule et met en évidence à la fois, c'est que les techniques de reproduction, avant de faire apparaître une perte de l'aura, sont la condition même de l'existence de cette aura: c'est, au moins en partie, parce que la photographie implique multiplication des images et des spectateurs, rapprochement de l'œuvre avec le public, perte d'authenticité de l'objet, que l'original a pu être perçu, en négatif, comme incarnation de l'unique, de la présence absolue (le concert vs le disque), de la distance maximum par rapport aux conditions réelles d'appropriation par la reproduction. (Heinich, 1983, p. 107)

Indeed, the aura of Alexandre Dumas' hero gives the Château d'If its authenticity as a literary place. As in the case of photographs of works of art, or recordings of concerts, examples cited by N. Heinich, the reproduction of the site in the numerous film adaptations of *The Count of Monte Cristo* only reinforces the authenticity of the place.

Rare privilege for an author, Alexandre Dumas had the opportunity to witness the legacy of his character and the beneficial economic consequences that his fame brought to Marseille and its population. In "Comment j'ai fait jouer à Marseille le drame des Forestiers" ("How I had the drama of the Foresters performed in Marseille"), a chronicle about another stay in Marseille in 1858, the novelist shares his experience when he takes his friend on another visit of the Château d'If. Even if, in 1834, the If Castle was especially known because of Count Mirabeau's cell, in 1858, 14 years after the release of *The Count of Monte-Cristo* " the false legend has taken the place of the true story " and Dantès escape has taken the place of Mirabeau's captivity (Dumas, 1861).

According to Dumas' account, the famous hole dug by Abbot Faria and Dantès between their cells was excavated anonymously a few years after the publication of the novel. It is one of the remaining testimonies of how the castle was almost instantly glorified into a literary location. In a more picturesque and almost premonitory way, Edmond Dantès himself, as has been said, makes the visit to his old dungeon, almost in the same conditions as a curious tourist:

Ever since the July Revolution, there had been no prisoners kept at the Château d'If. Look-outs posted there to discourage smuggling now had the guardhouse entirely to themselves. A janitor stood waiting for any visitors who came to show them round this monument of terror which had become a monument of curiosity.

He was taken to his old cell. Once more he saw the pale daylight filtered by the narrow airshaft. Once more he saw the place where his bed, now gone, had been, and behind it, now blocked up but still visible for being repaired with new stones, the hole made by the abbé Faria. Monte Cristo felt his legs turn to water. He pulled up a wooden stool and sat down.

'Are any stories told about the Château other than that of the imprisonment of Mirabeau?' asked the Count. '[...].

'This cell,' the janitor began, 'was occupied many, many years ago by a prisoner, a very dangerous man, so they say, and all the more dangerous for being very wily. Another man was held in the Château at the same time as him. But there was no harm in him. He was just a poor priest and quite mad.' [...]

'Were prisoners allowed to see each other?' asked Monte Cristo.

'Oh no, sir! That was strictly banned. But they got round the ban by digging a tunnel which went from one cell to the other.'. [...] (Dumas, 1846/2008, pp.1044-1046).

A bit further, a guide carries him to Abbot Faria's cell:

'It was here,' said the guide, 'that they put the mad abbé. And it was just there that the younger man would come to see him. (And he showed Monte Cristo the opening to the gallery which, on this side, had been left open.) Going by the discoloration of the stone,' he went on, 'an expert worked out that the two prisoners had been communicating with each other for ten years give or take. Poor fellows, they must have been very bored during those ten years.' (Dumas, 1846/2008, p. 1049)

When reading these lines, it becomes clear that the hole seen by Dumas during his 1861 visit must have been dug either by a passionate reader of his novel, or by a wise guide, who had the intuition of what an attractive location this place could become for the tourists. It is then the author, one could almost say, the character, who, through this imaginary hole, creates an actual hole. As it is often the case with literary location, the fictional imprisonment of Edmond Dantès is made authentic by an anonymous mark, still visible today on the walls of the castle (Figure 3).

Figure 3. The famous hole anonymously excavated between Dantès' cell and Abbot Faria's cells in the 19th century
Source: https://www.dreamstime.com/stock-photo-count-monte-cristo-cell-chateau-d-if-hole-carved-wall-jos%C3%A9-custodio-faria-marseille-prisoner-was-image44956698

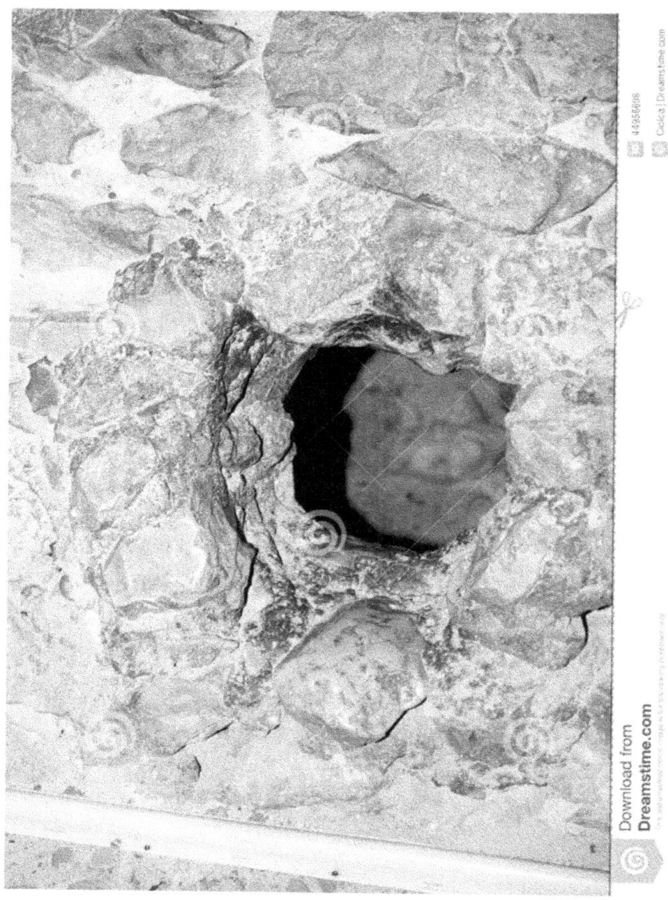

This is an obvious example of Samuel Taylor Coleridge's " willing suspension of disbelief ", here used for tourism, in the sense indicated by Rita Baleiro and Sílvia Quinteiro in their already quoted study:

This early nineteenth century concept refers to the notion that the unreal should be presented in such a way that it could be taken for real. However, currently, and as in the context of the tourism industry, the burden is not so much on the producer, but rather on the consumer. This means that the producer stages the imagined to make it more "real" and the consumer willingly chooses to temporarily engage in the illusion and suspend disbelief for the sake of the experience. (Baleiro & Quinteiro, 2018, p. 55)

Thus, creating his story, Dumas trusts fully the " aura " of the location to inspire him. In return, although the author does not always respect the verisimilitude of the story, the novel transmits to his reader an unshakable faith in his heroes, particularly the Count of Monte-Cristo. For the spectator of his cinematographic adaptation, what is expected is a bit different. The " intoxication " that Antonio Gramsci talks about follows through as long as the characteristics of the hero, and the story of injustice and revenge, remain. However, for those that practice literary tourism who would have read the book and visited the place, the most irresistible versions are the ones that are shot on location, in spite of the cheap technical effects, as seen in Henri Pouctal's adaptation, mentioned earlier.

Finally, in addition to the desire to believe, the readers and the tourists have in common the fetishist fondness related to the souvenirs of these locations. This practice was also planned by Dumas at the end of his visit to Dantès. After having received a generous tip from a mysterious traveller, the guide offers him an object, the most precious and authentic of all souvenirs:

'Sir,' said the guide, 'since you're so very generous, you deserve that I should offer you something too.'

'And what have you got to give me, friend? Shells? Models made of straw? No thanks.'

'None of those, sir, oh no, something that's connected to the story I told you just now.'

'Really?' exclaimed the Count eagerly, 'and what is it?' [...]

'A rope ladder and some tools.'

'Do you still have them?' cried Monte Cristo.

'No, sir. I sold them as curios to visitors. But I do still have one thing left.'

'What is it?' said the Count impatiently.

'I've still got a sort of book written on strips of cloth. [...]

The janitor was holding out the strips of cloth on which the abbé Faria had spread out all the treasures of his knowledge. [...]

The Count seized it eagerly [...]. Finding a safe place inside his coat for the relic which he had just recovered and which to him was worth more than the richest treasure, he hurried out of those underground vaults and, as he stepped into the boat, he gave the order:

'To Marseilles!' (Dumas, 1846/2008, pp. 1049 - 1051)

By a sort of wonderful *mise en abyme,* a writing device often used by the writer, Dumas seems to give the reader a reverse identification. The author, who maybe anticipated the curiosity that Dantès' adventures would create, gave the hero one last function, the one of the tourists, visiting the Château d'If. By placing himself in the shoes of the reader, the Count of Monte-Cristo is living eternally: dying and resuscitating, lurking in the shadow of the Château d'If.

By establishing the concepts of literary tourist, literary traveller and literary pilgrim, Sílvia Quinteiro and Rita Baleiro stated about the latter:

This proximity to the author's places and objects helps the literary pilgrim overcome the irremediable temporal distance between different orders of reality: that of the reader and that of the author (Westover, 2008: 67). For this reason, these literary places are often experienced as sanctuaries and the author's objects as relics. (Baleiro & Quinteiro, 2018, p. 94).

In the case of Dumas' hero, the " two orders of reality " refer to his past as Edmond Dantès and his present as the Count of Monte-Cristo, because he is a kind of author of the novel of his own life. Anyway, it is true that Edmond Dantès plays the role of a pilgrim rather than that of an ordinary tourist, and Abbé Faria's manuscript is more of a relic than a typical holiday souvenir. But isn't it the nature of popular art to give the general public access to aesthetic emotions? Who knows, a casual visit to the Isle of If might be enough to turn a mere literary tourist into a literary pilgrim.

CONCLUSION

The example of the Château d'If and its famous prisoner is thus at the confluence of reality and fiction, literature and history, as is often the case with the great sites of literary tourism. The question of their authenticity, as Rita Baleiro and Sílvia Quinteiro point out, is more related to public perception than to tangible evidence provided by history:

Therefore, we believe that when we discuss literary places, we should talk about gradations (or degrees) of authentic, of a game of replicas and staging, rather than authenticity (an impossibility, as we have previously mentioned). In this sense, the simulacrum constitutes a propeller for literary tourists and contributes to shaping places, as well as the way tourists feel and interpret them. (Baleiro & Quinteiro, 2018, pp. 60-61)

The consideration of the film adaptations of *The Count of Monte-Cristo* during the tourist development of the literary site also refers to another problematic, that of the sacralisation of the work of art. On the subject of literary tourists and their practices, Harald Hendrix, following Walter Benjamin, uses terms related to the sacred:

The fact that such practices oftentimes are coined as " rituals " and the tourists are called " pilgrims " underpins their culture significance. The performative perspective, moreover, logically opens up to aspects of place and location, as they determine the material context in which the performance is enacted. This again fits very well into the material-turn awareness that the physical dimensions of works of art, even of works of the imagination as literature usually is considered to be, need to be taken seriously when thinking about the ways aesthetic communication works: not only as objects of beauty, but also as tangible mediators between the physical world and the universe of the imagination. (Hendrix, 2014, pp. 20, 21)

By appealing to popular literature and the popular cinema that draws on it, literary tourism can undoubtedly answer the criticisms of those who see it as an elitist practice. Once again, the choice made by the site's designers to place on the same level the historical figures and the literary prisoners who have passed through its dungeons gives the Château d'If its ever-growing appeal.

As for the question of the authenticity of the site, it has recently been brought to light in a sort of contrast. On March 31, 2021, a Marseille website announced that Warner Bros. had bought the Ile d'If in order to make the Château d'If an avatar of Hogwarts Castle, the school of J.K. Rowling's famous hero, Harry Potter. Fortunately, it was an April Fools' joke, but the fake image created by the site's creators gives a clear idea of how inauthentic a tourist site can be. Admirers of Edmond Dantès must remain vigilant...

REFERENCES

Augros, J., & Kitsopanidou, K. (2009). *L'économie du cinéma américain: histoire d'une industrie culturelle et de ses stratégies*. Armand Collin.

Baleiro, R., & Quinteiro, S. (2018). *Key concepts in literature and tourism studies*. Universidade de Lisboa. Faculdade de Letras. Centro de Estudos Comparatistas.

Bouvier, J.-C. (2014). Toponymie des côtes insulaires. Les îles du Frioul à Marseille (Pomègues et Ratonneau). *Nouvelle Revue d'Onomastique*, *56*(1), 5–26. doi:10.3406/onoma.2014.1799

Candido, A. (1995). *On literature and society* (H. S. Becker, Trans.). Princeton University Press. (Original work published 1976)

De Laget, P. (1954). *Le Château d'If*. Imprimerie du Midi.

Delac, C. (Producer) & Pouctal, H. (Director). (1918). *Le Comte de Monte-Cristo* [Film]. Pathé frères.

DumasA. (1861). *Bric à brac*. https://www.dumaspere.com/pages/bibliotheque/chapitre.php?lid=m5&cid=7

Dumas, A. (1981). L'état civil du Comte de Monte-Cristo. In *Le Comte de Monte-Cristo* (pp. 1429–1436). Gallimard. (Original work published 1867)

Dumas, A. (2008). *The Count of Monte-Cristo* (D. Coward, Trans.). Oxford University Press. (Original work published 1846) doi:10.1093/owc/9780199219650.001.0001

Gramsci, A. (2011). *Prison notebooks* (Vol. 2; J. A. Buttigieg, Trans.). Columbia University Press. (Original work published 1950)

Heinich, N. (1983). L'aura de Walter Benjamin. *Actes de la Recherche en Sciences Sociales*, *49*(1), 107–109. doi:10.3406/arss.1983.2201

Hendrix, H. (2014). Literature and tourism: explorations, reflections and challenges. In S. Quinteiro & R. Baleiro (Eds.), *Lit&tour: Ensaios sobre literatura e turismo* (pp. 19–30). Húmus.

Lastécouères, C. (2014*). " L'Île-prison ": insularité, enfermement et pouvoir dans Le Comte de Monte-Cristo d'Alexandre Dumas (1844).* https://journals.openedition.org/criminocorpus/2866

Le Blanc, G. (2013). Ce qu'imaginer veut dire. *Entrelacs, 10.* https://journals.openedition.org/entrelacs/489

Chapter 6
Warfare, Oral Tradition, and Tourism:
Valorization of the Folk Narratives About the Gallipoli Campaign

Erol Gülüm
Bilecik Şeyh Edebali University, Turkey

ABSTRACT

Turkish folk narratives formed around the Gallipoli Campaign, which reflect the mental, psychological, and cultural attitude of Turks towards this war and hold an important place in Turkish folklore, also have the potential to make significant contributions to battlefield tourism of the region. The effective, creative, and innovative uses of the folk narratives conveying the mystical, supernatural, and miraculous events believed to have taken place in this war can be used in the enrichment and diversification of space, products, services, and experiences offered in battlefield tourism. The ultimate aim of the study is to discuss how authentic, creative, and innovative tourist attractions can be created by the valorization, remediation, and reenactment of intangible war heritage based on the example of the relationship between folk narratives about the Gallipoli Campaign and battlefield tourism in the Gallipoli Peninsula.

Sinan (the main character of the movie): *It* (book) *is all about life, people, the 'life culture' here. Çanakkale is in there, too.*

Ilhami (the side character that Sinan asks for financial support for the publication of his book): *You mean across the water?*

Sinan: *No, not the War Cemetery. To be honest, I don't think it's right to immediately associate Çanakkale with Troy or the War Cemetery. This place has its own life culture. But that's eclipsed by the cemetery.*

Ilhami: *But we're world-famous for that. It's the world's best-preserved battlefield. If the world values us for that, there must be a reason for it.*

Sinan: *True, but I see other things. There's an old man in a fedora, who sells fruit, sings, drinks wine. That's the kind of thing I write about. Tourists take photos of him, too.*

DOI: 10.4018/978-1-7998-8262-6.ch006

Ilhami: *Hold on! Are you comparing Wino Riza with our war martyrs?*
Sinan: *I didn't say that. I just...*
Ilhami: *Your "life culture" thing is everywhere. But is there a battlefield as well preserved anywhere? Why trust tourists? They snap away at everything they see. If they take photos of the old man, it's for a laugh. To be frank we're not on the same page here. If youngsters like you don't defend our martyrs and moral values, who will?*
Sinan: *Institutions and NGOs are all competing to support the War Cemetery. They're all using it to get work, an identity, even almost an existence. Don't you see that? There's a ton of identical books on Troy and the Memorial already. If you're that interested, get one and read it. "The Wild Pear Tree"* (Movie), (Nuri Bilge Ceylan, 2018).

INTRODUCTION

Tourism is generally considered a phenomenon that needs peace to thrive (Butler & Suntikul, 2013). However, as Smith (1998) remarked, war sites as touristic destinations that extend tourism beyond a calm and peaceful leisure activity possibly comprise the world's greatest single category of tourist attractions. Battlefield tourism is mostly considered within the concept of *dark tourism*, invented and defined by Foley and Lennon (1996) as the presentation and consumption of elements of death and destruction that have been made real and commercialized. People have been drawn to sights, attractions, or events associated with death, suffering, violence, or calamity since ancient times (Stone, 2005). Dark tourism, which is also considered a product of postmodernity, has reintroduced the reality of death into contemporary society, which has been downplayed by being confined between hospitals, morgues, and graves, by making it the object of touristic experience (Stone & Sharpley, 2008). Besides, the attractiveness of death sites and other manifestations of tragedy and brutality studies on dark tourism focus also on the aftermath of war and conflict (Butler & Suntikul, 2013).

Turning battlefield visits into planned and organized touristic events took place at the beginning of the twentieth century. Travel to renowned battlefields became an important commercial activity for tourism, especially at the outbreak of the Great War, which is considered one of the greatest disasters in human history. Immediately after World War I, a few travel agents organized tours for an increasing number of European travelers and, as a result, the fundamental characteristics of battlefield tourism emerged. Interest in battlefield tourism increased further after World War II. Tourists visited battlefields not only to remember relatives who had fought in one area and lost their lives, or to organize various commemorative events, but also to better understand human conditions, cultural responses to pain and grief, and turning points in global and national history and politics. In this process, the forms, and dimensions of war-related tourism activities were dramatically diversified, and historical battlefields swiftly became one of the world's most intensive tourist destinations.

Famous battlefields such as Waterloo, Gettysburg, Gallipoli Peninsula, Pearl Harbor, and Normandy are regarded and experienced as both cultural heritage and memory sites. Stone and Sharpley (2008) pointed out that sites associated with war and atrocities have long been viewed in the broader context of heritage tourism while Misztal (2003) argued that battlefield tourism allows for creating cultural memory sites that institutionalize memory through cultural mediators such as memorial ceremonies, monuments, and museums. As Hunt (2010) stated the battlefields are examples of how there is a development from memory (personal and collective) to history to heritage, which is where events in the relatively distant past become fixed and unchangeable according to the ideas of the present. Battlefields, as sites of collective trauma, discussed also within the concepts of difficult heritage or difficult past, is the focus of cultural memory and heritage activity. By visiting these locations, tourists might feel as if they have connected to these traumatic occurrences and, as a result, have received a sense of authenticity (Sturken, 2007).

Integrating the war into regional tourism primarily depends on the geography of war traces, which is dictated not only by the extension of the battlefront but also by the geography of war traces in establishing a heritage (Hertzog, 2012). Thus, the major attractions offered in battlefield tourism are geographical, and tangible elements of the war legacy. The spatial and built elements left over from wars such as the castles, walls, mausoleums, cemeteries, army roads and routes, headquarters, trenches, arsenals, hospitals, seaports, factories producing war equipment, areas declared national parks and military or war museums, and military or war museums, are among the main attractions in battlefield tourism. Tangible dimension of war heritage; a) bears the physical traces of a war that took place on a certain date, b) includes spatial attractions such as trenches, fronts, headquarters, hospitals, c) is protected in various ways because of their historical-cultural importance, d) continues to have traumatic effects on societies, e) experienced as a collective memory, national identity, and consciousness sites and f) contributes to tourism through recreations, restorations, and interpretations (Hall & Basarin, 2009).

Another significant source of attraction that can be evaluated in battlefield tourism is the intangible manifestations of war folklore that emerged in the famous battlefields. The essential component of the intangible dimensions of war folklore includes the cultural practices, representations, and expressions of a society's collective wartime psychic. Oral war narratives comprise the most significant aspects of intangible war folklore. Folk narratives that reflect the heroic deeds, suffering, and drama that took place on the battlefield contain extraordinary details about the community's collective attitudes towards times of severe trauma and the cultural expression of war psychology. In this sense, World War I is a unique laboratory for war folklore, oral war narratives in particular (Davies, 2018). The Gallipoli Campaign, which reflects all the characteristics of WWI, holds a distinct position in terms of war folklore and narratives. It is not possible to truly comprehend the Gallipoli Campaign or experience the Gallipoli Peninsula without considering these narratives, which is the folkloric expression of the mental and behavioral attitude of Turks towards this war. Folk narratives formed around the Gallipoli Campaign, which have an important place in Turkish folklore, have also the potential to make significant contributions to the battlefield tourism of the region. The effective, creative, and innovative uses of oral war narratives conveying the mystical, supernatural events believed to have taken place in this war and can make important contributions to the enrichment and diversification of space, products, services, and experiences offered in battlefield tourism. The ultimate aim of this study is to discuss how authentic, creative, and innovative tourist attractions can be created by the valorization, remediation, and reenactment of intangible war heritage based on the example of the relationship between folk narratives about the Gallipoli Campaign and battlefield tourism in the Gallipoli Peninsula.

HISTORICAL AND THEORETICAL BACKGROUND

The tensions between the major powers in Europe turned into a war when Austria-Hungary declared war on Serbia on July 28, 1914. First, the Ottoman Empire managed not to intervene in this war for a certain period. However, after the Ottoman administration realized that this situation was not sustainable, it first made various negotiations with the allied states with the motivation to protect its borders and to reclaim the lands it had recently lost. The major allied powers, which made plans to divide the Ottoman legacy, did not accept the Empire as an ally and also did not give a guarantee that their borders would not change after the war. Recognizing the aims of the allied states on their territory, the Ottoman administration intervened in the war by establishing a defensive alliance with Germany, which was in a difficult position against this powerful bloc. Thus, the war that started in Europe turned into a world war by spreading to other continents with including the Ottoman Empire, which also had territories in Asia and Africa (Aliagaoglu, 2008).

The battles on the Gallipoli Peninsula started with the attack of the British and French navies on the strait on 19 February 1915. The main purposes of this attack were to seize Istanbul, relieve pressure on the Russian army, and urge additional Balkan governments to join the Allies or at the very least not align with Germany. Churchill thought it would provide a back door to assault the Central Powers and save the troops on the Western Front (Hart, 2011). The Gallipoli Campaign, carried out by the allied powers intensely and uninterruptedly, reached its peak on March 18, 1915. The allied powers, who saw that the intense sea attacks that lasted for about a month did not give any results, made plans to capture the Dardanelles Strait by land. As part of this plan, British and French forces staged land assaults from 6 different locations on the Gallipoli Peninsula on April 24, 1915. Thus the process of ground wars, which continued for about eight and a half months, began. On 19-20 December 1915, which coincided with the 236th day of ground warfare, the Northern Group of the allied forces was removed from the Conkbayiri and Anafarta Hills. Following this incident, the Southern Group had to withdraw from Gallipoli on January 8-9, 1916 The other allied forces in the Gallipoli Peninsula had to retreat, both because of the unbearable levels of their casualties and the breaking of the belief that they would win the war and the determination of the Turkish army (Nihad, 2012).

Immediately after the withdrawal of the allied powers from the Gallipoli Peninsula in early 1916, the Turkish General Staff ordered that the battlefield, where thousands of Turkish soldiers lost their lives, be urgently mapped to future generations without being affected by physical and human factors. Under this order, the combat area was mapped at the end of 1916, which is currently known as Sevki Pasha's maps. The tangible and spatial elements left over from the war such as collision sites, trenches, scaffolding left behind by fighting forces, shipwrecks, water pipes, barbed wire, gun locations, warehouses, roads, barracks, support facilities, military hospitals, and cemeteries, which are evaluated within battlefield tourism today are documented with these maps (Aliagaoglu, 2008 citing Bademli et al., 2002). And then, the process started with The Treaty of Lausanne signed on July 24, 1923, to identify the martyrs on the Gallipoli Peninsula, mark their locations, and create monuments and cemeteries was completed in 1926 (Dore, 2006). Gelibolu Peninsula gained the status of a national park with the decision of the Council of Ministers in 1973. The international importance of the Gallipoli Peninsula Historical National Park, which spreads over an area of approximately 33,000 hectares, has gradually increased in 1997 when it was included in the protected areas list by the United Nations World Protection Organization (Aliagaoglu, 2008).

The name of the Gallipoli Peninsula Historical Park was redefined as the Gallipoli Peninsula Peace Park with a joint project carried out in cooperation with the governments of Turkey, Australia, and New Zealand (Dore, 2006). Because of these processes, the Gallipoli Peninsula has turned into a place where communities that came face to face in the war come side by side with various commemorative events held on certain dates every year while sharing their common pain and sorrow on the one hand and sharing and reinforcing their feelings of pride and victories on the other. The crucial importance of Gallipoli for Australians and New Zealanders is that despite being thousands of kilometers away from their lands, this place marks the psychological birthplace of the nationalization process of the two communities. (Slade, 2003). Every March 18 is celebrated annually by the Turks as The Victory of Canakkale and Remembrance Day for the Martyrs, while April 25 which was recorded as the date (April 25, 1915) when Anzac forces landed on the Gallipoli Peninsula is celebrated as Anzac Day with various commemorative ceremonies attended by citizens from Australia and New Zealand.

The ultimate paradox of battlefields as a touristic destination like the Gallipoli Peninsula, as Prideaux (2007) stated, is the freedom of the tourist to wander through a once dangerous place where the agony of the combat has given way to the tranquillity of peace. However, battlefield tourism is more than just visiting places where tragic events occurred at a specific point in history or consuming spatial and material remnants of wars with a 'touristic gaze' (Urry, 1990). Tourists visit the battlefields not only to see the very few things leftovers from the destructive nature of war but also to experience the war space in its sensory, and imaginary dimensions. Tourists come to seek meaning through a process of engagement with the memory and heritage of war (Miles, 2017). Therefore, the senses play an essential role in this process since contact with memory and heritage in an interpretative context may be achieved through multisensory and embodied performances, as well as a sensory and emotive contextualization that leads to an "internal understanding" and empathy (Schorch, 2014) creative representations and presentations of elements that can stimulate the senses of hearing, touch, and even smell plays an important role to provide satisfying experiences to tourists in battlefield tourism (Iles, 2006).

Touristic imagination and destination images have an important role in the design of cultural activity that enables the emergence of touristic experiences based on the combination of sensation, vision, and emotion. Reijnders (2011), who analyzed this issue in terms of the concept of "places of the imagination" defined the touristic movements towards the destinations that can turn into an execution context with their imaginary dimensions as the visits made by the tourist to his/her imagination. According to McGinn (2004), who defines this phenomenon with the concept of "imaginative seeing", in such experiences, the image precedes the sensation, penetrates it, reaches it, and dresses it up. The tourist who can perform imaginary seeing can enrich their experiences of the destination by going beyond the physical dimensions of the place. Hence, as Harvey (1989) put it clearly, tourism is fundamentally part of the image production industry because the identity, aura, and attractiveness of destinations considered as part of cultural tourism are invented, created, or produced using images.

The folk narratives are one of the most functional sources that can be used in producing sensory, emotional, and imaginative experiences. The narratives, which have been used in schematizing, interpreting, and making sense of the chaotic reality that surrounds the human species since ancient times, also have a function that allows spaces to be experienced with their imaginary dimensions. According to Gottschall (2013), narratives are also a highly functional mechanism for attributing stable identities to variable elements through the establishment of random connections between individual events and the addition of form and color to the imagery. Words, to Jenkins (2019), are visual and evocative in terms of the 'mind's eye' and can act similarly to a picture. The words enliven and contextualize the narrative to give an additional dimension to the locations. As a result, words have the stimulating power to encourage people to visit both actual and imaginary locations.

The role of the relations between spaces and narratives in the emergence of touristic experiences that include imaginative, sensory, and emotional elements can also be evaluated in terms of the relationship between narrative and space. According to Reijnders (2016), narratives consist of a causal combination of meaningful events and each event takes place in one place, therefore it can be claimed that there are organic relationships between narratives and spaces. Narrative space is described by Buchholdz and Jann

(2005) as environments and spaces that are part of literary and cinematic fiction and in which fictional characters live and interact. However, to Hovi (2016) the narrative space is the space on which various narratives are produced, performed, and consumed. Narrative spaces are spatial organizations that are a part of the enacted story and may be turned into an experiencing place in performance. Lefebvre (1991) has already suggested that all places are built socially, that their meaning depends on how people act within them and how we communicate about them. The narratives can also serve as a framework for appreciating the dynamic and multifaceted nature of the places (Lichrou, O'Malley & Patterson, 2008). Especially folk narratives may also help communities make sense of their surroundings while expressing shared values and rules (Boje, & Hillon, 2008).

An important part of the touristic activities is composed of the mobility of the experience hunters who pursue cultural spaces that are imagined, mythicized, idealized, and exoticized through narratives. According to Reijnders (2016), who also analyzed the connections between touristic destinations and imaginative narratives using the concept of topo-phobia, the narratives about these destinations where traumatic, frightening, or scary events occur are the primary source of touristic attraction. The symbiotic relationships between tourism, imagery, and narrative become more apparent when the interest in topophobic spaces created with folk narratives has a scope that can make up a touristic activity defined by the concept of "legend-tripping" (Kinsella, 2011). Dark places such as abandoned buildings, haunted houses, distant bridges, spooky caves, uncanny tunnels, unoccupied country roads, cemeteries, forested areas are among the places evaluated in legend tripping activities. What makes even these idle and harmless places uncanny and makes it possible to be experienced in dark tourism is the imaginary reality of the destination constructed through narratives. Narrative spaces, in this respect, are experienced in simultaneous integrity of both physical and imaginary dimensions. Stories create fictional worlds in imagination and frame a particular literary tourist gaze that structures tourist behavior during the trip and shapes the perception of the place (Charapan & Mikulich, 2019).

The researchers, who suggest that tourism should be also questioned in terms of the performance paradigm took these claims one step further and argued that the touristic event is not just something similar to the performance, but the performance itself. Kirshenblatt-Gimblett (1998), one pioneer of this trend, revealed that various activities based on the interaction of objects and people in destinations are essential kinds of cultural performances. Edensor (2001) claimed that all the components of the touristic performance modes are constantly redefined under the characteristics of the scene on which it takes place and interactively. If the most effective way to reveal the potential modes of performance that are expected to occur in a destination is to carefully plan, produce and manage the scenes that allow these performances, then a destination where folkloric content is tried to be a part of touristic performances and experiences should be arranged in a way that allows such performances to emerge. According to Edensor (2001), what a folkloric scene arranged as a touristic area promises is that it allows the tourist to comprehend and experience the culture of the "native" or the "other" by being involved in the performance. In such contexts, the tourist may play the roles of audience or performer of the folkloric performance being produced, depending on the level of interest, curiosity and desire, and the possibilities offered by the scene. The success of touristic products and services based on folkloric scenes and performances depends on their aesthetic pleasures by offering creative, artistic, authentic, and mystical experiences equipped with special codes of artistic nature, symbolic and semantic frameworks.

TURKISH FOLK NARRATIVES ON THE GALLIPOLI CAMPAIGN: CLASSIFICATIONS AND SAMPLES

Turkish written literary productions about the Gallipoli Campaign are very few when compared to Turkish historical sources and the place of the war in Australian and New Zealand literature (Horzum, 2016). The first criticisms about the importance and function of war literature and the lack of Turkish literature on this subject were mentioned by Mehmet Talat and İsmail Hakkı in the special issue of Yeni Mecmua, titled *Çanakkale Nüsha-yı Fevkaladesi,* published in 1918 (Koroglu, 2007). These two authors point to the fact that war literature is not only a medium of representation and transmission, but also a site of memory, remembrance, and commemoration. Even during the war, significant efforts were made to create artistic representations of the wars in Gallipoli and literary representations in particular and to use them as a propaganda tool. For this purpose, even a committee of 17 people, under the premise of the General Staff Headquarters Intelligence Office, was established, and this committee visited the front between June 15 and June 22, 1915 (Koroglu, 2007). The goal of the visit was to allow Turkish intellectuals to experience the war and the battlefield first hand, and to produce written, visual and audio content based on the impressions gained from these experiences. However, despite such efforts, satisfactory Turkish war literature about the Gallipoli War could not be created.

Some Turkish writers and literary historians such as Ahmet Mithat Efendi and Nihat Sami Banarlı attributed the weakness of Turkish war literature to a superior reason by arguing that the Turks were a nation capable of creating epics rather than writing them (Kocak, 2013). In fact, this ironic idea has a share of truth because Turkish war folklore exhibits a quality that can fill the gap in written war literature. Turks, since ancient times, have represented collective traumatic experiences that have had a profound impact on society through the cultural content they have produced through their deep-rooted and strong oral traditions. Turkish oral tradition has been the most common means of expression, communication, and representation in Anatolia until the middle of the 20th century, including the period of the Gallipoli Campaign. Therefore, the Gallipoli Campaign occupied an important place in Turkish folklore and oral literature. The Gallipoli Peninsula has also authentic, and rich war folklore that reflects dimensions of the war trauma taking place in this area from psychological, social, and cultural perspectives.

Folk narratives that present the reflections of the Gallipoli Campaign in the collective imagination and memory of the Turks, was mostly produced in the genres of legends, legends, rumors, stories, and memorates. Yakici's (1994) study on the classification of folk narratives formed around the Gallipoli Campaign provides valuable information about the generic, contextual, and functional characteristics of these narratives. This classification, shows that even oral narratives produced only in a legend type (also called menkabe) have a richness and diversity that can form 7 basic categories:

1. Legends about people who show extraordinariness,
2. Legends about place names,
3. Legends formed around the cult of trees,
4. Legends formed around dream contents,
5. Legends about the immortality of martyrs,
6. Legend about protective aspects of Islam,
7. Legends formed around the cult of Khidr-Elijah[1]

A more comprehensive and detailed study on the classification of folk narratives about the Gallipoli Campaign was carried out by Yilmaz (2015). In this classification, 9 basic categories were determined in 133 folk narratives produced from different narrative types:

1. Narratives about The Prophet Muhammad,
2. Narratives about the cult of Khidr,
3. Narratives about the cult of saints,
4. Narratives about natural events,
5. Narratives about martyrs (war period),
6. Narratives about martyrs (post-war period),
7. Narratives about dreams,
8. Other narratives (war period),
9. Other narratives (post-war period)

The most striking feature of the folk narratives formed around the Gallipoli Campaign is the diversity and richness in categories that include supernatural, extraordinary, or supernatural persons, entities, and events. In this context, it is especially noteworthy that the narratives about the divine help of the Prophet Muhammad to the Turkish army are rich enough to form a category. The legend titled "Hz. Muhammed Canakkale'de" (the Prophet Muhammed in Canakkale) is the most known of the narratives in this category: *A person from Indian Ulama who went from the Indian ulema to hajj in 1915 visits the tomb of the Prophet Muhammad. During his days in Medina, this person spiritually senses that the Prophet Muhammad is not there shares this feeling with the custodian (tomb keeper) of Al-Masjid an-Nabawi (known in English as The Prophet's Mosque). After this event, the custodian sees the Prophet Muhammad in his night dreams. The Prophet said to him, "Yes, what is felt is true. I am not in Medina right now. I am in Çanakkale. I was not willing to leave my troubled soldiers alone. I help them."* (as cited in Yilmaz, 2015: 49).

Another narrative category about the Gallipoli Campaign was formed around the cult of Khidr-Elias. The narrative containing the specific beliefs, practices, and motifs of the cult of Khidr-Elias, defined as saviors with supernatural hidden powers who are believed to have come to the aid of those who are in difficult situations especially at sea and on land, is: *As the fighting continues in Eceabat, a soldier notices a tall, white-bearded man walking over the sea towards him. This man approaches the soldier who notices him, salutes him, and says some things. The soldier, who carefully examines the person who comes to him, realizes that his feet are not wet even though this person comes from the sea. After the conversation is over, this person walks away, mentioning the names of Allah.* (as cited in Yilmaz, 2015: 69). The beliefs, practices, and motifs formed around saints, which are closely related to the Khidr-Elias cult and believed to be sent by God to help the desperate, are also frequently encountered in war narratives. The narrative that can be considered in this category is: *A saint named Câhidî Sultan, who died about two hundred and fifty years before the war, was resurrected on March 18, 1915, when the enemy ships passed the Dardanelles Strait, and he caused them to sink by scattering the grains of his rosary on these ships* (as cited in Yilmaz, 2015: 78).

There are also frequently encountered narratives with supernatural and fantastic content about the aid of Turkish soldiers by air forces and events such as fog, cloud, and wind in the wars on the Gallipoli Peninsula. War narratives, which deal with supernatural events caused by natural elements, are woven with belief motifs such as divine help and divine grace. The most familiar narratives in this category are about the cloud that engulfs the enemy battalion. According to the legend, when the Sandringham Battalion of the Norfolk Regiment, one of the British forces that landed on August 12, 1915, attacked the Sülecik District of Kücükanafarta, a cloud descends from the sky, swallows the entire battalion, and slowly disappears again. Some official research and explanations have been made about the incident. However, it was believed that news was received from the missing soldiers during and after the war. This event, which was opened for discussion with the name of cloud event on the 50th anniversary of the Gallipoli Campaign, evolved into another dimension with the notarized statements of three soldiers. The soldiers stated in their statements that although the weather was clear during the attack on August 12, when the incident took place, a cloud 250 meters long and 60 meters wide suddenly landed on the ground and disappeared, including a battalion of soldiers (Smith, 2013; Gariepy, 2014; Yurdal, 2018). The different perspectives, arguments, and stories about the disappearance of the Sandringham Battalion were also remediated across various media cultures. The movie titled *All The King's Men* (1998) adapted from Nigel McCrery's book (same title) and the novel titled *Uzun Beyaz Bulut - Gelibolu* (Long White Cloud - Gallipoli), written in 2002 by Turkish writer Buket Uzuner is some of the prominent remediations of this event.

Figures 1. The scenes from All The King's Men
(HTTP://YEOLDECYBERSHOPPE.COM/MALAHIDE/PHOTOS/PATRICK-MALAHIDE-CAPT-CLAUDE-HOWLETT-KINGS-MEN/ATTACHMENT/ALL-THE-KINGS-MEN-070/).

Figures 2. The scenes from All The King's Men (HTTP://YEOLDECYBERSHOPPE.COM/MALAHIDE/PHOTOS/PATRICK-MALAHIDE-CAPT-CLAUDE-HOWLETT-KINGS-MEN/ATTACHMENT/ALL-THE-KINGS-MEN-070/).

War narratives dealing with supernatural events caused by the forces of nature are closely related to mythical thought. The presence of motifs specific to pre-Islamic Turkish faith ecology and mythology in war narratives reinforces this claim. The war narratives, including the tree cult, the cult of Khidr-Elijah, place names, natural events, and dream motifs, show that the Turkish mythological memory is also reproduced in these narratives. The narratives that convey the fantastic experiences related to the help of mythological and divine powers are quite functional to protect the psychological endurance of the soldiers. Davies (2018) has found that the narrative and iconographic representations of such experiences are also highly influential in the armies comprise of Catholic and Orthodox soldiers. In the war narratives produced and circulated on the Western Front of the First World War, the contents about the visits of Jesus, Mary, and guardian angels to the battlefields and their divine help show the universality of the motif in question. Such visions and experiences emerge because of the effects of powerful emotions, physical stress, insomnia, extreme fatigue, uncertainty, and helplessness on the body, perception, and mind mechanisms that form the basis of war psychology. Perhaps the psychology of war led to the emergence of a mythological, magical and religious vision that could activate the collective unconscious, based on narratives containing animistic, occultist, magical, and religious images and motifs.

The narratives in which the experiences of the spirits or ghosts of the martyred soldiers are conveyed have an important place in the Gallipoli war narrative corpus. The concept of martyrdom is often used in religious sources to describe those who died/were killed in the way of God. The reflections of the concept of martyrdom in Turkish folk culture show that the meaning of this concept has been expanded to include those who sacrificed their lives for the sake of the country and the nation. Folk narratives about the Canakkale martyrs, which were produced and circulated after the war, are mostly specific to legends and memorate types in terms of their creation context, content, motif structure, and functions. For example, the narrative titled *Sehitlerin Görünmesi* (The Appearance of Martyrs) is described how in the years after the war two people encountered a group of fallen soldiers in Eski Hisarlık (Old Fortress) in Gallipoli: *The martyrs came to the beach on their horses and performed ablution and morning prayer together. After the prayers, the martyrs went to the direction of Kanlıdere by riding their horses again* (as cited in Yilmaz, 2015: 153). The narratives about the martyrs guiding or accompanying those who visit the battlefield and martyrdom are also striking. For example, in the narrative titled *Sırlı Bir Çanakkale Şehitlik Gezisi* (A Glazed Canakkale Martyrdom Tour) a group of students is told about the mysterious events that happened to them while visiting the Çanakkale Martyrdom. According to the narrative, when the group of students who visited the martyrdom could not reach the person who would guide them, a man they met there offered to accompany them. This mysterious person who guided the student group throughout the day suddenly disappeared when the trip was over. Visitors believed this person was a martyred soldier who fought in Gallipoli (as cited in Yilmaz, 2015: 158).

Australian and New Zealand folk culture as well contributed very much to the make of Gallipoli into a legendary narrative place. Seal (2004) claims that the Anzac myth was created as an intentional ideological construct that functions within an Australian culture in collaboration with the digger tradition. In the Anzac narrative, the digger appears as a brave, resourceful fighter who answers the call of duty and sacrifices himself unquestioningly for the good of his country in Gallipoli. According to Seal (2010), it can be nearly pinpointed the time, location, and conditions that gave rise to Anzac customs and its central figure, the digger. But later, they branched into autonomous but interwoven traditions, which together formed the great national mythology known today as Anzac. The digger, as a war hero, is a descendant of the older character of the Australian bushman who was also a heroic worker who liked to fight, drink, curse, and gamble, and who was anti-authoritarian, egalitarian, and resourceful. As soon as WWI was declared in 1914, the bushman was transformed into the figure known as the digger (Seal, 2010). During and after the war, the diggers quickly became a separate folk group that mirrored much of the bush tradition while also linking it, via the experience of war, to an explicit sense of national identity. This was accomplished by the integrating of digging culture's esoteric and exoteric components into a war legend that emerged around the Gallipoli Campaign (Seal, 2004).

In *The Anzac Book* edited by C.E.W. Bean and first presented in 1916 Anzac soldiers are represented as strong, resilient, inventive, good-humored, and duty-bound. This is strikingly similar to the archetypal Australian bushman of the late nineteenth and early twentieth century. It also shows how the bushman's characteristics as an Australian folk hero became a model for producing an idealized image of Anzac soldier following the Gallipoli landing (Seal, 2004; 2010). The historical, geographical, and cultural parallels with the Trojan War and Gallipoli Campaign also play an important role in the mythification of the struggles and adventures of Anzac in Gallipoli. The proximity of Gallipoli to Troy drew the attention of journalists and historians, and the region's mythological history was referenced in Australian legends and histories. Especially, Bean ensured that the Greek context would have a part to play in how the Australian sojourn at Gallipoli would be remembered (Midford, 2010; 2011). According to Mackie (2017), no one among the Australians was more conscious of the ancient Greek context of Gallipoli than Bean because he had studied Classics as a child in Australia and Britain, and then went to Oxford where he studied classics. No one could use it to such a beneficial effect. He is the master of memorializing soldiers in the Greek way, but without explicit reference to classical borrowings. In this respect, it is possible to consider Anzac narratives among the remediations of "Homeric memory" and "Homeric genea-logic" (Erll, 2018). That the Anzac narratives were produced, especially regarding the narrative scheme of the Iliad, supports Erll's (2018) argument that Homeric narratives are also a powerful source of premediation. It can be concluded that the representations of the Troy War war premediated the experience of Anzac in the Gallipoli Campaign, and Homer's epics premediated Anzac narratives.

The Anzac narrative, which became more popular with the British journalist Ellish Ashmead Bartlett's report of the landings of the Australian Imperial Force (AIF) at Anzac Cove, also inspired the silent film *The Hero of the Dardanelles,* shot by Alfred Rolfe and released on 17 June 1915. In the 1980s, Anzac cinema and television played a significant part in creating the Anzac legend, combining with political and social elements to promote a version of the legend that was best suited to the times' patriotic ambitions. The most prominent among the films shot in this period is Peter Weir's Gallipoli (1981). *Gallipoli* presented to Australian audiences a readily acceptable version of the Anzac story and remains to this day its best-known representation. It is probably the most widely discussed Australian film of all time and is the subject of a formidable body of work, both scholarly and popular. It affected many Australians' understanding of the Gallipoli campaign, as evidenced by Prime Minister Bob Hawke's visit to the Nek on the Gallipoli Peninsula in 1990 (Reynaud, 2014).

SOME METHODOLOGICAL SUGGESTIONS ABOUT VALORIZATION OF FOLK NARRATIVES IN BATTLEFIELD TOURISM

The concept of valorization, which is used to mean assigning a value to something, making something valuable for economic purposes has been used in cultural heritage management recently. Value, on the other hand, can be defined as the degree of usefulness or desirability of something, especially in comparison with other things, and is by definition subjective. Things can be valuable because they are beautiful, pleasing, or in other ways desirable, which is why the term desirability is included in the definition. Usefulness and desirability are not mutually exclusive. However, what is useful to someone does not have to be equally useful to somebody else. This approach lets us conceptualize cultural heritage as a 'thing', 'knowledge', 'resource' or 'capital' that can be created, stored, transferred, sold, and used through various revitalization practices (Andriessen, 2004). Therefore, the valorization of cultural heritage is the process of activating the developmental potentials of cultural heritage as knowledge and capital based on its values. By embedding cultural capital in the regional economy, quality of life and economic competitiveness is enhanced, and the regional identity is strengthened (Graham, 2002). In this context, it is also important to analyze the process of valorization of intangible cultural heritage to provide effective results for attempts to capitalize folk narratives to lead to performative, imaginative, and narrative experiences within battlefield tourism. The valorization steps of the intangible cultural heritage elements such as oral narratives outlined above can also be embodied in the visual representation of the 5-stage CCC methodology produced within the transnational project titled Cultural Capital Counts, which was funded by the EU's Central Europe program between 2007-2013:

Figure 3. Visual representation of the 5-step CCC Methodology (Krotscheck & Fend, 2014).

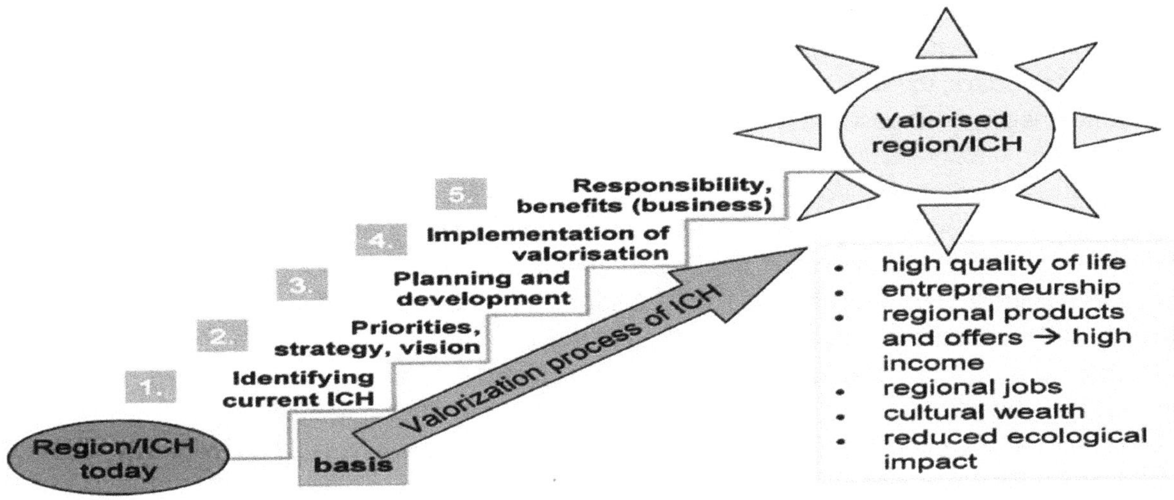

This methodology can be used to achieve a more systematic, manageable, and sustainable integration of the folk narrative tradition formed around the Gallipoli Campaign and the memory of images embedded in them into battlefield tourism in the Gallipoli Peninsula. An alternative action plan for tourism valorization of folk narratives in Gallipoli Peninsula that designed based on the above methodology can be formulated in five basic stages: First, the strategic importance of the integration between cultural memory and tourism in general, and war narratives and battlefield tourism in particular, in achieving social, cultural and economic development goals should be evaluated with its theoretical and practical dimensions. At this stage, it also should be put forward with its theoretical dimensions that oral literature is one channel to access content that can create added value in battlefield tourism, and a practical, functional and flexible narrative identification, documentation, and classification methodology should be developed. Later, it is necessary to determine the priorities, short- and long-term visions, and strategies for the creative and innovative use of war narratives in battlefield tourism. In the fourth stage, management, capitalization and implementation plans that enable the use of folk narratives in battlefield tourism should be developed. In the last stage, the right and responsibility of the management, design, and application models developed and tested with pilot applications should be transferred to the relevant agencies/intermediaries.

As also seen in the methodology, valorization of cultural heritage is a planned, systematic, and manageable process that comprises practices such as "defining, estimating, attributing, changing, confirming and even denying the value of a heritage content" (cited in Throsby, 2010: 17). While archaeologists, art historians, museologists, anthropologists, and folklorists use valorization tools for clarifying and highlighting the historical, cultural, spiritual, and aesthetic values of cultural heritage, tourism professionals, professionals directing the cultural and creative sectors, individual entrepreneurs, and local governments resort to valorization to actualize the instrumental value potentials of cultural heritage that can be converted into economic gain. Labeling the first group as culturalists and the second as economists, and who thinks that both alone are not sufficient for an effective valuation, Klamer (2013: 421), argued that effective valorization of cultural heritage can only be managed by cultural economists who emerge from the balanced union of these opposite poles. Cultural-economic valorization of heritage for tourism is the prerequisite for producing cultural creative products, services, and experiences by adapting the lifestyles, practices, knowledge, techniques, abilities, and images that make up the folk culture to the desires, fantasies, needs, and demands that shape the global cultural market. As Lowenthal (2009: 148) succinctly states, the sectors and professionals that shape the cultural market transform the scenes, actors, and images of the past and traditions into consumable cultural content by updating, improving, or excluding production techniques.

One issue that needs to be addressed in this context as it is directly related to the interpretation, mediation, reproduction, rewriting, and commercialization practices that make up the tourism valorization of folkloric resources is the phenomenon of folklorism/folklorismus, which is also defined as "second-hand folklore" by Hans Moser. While Moser defines folklorism as a form of cultural production based on the imitation of folk culture contents (Bendix, 1997: 177-178), Bausinger (1990: 152) marks modernity, and the culture industry, as the lever of folklorism, which takes folklore out of its authentic form and drags it into a foreign context. According to Bendix (1997: 176-178), the commodification of expressive culture for economic purposes and its effective use for political manipulations have been effective in the cultural content defined by this concept to become an integral part of daily life. Roginsky (2007: 44) argued that authentic and reified forms of folklore (folklorism) live side by side or intertwined synchronously in the dynamic, interactive, and relational contexts of contemporary everyday culture, where the sharp lines thought to exist between folklore and folklorism have faded. This argument takes a step forward with the debates taking place around the concept of "folkloresque" which Foster and Tolbert (2016) invented to describe cultural content that appears to its consumer as derived from ancient folkloric traditions, but which is mostly produced for commercial, which can neither be fully incorporated into folk culture nor popular culture has been moved further. Such cultural productions are formed by combining folkloric elements got from specific sources that have the power to create nostalgic and authentic connotations, and hybrid popular culture elements in new contexts, in creative and conscious ways.

Folklorism-based tourism valorization comprises the stages of identifying folkloric resources, clarifying their touristic value, turning them into touristic attractions, and finally into touristic products. This process determines which areas of destination folklore the profile of the tourist addressing may be interested in. After the areas of interest and patterns are determined, authentic or second-hand folkloric sources, which continue their existence passively or actively as a part of the destination culture, are determined. The resources got are reinterpreted under the tourism valorization and transformed into touristic products and services, while the infrastructure required to perceive the consumption of these products and services as a cultural experience and execution is prepared. Folkloric elements, whose touristic value is clarified and the performance infrastructure is prepared, turn into one of the major attractions of the destination. With the marketing and promotion of the destination's folkloric attractions, these elements turn into touristic products or services. A visualized representation of the entire process is as follows (Mokras-Grabowska, 2013):

Figure 4. The process of commercialization of folk culture resources
(Mokras-Grabowska, 2013 based on A. Kowalczyk, 2009).

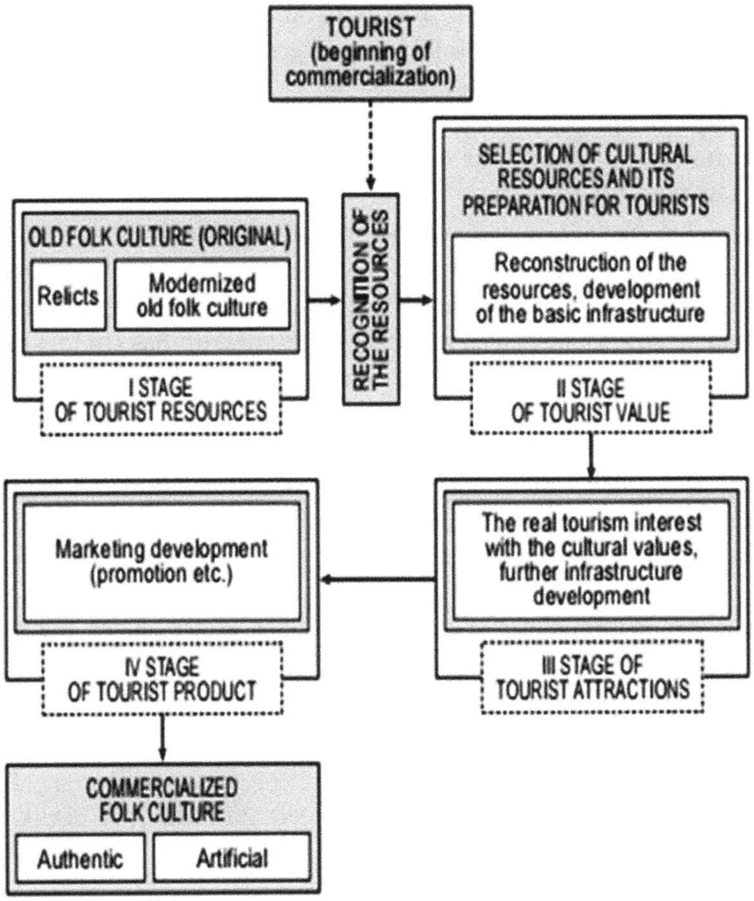

One of the most effective strategies that can be used in tourism valorization of folk narratives in battlefield tourism is storytelling. Integrating narratives into destinations with their authentic backgrounds and natural performative attitude will enable the production of participatory, performative, imaginative, experiential, and creative touristic services and events. According to Korez-Vide (2017), when storytelling is used in effective and creative ways in tourism, it can contribute to creating touristic products, attractions, and tourism forms with sustainable qualities; to highlight and; promote the historical, cultural, and traditional essence of a destination. Storytelling can also be used to share specific place goals; spread and justify place values; motivate tourism visitors and residents to visit a place; maintain the memory, ensuring a continuity of knowledge and guidance of behavior; create trust, confidence, and a sense of belonging; share tacit knowledge; share norms and values; reformulate place stories and re-engineer place image narratives; provide value categories and interpretative patterns (Bassano et al., 2019 citing Boje, 1991 et al.).

Professional tourist guides play an important role in performing different storytelling practices in tourism. Such as, the guides working in the Gallipoli Peninsula and defined (Atay 2008) as *Field Guides* tell the history of the Gallipoli Campaign, the cemeteries, cemeteries, and the events that took place during the war or were considered having happened to the visitors. Field Guides, which accompany the visits in the field and play the role of narrator, also undertake important tasks in terms of the continuity of the folk narratives evaluated within war folklore and heritage. During the guidance activities, Field Guides intensely and effectively make use of diaries, memoirs, military archival documents, and letters, as well as folk narratives. In this respect, Field Guides contribute to the performance, reproduction, and circulation of the narratives in question, as well as the touristic uses of the folk narratives about the Gallipoli Campaign (Birtane, 2020).

The other most popular storytelling and remediation strategy that can be used in battlefield tourism is cultural reenactment. Reenactment is a crucial factor in the cultural responses to traumatic events within this economy of survivors and consumerism (Sturken, 2007). According to Prideaux (2007), the hunger of the tourism industry for new sites to attract visitor interest has led to a new understanding of the concept of the battlefield that encompasses the 'real' and increasingly the 'play'. The 'real' refers to the actual place of battle previously described, while the 'play' is a new form of battle that involves the re-enactment of past battles and, more recently, battle scenarios based on alternative histories. The Turkish Red Crescent Hospital Exhibition 1915, which was realized in 2018 in the Alcıtepe by the Presidency of the Historic Site on the Gallipoli Peninsula, reveals the importance of stimulating the senses in battlefield tourism. This open-air exhibition, designed with dramatic elements that appeal to the tourist's senses of sight, hearing, touch, and smell, can be considered as a successful example of activating the imagination and emotions of the tourist. The reenactment area, which contains various details of a hospital behind the facade, offers an impressive experience that allows visitors to feel the atmosphere with its audio reenactments of the turmoil in the bandaging area bombed by enemy planes:

Figure 5. Hospital Exhibition
(HTTPS://WWW.KULTURPORTALI.GOV.TR/TURKIYE/CANAKKALE/GEZILECEKYER/CANAKKALE-1915-HILAL-I-AHMER-HASTANESI-CANLANDIRMA-ALANI)

Figure 6. Hospital Exhibition
(HTTPS://T24.COM.TR/HABER/SAVASIN-ACI-YUZU-CANAKKALE-HASTANE-SERGISINDE,602630)

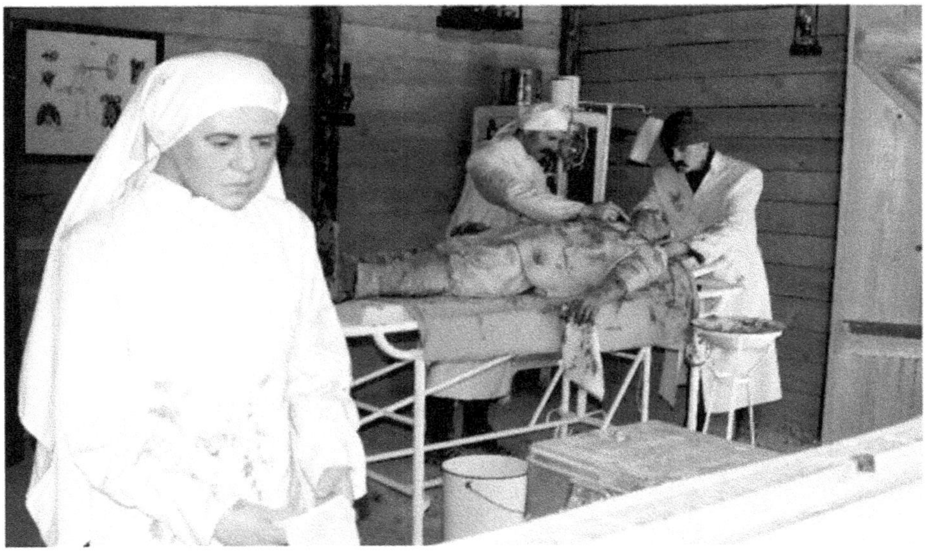

One of the war narratives and images that have been successfully associated with the Gallipoli Campaign and the battlefield through re-mediation and reenactment practices is Turkish Corporal Seyit Cabuk. Seyit Cabuk was drafted in 1909 and joined the Balkan Wars in 1912 when he was a three-year soldier and joined the army. After the Balkan wars, he was assigned to the Canakkale front. Corporal Seyit was assigned as a gunner in the Mecidiye bastion on the Rumeli side on the Canakkale front (Torun et al. 2020, citing Ozdemir, 2015). The historical event, which is also the foundation of Seyit Corporal's narratives, is mostly associated with the capture of the Mecidiye bastion by the British battleship Ocean on March 18, 1915. According to the narrative, some soldiers who served here took refuge in the trenches during the bombardment. Corporal Seyit was wounded in the bombardment, but could still stand up. Seyit noticed that the enemy ships were getting closer to the land and the guns in the battery were damaged. One of the guns is intact, but the winch that took it to the muzzle has malfunctioned. Seyit single-handedly carried the 275 kg bullet into the barrel. The bullet that Seyit put into the barrel hit the rudder of the battleship Ocean and the armorer was heavily damaged (Taşkıran, 2009).

After this incident, the commander of the said unit, Cevat Pasha, requested that Seyit Corporal's photograph be taken while he was carrying a bullet to announce his heroism. However, Corporal Seyit could not lift the cannonball despite several attempts. Thereupon, a wooden cannonball was prepared to immortalize this heroism, and the picture of Corporal Seyit that has survived to the present day was taken (Taşkıran, 2009). Besides, there was no gun in Rumeli Mecidiye Fort that could fire a 275 kg shell, and military reports and historical records also state that the battleship Ocean was out of the effective range of the Mecidiye Fort where Seyit served (Yurdal, 2018). Despite these facts, Harp Mecmuası (War Journal), the most popular Ottoman magazine during the First World War, covered Seyit's photo in December 1915 issue (Figure 7) with the footnote of *"A strong, gallant soldier in Canakkale fortification who carrying a shell on his back, weighing 215 kıyye. Seyyid, son of Mehmet. It is an example of our army's love of war"*. Harp Mecmuası was widely disseminated among both civilians and soldiers and aimed at convincing the public of the idea that the Ottoman Empire had sufficient power to maintain the war. Corporal Seyit's this photo must have been used as a cover for the same aim (Yilmazer, 2020):

Figure 7. The photo of Corporal Seyit on the cover of Harp Mecmuası
(H.M, n. 2)

Regardless of the accuracy and validity of this historical moment, and the arguments about it, the reenactments and remediations of the narrative of Corporal Seyit have turned into a prominent focus of attraction in battlefield tourism. The interest of local and foreign visitors to Corporal Seyit's statue, which was erected at his mission in the Rumeli Mecidiye Fort and depicted while carrying cannonballs, can be regarded as a simple but successful example of the interactions between war narratives and battlefields:

Seyit Corporal's narrative and image have been animated in the 4th Camber named "Mecidiye Fort" in the Simulation and Information Center for the Legend of Gallipoli. The Simulation and Information Center for the Legend of Gallipoli, which was established in 2012 aims to reenact the war through various technologies. The competitors could use pictures, statues, movies, 3-D movies, and simulation and animation techniques separately, and they could also use some of them in one project. The primary aim of this center is to make visitors feel as if they were really in The Battle of Dardanelles and to let them learn as much as possible about the Battle of Dardanelles. As Stankova and Vassenska (2015) have argued and also it can be seen from the below examples, the re-enactments based on folkloric elements also are said to encourage the transformation of memory, history, and narrative into a visible leisure performance, where individuals draw on nostalgic symbolism to recollect and tie together their past and present lives:

Figures 8. Rumeli Mecidiye Fort
(Republic Of Turkey Ministry Of Culture and Tourism, 2016)

Figures 9. Rumeli Mecidiye Fort
(Republic Of Turkey Ministry Of Culture and Tourism, 2016)

Figures 10. 4th Camber "Mecidiye Fort"
(Republic of Turkey Ministry of Forestry and Water Affairs, 2012)

Figures 11. 4th Camber "Mecidiye Fort"
(Republic of Turkey Ministry of Forestry and Water Affairs, 2012)

In recent years, the role and influence of media technologies and cultures in the clarification, popularization, and iconization of the image and narrative of Corporal Seyid have gradually increased. Such examples are examples of how the cultural memory of a community is embodied, transmitted, and consumed through primary (bodily and verbal media) or secondary media technologies (printed, visual, audio-visual, and digital media) that enable synchronic and diachronic communication between members.

Therefore, it can be said that mediation practices, technologies, and ecologies have strategic importance in terms of re-functionalizing, popularizing, iconizing, and circulating cultural memory images specific to the Gallipoli Campaign:

Figure 12. Seyit in Çanakkale 1915 (Movie)
(HTTPS://YOUTU.BE/1ZNRST0SKUE)

Figure 13. Seyit in Çanakkale Geçilmez (Animation)
(HTTPS://YOUTU.BE/8IZNYQYRD_8)

Figure 14. Corporal Seyit in 3d digital art
(HTTPS://WWW.ARTSTATION.COM/ARTWORK/OOMZY4)

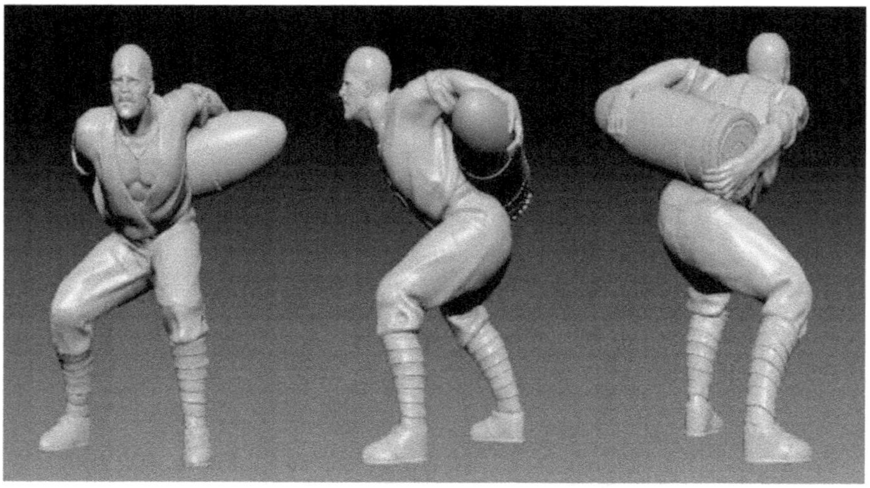

Figure 15. Corporal Seyit in Nusrat (Digital game)
(HTTPS://YOUTU.BE/EZQKPPUXVIW)

Each remediation of cultural narratives and images produced and transmitted in the oral tradition, in audio-visual and digital media technologies, as in the examples above, contributes not only to the continuity or sustainability of folk narratives but also to reveal the potential to produce added value in different fields. Memory images, which become stronger to mean, function and impact as they interact with the intra- and inter-media dynamics of mediation, can even infiltrate the subcultures produced by these media agents and diversify and expand the audience. The representations derived from the mediations of the narratives in different media technologies and cultures also render the cultural memory functional and functional as it also refers to the main image in every context in which it is positioned because of its analogous and associative nature. With the analyzes of these reception contexts, as Erll (2008) draws attention to, both the understanding of the intermedia dimensions of cultural remembering and how narratives and images are crystallized, stabilized, iconized, and popularized through remediation can be revealed.

FUTURE RESEARCH DIRECTION

Touristic experiences based on the combination of sensation, imagination, and emotion are possible with the construction of cultural performances. The source of the pleasure created by being an active participant in a touristic performance is that it allows the culture of interest to be observed with a sincere and empathetic sense and to be experienced with sensory intimacy and present presence. Performative events that can draw the visitor into a different world for a certain period, bracing the moment of experience, distract from the rhythm of daily life, build a simulated reality, and create intense emotional effects can contribute to the acquisition of authentic experiences in the battlefield tourism. Therefore, more detailed studies must be done to reveal what kind of organic relations exist between cultural performances and oral narrative worlds and on how to turn folk narratives into a participatory performance to attempt a questioning about how the narratives can be evaluated in creative and innovative ways in increasing the tourism potential and enriching the touristic experiences.

CONCLUSION

Folk narratives, which shed light on many aspects of the Gallipoli Campaign that cannot be revealed by historical information, documents, and research, also have the potential to bring new perspectives to those interested in the war and battlefield. The identification, clarification, and interpretation of the folk narratives formed around the Gallipoli Campaign and their association with the battlefield in creative and innovative ways may also contribute to enriching the performative, imaginative, and narrative dimensions of battlefield tourism in the Gallipoli Peninsula. Integration between the tangible and tangible heritage elements left over from the Gallipoli Campaign can enable the establishment of contexts that can allow an imaginary journey in this field, and an empathic understanding of the war, and the experience of it with performative participation. Such an orientation can also make significant contributions to the clarification, preservation, interpretation, and re-functionalization of Turkish war folklore, war heritage, and war memory. The sustainable capitalizing on the war folklore, heritage, and memory formed around the Gallipoli Campaign in tourism also contributes to a) the development of the war-related tourism potential in this region, b) the presentation of an imaginative, performative, and creative tourism experience to the visitors, c) the protection, interpretation, and functionalization of the elements evaluated in the intangible cultural heritage category with sustainable practices, d) the iconization of cultural contents that are brought to the fore within battlefield tourism and put into global circulation by mediating them in other cultural and creative sectors.

REFERENCES

Aliagaoglu, A. (2008). Savaş alanları turizmi için tipik bir yer: Gelibolu Yarımadası Tarihi Milli Parkı [A typical place to battlefield tourism: Gallipoli Peninsula Historical National Park]. *Milli Folklor, 20*(78), 88–104.

Andriessen, D. (2004). *Making sense of intellectual capital: Designing a method for the valuation of intangibles*. Routledge. doi:10.4324/9780080510712

Atay, L. (2008). Alan Kılavuzluğunun Çanakkale ilindeki uygulaması ve ilgili turizm mevzuatı açısından değerlendirilmesi. *Anatolia: Turizm Araştırmaları Dergisi, 19*(2), 169–176.

Baldwin, F., & Sharpley, R. (2009). Battlefield tourism: Bringing organized violence back to life. In R. Sharpley & P. R. Stone (Eds.), *The darker side of travel: The theory and practice of dark tourism* (pp. 186–206). Channel View Publications. doi:10.21832/9781845411169-011

Bassano, C., Barile, S., Piciocchi, P., Spohrer, J., Iandolo, F., & Fisk, R. (2019). Storytelling about places: Tourism marketing in the digital age. *Cities (London, England), 87*, 10–20. doi:10.1016/j.cities.2018.12.025

Bausinger, H. (1990). *Folk culture in a world of technology* (E. Dettmer, Trans.). Indiana University.

Bendix, R. (1997). *In search of authenticity: The formation of folklore studies*. The University of Wisconsin.

Birtane, Y. (2020). Sözlü ve yazılı kültür bağlamında Çanakkale muharebe anlatıları ve alan kılavuzları. In B. Borlat & B. Sayılır (Eds.), 105. Yılında Çanakkale Muharebelerine Bakış (pp. 475-483). T.C. Kültür ve Turizm Bakanlığı Çanakkale Savaşları Gelibolu Tarihi Alan Başkanlığı.

Boje, D. M., & Hillon, M. (2008). Transorganizational development. In T. Cummings (Ed.), *Handbook of organization development* (pp. 651–654). Sage.

Buchholz, S., & Jahn, M. (2005). Space in narrative. In D. Herman, M. Jahn, & M. L. Ryan (Eds.), *Routledge encyclopedia of narrative theory* (pp. 551–555). Routledge.

Butler, R., & Suntikul, W. (2013). Tourism and war: An ill wind. In R. Butler & W. Suntikul (Eds.), *Tourism and war* (pp. 1–11). Routledge. doi:10.4324/9780203107706

Ceylan, N.B. (Director). (2018). *The wild pear tree* (Movie).

Charapan, N., & Mikulich, H. (2019). 'Magic realism' of Vilnius: Conceptualization of literary tourist gaze based on Tales of the Old Vilnius by Max Frei. In I. Jenkins & L. A. Lund (Eds.), *Literary tourism: theories, practice and case studies* (pp. 15–27). CABI. doi:10.1079/9781786394590.0015

Davies, O. (2018). *A supernatural war: Magic, divination, and faith during the First World War*. Oxford University Press.

Dore, L. (2006). Gallipoli: A visitor profile. *The Historic Environment, 9*(2), 46–51.

Edensor, T. (2001). Performing tourism, staging tourism: (Re)producing tourist space and practice. *Tourist Studies, 1*(1), 59–81. doi:10.1177/146879760100100104

Erll, A. (2008). Literature, film, and the mediality of cultural memory. In A. Erll & A. Nünning (Eds.), *Media and cultural memory* (pp. 389–398). Walter de Gruyter.

Erll, A. (2018). Homer: A relational mnemohistory. *Memory Studies, 11*(3), 274–286. doi:10.1177/1750698018771858

Foley, M., & Lennon, J. (1996). Heart of darkness. *International Journal of Heritage Studies, 2*(4), 195–197. doi:10.1080/13527259608722174

Gariepy, P. (2014). *Gardens of Hell: Battles of the Gallipoli Campaign*. University of Nebraska Press. doi:10.2307/j.ctt1d9nkq4

Gottschall, J. (2013). *The storytelling animal: How stories make us human*. Mariner.

Graham, B. (2002). Heritage as knowledge: Capital or culture? *Urban Studies (Edinburgh, Scotland), 39*(5/6), 1003–1017. doi:10.1080/00420980220128426

Hall, J., & Basarin, V. J. (2009). Motives for attendance at Anzac Day commemoration in Turkey in tourist experiences. In *Meanings, motivations, behaviours conference*. The University of Central Lancashire Press.

Hart, P. (2011). Gallipoli. Profile Books.

Harvey, D. (1989). From managerialism to entrepreneurialism: The transformation in urban governance in late capitalism. *Geografiska Annaler, 71*(1), 3–17. doi:10.1080/04353684.1989.11879583

Hertzog, A. (2012). War, battlefields, tourism, and imagination. *Tourist Imaginaries, 1*, 1-13.

Horzum, S. (2016). Reflections on the Gallipoli Campaign in Turkish literature. In M. Gurcan & R. Johnson (Eds.), *The Gallipoli Campaign The Turkish perspective* (pp. 206–216). Routledge.

Hovi, T. (2016). *Finding heritage through fiction in Dracula tourism*. Suomalainen Tiedeakatemia/ Academia Scientiarum Fennica.

Hunt, N. (2010). *Memory, war, and trauma*. Cambridge University Press. doi:10.1017/CBO9780511845017

Iles, J. (2006). Recalling the ghosts of war: Performing tourism on the battlefields of the western front. *Text and Performance Quarterly, 26*(2), 162–180. doi:10.1080/10462930500519374

Jenkins, I. (2019). Practical applications of literature to the tourist experience and tourist industry. In I. Jenkins & L. A. Lund (Eds.), *Literary tourism: theories, practice and case studies* (pp. 71–72). CABI. doi:10.1079/9781786394590.0000

Kinsella, M. (2011). *Legend-tripping online: Supernatural folklore and the search for Ong's hat*. University of Mississippi Press. doi:10.14325/mississippi/9781604739831.001.0001

Kirshenblatt-Gimblett, B. (1998). *Destination culture: Tourism, museums, and heritage*. University of California Press.

Klamer, A. (2013). The values of cultural heritage. In I. Rizzo & A. Mignosa (Eds.), *Handbook on the economics of cultural heritage* (pp. 421–437). Edward Elgar. doi:10.4337/9780857931009.00035

Kocak, A. (2013). Savaşı şiirden okumak: Çanakkale zaferinin Türk şirinde akisleri, *Uluslararası Gelibolu Sempozyumu*, 467-479.

Korez-Vide, R. (2017). Storytelling in sustainable tourism management: Challenges and opportunities for Slovenia. *Journal of Advanced Management Science*, *5*(5), 380–386. doi:10.18178/joams.5.5.380-386

Koroglu, E. (2007). *Ottoman propaganda and Turkish identity: Literature in Turkey during world war I*. I.B.Tauris.

Krotscheck, C., & Fend, M. (2014). The CCC method in brief. In Cultural Resources Strategy for Sustainable Regional Development-The CCC Method. Verlag.

Lefebvre, H. (1991). *The production of space*. Blackwell.

Lichrou, M., O'Malley, L., & Patterson, M. (2008). Place-product or place narratives(s)? Perspectives in the marketing of tourism destinations. *Journal of Strategic Marketing*, *16*(1), 27–39. doi:10.1080/09652540701794429

Lowenthal, D. (2009). *The heritage crusade and the spoils of history* (7th ed.). Cambridge University.

Mackie, C. (2017). Further investigation: Gallipoli's rich history of conflict started well before 1915 (conversation). In Gallipoli, Anzac and the Great War (pp. 36-37), LaTrobe University Press.

McGinn, C. (2004). *Mindsight: Image, dream, meaning*. Harvard University Press.

Midford, S. (2010). From Achilles to Anzac: Heroism in the Dardanelles from antiquity to The Great War. *ASCS 31 Proceedings*, 1-12.

Midford, S. (2011). Constructing the Australian Iliad: Ancient heroes and Anzac diggers in the Dardanelles. *Melbourne Historical Journal*, *2011*(Special Issue), 59–79.

Midford, S. (2017). Gallipoli, Anzac and the Great War. LaTrobe University Press. doi:10.26826/1002

Miles, S. (2017). Sensorial engagement in tourism experiences on the Western Front. In N. J. Saunders & P. Cornish (Eds.), Modern conflict and the senses. Routledge. doi:10.4324/9781315682228-6

Ministry of Forestry and Water Affairs. (2012). *Simulation and information center for the legend of Gallipoli*. Author.

Misztal, B. A. (2003). *Theories of social remembering*. Open University Press.

Mokras-Grabowska, J. (2013). Folk culture resources as a component of tourism space. *Tourism (Zagreb)*, *23*(2), 43–47.

Nihad, M. (2012). *Büyük harpte Çanakkale seferi [The Gallipoli Campaign in the Great War]*. Türk Şehitlikleri İmar Vakfı.

Ocak, A. (1998). *Hıdırellez. In TDV İslâm ansiklopedisi* (Vol. 17). TDV İslâm Araştırmaları Merkezi.

Prideaux, S. (2007). Echoes of war: Battlefield tourism. In C. Ryan (Ed.), *Battlefield tourism: History, place, and interpretation* (pp. 17–27). Elsevier. doi:10.1016/B978-0-08-045362-0.50007-5

Reijnders, S. (2011). *Places of the imagination: Media, tourism, culture*. Ashgate.

Reijnders, S. (2016). Stories that move: Fiction, imagination, tourism. *European Journal of Cultural Studies*, *19*(6), 672–689. doi:10.1177/1367549415597922

Reynaud, D. (2014). National versions of the Great War: Modern Australian Anzac cinema. In M. Loschnigg & M. Sokolowska-Paryz (Eds.), *The Great War in post-memory literature and film* (pp. 289–304). De Gruyter.

Roginsky, D. (2007). Folklore, folklorism, and synchronization: Preserved created folklore in Israel. *Journal of Folklore Research*, *44*(1), 41–66. doi:10.2979/JFR.2007.44.1.41

Schorch, P. (2014). Cultural feelings and the making of meaning. *International Journal of Heritage Studies*, *20*(1), 22–35. doi:10.1080/13527258.2012.709194

Seal, G. (2004). *Inventing Anzac: The digger and national mythology*. University of Queensland Press.

Seal, G. (2010). Folklore, history, and myth at an Anzac memorial. *Australian Folklore*, *25*, 171–181.

Slade, P. (2003). Gallipoli thanatourism: The meaning of ANZAC. *Annals of Tourism Research*, *30*(4), 779–794. doi:10.1016/S0160-7383(03)00025-2

Smith, S. (2013). *And they loved not their lives unto death: The history of Worstead and Westwick's war memorial and war dead*. Tommies Guides.

Smith, V. L. (1998). War and tourism: An American ethnography. *Annals of Tourism Research*, *25*(1), 202–227. doi:10.1016/S0160-7383(97)00086-8

Stankova, M., & Vassenska, I. (2015). Raising cultural awareness of local traditions through festival tourism. *Tourism & Management Studies*, *11*(1), 120–127.

Stone, P. (2005). Dark Tourism: An old concept in a new world. *Tourism Magazine*, *4*(25), 20.

Stone, P., & Sharpley, R. (2008). Consuming dark tourism: A thanatological perspective. *Annals of Tourism Research*, *35*(2), 574–595. doi:10.1016/j.annals.2008.02.003

Sturken, M. (2007). *Tourists of history. Memory, kitsch, and consumerism from Oklahoma City to Ground Zero*. Duke University Press.

Taşkıran, C. (2009). 18 Mart Canakkale deniz savaşı. *Atatürk araştırma Merkezi Dergisi Çanakkale Savaşı Özel Sayısı*, *73*(25), 25-38.

Throsby, D. (2010). *The economics of cultural policy*. Cambridge University. doi:10.1017/CBO9780511845253

Tolbert, J. (2016). Introduction. In M. D. Foster & J. A. Tolbert (Eds.), *The folkloresque: Reframing folklore in a popular culture world* (pp. 37–41). University Press of Colorado.

Torun, F., Doğan, Y., & Gülüm, K. (2020). Seyit Onbaşı'nın Çanakkale savaşlarındaki rolüne yönelik öğrencilerin ve öğretmen adaylarının metaforik algıları: Boylamsal bir araştırma. *Balıkesir Üniversitesi Sosyal Bilimler Enstitüsü Dergisi*, *23*(43), 63–86.

UNESCO. (n.d.). *Spring celebration, Hıdrellez*. https://ich.unesco.org/en/RL/spring-celebration-hdrellez-01284

Urry, J. (1990). The Tourist gaze: Leisure and travel in contemporary societies. *Sage (Atlanta, Ga.).*

Yakici, A. (1994). Canakkale savaşları etrafında oluşan menkıbelerin Türk folkloru içindeki yeri [The place of the religious legends (menkabe/menkibe) formed around the Canakkale wars in Turkish folklore]. *Atatürk Araştırma Merkezi Dergisi, 10/30*, 599–608.

Yilmaz, B. (2015). *Çanakkale savaşları etrafında teşekkül eden halk anlatıları–Çanakkale örneği* [Folk narratives occurred around the Gallipoli Campaign: The case of Canakkale] (PhD thesis). Çanakkale Onsekiz Mart Üniversitesi, Sosyal Bilimler Enstitüsü.

Yilmazer, T. (2020). Cabuk, Seyit Ali. *International Encyclopedia of the First World War.* https://encyclopedia.1914-1918-online.net/article/cabuk_seyit_ali

Yurdal, M. O. (2018). Efsanelere kurban edilen savaş [War Sacrificed to Legends]. *Tarih, 53*, 28–33.

ADDITIONAL READING

Ashworth, G., & Hartmann, R. (Eds.). (2005). *Horror and human tragedy revisited: The management of sites of atrocities for tourism.* Cognizant.

Clarke, D. (2002). Rumors of Angels: A Legend of the First World War. *Folklore, 113*(2), 151–173. doi:10.1080/0015587022000015293

Davies, O. (2018). *A supernatural war: Magic, divination, and faith during the First World War.* Oxford University Press.

Iles, J. (2006). Recalling the ghosts of war: Performing tourism on the battlefields of the western front. *Text and Performance Quarterly, 26*(2), 162–180. doi:10.1080/10462930500519374

Johnson, R., & Gürcan, M. (2016). *The Gallipoli Campaign: The Turkish Perspective.* Routledge.

Krotscheck, C. (2014). *Cultural Resources Strategy for Sustainable Regional Development-The CCC Method.* Verlag.

Reijnders, S. (2011). *Places of the imagination: Media, tourism, culture.* Ashgate.

Sturken, M. (2007). *Tourists of history. Memory, kitsch, and consumerism from Oklahoma City to Ground Zero.* Duke University Press.

Whitehouse, A. (1964). *Epics and legends of the First World War.* Frederick Muller.

KEY TERMS AND DEFINITION

Cultural Reenactment: Dramatic narrations or performances of culturally specific historical events for education and entertainment.

Folk/Oral Narrative: Cultural narratives that are transmitted orally from generation to generation and determined by the public as references on which behavior, attitudes, norms, and beliefs are based.

Gallipoli Campaign: Naval and land wars that took place on the Gallipoli Peninsula between 17 February 1915 and 9 January 1916 as part of the First World War.

ICH Valorization: The practices and processes of the increasing value of intangible cultural heritage resources.

Intangible War Heritage: Cultural representations, expressions, and beliefs formed around a war comprise a society's collective wartime psychology and behavior patterns.

Oral Tradition: A cultural system comprising traditional knowledge, skills, practices, and expressions transmitted from generation to generation through oral communication techniques and contexts.

Remediation: The practice of reproducing, embodying, iconifying, and transferring cultural memory contents through all available media technologies, cultures, and ecologies.

ENDNOTE

1 The cult of Khidr-Elijah was formed around the folk beliefs and narratives about the meeting of the prophets Khidr and Elijah once a year, who were believed to have attained the secret of immortality and helped those in distress, one on land and the other in the sea. In a significant part of the Islamic world and by the way among the Turks, the memory of the day Khidr and Elijah were united has been celebrated in different days and forms since ancient times. Hıdrellez, celebrated on May 6 among the Western Turks, whose center is especially formed by Anatolia and the Balkans, Crimea, Iraq and Syria which consists of the form that the names Hızır and İlyas is the name of a folk celebration whose root is based on the old Central Asia, Middle East and Anatolia (Ocak, 1998). Hıdrellez inscribed in 2017 on the Representative List of the Intangible Cultural Heritage of Humanity with the joint multinational file "Spring Celebration: Hıdırellez", prepared by Turkey and Macedonia (UNESCO).

Chapter 7
The Literary Geography of İstanbul and the Ahmet Hamdi Tanpınar in the Framework of the Narration and Tourism Relationship

Zeliha Öztürk
https://orcid.org/0000-0001-7900-4108
Bilecik Şeyh Edebali University, Turkey

ABSTRACT

This chapter aims to evaluate the works made under the title "the literary geography of İstanbul" concerning literary tourism in the activities carried out by the Ahmet Hamdi Tanpınar Research and Application Centre regarding Ahmet Hamdi Tanpınar and the novel Huzur. In this framework, the author examines the tourist experiences of literary tourists in actual and fictitious localities, their psycho-aesthetic experiences, the relationships they build with the city, the method used by the research center in these touristic activities, as well as the consequences of these experiences for literary tourism. The potential of Turkish literature in determining the future of cultural tourism and literary tourism will also be evaluated.

INTRODUCTİON

As in all other fields, postmodernity also paved the way for experiences oriented toward the individual in tourism and gave rise to the emergence of the special interest tourism which is a new field of experience in addition to mass tourism. In this alternative tourism, the desire and pleasure object of the individual moved towards his own experience fields, rather than collective experiences. In this new touristic experience shaped according to the individual and intellectual needs of the person, the concepts of "identity" and "desire" started to come into prominence and thus the subject of "diversity" in tourism became a matter that should be given thought. From health to sports, from science to arts many needs or interests

DOI: 10.4018/978-1-7998-8262-6.ch007

have oriented touristic experiences under this diversity. At this point, it is clear that one of the most important factors shaping the touristic experiences of individuals is their intellectual experience. Since the intellectual experiences of individuals affect their touristic activities while their touristic experiences affect their intellectual accumulation; there is a reciprocal and inseparable relationship between tourism and intellectual experiences. Among these intellectual experiences, literature started to affect, change and diversify the touristic activities of individuals as one of the most prominent areas when it comes to diversity in tourism and alternative tourism. Many subjects from literary texts (narrations, travel writings) to literary personalities (writer, fictitious personalities), from literary geography (actual-fictitious localities) to literary activities provide individuals an aesthetic pleasure that would be incorporated into their intellectual experience together with their desire to re-experience in the course of the touristic activity. At this point, all the activities experienced by individuals in the field of literature as tourists also turn into an artistic activity as much as they are a touristic activity. The psychological state of the person in relation to an artistic object and the psychological state created by a touristic activity he/she experiences in the field of arts and literature come close to each other. Since all the details concerning writers, works and readers that you might make mention of within the field of literature have the potential of turning into a touristic motivation sociologically as well as psychologically, at this point, it has become necessary to talk about literary tourism on its own. Literature, which is one of the fields for experiencing a city within the scope of cultural tourism, also turned into special interest tourism as a field having a touristic experience diversity. Literary tourism is an important and growing sector of the tourism industry, associated with places celebrated for literary depictions and/or connections with literary figures (Squire, 1996). Tourism then originates when the popularity of a literary depiction or the stature of an individual author is such that people are drawn to visit the places that he/she wrote about or was associated with (Busby & Klug, 2001, p. 319). Since men of letters and literary works which form an integral part of the city identity have the power to form a feeling of identity and sense of belonging for the tourist, it is necessary to consider literary tourism on a multidimensional basis with respect to tourism as well as artistic activity.

First literary localities draw attention in literary tourism. A literary or artistic place can be regarded as a place to which visitors attach meaning and it is the value of this meaning to them which draws them there. Meaning can be derived from reading a novel or seeing a work of art as much as from knowing about the life of the writer or artist (Herbert, 1996, p. 78). Although the localities where writers lived, produced their works or those they mentioned in their works are the first routes drawing the attention of the modern tourist in a touristic activity, they have also been the foremost localities for understanding the identity of the city and participating to it. Literary tourism works to put these localities into the service of the tourist while it also works to ensure that these localities contribute to the local economy. Nevertheless, today literary tourism also has to think on the authenticity and attractiveness of fictitious and real localities, their suitability to tourism, their power and efficiency in transferring cultural heritages and the bond the modern tourist establishes with the literary work, the image it produces over it and his expectation from literary tourism. All these values produced over the locality give the tourist the possibility of forming a literature destination. On the other hand, literary festivals, publishing fairs, reader's meetings and activities that can be mentioned under the writer, reader, work and publisher axis are also considered as activities appealing to the tourist within the scope of literature (Çevik, 2017, p. 158). While a big majority of today's world capital cities try to keep today's literary tourism dynamic with writers and literary works identified with the city; they continue to design creative and original contents and activities in this special field and create classical and new media that would experience them. Nowadays, it is possible to talk about literary tourism and experience reduced to much more special interests beyond the museum houses of writers and literary festivals which became classic.

İstanbul, which stands on the rich cultural heritage of the Byzantine and Ottoman states, is a city with a rich literary tourism potential as it hosted many men of letters from different languages and cultures throughout history. After the establishment of modern Turkey, literary tourism in İstanbul has been generally shaped by turning the houses of writers into museums and exhibiting the literary works of writers, their documents and various objects they used in their lives and has been perceived in a classical manner for long years. While the literary tourist was only an observer in this classical experience, some adventures in which the literary tourist will participate started to get designed and presented in accordance with the interests and requirements of the modern tourist. In the scope of this new literary tourism which started in Turkey with private and boutique enterprises; fictitious and actual locality experiences, the creation of inner-city travel routes based on fiction, the objects used by fictitious personalities in their lives, city experiences, the creation of city maps in parallel to these experiences and their sharing with tourists started to increase considerably and also attract the attention of the new tourist much more.

The fact that literature has also been transformed into a marketable touristic element in today's world where each tangible and intangible cultural heritage has the potential of getting transformed into a product pushed the institutions directing cultural policies of states to think on the subject and develop policies. By the year 2007, the Ministry of Culture and Tourism of the Republic of Turkey gave place to the title *The Strategy of Tourism Diversification* in its planning entitled *Tourism Strategy of Turkey-2023* and attached a special importance to reduce touristic experiences of local and foreign tourists into quite specific fields and to create touristic experience fields for tourists having special interests and it started shaping its activities in this direction.http://www.sp.gov.tr/upload/xSPTemelBelge/files/J4WC8+20070302-17-1.pdf At this point, literature has also been considered as an important cultural heritage referred in this tourism diversification strategy.

As a result of the area of the modern Turkish literature centred especially in İstanbul, each one of the ancient quarters of İstanbul has been known and named after a man of letters. At this point, the Beyoğlu quarter comes to the forefront as it was the one where especially the publishing sector pursued intense activities, where men of letter lived and frequently cited in their works. In its project entitled "The Literary Geography of İstanbul", the Mimar Sinan Fine Arts University located in İstanbul launched its project by planning to offer readers and thus tourists an experience and route that will give them the chance to trace the writers and works they read willingly within the city after determining the localities where men of letter who lived and wrote in İstanbul left some traces. The University and the Beyoğlu Municipality launched their project from Beyoğlu where many writers grew up and many works have been produced and started to make efforts within the field of literature as well as within the field of tourism by determining the houses of writers, the localities where well-known works have been written and hanging out signboards for creating some awareness on the subject. Beyond these works, by means of the research centre named The Tanpınar Literature Research and Application Center it has established within its body, the University also targeted to shape a new İstanbul experience by combining it with an aesthetic experience and offer it to interested individuals based on the biography as well as the works of Ahmet Hamdi Tanpınar who is known as the writer best describing İstanbul in the Turkish literature.

The Tanpınar Literature Research and Application Centre established in the year 2017 had for objective to offer a brand new writer-work-reader experience by starting to use the physical means and archives of the Ahmet Hamdi Tanpınar literature museum and library established in 2011 by the Ministry of Culture and Tourism of the Republic of Turkey in Alay Palace (19th century) situated in Gülhane Park within the historical peninsula which is one of the most popular localities of İstanbul for local as well as foreign tourists. The centre was aware that it was a research and application centre dealing with an artist and works of art and it also started to carry out activities especially for those who were the readers of Tanpınar, wished to see and experience İstanbul from an artistic point of view and reach the identity and sense of belonging over these experiences. At this point, the centre revised all the archives based on the artistic identity of Ahmet Hamdi Tanpınar and started to create a number of touristic experience fields equivalent to the aesthetic pleasure the reader takes as he/she consumes the work of art and also can become a participant. The centre which designs activities based on the inventory it has made for the artistic materials mentioned in all the works of Ahmet Hamdi Tanpınar, offers small groups of interested persons the possibility to experience the films spoken highly of in the works of Tanpınar, musics and certain well-known paintings he made mention of in his novels with renewed activities in the world of thought of Tanpınar.

Tanpınar Literature Research and Application Centre attaches a special importance to Huzur (Peace), the most known and loved novel of Tanpınar in Turkish which places the love of İstanbul to its centre. Huzur, which is a novel placing İstanbul to its centre around the adventure of two lovers, tells about the love the two lovers feel for İstanbul beyond each other. The two lovers walk around the different quarters of İstanbul, just like tourists and make have talks on the emotional state these walks create on them. The Centre brings this fictitious İstanbul experience in the novel to a real experience and draws the İstanbul map of the Huzur novel; It presents the routes followed by the two lovers during their talks and turns them into a touristic tour plan. In this context, the Research Centre which realizes quite attractive touristic activities for the literary tourist with the route it has set based on a fictitious text, offers an alternative İstanbul tour for the tourist based on a novel accepted as the best İstanbul novel in Turkish. This tourism which offers a new and exciting experience to the literary tourist and has been tried by just a few museums and enterprises in İstanbul for the time being, presents touristic activities worth being examined with respect to the development of literary tourism and gives some clues on the future of the literary tourism.

This chapter aims to evaluate the work developed under the title *the literary geography of İstanbul* with respect to literary tourism, within the activities the Ahmet Hamdi Tanpınar Research and Application Centre carries out regarding Ahmet Hamdi Tanpınar and the novel *Huzur*. The touristic experiences of literary tourists in actual and fictitious localities, their psycho-aesthetic experiences, the relationships they build with the city, the method used by the research centre in these touristic activities will be examined. Within the scope of this analysis, the potential of the Turkish literature in determining the future of cultural tourism and literary tourism will be evaluated.

BACKGROUND

The definition and placement of literary tourism are quite important in the acceptance of literary tourism as a systematic tourism field. In general, literary tourism centers the literary field that can be mentioned within the written and oral cultural heritage specific to a population, nation or humanity, and shapes the product and event accordingly. In this respect, literary tourism is a field that is expected to be explained and governed within the framework of cultural tourism. According to Robinson and Andersen (2002), literary tourism is a sub-type of cultural tourism from an anthropological point of view when it involves visitors and tourists exploring, defining and creating the identifiers of cultural values. There are different opinions in the literature on which tourism type literary tourism should be evaluated in. In studies of some people such as Squire (1996), Herbert (2001), and Smith (2003), it is seen that literary tourism is discussed within heritage tourism (Çevik, 2017, p. 156). Literary tourism is a tourism type that can also be evaluated within the special interest tourism category, as a tourism type shaped in accordance with personal, intellectual interests and needs. In this sense, in our study, literary tourism and the aforementioned example has been positioned and examined through special interest tourism based on culture.

The first aspect that should be carefully addressed in the literary tourism field is the topic of "literary place". The places where the popularity of the author/poet or work creates value, are referred to as literary place within literary tourism. It is seen that those who work on literary tourism and design touristic events think about the reasons why these places turn into literary places, how a literary place can be redesigned from a touristic aspect, and the type of literary places that are produced based on touristic trends. Squire (1994), and Herbert (1995, 1996) discuss the literary place concept as two main topics. The first of these places is the actual places related to the life of the author/poet, the other is the fictional place that emerges contingent to the work/fiction. Urry (1990, 1995) emphasizes the effect of cultural heritage and touristic trends in assigning value to the place and creating places, and remarks that these should be highlighted in touristic place design. In this sense, understanding how the literary tourist perceives the literary object (work-author's life) in the creation and design of the literary place from a touristic perspective is also important in revealing the quality of the literary place. Literary places are also quite effective in the routes of literary tourism and by extension, diversification of literary tourism. Literary tourism regarding actual and imaginary/fictional places has currently expanded by absorbing many other touristic activities centering literature. With the contributions of Busby and Klug (2001), Busby and Laviolette (2006), and Mintel (2011), the types of literary tourism have spread to a wide area that includes, trips to actual places, trips to fictional places, trips to destinations that appeal to literary figures, trips to destinations based on the popularity of an author or work, travel writing, literary tourism due to movies, literary festivals, bookstore tourism (Çevik, 2017, p.157).

The quality, definition, behavioral practices and motivation that shapes literary creates another important field of study about the definition and sustainability of this type of tourism. It is a reality that the life conditions of modern life created a new type of tourist and sense of tourism. The transformation of the concept of travel into the concept of tourism with industrialization has led to the criticism that the "tourist" has moved from an active position to a passive position, and that the situation in tourism, which has become a commodified activity, has changed. In the face of this industrialization, it has become necessary to re-examine and redefine the tourist, who is after the unique and experiential. At this point, Boorstin (1961) discussed the concepts of tourist and traveler separately, defined the tourist as those who consume the ready and attractive content of the tourism industry from a pleasure focused and passive position, while defining the traveler as those who discover, are focused on adventure and thus, in an active position. (Sağıroğlu, 2019, p. 93). In the work titled *The Tourist*, where MacCannell (1976) discusses the tourist concept in detail, it is stated that tourism is the sacred journey of the modern times, and that the modern person/tourist, who has severed the ties between their own life and reality, pursues unique experiences for seeking shelter. This situation also provides data on the motivation for travel of the literary tourist who is after unique experiences. Gentile and Brown (2015) reached the conclusion that the feeling of nostalgia and the drive for escape are the markers of the literary tourist's motivations (As cited in Çevik, 2017, p. 159). The place that the literary tourist escapes to is the imaginative world that literature creates. The literary tourist romantically creates an image in their mind with this imaginative world and pursues this image in their travels.

According to Reijnders (2011), the tourist travels to compare real places to the images they created in their minds. By trying to establish a connection between the literary work or the fictional character and the ideal world that the author mentally creates in their image, the literary place, touristic route and daily reality, the literary tourist carries this tourism into a field of experience. These expectations and comparisons of the literary tourist expose yet another problem regarding literary places. Understanding and meeting the tourist's authenticity expectation from the place and experience should be taken into account as a significant factor that will contribute to their approach and adoption of the touristic experience. Jia (2009) examined authenticity in literary tourism by considering the tourist's self, the relationship between reality and the displayed object and the presentation and configuration of the displayed object in the literary tourism sector. As a result of this examination, they discussed authenticity under the titles of objective authenticity, constructive authenticity, postmodern authenticity and existential authenticity in literary tourism. In our study, especially postmodern authenticity was examined due to its suitability to our topic.

An important topic examined within literary tourism is the quality and characteristics of the literary tourist. As a result of these literature centered touristic activities, it stood out that those who are interested in this type of tourism consisted of groups that have particular and similar characteristics. Specifically, it is clear that these groups are a community with above-average intellectual tools and at least one specialization, with more exclusive and original expectations from tourism, and literary tourism takes these expectations into account when structuring and presenting a touristic value. In their studies on the quality of literary tourists, Squire (1994) and Herbert (1995, 1996, 2001) revealed that literary tourists consist of tourists who have high education and income, awareness about literature and literary tourism, who are over the age of 35 and mostly women. The way the literary tourists, whose common characteristics are gathered under these titles, approach literary tourism has a state of mind and expectation in line with these qualities. Cohen (1979), who tried to make a classification on the state of mind and expectation of the tourist approaching the touristic event, took criteria such as entertainment, rest and the pursuit of authenticity in touristic experiences into consideration and determined touristic temperaments such as; recreational tourist mode, diversionary tourist mode, experiential tourist mode, experimental tourist mode and existential tourist mode. Due to the general quality and expectation of the literary tourist, the search for originality and the desire to experience the mental image created through literature is prominent, and it is possible that the literary tourist generally approaches the experiential mode.

It is seen that academic studies on literary tourism are also closely interested in the realization process of literary tourism. This process is not only an activity that can be limited to the moment that the touristic activity is performed, but it is a process that spreads over a wide period of time, and in this process, the psychological state and readiness of the tourist also affects the perception of the performed activity. To have knowledge about the author, works and literature in general can be evaluated as literary awareness in literary tourism; being aware of the fact that a literature-based touristic event is an independent type of tourism and traveling accordingly can be considered as literary tourism awareness. This awareness that the literary tourist will have stands out as an important factor in terms of its potential to differentiate the experience in the realization process of literary tourism. All of the accumulations regarding the image, myth and imagination created by the literary tourist with this knowledge may have the power to direct them to tourism. This can take the realization process and beginning of literary tourism back even further. Amirou (1999) discusses the realization process of tourism as "past, present, after". The images created by literature (author's biography or reading a novel, etc.) in the tourist and the literary tourism that starts to take place from the moment they set off, continues in the future due to the main experience offered being lived through and that this experience continues for longer periods of time with effects such as feelings of belonging, relaxation, sympathy, discussion and questioning in the tourist. This expands the realization process of literary tourism.

In addition, today, literary tourism is an area that continues to examine and discuss sociological cases such as the impact of this type of tourism in creating urban culture, and to examine the potential and application methods of this type of tourism implicated in cities and opens up to broad academic issues.

AHMET HAMDİ TANPINAR AND HUZUR

Ahmet Hamdi Tanpınar is one of the prominent names in modern Turkish literature. Tanpınar, born in 1901, witnessed the downfall of the Ottoman Empire and establishment of the Republic of Turkey, and is a writer who was deeply affected by the political and cultural implications of this fracture. Tanpınar, who was educated in the literature department, gave literary history and aesthetic classes in various universities. Both his academic writings and his literary works rely on a wide academic accumulation including mythology, aesthetic, history, culture and politics. In Tanpınar's understanding of art, the literature, plastic arts and music of the West are at least as effective as the local culture. This situation enables him to put forward an attitude that constantly questions this dualism in his literary works and is sometimes nourished by it. Okay (2010) states that Tanpınar, who understood the last periods of the Ottomans and still experienced its effects, is a part of a generation that has mystical and metaphysical tendencies in terms of temperament. Concepts such as time, eternity and dreams, which Tanpınar frequently uses in both his novels and poems and which have become his aesthetic, are nourished by this mystical and metaphysical tendency. Saying that Tanpınar was influenced by Proust especially in his perception of time in his novels, Okay (2010) emphasizes that he mostly pursues the past times. Tanpınar, who cannot remain indifferent to the socio-cultural issues of the civilization, country and city he lives in, takes an attitude in favor of the ideas of composition, movement, change, renewal and continuation while examining these issues in his works. In particular, the concept of the city often occupies Tanpınar's thought and works as the concept that best reflects a mass change or ongoing culture. The city that the author, who has many essays on the city's architecture, life culture, and human types, thinks about the most, is undoubtedly İstanbul. The author, who examines the changing life culture of İstanbul within a civilization problem, takes a careful look at the city life of İstanbul in his novels. In particular, the novel Huzur attracts attention as a work in which İstanbul has turned into a decoration and moreover, a novel hero.

The novel Huzur, published in 1949, is a novel that has become famous as the most beautiful romance novel and the most beautiful İstanbul novel in Turkish. The novel, in which the love for each other and the love for İstanbul of the heroes, Mümtaz and Nuran, who explore İstanbul together, intersect, creates an atmosphere in which İstanbul is visited and depicted from district to district. The novel, which is basically about one day the main character named Mümtaz spent leaving the house on August 31, 1939, expands to important dates such as 1911, 1922, 1937-1938, 1938-1939 in the personal story of the main protagonist, with various flashbacks. The novel ends with the hero Mümtaz, who returned after completing one day, hearing the news that the second world war had begun. Kaplan (1962) mentions that Tanpınar has the "actual time" of the novel as twenty-four hours and takes James Joyce's "Ulysses" as an example in doing so and adds: The novelist read this book when he wrote his work. As a matter of fact, there is a short reference to him somewhere in the novel. It can be argued that Tanpınar was under the influence of James Joyce in the way he walked his hero around the city, the way he portrayed the life and decor as a landscape of confusion and ruin, and on some details. (p. 43). Also, in respect to turning the city into a travel route, reflecting on certain points of the city and works on flashbacks, intellectual anxiety and socio-cultural problems, Huzur has similarities with Ulysses. Okay (2012) states that Huzur is a novel of rediscovering İstanbul with its nature, districts, history and works of art. He says that İstanbul entered this novel as a natural decoration of music, painting, calligraphy, architecture and all aesthetic and cultural accumulations to the love and relationship of the individuals. While the Bosphorus creates a motif that prompts the love between two young intellectuals in the novel, the city's slums cause them to focus on social issues such as cultural conflicts, the wreckage of the empire, and the newly established republic's methods of coping with it, and thus, İstanbul forms the basis of another theme in Huzur. The novel takes place in 4 main sections, where the names of the main protagonists Mümtaz and Nuran, İhsan and Suat are given. Nuran becomes prominent as the female character with whom Mümtaz is in love. Okay (1998) states that the love that starts between Mümtaz and Nuran develops within the historical and natural beauties of İstanbul. Thus, in every aspect, İstanbul becomes both an effective decor and a supporting figure during the relationship of these two lovers, and this brings the novel's transformation into an İstanbul novel.

THE TANPINAR RESEARCH CENTER AND TOURS TO İSTANBUL WITHIN THE FRAMEWORK OF HUZUR

The Tanpınar Literary Research and Application Center, which was opened in 2017 in order to examine the poet and writer Ahmet Hamdi Tanpınar, especially in an academic sense, aims to create a very wide archive from Tanpınar's biography to his works and what is written about him in the literature. In addition, this research center attempts to be shaped according to the social expectations of the literature readers and researchers. At this point, besides organizing symposiums and panels about Tanpınar and creating reading maps through novels, The Tanpınar Research Center also plans touristic activities, especially based on the relationship of the novel Huzur with the city of İstanbul. In line with the content of the novel, the research center, which creates a step-by-step map of İstanbul where the main character Mümtaz and his lover Nuran travel, designs events that realize the transformation of İstanbul into a travel route, which is only noticed during literary studies on the novel Huzur, in the field of tourism and present these events for literary tourists to experience. Therefore, these activities and efforts that work on the city and the novel together, evaluation and structuring of these, provide new and interesting possibilities regarding the literary tourism potential of İstanbul.

As mentioned in the novel Huzur, The Tanpınar Research Center has created four different travel routes in line with the places visited by the characters of the novel, taking into account the transportation and proximity of these places. Most of the places mentioned as historical in the novel are real places. However, these real places in the novel where especially Mümtaz and Nuran interact, walk, chat, and discuss are also fictional places as they are recreated through these characters and reconstructed through the discourse of the novel. Real and fictional spaces come together in the novel, and this combination creates places that are attractive for literary tourists and creates high expectations of authenticity. The direct relationship of these fictional places in the novel with reality, while creating a travel route, caused a design through the expectation of real-life transportation convenience, such as the proximity of these places to each other and their accessibility. Each route is formed by choosing 8-12 places close to each other; these places can be visited mostly on foot and each of them can be accessed during the day. Each route is also located on interesting touristic routes other than literary tourism. A large number of literary places, including almost all of the historical parts of İstanbul described in the novel, were divided into four main routes, and therefore, four different travel alternatives were created. According to Akkaya et al. (2015), this route planning is based on the principle of from upper scale to lower scale. Then, while creating the route, the principle of holistic and thematic thinking was again acted on, in a way that tightly preserves the connection of these routes with the fiction and narrative of the novel. These four routes are as follows: 1. Historical peninsula - Beyazıt square and its surrounding; 2. Üsküdar; 3. Emirgan - Kanlıca - Bosphorus Line; 4. Fatih district.

Figure 1.
Source: http://www.tanpinarmerkezi.com/roman-haritalari/

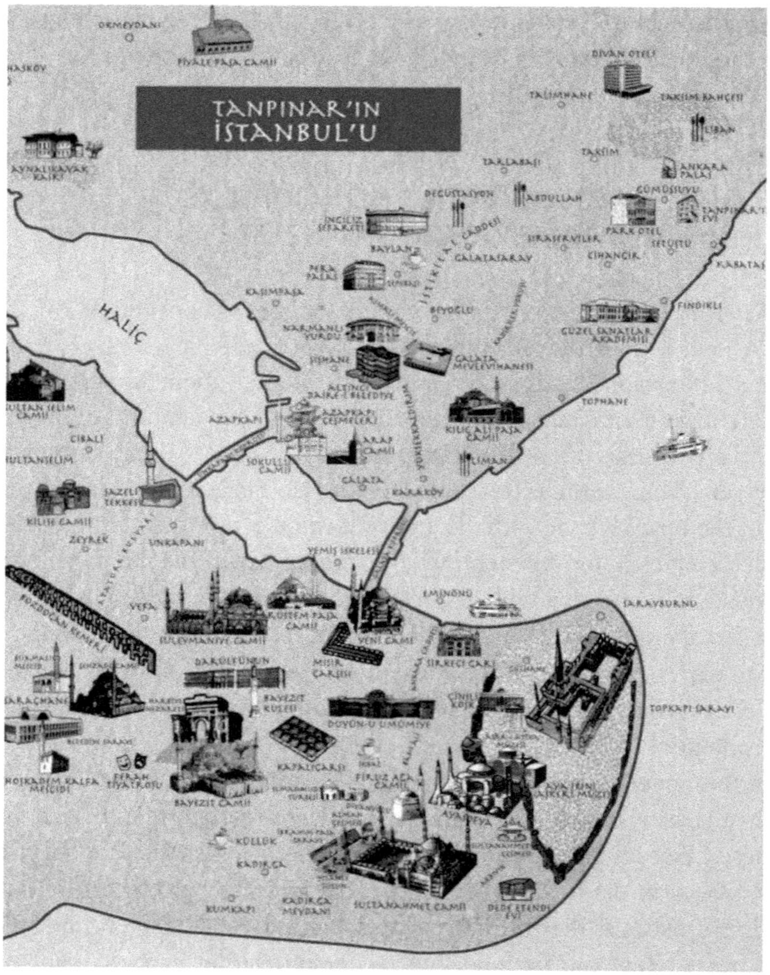

Route I: It consists of 12 places located in the most popular touristic area of İstanbul between Beyazıt Square and Eminönü. These places, which consist of mosques, squares, bazaars, coffee houses and bookstores, are mostly mentioned in the first chapter of the novel. While looking for a doctor for İhsan, the main character Mümtaz grew up with, these places not only appear as stops, but also as places that carry the memories of Mümtaz's past experiences. Mümtaz goes into a pessimistic mood as a result of the pain of breaking up with the girl he loves and İhsan's illness. This pessimistic mood affects his view of the places he visited on this route. These places have two main thematic aspects in the novel. Firstly, these places cause a sense of belonging to his culture for Mümtaz (square, coffee shop, booksellers); secondly, these are the places where the cultural, political, and economic problems experienced during the transition from the Ottoman Empire to the Republic are reflected. According to Mümtaz, these places, which consist of streets, bazaars, and small shops, are tired, untidy, noisy, and places where piles of goods accumulate in a chaos.This narrative will inevitably affect the perspective of the readers and the tourists who will visit this route towards the places and things; it is also a highly critical and very effective in terms of style. The Tanpınar Research Center has determined these places, which are within walking distance of each other, as the Route I in line with the places mentioned mostly in the first part of the novel. The research center has prepared an animation map that can be used by literary tourists during the trip, where all the stops on the route can be seen together, and shared this map on the website of the center. The center, which prepares animation maps for the other three routes in the same way, has shaped each stop on the route on these maps by quoting from the related parts of the novel. These maps stood out as an element that increased the reader's commitment and motivation to both the text, the route, and the literary places, and used effectively during the trip.

Figure 2.
Source: http://www.tanpinarmerkezi.com/birinci-rota/

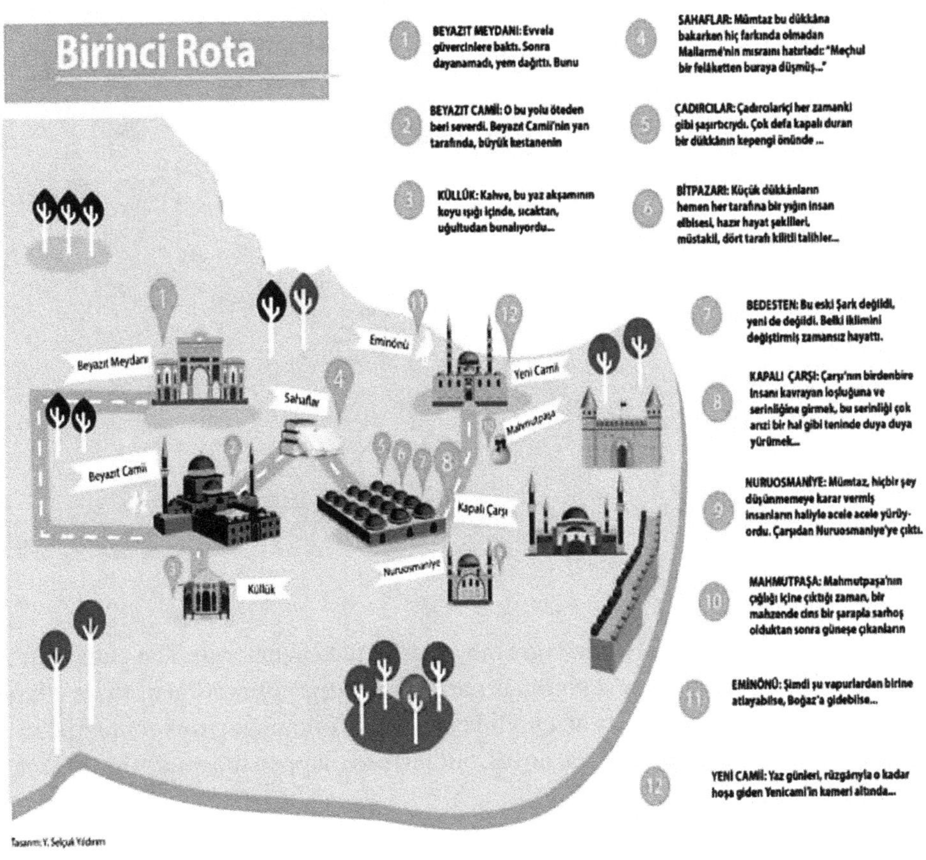

Route II: It covers the prominent locations of Üsküdar district, which is like a treasure according to Mümtaz and Nuran (Anar, 2020). The research center has determined fourteen locations for this route based on the novel. The majority of the places consist of mosques; however, dervish lodges, cemeteries and recreation areas are also included in the route. Not all locations are within walking distance of each other, but some places are quite close to each other and can be visited and experienced in a single trip. This route mostly bears the traces of the love and intellectual sharing of the novel characters Mümtaz and Nuran. In the novel, Mümtaz provides historical information and architectural interpretations of the places. It can be said that he did this mostly for the places on this route; Mümtaz sometimes answers Nuran's questions about places, just like a travel guide. It is noteworthy that the places he visited during these happy times are mostly places such as mosques, dervish lodges and cemeteries. These places, which mostly have an emphasis on eternity in Turkish culture, are places that are considered to have high spiritual-mystical value in terms of the civilization. This is a route for literary tourists, where a romantic love story is shared with their fans and a sense of empathy and belonging to a civilization is created. At the same time, it is a route that provides the opportunity to get some information about real places through Mümtaz's stories about places, so it is expected to attract the attention of the readers more than other routes.

Figure 3.
Source: http://www.tanpinarmerkezi.com/ucuncu-rota/

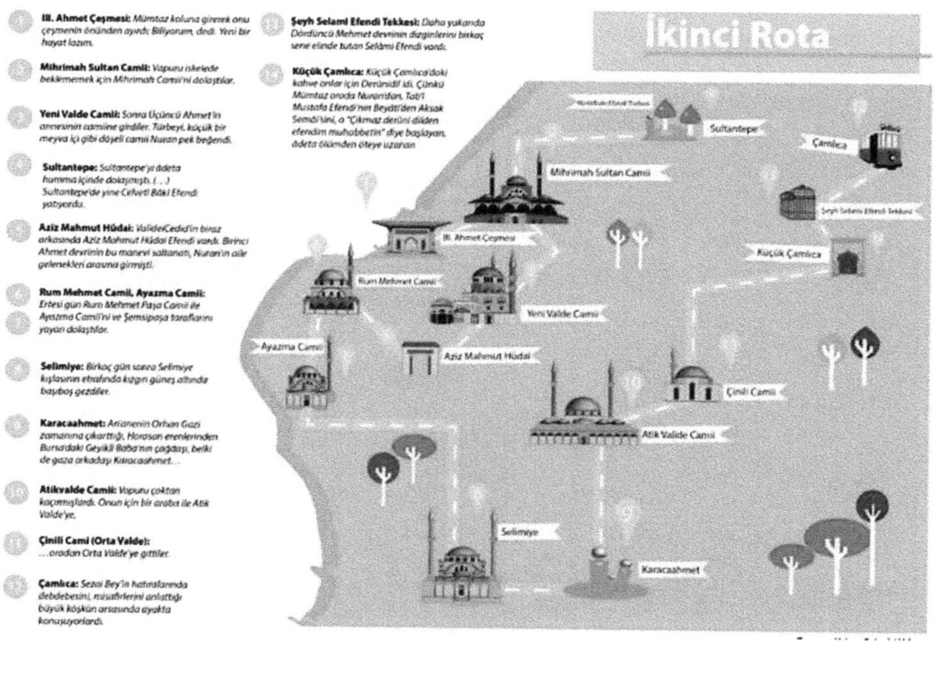

Route III: It consists of districts located on both sides of the Bosphorus. The places on this route are the places where the love of Mümtaz and Nuran is narrated and the admiration of these characters for the Bosphorus is experienced by the tourists at a high level. While organizing this route, the research center did not focus on the districts described in the novel, but instead mapped the roads that Mümtaz and Nuran crossed, sometimes on foot and sometimes by boat, and created the travel route based on these paths. 11 places from the novel are included in this route. These districts, many of which were known as fishing villages until the 18th century, are places preferred by İstanbul's prominent families and built a lifestyle and culture in the form of being a native of Bosphorus. As a matter of fact, the houses of Mümtaz and Nuran are located in these districts as well. The elements highlighted in this route, which Mümtaz and Nuran travel in the novel, are İstanbul's coastal districts on the Bosphorus, the sea and nature. Here, both the Bosphorus itself and the views of İstanbul seen from the Bosphorus create an atmosphere that complements the love of these two lovers; Mümtaz and Nuran's love for each other and their love for the Bosphorus and İstanbul merge. Because of this romantic story, this route seems to be the most attractive route to attract literary tourists.

The research center has remarkably emphasized the time factor when structuring and mapping this route. The love of Mümtaz and Nuran lasted about a year. During this period of time, they mostly visited the Bosphorus. For this route, the research center has created a map that uses the connection of some special dates such as May 6, May 11, August 13, which are the dates of the travels in Bosphorus districts according to the novel, and seasons such as spring and autumn or months such as June and September. Accordingly, they encourage the route to be visited at the mentioned dates and seasons, with references to these dates. This application provides the perception of time and space, which are the elements of the aesthetic structure of Huzur, together and embedded. Thus, it presents a holistic approach to the expectation of authenticity and the desire for experience of the readers and tourists.

Figure 4.
Source: http://www.tanpinarmerkezi.com/ucuncu-rota/

Route IV: It consists of certain locations in the Fatih district of the European side of İstanbul. Fatih, one of the historical districts of İstanbul, is the settlement of places visited to connect with the past, according to the novel. Especially the traces of the conquest of İstanbul and the reign of Mehmet the Conqueror shape the travels of Mümtaz and Nuran. This route includes nine places consisting of mosques, tombs, madrasahs, cemeteries, and historical walls. Mümtaz and Nuran enjoy reading Ottoman inscriptions in these places, witnessing the stories of mystical figures integrated with the history of the city in the tombs, and experiencing the traces of old lives. This route is similar to Route II on the Anatolian side of the city in terms of formation and purposes. In both routes, travelers who follow the route, similar to the characters of the novel, have the opportunity to experience the history of İstanbul, especially the post-conquest, and the architecture and culture in which the Turkish-Muslim identity was built, within the framework of nostalgic expectations. The idea presented in the novel Huzur is that despite all the modernization winds, it is necessary to renew the local culture by maintaining its essence. Every place seems to serve this continuation idea with every historical detail. This route has been shaped with the aim that the tourists who will participate in the tour experience the feeling of "continuation of civilization" that the novel characters and author wanted to convey. Thus, the novel, which questions the chaotic gap created by modern life between the individual and the past, provides the opportunity to close this gap for tourists who feels the same chaotic gap through these touristic experiences.

Figure 5.
Source: http://www.tanpinarmerkezi.com/dorduncu-rota/

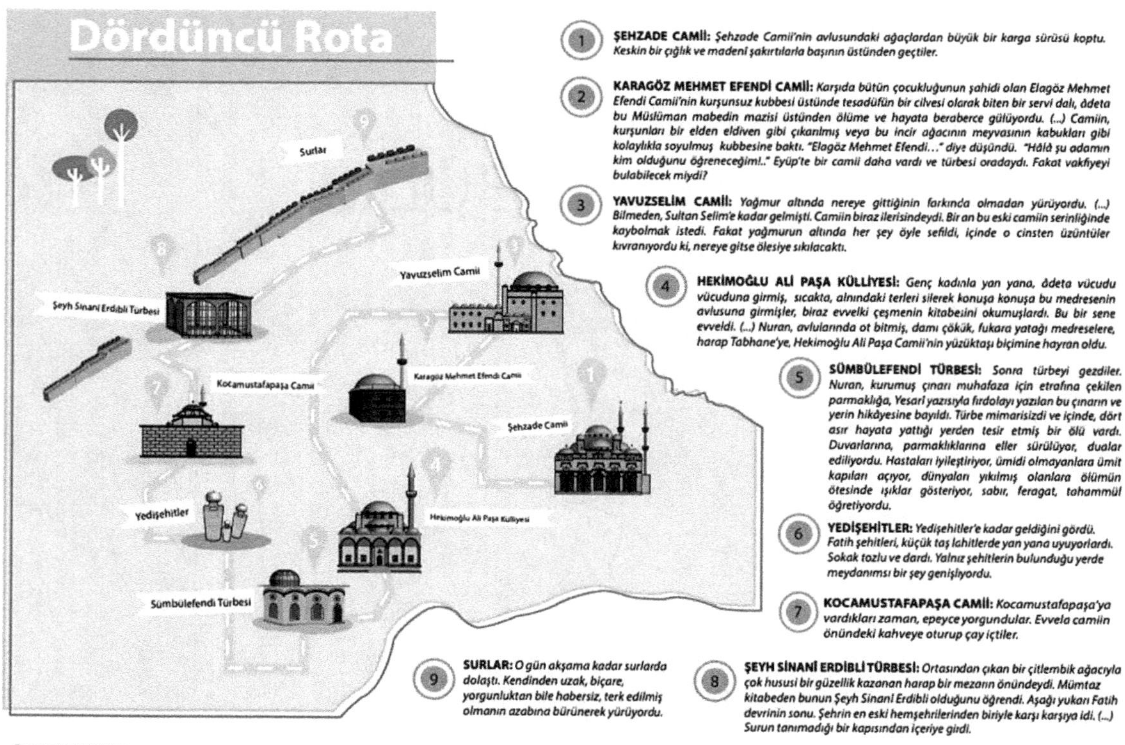

The Tanpınar Research Center does not open the travel route of Huzur to experience only through places; the research center also adds other experiences related to cultural and daily life of İstanbul in the design through these trips. Use of trams and ferries, which are transportation vehicles used in İstanbul, especially in historical districts, are a part of these trips. In some of the trips, the group gets on the ferry, which is a part of the love of Mümtaz and Nuran according to the novel, and experiences it as a part of the trip. During the ferry ride, the tour group shares their opinions and quotes on the fiction, characters, and literary places of the novel. Thus, this ferry journey turns into a touristic experience, unlike the daily ferry journey.

The Tanpınar Research Center specifically includes quotations from the novel in these trips. The research center uses the passages related to the places mentioned in Huzur in the maps it creates. In addition, reading the passages in the novel related to the places visited during the trip by the guide or the participants was added to this trip as an activity. After seeing the real places in the trip, their relationship with the novel is evaluated by the group. These evaluations are about whether the present reality of the place and its fictional and historical reality match. The most interesting element in this activity for literary tourists is the short stories about the history of these places that they read about in the novel. These stories belong to places that can be considered historical within the actual time of the novel. But these are the stories that are told based on Mümtaz's interest in history, culture and architecture and his intellectual background. The content of these stories consists of narratives such as the formation of the places, their relationship with historical figures, their place and importance in terms of culture and civilization. It is the content that connects both readers and literary tourists to the text the most and pushes them to the touristic activity the most. Readers or literary tourists who like to read and listen to stories desire to experience literary tourism to compare these images they have witnessed in the novel with the existing reality.

One of the important aspects added to these trips is music. Some works and composers of both classical Turkish and classical European music are mentioned in the novel; and there are musical pieces accompanying some of the romantic and critical scenes. Dede Efendi, Beethoven, Wagner are the prominent ones among these names. The research center has made this emphasis on music, which is intensely presented to the reader in the novel, as a part of the tour. For this, the musicians mentioned in Huzur were listed on the website of Tanpınar research center, and a playlist prepared from the songs mentioned in the novel was shared through the online music platform Spotify. In addition, the research center includes the music group Boğaziçi Mehtapları, which perform the compositions in the novel, in some of its trips, so that these events and their participants could relate themselves to the novel as a whole through the live performance. http://www.tanpinarmerkezi.com/huzurdan-muzikler/.

Mapping efforts are important in terms of the organizing process of these trips. These maps, some examples of which are shared for the routes, are both accessible on the website of The Tanpınar Research Center and are used as printed papers to strengthen the connection of the literary tourists with the event. These maps can be seen as novel and literary maps prepared for the activities of the research center to market destinations. In addition to these route maps explained in detail above, the center has created another map that marks the current and real locations of the places mentioned in the novel through the Google Maps application. Clicking on the marked points on this map opens relevant passages from the novel and offers tourists an independent guidance. In addition to these tours, which are mostly experienced in groups and accompanied by a guide, an opportunity has also been created for tourists who want to experience the İstanbul of Huzur individually through this map.

Figure 6.
Source: http://www.tanpinarmerkezi.com/roman-haritalari/

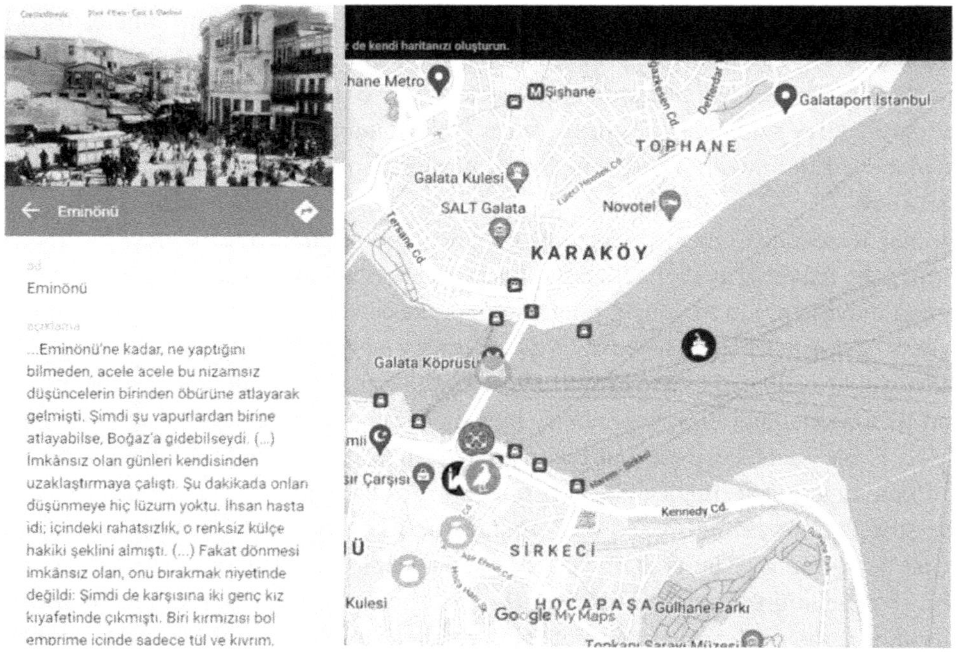

The Tanpınar Research Center effectively uses social media platforms Twitter, Facebook, and Instagram in order to share its academic and touristic activities with the people. With the same username, The Tanpınar Research Center shares posts, stories and tweets via these social media platforms to announce academic meetings, some archival documents of the author Tanpınar, and İstanbul trips guided by Huzur. The center not only announces the planned trips, but also shares multimedia related to these touristic events with photos of the participants and places. Many of these posts are shared using the relevant passages in the novel along with the picture of the route or places visited; thus, a relationship is once again established between the trip and the novel. These posts have a significant contribution to the impact of this touristic activity over a long period of time (before, during, after). As a result of the reading of the novel, the imagining process and the pre-event are built, while the trip plan is announced via social media and multimedia is shared from previous trips contributing to it. This process greatly affects the mood during the trip as well. Moreover, these posts also contribute to the post-trip period. Photos shared after the trip, comments conveying feelings and thoughts, and passages quoted from the novel ensure that these effects last for a long time and increase the feeling of satisfaction.

Through these holistic activities organized by The Tanpınar Research Center for the novel Huzur, one of the most qualified organizations of the touristic activities that can be designed using the fictional text has been achieved. And in terms of literary tourism in İstanbul and Turkey, these trips set an example of literary tourism that reveals the potential for spreading touristic activities to almost the whole of İstanbul without being limited to a single museum.

Figure 7.
Source: https://twitter.com/tanpinarmerkezi/media

There are a number of other programs created by The Tanpınar Research Center based on real places related to literature. As of 2019, The Tanpınar Research Center has started a project titled The Literary Geography of İstanbul in order to transform real places, which are related to literary figures who lived in İstanbul, into a route in terms of literature and tourism, to raise awareness on this issue and to bring these places into urban culture. The research center started its efforts primarily in the Beyoğlu district. Together with the municipality, it aims to identify the buildings where the literary figures live, to formally register them, and to contribute to the city culture via road signs. Among the activities of the research center, there are activities such as inviting literature lovers to meetings related to the registration and sign works of these buildings, commemorating the relevant writers or their works by organizing trips to these buildings at various times, and raising awareness about literary tourism in the city. For this new and ongoing project, Ahmet Hamdi Tanpınar's and Oğuz Atay's houses were identified, and signage works were done. https://www.aa.com.tr/tr/kultur-sanat/oguz-atayin-evine-bilgilendirme-tabelasi-yerlestirildi/1673158

The house in the Beyoğlu - Gümüşuyu neighborhood, where Tanpınar lived for a while, has been provided to İstanbul's literary tourism and literature lovers, with an information sign on it stating that Tanpınar lived in that place. Tanpınar's house can be included in the tour within the scope of the Literary Geography of İstanbul project for touristic trips to the literary places mentioned in the novel. In this way, a touristic experience will be created that is related to both the author and the novel, the actual place and the fictional place; these events, which will provide a holistic approach, will create a more satisfying literary tourism experience in terms of expectations and motivation of tourists.

AN EVALUATION ON THE TOUR IN LIGHT OF HUZUR AND THE LITERARY GEOGRAPHY OF İSTANBUL REGARDING LITERARY TOURISM

Touristic activities within the scope of the project titled Huzur Novel and Literary Geography of İstanbul are carried out by The Tanpınar Literary Research and Application Center, a research center affiliated with the Mimar Sinan Fine Arts University. Therefore, the primary purpose of organizing these activities is for academic and intellectual expectations. The Tanpınar Research Center is a non-profit organization that designs and presents these kinds of trips and activities. The research center is financed by cooperating institutions such as the municipality and the university it is affiliated with. Therefore, it is difficult to evaluate the touristic activities designed and provided by The Tanpınar Research Center based on the "Huzur Novel and Literary Geography of İstanbul" project in terms of industry, market, and marketing within the concept of tourism.

The first noteworthy aspect of the trip, which was designed based on the locations in the Huzur, is the route design of the tour. Route design should be based on the processes of understanding the region, evaluating it and building the setting (Akkaya, et al., 2015, p.61). According to Akkaya et al. (2015), for route design, it is very important to analyze and reflect the local elements well. If the route in question is a cultural route, the historical and cultural context of the content to be used and the symbolic values to which it is connected should also be taken into account. İstanbul of Huzur trips have been made in accordance with the elements of the novel; organizers have also carefully analyzed the literary places in the novel and the urban life of İstanbul with which it is related, and successfully divided the mentioned places in four routes. In this division, it is striking that three themes presented by the novel shaped the design: 1. Current problems of civilization, 2. İstanbul's historical and natural beauties, 3. The romantic love story. As explained before, the first route deals with the problems of civilization; the second and fourth routes are related to the historical values of İstanbul, the integration of the city and the people, and the sense of belonging; and the third route deals with scenes from a romantic love story. It is a design in which factors such as the participants, the type of trip and the duration of the route are taken into account, apart from the theme. Most of the routes are designed to be experienced on foot, and route selections are made accordingly. While designing these routes, The Tanpınar Research Center aimed to highlight the idea that İstanbul can be visited within the framework of the passion for literature and the novel. Therefore, the organizers developed a sensitive attitude that prioritizes the authentic values of İstanbul as a shared attitude with Huzur.

The realization processes of literary tourism, organized in line with the novel, include details about the process success of this event. According to Herbert (1996), the social construction process of literary heritage tourism has four stages. These are the generation, text, reading, and impact stages. At the generation stage, decisions are made on promotion, image and development, these decisions are placed in the space-setting at the text stage; and the reading stage consists of the use of the space and the way visitors interact with their environment. The impact stage consists of the effects of the visit on the behavior and value judgments of the visitors (A Cited in Narlı and Aliağaoğlu, 2012, p. 6). Accordingly, during the generation phase of the trip, there is an image that the readers have created in their mind through the novel, as well as an image created in the minds of the tourists via the Tanpınar research center's efforts and invitations to encourage İstanbul of Huzur trip. This image, which begins with the reading of Huzur, is mostly shaped by the expectation of understanding İstanbul through empathizing with the feelings and the emotional world of the romantic protagonist Mümtaz. The decision of The Tanpınar Research Center to place the images created by the novel in places of the route, as has been clearly described in the sections discussed in detail above, consisted of two main topics: love and civilization. The selected places are presented together with the passages of the novel as a civilization problem, a sense of belonging or a part of love. The most important question in this process is whether the expectation created by this touristic trip has been met in line with the image that literary tourists have created in their minds as a result of reading the Huzur. The common expectation for the images in the thoughts of those who participated in the trip is to visit İstanbul from district to district, like the main character Mümtaz. This expectation was met to a large extent during the trip, and the image created in the minds of the readers and the experience provided converged at a certain point. The impact of the tour continued for a long time, especially on the tourists who participated and experienced all the routes of the trip. Some of the participants are known to made comments on the ongoing effects of the trip under their related posts on social media platforms. This experience had different effects, especially for the participants. For example, it brought a different perspective to İstanbul, the city they lived in, and built a new sense of identity with this perspective. All these can be listed among long-lasting impacts.

The concept of space-setting is the most carefully considered issue in terms of tourism, in this literary trip led by the footsteps of Huzur. Kaplan (1962) emphasized that the novelist Tanpınar and his protagonist Mümtaz are very fond of nature and architecture, so there are a lot of depictions of places and objects in the novel. In the novel, the natural and historical places that Mümtaz contacted with were chosen from his own memories and perspective. These are fictional spaces, but most of these spaces are also real spaces that belong to the non-fiction world. For this reason, it is not possible to define this trip as visiting only real places or only fictional places. There are real places in the novel that are in line with the historical reality and still exist today. These places are also considered as fictional spaces set as a background for the adventures of fictional characters. However, these places, which are mentioned in the novel and still exist in today's life, due to the touristic trips organized in line with novels and novel itself, have turned into literary places for literary tourism.

The most important extension of the literary place is the expectation of tourists from literary places. Especially fictional literary places that exist in real life or have a very high relationship with reality create an expectation of authenticity for tourists. According to MacCannell (1976), tourists are adventure-seekers who try to find authentic elements and spaces staged by cultural tourism organizers (as cited in Çevik, 2017, p. 158). Therefore, the tourists who participate in the touristic trips organized in the direction of the novel want to experience a route and place equivalent to that of the main character Mümtaz; they want to see a presentation of a place that will fit with both historical and daily reality. This takes place mainly through the design of the route and literary places of the trip. Especially with the leadership of the tour guide, the narrative of real places, the position of these places in the novel, and their place in the world of the main character are presented together with the emotions they create; the main character converges to the literary places with a bond of authenticity. This type of authenticity is similar to postmodern authenticity, which is one of the definitions of authenticity that Jia (2009) points to in related study on literary tourism. Postmodern authenticity means rebuilding authenticity and destroying the boundary between the imitation and the original. The romantic universe created by the novel and the passion for İstanbul, which is at the center of it, is not a direct presentation of reality, neither actual nor in past. On the other hand, the way of presenting and perceiving the trip goes beyond an imitating activity presented to them in the actual time with holistic activities. Thus, it suspends for a while the idea that both the novel and the travel are fictional. This, in accordance with the postmodern attitude, means fascination of the disenchanted İstanbul both through this trip and the images created by the tourists.

The qualifications of the literary tourists participating in the trip match the qualifications of the tourists oriented towards special interest tourism. It can be seen that travelers who participate in novel-guided trips are middle/upper aged, professionally or intellectually interested in social sciences and/or arts, and have a specialty. These participants are usually academicians, philology students, TV programmers, translators, and writers in terms of their professions. Those who read the novel and participate in the touristic trip guided by the novel are very interested in the culture and identity of the city, and enjoy experiencing literature-centered activities. In addition, it can be seen that most of the participants who had literary awareness about both the novel and world literature before the trip they attended, also had awareness about literary tourism. The element that creates the motivation of those who participate in these trips is a novel. Considering this; designing the trip according to this element, the quality and intellectual background of its participants requires that this trip be evaluated separately within the scope of special interest tourism as well.

The aim of the travelers who participate in the tours is to experience the images and literary places created by the novel, to interact with them, and to reposition the relationship between the city and themselves. The participants of the tour desire having a unique experience in the city of İstanbul and finding the traces of the world they have built in their intellectual world, especially in light of the literature, while they have already exhausted the opportunities provided by İstanbul. From this point of view, İstanbul travelers following the footsteps of Huzur are included in the experiential tourist mode, which focuses on the search for originality among the tourist modes, according to Cohen (1979), who evaluates tourism from a sociological perspective. Cohen (1979) defines experiential tourists as those who have lost their own spiritual centers and original lives, trying to reconstruct the meaning in their own lives by experiencing the reality of other lives. Travelers who participate this trip desire to purify their spiritual world, which has been severely suppressed by modernism in a metropolis like İstanbul. They seek to join the universe of the civilization in the novel and the world of the romantic main character Mümtaz. Just like Huzur, which emphasizes the pure, unspoiled and untouched state of the reality of the past, travelers guided by the novel seek the lost purity inside modern İstanbul. This is a nostalgic bond with the past and a reflection of a romantic point of view.

The travelers of Huzur want to establish a semi-spiritual connection with the mostly historical places they come across. They desire mostly individual experiences such as creating meanings from the travel and places with their own images and transferring these to their stuck lives. These travelers are the type of tourists who have a romantic point of view among the tourist perspectives that Urry (1995) has classified and defined. They especially seek cultural originality, a lost meaning, and purity in their experiences. The same perspective is valid for the İstanbul trips of the main character in the novel. Both the traveler, who is the main character of the novel, and the actual traveler who follows him are in pursuit of an İstanbul where they believe and try to find the lost values buried.

CONCLUSION

The design of this tourism route, guided by the novel Huzur, has become a new example for Turkish tourism in terms of travelers participating in these tours. In the route design process of the tourism type in question, both the fictional and thematic dynamics of the novel, the symbolic values of the cultural circle to which it belongs, and the daily life of İstanbul are meticulously included in the design. This design, in line with the main idea of the novel, contributes to the continuation of the values of the civilization it is involved in and the urban culture, which is an integral part of it, both touristically and intellectually. In addition to these trips, initiatives on the Literary Geography of İstanbul also contributed to raising awareness of literary tourism in İstanbul's urban life. They have almost frozen a part of time for new tourist groups and enabled the participants to realize that there can be alternative literary tourism tours against museum visits where the participants are in a passive position. Dozens of different tourism types and routes have been designed for İstanbul, which has a rich cultural heritage. On the other hand, these new types of events aimed to actively add Turkish literature to İstanbul's tourism and succeeded to a certain extent. Promising to experience İstanbul, to feel it in a romantic perspective, these trips once again reflected the existence of a new type of tourist to the sector as a unique demand. Demands for Huzur and the Literary Geography of İstanbul, which has been experienced with a few new boutique initiatives in Turkish tourism and with a few similar events, have been revealed that literary tourism should be evaluated and defined independently within national tourism policies, and that policies should be planned in accordance with the expectations of this new tourist type.

FUTURE RESEARCH DIRECTIONS

These trips and their configurations, being presented as touristic, brought forth the necessity of new viewpoints and activities both regarding the future direction of literary tourism and the tourism strategies of the Republic of Turkey Ministry of Culture and Tourism. In this direction, the activities listed in detail below can be carried out.

1. The Ministry of Culture and Tourism of the Republic of Turkey should expand and elaborate on the clauses indicated as special interest tourism in its tourism strategy, acknowledge the independent presence of literary tourism and add goals regarding encouragement for literary tourism to its strategies. Alongside this, it should diversify literary tourism with creative perspectives by bringing its literary tourism activities, structured with a mass tourist viewpoint, closer to individual touristic expectations.
2. Private enterprises in the tourism sector in Turkey should add the discipline of literature to touristic events, design new routes regarding literary tourism to produce more new products and services regarding local and foreign literary tourists.
3. Publications regarding the literary tourism potential of Turkey, reports on this touristic event should be produced systematically and continuously, and shared publicly.
4. Brochures, city maps centered around literature, literature festivals based on authors, works, fiction, marketable souvenirs that would make literary tourism more attractive should be designed, and these should be supported with an image and visual culture in accordance with the literary tourism audience.
5. Specific to the novel Huzur, a special day should be determined where the traditional life of İstanbul can be experienced in a touristic way. Efforts should be made for the people of the city to embrace this day as a mass and turn it into a tradition.

REFERENCES

Akkaya, B., Karataş, E., Sarı, O., Seten, S., & Zilif, S. (Eds.). (2015). Kültür rotaları planlama rehberi. Çekul Vakfı Tarihi Kentler Birliği Yayınları.

Amirou, R. (1999). Les nouvelles mythologies du tourisme. *Sciences Humaines*, *90*, 22–25.

Anar, T. (2020). *Huzur atlası – Ahmet Hamdi Tanpınar'ın Huzur romanını okuma kılavuzu*. Kapı Yayınları.

Boorstin, D. J. (1961). *The image: A guide to pseudo-events in America*. Atheneum.

Busby, G., & Klug, J. (2001). Movie-induced tourism: The challenge of measurement and other issues. *Journal of Vacation Marketing*, *7*(4), 316–332. doi:10.1177/135676670100700403

Busby, G., & Laviolette, P. (2006). Narratives in the net: Fiction and cornish tourism. *Cornish Studies*, *14*(1), 142–163. doi:10.1386/corn.14.1.142_1

Çevik, S. (2017). Kültürel miras kapsamında edebiyat turizmi deneyimi: Sait Faik Abasıyanık'ın izinde otoetnografik bir çalışma. *Turkish Studies*, *12*(29), 51–168. doi:10.7827/TurkishStudies.12404

Cohen, E. (1979). A phenomenology of tourist experiences. *Sociology*, *13*(2), 179–201. doi:10.1177/003803857901300203

Demirer, D. (2015). *Kültürel mirasın sürdürülebilirliği kapsamında edebiyat turizminin yönetici ve tüketici perspektifinden değerlendirilmesine yönelik bir araştırma* [Unpublished master dissertation]. Düzce University, Akçakoca.

Gentile, R., & Brown, L. (2015). A Life as a work of art: Literary tourist's motivations and experiences at il vittoriale degli Italiani. *European Journal of Tourism, Hospitality and Recreation*, *6*(2), 25–47.

Herbert, D. (1995). *In heritage tourism and society tourism leisure and recreation*. Mansell Publishing.

Herbert, D. (1996). Artistic and literary places in France as tourist attractions. *Tourism Management*, *17*(2), 77–85. doi:10.1016/0261-5177(95)00110-7

Herbert, D. (2001). Literary places, tourism and the heritage experience. *Annals of Tourism Research*, *28*(2), 312–333. doi:10.1016/S0160-7383(00)00048-7

İstanbul'un edebiyat coğrafyası, Ahmet Hamdi Tanpınar. (2020, May 20). *İbb Tv*. https://ibb.tv/haberler/detay/evine-tabela-asildi

Jia, H. (2009). The construction of literary tourism site. *Tourism (Zagreb)*, *57*(1), 69–83.

Kaplan, M. (1962). Bir şairin romanı: Huzur. *İstanbul Üniversitesi Edebiyat Fakültesi Türk Dili ve Edebiyatı Dergisi, 12,* 33-86.

Kültür ve Turizm Bakanlığı. (2007). *Türkiye turizm stratejisi 2023*. http://www.sp.gov.tr/upload/xSPTemelBelge/files/J4WC8+20070302-17-1.pdf

MacCannell, D. (1976). *The tourist: A new theory of the leisure class*. Schocken Books Inc.

Mintel. (2011). *Literary tourism – international – September 2011*. Mintel International Group.

Narlı, M., & Aliağaoğlu, A. (2012). Edebi miras turizmi ve Türkiye'de edebî mekânlar. *Bilig*, *60*, 1–18.

Okay, O. (1998). Huzur. In *TDV İslam Ansiklopedisi* (Vol. 18, pp. 439–441). Türkiye Diyanet Vakfı Yayınları.

Okay, O. (2010). Ahmet Hamdi Tanpınar. In *TDV İslam Ansiklopedisi* (Vol. 39, pp. 567–570). Türkiye Diyanet Vakfı Yayınları.

Okay, O. (2012). *Bir hülya adamının romanı*. Dergah Yayınları.

Reijnders, S. (2011). Stalking the count: Dracula, fandom and tourism. *Annals of Tourism Research*, *38*(1), 231–248. doi:10.1016/j.annals.2010.08.006

Robinson, M., & Andersen, H. C. (2002). Literature and tourism: Reading and writing tourism. *Continuum*.

Sağıroğlu, M. (2019). Turistik mekanda gerçekliğin algisal değişimi: Modern ve post-modern turizm deneyimlerinde özgünlük. *Planlama*, *29*(2), 90–101. doi:10.14744/planlama.2019.42204

Smith, K. A. (2003). Literary enthusiasts as visitors and volunteers. *International Journal of Tourism Research*, *5*(2), 83–95. doi:10.1002/jtr.419

Squire, S. J. (1994). The cultural values of literary tourism. *Annals of Tourism Research*, *21*(1), 103–120. doi:10.1016/0160-7383(94)90007-8

Squire, S. J. (1996). Literary tourism and sustainable tourism: Promoting Anne of Green Gables'in Prince Edward Island. *Journal of Sustainable Tourism*, *4*(3), 119–134. doi:10.1080/09669589608667263

Tanpınar, A. H. (2001). *Beş Şehir*. Dergah Yayınları.

Tanpınar, A. H. (2013). *Huzur*. Dergah Yayınları.

Tanpınar Edebiyat Araştırmaları ve Uygulama Merkezi. (2017). *Huzur'un İstanbul'u* [Data set]. http://www.tanpinarmerkezi.com/roman-haritalari/

Tanpınar Merkezi. (2013). https://twitter.com/tanpinarmerkezi/media

Urry, J. (1990). *Tourist gaze: Leisure and travel in contemporary societies*. Sage Publications.

Urry, J. (1995). *Consuming places*. Routledge.

KEY TERMS AND DEFINITIONS

Authenticity: The case of being unique, existing and being regulated while being faithful to the original.

Experiential Tourist Mode: The tourist mode that focuses on reestablishing and understanding their own spiritual center as a result of touristic experiences.

Literary Place: Places that emerged in a literary tourism event based on a literary figure, literary work, or its popularity, and has the potential to turn into a touristic place.

Literary Tourism: A trip to places that were given value to by a literary figure or a literary work and events centered around literature.

Literary Tourism Awareness: The case of having theoretical and practical information about literature/literary tourism and travelling based on this information.

Romantic Tourist View: The tourist view that approaches touristic trips and touristic products through individual and spiritual experiences and has expectations from tourism in this direction.

Chapter 8
Porto, Gaia, and the Sant'Ana Arch:
Places in Almeida Garrett's Literary Work

Isabel Vaz de Freitas
Portucalense University, Portugal

Helena Albuquerque
https://orcid.org/0000-0003-3064-1514
Portucalense University, Portugal

ABSTRACT

This study aims to analyse the novel O Arco de Sant'Ana, by Almeida Garrett, one of the most important Portuguese writers of the 19th century. O Arco de Sant'Ana is a historical novel that describes a medieval narrative that is used as a context and emphasis for the presentation of the author's liberal ideas of his time. Using geographical information system as a methodological tool, a literary cartographic analysis will be conducted by identifying places, streets as well as tangible and intangible heritage, described in the novel. Several analyses will be performed to pinpoint the places where the medieval narrative occurs, transposing them to the current urban map. In this way, it should be possible to overlay the literary landscape onto the present map of Porto to offer the tourist a new product based on a journey through time based on the writer's literary work.

INTRODUCTION

Literary tourism gives us an opportunity to interact with the geographical background presented in novels. Seen like this, rather than simply being a closed universe, a novel describes places and environments allows readers to enter and visit this world of knowledge. Reading a novel is, thus, an experience to (re) visit places, feel the sense of place, understand the symbolism, the characters' impressions and emotions. Knowing the significance of the places depicted in the novel also allows a better understanding

DOI: 10.4018/978-1-7998-8262-6.ch008

of the author's narrative. In this way, literary tourism is identified with the notion of finding reality in a literary place, which is not normally associated with reality (Kaczmarek, 2020).

Literary places can also provide an opportunity to connect with cultures, habits and different communities, thereby linking culture, history and tourism. Furthermore, they can make the experience rich from the point of view of the aesthetic value of nature linked with sensorial stimulation, thus enriching tourists' perceptions of nature (Zhang and Xu, 2020). What we are looking at is a literary landscape, which is a discovery of time and space as an individual and personal involvement within the scope of touristic experience (Kaczmarek, 2020). Even the imaginative places that are depicted in novels are significant for tourists that search for the real place through the figurative and are motivated to try to find those potential places (Jiang and Xu, 2016).

In Portugal, literary tourism can increase positive perceptions of the country's culture on an international level and is a very promising area for cultural tourism considering the enormous potential of Portuguese literature (Cacilhas, 2020). According to the Instituto Camões (2021), the Portuguese language is the 4th language in the world, given that 260 million people speak Portuguese in the 5 continents and, in 2050, there will be almost 400 million people speaking Portuguese, according to the estimates of the United Nations (Instituto Camões, 2021). With those numbers, it is a real opportunity to increase literary tourism in the near future.

For historical cities, literary tourism is an interesting approach to develop their cultural heritage and it is also an opening to cultural innovation and a challenge to develop the city's potential, in terms of culture. In turn, it is a golden opportunity to think about tourism planning and to decentralise tourism in some historical areas of the city. Tourism is an excellent method to create new images of places and to share knowledge and, as Vitić-Ćetković, Jovanović and Potočnik Topler (2020) argue, tourism is very important to promote cultural and historical heritage, to preserve destinations in the context of sustainability and furthermore could promote interesting local storytelling. Baleiro and Quinteiro (2019) highlighted the point that literary tourism could be an important alternative to mass tourism, and might give rise to a more responsible, ethical and sustainable tourism experience, while at the same time provide a fundamental means of education in terms of history and culture. Arcos-Pumarola, Osácar Marzal and Llonch-Molin (2018) emphasise the possibilities of literary tourism in order to diminish the impact of overcrowded places by creating new points of interest in cities to which tourists could flock. Besides this they refer to the fact that the literary landscape, as a place interpretation, is still largely being ignored.

Thereby, this study aims to analyse the novel *O Arco de Sant'Ana*, by Almeida Garrett, identifying the descriptions of tangible and intangible heritage. A literary map of *O Arco de Sant'Ana* will be developed based on the analysis of the novel, using a geographical information system as a methodological tool.

GIS are important methodological tools in literary tourism, although their use is still very recent. As Murrieta-Flores, Donaldson, and Gregory (2017) point out, the use of these tools has occurred because of two reasons: the growth of interest in the study of space, place, and landscape; and the opportunity of using computational approaches in literary research. Thus, these tools become essential in developing itineraries built by reading works and their interpretation. The combination of reading novels with GIS tools can play an important role in the definition of a destination's tourism content. It is possible, through this combination, to create itineraries where memories of an era, descriptions of physical spaces, places and society are presented, through the way authors tell the story, or through their own experience and the social context in which they write (Arcos-Pumarola, Osácar Marzal, & Llonch-Molina, 2018).

The process of converting a book into an itinerary and, eventually, into a tourist product requires intense research, compilation and editing of information before the mapping process itself, but in the end, facilitates the reading and understanding of the locale, presenting it in an appealing way to the visitor and, at the same time, becoming a base of work for both academics and managers (Hladnik and Fridl, 2012). Having this in mind, the expected result will be a literary map of *O Arco de Sant'Ana* novel, highlighting the tangible and intangible medieval heritage, which was inventoried and described by the writer in the 19th century. In this cartographic description, streets, monuments, places, events, and cultural landscape will be mapped and described, as well as emotions, feelings, and other sensorial information. Several analyses will be performed to understand the changes that have occurred between the medieval time, 19th century, and present time. In this way, it would be possible to overlay the literary landscape components with the current map of Porto, to offer the tourist a new product based on a journey through time and through the writer's literary work.

This chapter is divided into several parts. After the introduction, the authors present a theoretical background for destinations regarding literary tourism, with the main focus being on Porto and Gaia (Portugal); followed by the use of GIS as a methodological tool for literary tourism, and finally, the example of Almeida-Garrett as a writer. This is followed by a section dedicated to the analysis of the case study, the novel *O Arco de Sant'Ana*, where the context of the novel and the development of itineraries are described. Lastly, in the concluding remarks, a summary of the main findings and the limitations are described, as well as future research that the authors intend to develop.

BACKGROUND

Literary Tourism Destinations

Herbert (2001) points out that literary places attract heritage visitors and considers that those literary places are part of heritage places. Ferreira, Alén, Liberato and Liberato (2020) consider literary tourism to be a rising tendency in cultural tourism and believe it can have an important impact on tourism activity. UNESCO (2003) considers the performing arts, expressions and representations as forms of immaterial heritage. In the same way, Marques and Cunha (2013) understand literary tourism as an emotional relation with spaces and sites where the connection with a novel, a character or with the author occurs.

Literary tourism is an opportunity to explore the daily life of the chosen writers, including their places of residence, birthplaces, graves, among other biographic places of interest. These are all places of memory that could be valorised to establish emotional relation with the writers, when interpreted as heritage (Menezes, 2020). The place of memory is not only a place of a biographic relation but also a place of literary encounters.

To promote the imagined landscape in a city is also a way to create new destinations in a city (Arcos-Pumarola, Osácar Marzal & Llonch-Molin, 2018) since literary places are "consumed" by city visitors (Hoppen, Brown and Fyall, 2014) and by "literary pilgrims", as Smith (2003) identified literary tourists. Herbert (2001) explains that "literary pilgrims" are cultural tourists who appreciate literature. They are not only cultural visitors but tourists that search for places linked with writers and the places they describe in novels. For them, these places offer genuine touristic experiences.

According to Beigi, Hossein and Azani (2015), literary tourists look for different types of places: they can look for actual places, imaginary places that are expressions of the novels, plays or poetry, or other attractions in places related to moments in the author's life.

Literary novels of the 19th century are an opportunity to explore the history and heritage of a city and develop historical knowledge about its immaterial and material heritage as well as promote its conservation. As Jiang and Xu (2017) concluded, literary tourism is a significant counterpart for historical cities tourism, creating place links between all related aspects: literature, historical places and literary appreciators, and also playing a significant role in local development.

Tourists are attracted by the significance of the heritage and culture that is possible to find in those literary descriptions. Nineteenth-century authors can also be a living book of history and past emotions and feelings that could be preserved in the same way as cultural and heritage, built structures, natural environment, landscape and traditions (Taylor and Pouliot, 2020). In this context, Wang and Zhang (2017) concluded that literary tourists consider the authenticity of scenarios in which historical architecture and landscape are depicted to be more important than their literary value. This important conclusion could help the tourism industry rethink places and their role in literary tourism.

The literary place is a mixture between the real and the imagined world where emotions and sensorial perceptions exist and permit the reader to understand all the actions and sceneries (Jiang andYu, 2020). It is well known that the tourism experience has a strong connection between literary description and observed reality (Machado et al., 2020).

Porto and Gaia as Literary Tourism Destinations

Porto and Gaia are two municipalities of the Metropolitan Area of Porto, geographically located on the North coast of Portugal, being divided by the Douro River.

The municipality of Porto is considered to be the capital of the Metropolitan Area of Porto and the North region, having a high concentration of the most demanding activities from the point of view of qualifications, creativity, and knowledge incorporation (DMPU, 2018). In addition, tourism, and all related activities, from commerce to creative industries, from culture to gastronomy, from transport and accessibility to international connectivity by the presence of Sá Carneiro Airport, constitute some of the municipality's attractions. These factors have created conditions for the emergence of new dynamics of urban regeneration, new spaces for entertainment, the increase of tourist accommodation, and the offer of a diverse set of services aimed largely at visitors. Furthermore, the presence of a diverse set of prestigious universities, research centres with international projection, and favorable quality of life indicators, contribute to the growing attractiveness of Porto (DMPU, 2018).

The municipality of Vila Nova de Gaia, or just Gaia, as it is known, is inseparable from the municipality of Porto, both by the geographical factor as well as the presence of the Douro River that divides them, and by the transport networks and infrastructures that connect these two municipalities. It is known especially for its Port and Douro wine companies and cellars. In addition, it boasts a long history that stretches back to medieval times.

In terms of population, it is the municipality with the third highest number of residents at a national level (INE, 2021), and a large sector of this population commutes daily to Porto, either to work or to study.

These two adjacent cities on either side of the Douro, Porto and Gaia, which are both visited by a large number of tourists, share a common denominator: they are both cities of writers and the arts. Porto and Gaia have a connection in tourism linked by the old iron bridges, the boats that cross Douro River, the wine cellars and the splendid view of Porto from Gaia. Besides that, the streets, monuments, old cafés, historical theaters, and cinemas, among others, are a live museum of urban art, art history,

history of theatre, cinema, literature and poetry. In Porto and Gaia, heritage and culture can be visited and experienced.

As Porto and Gaia are cities of writers and artists, and given that literature can arouse interest in its readers to investigate and discover the places or events depicted in the books (Oliveira, 2017), the opportunity arises to develop a product that has been poorly explored and/or disseminated. Oliveira (2017) has proposed a literary route as a resource material to disseminate literary tourism in Porto. Her idea was to identify literary places, based on the life and work of the main writers and at the same time discover and experience the atmosphere, landscape, people, colours, cafés, gardens and typical sounds of the city that inspired the cultural legacy that they have left (Oliveira, 2017).

In addition to this, creating specific itineraries based on those writers' novels can also help visitors to discover Porto and Gaia cities in a different way. In summary, literature can play an important role in creating the image of a place. Its use is a pretext for storytelling, which can itself be transformed into a tourism product. (Arcos-Pumarola, Osácar Marzal, & Llonch-Molina, 2018).

GIS as a Methodological Tool for Literary Tourism

Geographical information systems (GIS) are powerful tools that allow the aggregation of georeferenced data, being essential in the process of decision-making, namely concerning tourism planning (Farsari and Prastacos, 2008, Bahaire and Elliont-White, 1999, Mcadam, 1999, Engelmo-Moriche, et al., 2021, Jurkus, et. al., 2021).

For this reason, several tourism projects and studies have been developed using GIS as a methodological tool, to produce inventories of tourism resources, to identify the most suitable location for tourism development, to measure the impact of tourism or to analyse relationships between different resources and develop destination marketing strategies (Bahaire and Elliot-White, 1999; Albuquerque, Costa, Martins, 2018). As Sousa and Fernandes (2007) point out, the use of GIS can encourage community participation in tourism planning. Cruz (2009) adds the importance of GIS in the management of tourists and in the analysis of the impacts of tourist activity.

GIS have been used in tourism since the 1990's (Farsari and Prastacos, 2008), but their use in literary tourism is still scarce. As Murrieta-Flores and Howell (2017) highlight, there are two characteristics that texts from novels must have to use GIS as an analysis tool: they must have a geographical nature and it must be possible to identify the geographical coordinates of the places mentioned in the novel. However, despite having a geographical nature, some of the places mentioned in many of the novels are either imaginary, or no longer exist or else are transformed by the writers, making it difficult to locate them and therefore also making it difficult to create real maps based on those novels (Hladnik and Fridl, 2012, Murrieta-Flores and Howell, 2017).

Nevertheless, it is possible to identify some studies that have used GIS in literary tourism. Such is the case with Hladnik and Fridl (2012), who have used a database of literary heritage of Slovenia where several real and imaginary places are identified to propose presentations of the narratives in web maps. Donaldson, Gregory and Murrieta-Flores (2015) presented a study about literary tourism in Victorian Lakeland, using GIS to create maps identifying literary locations in the Lake District. The same authors have also developed another study based on the travels of three British writers (the poet Thomas Gray (1716-1777), the naturalist Thomas Pennant (1726-1798) and the agriculturist Arthur Young (1741-1820), using GIS to determine spatial models based on their travel writings (Murrieta-Flores, Donaldson and Gregory, 2017). In this way, it is possible to conclude that GIS allows the combination of literary novels, tourism and the definition of itineraries, through the development of promotional content that is simple, easy and interactive for users and by carrying out analysis that provides valuable indicators to destination managers.

The Example of Almeida Garrett

Garrett is an important (and highly respected) Portuguese writer. As was noted previously, literary landscapes and writers' places can be considered as tangible and intangible heritage. Thus, if sufficient merit and value is bestowed on them, this might allow the development of new attractions in those areas.

It is well known that those two worlds of the narrative literature (biographic spaces and literary places) could have a positive effect on the number of tourists that want to visit places, monuments and sites connected with writers or novels of personal preference. If the writer has a high international profile, the numbers could be even higher and could be a real tourist phenomenon.

As a case study, Almeida Garrett, or João Batista da Silva Leitão de Almeida Garrett, to give his complete name, is an interesting writer for those wishing to understand the connection between the sense of place and the concern of literary landscape as heritage. Almeida Garrett, a Portuguese writer from the beginning of the 19th century, was also a government minister and honorary secretary of state. He was one of the most important personalities of the Portuguese literary scene and was seen as a dynamic force in Portuguese culture at that time. However, his impact has mostly been national, and confined to the city of Porto, where he was born, on 4th February 1799. His house in Porto is located in Calvário Street. He was born in a family of merchants with trade relations with Brazil and baptised in Santo Ildefonso church in Porto (figure 1).

Figure 1. Places where Almeida Garrett lived in Porto and Gaia.
Source: authors, based on Map data ©2015 Google (https://www.google.at/permissions/geoguidelines/attr-guide.html)

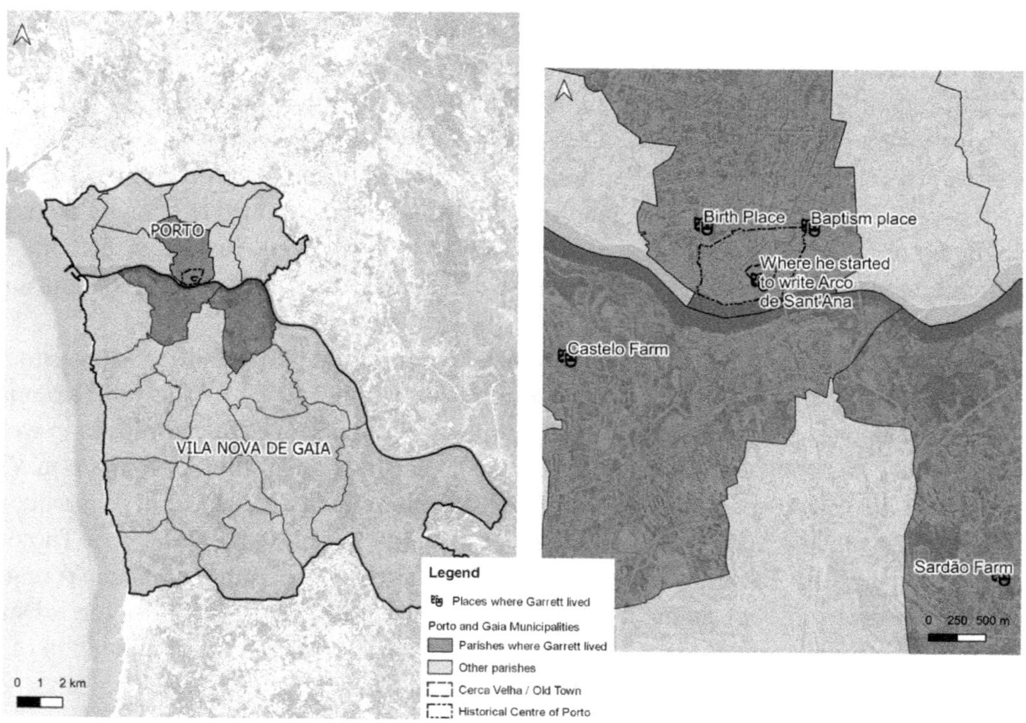

The writer was the first to introduce the romantic period in Portugal, and he was also an important figure in theatre and dramaturgy. As a politician, he was a liberal that opposed the Portuguese absolutists. In his time, he experienced a historic conflict that occurred after the French invasions at the beginning of the 19th century. The Portuguese king was exile in Brazil, and Portugal was governed by a regency, led by General Beresford, an Englishman. This critical situation gave rise to the creation of a new political movement, which gained momentum with the collaboration of a Portuguese intellectual group (Ribeiro, 2020) that opposed the existing political system and introduced a new liberal perception. The liberal movement was deeply embedded in the new mentality of the era, and Porto freely collaborated with the Liberal project established between 1832 and 1834. Partly as a result of this, Porto was, for Almeida Garrett, a setting for what happened in real life and subsequently for his writing.

His biography has been studied by several authors, such as Amorim (1881-1884) and David (2007). He lived in Porto, but in 1804 the family moved to "Castelo Farm" to the south of the Douro, located at Candal, Santa Marinha, near Gaia castle. Some years later he moved with his family to the Sardão farm in Oliveira do Douro, a rich property well known for its water aqueduct. In 1809, with the Napoleonic invasions, the family moved to Lisbon and then to Angra do Heroísmo, in the Azores. In 1816 he went to Coimbra to study law.

In 1820, he returned to Lisbon where he married Luisa Midosi and remained with her until 1836. In 1820, on August 24th, the liberal revolution began in Porto. In 1821, Garrett helped the liberals to implement the liberal constitution in Azores. In 1823 he was exiled in London after the Vilafrancada conflict. In 1826, after the death of King João IV, the new King Pedro IV abdicated the thrown of Portugal in favour of his daughter D. Maria and acquired the title of Emperor of Brazil. Garrett returned to Portugal in the year 1826.

In 1828, the brother of King D. Pedro, D. Miguel, returned to Portugal from Brazil with the intention of imposing an absolute monarchy by assuming the crown. Garrett had to return to exile in London. In 1832, he joined the liberal army led by D. Pedro, who came from Brazil to help the liberal cause. During this period, in the Siege of Porto (*Cerco do Porto)*, he was with the liberal army in the *Seminário Maior de Nossa Senhora da Conceição*, where he began to write *O Arco de Sant'Ana*. After the liberal victory he became a congressperson and Ministry of Foreign Affairs, chronicler of the king, among other distinctions. He also received the title of "*Visconde*" (Viscount).

In 1838, his passion for culture and art, as well as being vice-president of Lisbon's Royal Conservatory and General Inspector of Theatres, led him to present to the government,, a regulation for three schools: oratory, dance and music. With these schools, Garrett intended to improve the level of artistic and theatre education. (Ferreira, 2008). As General Inspector, he developed a national repertoire for theatre, choosing a medieval author, Gil Vicente, as the father of theatre in Portugal.

Garrett died on 9th December 1854 after a very troubled political period in which he was an active participant. He had 4 children, but only one, Maria Adelaide, daughter of his relationship with Adelaide Pastor, reached adult age. Before he died, he fell in love with a lady named Rosa de Montufar, Visconti of Luz. His last book of poems, published anonymously, "Folhas Caídas" (Fallen Leaves), was dedicated to her.

His political interests and social status are always present in his work. As a writer, he searched for inspiration from the late Middle Ages to Modern times, namely in Bernardim Ribeiro, Gil Vicente, Frei Luiz de Sousa, and also in Camões. (Alves, 1999).

METHODOLOGY

The methodology used to develop this case study was based in three main parts, described in figure 2.

Figure 2. Methodological framework

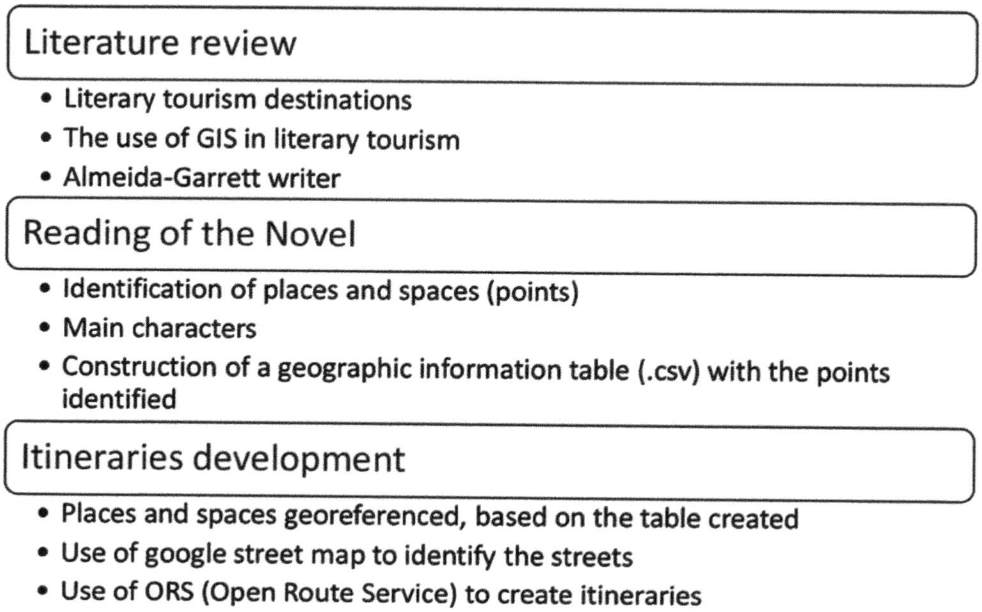

First, the authors of the article carried out a literature review concerning the three main topics: literary tourism destinations, the use of GIS in literary tourism, and Almeida Garrett, the writer. After the literature review, the novel was analysed, and commentaries were written to clarify the meaning of the novel, the historical facts, the real and imaginary sceneries as well as the main characters.

From the information in the novel and the literature at that time, a .csv table with all the places identified (points) was constructed, containing geographical and descriptive information. Several aspects of these points were analysed including name, location, pages, chapters, number of references and geographical coordinates (latitude and longitude). It was necessary to research several different types of references, namely historical documents, archives, images and maps, documents from the directorate-general for cultural heritage, dissertations and websites. This was due to the fact that some of the points or locations no longer exist or can be considered imaginary. In this way, it was possible to determine the approximate location of the places identified by Almeida Garrett. It should be said that some of the points are just a reference to the area that is mentioned in the novel, since Garrett describes places, streets, areas, and not just specific points.

QGIS v. 3.16 software was used to map all the information collected. The Official Administrative Map of Portugal (CAOP – [Carta Administrativa Oficial de Portugal], 2018) was used to identify Porto and Gaia parishes, and Google Street maps to identify the streets. In addition, a map of the city of Porto from the 19th century was overlapped on the Google Street map to try to contextualise the identified points of the novel on a map from that century.

Using QGIS v. 3.16 software a dot map was produced, with the locations identified by Garrett in the novel. The points collected in the .csv table were imported to QGIS software, using WGS84 as coordinate reference system (CRS).

This dot map enabled us to understand where the action of the novel takes place and to see how the city of Porto was organised during Almeida Garrett's time. Then, using ORS (Open Route Service), with the travel mode "foot-walking" and travel preference "shortest", the itineraries were created, linking the points identified in the novel. To link the itineraries in Gaia and Porto, by crossing the river, the tool "add element" was used, which creates a new line. Through the creation of these itineraries, the different parts of the novel and several maps were produced with the itineraries of the characters. In these itineraries, the most important parts of the novel *O Arco de Sant'Ana* are described.

After this first analysis, a project about Almeida Garrett´s vision of Porto is expected to be developed, by comparing existing maps of the medieval era, the 18th century with the current ones. This project is expected to present an innovative offer in Porto, based on a journey through time and on a journey in the writer's literary work. In this context, Porto could explore the liberal city image brought by Garrett's work and improve the historical and cultural dimension of the city. This is a tremendous opportunity for decision makers to act.

O ARCO DE SANT'ANA NOVEL

The Context of the Novel

O Arco de Sant'Ana is a historic novel, based on a medieval narrative which describes a historical period with many similarities with the Portuguese liberal war period during Almeida Garrett's time. The importance of the writer and the historical singularity of this novel are enough to be considered as literary heritage.

It is interesting to notice that the writer recreates the Medieval Age from his own reality. In this context, it is important to understand the places where narratives were developed, and to appreciate the place as a medieval description of the 19th century. *O Arco de Sant'Ana* is an imaginative description of the territory in the Medieval era, but also where the author identifies common and actual places of his time. This literary landscape is a tangible and intangible heritage that needs to be inventoried, protected, and disseminated as a legacy to be visited and safeguarded.

O Arco de Sant'Ana was published in two volumes, in 1845 and 1850 and represented two of Porto's historical periods. It was written in a post- conflict era, which opened up a liberal period of great dynamism in literary works in Portugal. These works address a new ideological structure that presents actions of economic, social and political regeneration, based on historical facts, but with implicit romantic ideals. (Ferreira, 2008).

Based on a medieval manuscript from the 14th century found at the library in Grilos church entitled "Cronica Portuense" (Porto Chronicle), the novel *O Arco de Sant'Ana* can be seen as a historical narrative that also seems to contain humoristic historical descriptions. However, it follows the dominant literary structure of a historic novel of gothic inspiration with plenty of drama (Marques, 2012). The manuscript is described by the author as a true story of the Medieval period, containing both historic truth and historic invention. The medieval period is evoked as a reminder of the historic nature of the novel's creation. Nevertheless, the historical validity and truth are interrupted by the narration of the novel itself. The author, constantly using anachronisms, introduces ironic facts and contexts from his own time.

The question of historical truth is debated by other authors that studied *O Arco de Sant'Ana*. Lopes (1981), for example, underlines the humoristic character of this novel based on the fact that the author tried to pass off as a true story what was found in the original manuscript. This discovery itself is a literary scenery that gives authenticity to this story.

The streets, buildings and places are described as the author can saw them in his own time. This is also observed in the description of the places and in the main characters, which are described by the author as medieval, despite the fact that its characteristics are based on the 19th century background. This is the case of the Episcopal Palace, the Porto statue and the medieval social rebellion described more from the point of view of a liberal age scenery. None of the characters are described as medieval. The lady in the arch, Vasco as well as others are contemporary of the author. To go back to Medieval times, the author introduces the witch and the secret hatches of the episcopal Palace (Santana, 2004). The authors of the chapter agree with Santana (2004) that refers that this novel is a19th century description that introduces the medieval and gothic scenery as a common contemporary literary approach, and that criticises other novels that describe a fake Medieval Age. The author's notes too, which are always brimming with criticism and humour, when he describes the Medieval Ages, are a sign that he is not in fact describing the 14th century. In fact, he creates a novel based on a 14th-century story (true or false) to criticise the contemporary political period. His action while narrator of the story is a useful way of criticising his epoch and takes his stance as a critic of politics and anticlerical behaviour a step further.

Lopes (1981) considers that *O Arco de Sant'Ana* is not only a historical novel but a novel that relaunches the medieval criticism of the attitude of the clergy and nobility and takes up the question of the rapid enrichment of the bourgeoisie. This scenario has its parallel in the time in which Almeida-Garrett lives and those ideas are contested by him as a liberal. The writer is very explicit that the novel is written through the eyes of the author in the 19th century (Lopes, 1981). Garrett prepares the readers for this encounter with a different Medieval Age view and with a style of writing which is neo-gothic. In this way he introduces a revival of the medieval aesthetic that promotes the approach to Medieval Times, reinforcing an aesthetic approach that is far removed from the real medieval aesthetic. In this context, he presents an interesting vision about Porto in the Liberal age and a more sketchy portrait of Porto in the Medieval period.

Mukhametzyanova, Yao, Emanova, & Pozdnyakova (2017) give the example of the neogothic in England, where Garrett lived and was inspired. These authors point out that the medieval inspiration of the neogothic in England is an antithesis of the mid- 19th-century lifestyle. They emphasise the fact that the social attitude of the 19th century is far away from the habits and culture of the Middle Ages. It is possible to find a similarity between these authors' explanations and *O Arco de Sant'Ana*. On the one hand, the novel is an approach to the Medieval Era and on the other hand, is a straightforward criticism of that epoch, as the focus of comparison is more about the author's contemporary lifestyle. In these two points of view, Garrett develops the action in a way that appeals to humour and criticism in a location that is intentionally imaginary but which bears some resemblance to his contemporary milieu.

Creating Itineraries Based on O Arco de Sant'Ana Novel

In the novel, Garrett identifies several places, some of which still exist nowadays, others that have completely disappeared as a result of the city's growth outside the walls and/or have even been destroyed. A few other places turn out to be completely imaginary or else are described according to what the author perceived them to be back in the medieval era.

With this in mind, the objective was to geo-reference the places identified in the novel and create an itinerary based on those places.

Most of the action in the novel takes place in the centre of Porto, inside the walls, near the Arco de Sant'Ana, on some of the main streets of that time and in the area of the *Paços da Sé*. Wishing to express criticism of the power that the bishop, at the time, had over the city, the author created a novel in which these places were given a great deal of respect and value.

Yet, the main character crosses the Douro River, to Gaia, where some important events are reported, but always in a critical and satirical tone regarding the times in which the novel is written. All the places identified were geo-referenced and are shown in figure 3.

Figure 3. Places identified in the novel.
Source: authors, based on Map data ©2015 Google (https://www.google.at/permissions/geoguidelines/attr-guide.html)

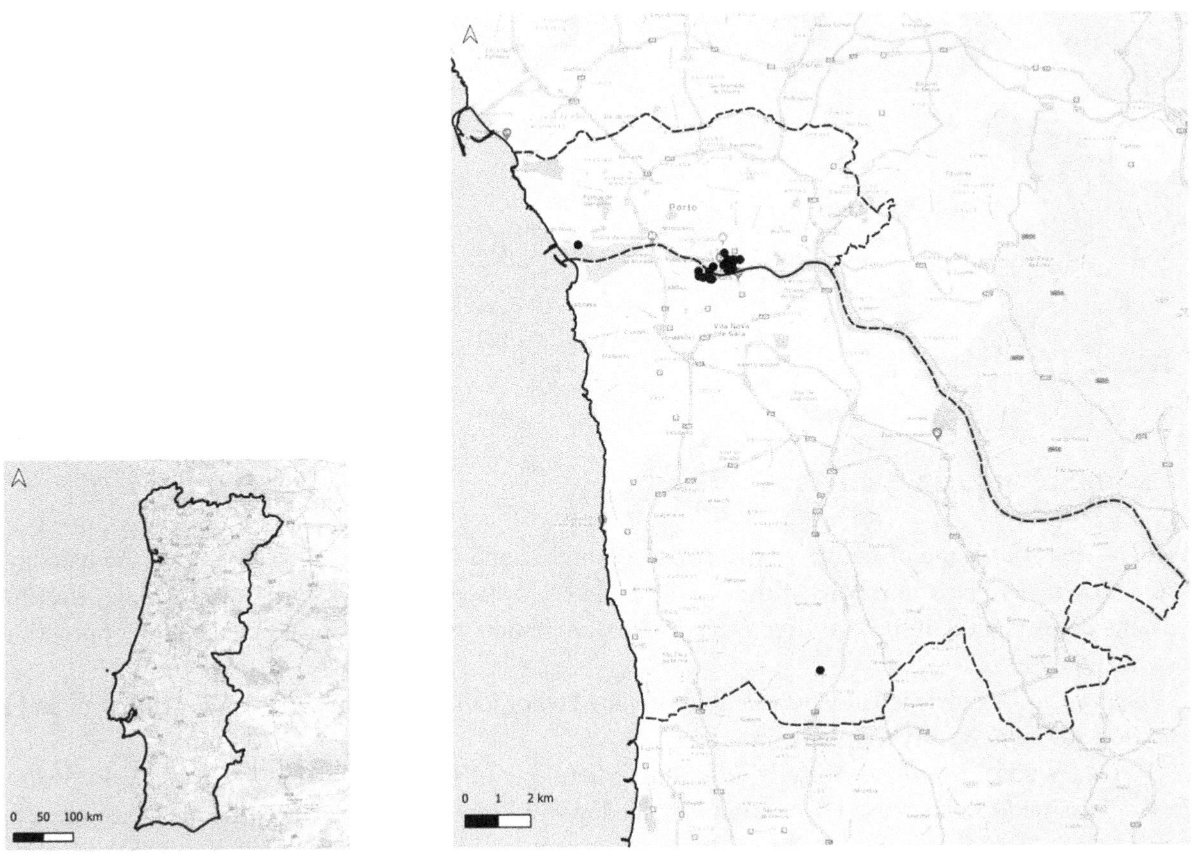

Several itineraries were produced, allowing an interpretation of the different parts of the novel (figure 4). These itineraries are Vasco on his way to Gaia, the narrator's itinerary, the Mutiny itinerary, the Procession itinerary, and Vasco's journey to Porto.

Figure 4. Itineraries of the novel O Arco de Sant'Ana.
Source: authors, based on Map data ©2015 Google (https://www.google.at/permissions/geoguidelines/attr-guide.html)

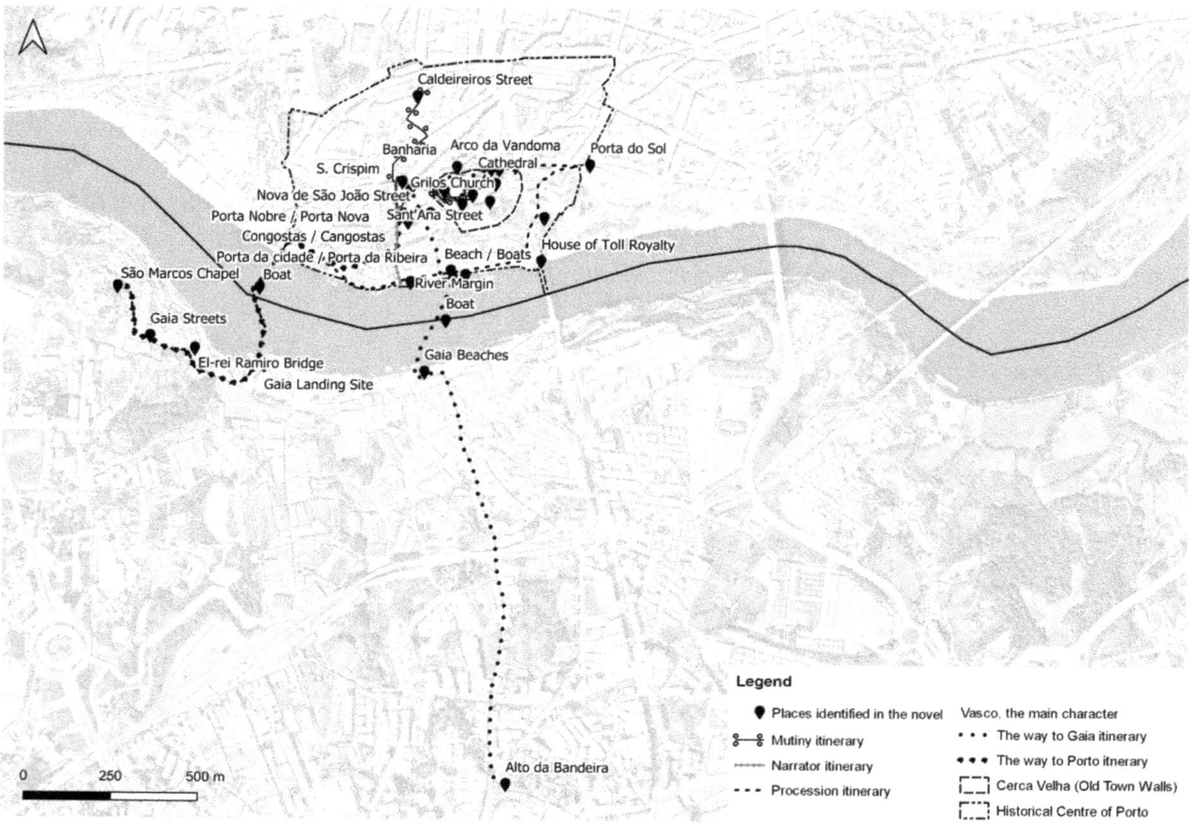

The Area of Arco de Sant'Ana

Arco de Sant'Ana is the place where two friends, Aninhas and Gertrudes, live and talk on the balcony. The action takes place in medieval times (14th century), between the city of Porto and Gaia, starting with the conversation of the two friends about the domination of the bishop and the state of oppression in which they live.

Meanwhile, Aninhas is kidnapped and imprisoned by order of the bishop. Gertrudes finds out and a rebellion starts in the city to free Aninhas.

The story unfolds around *Arco de Sant'Ana*, where one of the gates of the so-called *Cerca Velha*, Porto's first medieval wall used to be (figure 5). Today, in *Sé (the Cathedral)*, it is possible to see some of the remains of this first wall of the city. The other settings of this novel are developed inside the second Porto wall called *Muralha Fernandina*, constructed in the 14th century during the reign of Fernando I

(1370). This wall is still visible in some places of the city centre. The streets and places of the medieval city centre of Porto are the main scenarios of the novel.

Garrett refers to this location as *Arco de Sant'Ana*, a name given to this door in the 16th century, due to the existence of an image of *Nossa Senhora de Sant'Ana* and the veneration that took place there (Oliveira, 2013). However, Oliveira (2013) mentions that previously, this arch or gate was called "*Portal*", which is how it appears in the council minutes of 1449 (Oliveira, 2013). This was a medieval wall door located in the medieval street called first *Penaventosa* street, but whose name has been changed to *Aldas* street and then to *Sant'Ana* street. This street began in a street staircase (Basto, 1962). The street and staircase were located near the *Grilos* church and the college described by Garrett. However, those buildings date from the 16th century, so they couldn't have existed in the as a scenario of a novel set in of the Medieval Age (Basto, 1962). This indicates clearly that Garrett describes the city from the point of view of his own times (Lopes,1982).

Figure 5. The Area of Arco de Sant'Ana.
Source: authors, based on Map data ©2015 Google (https://www.google.at/permissions/geoguidelines/attr-guide.html)

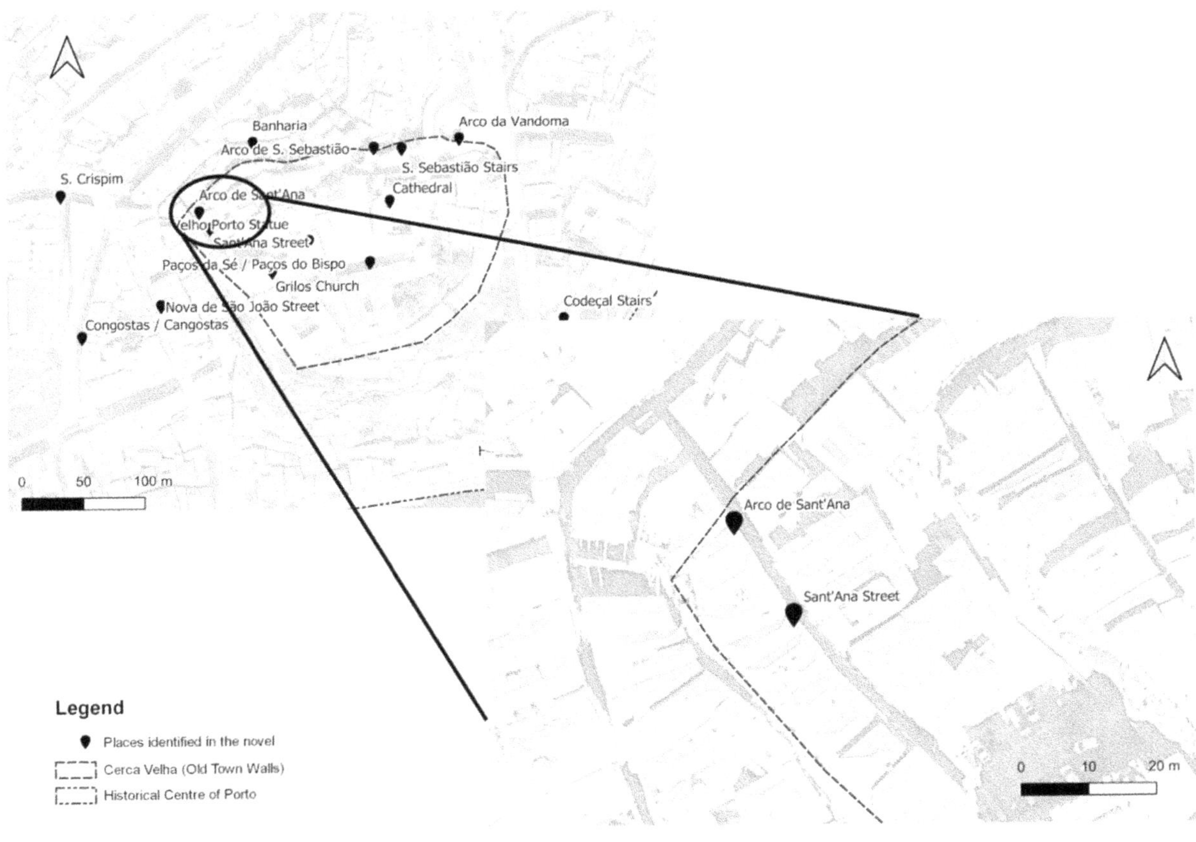

Vasco, the Main Character

Vasco, the main character of the story, a student who is under the protection of the Bishop of Porto (and who, as the novel unfolds, is discovered to be his son), is in love with Gertrudes and does everything to win her love. Vasco is a medieval character, but, in the descriptions of Garrett, seems to display liberal traits. In the novel, Vasco is encouraged by Gertrudes to leave Porto in an attempt to meet D. Pedro I, to report that the bishop is oppressing the people of Porto, and that her friend Aninhas had been kidnapped. This act had already prompted a revolt against the power of the bishop (figure 6). This conflict between the city of Porto and the bishop is well documented by historians of the city (Coelho, 2003; Sousa, 1994; Basto, 1940).

Figure 6. Vasco on his way to Gaia.
Source: authors, based on Map data ©2015 Google (https://www.google.at/permissions/geoguidelines/attr-guide.html)

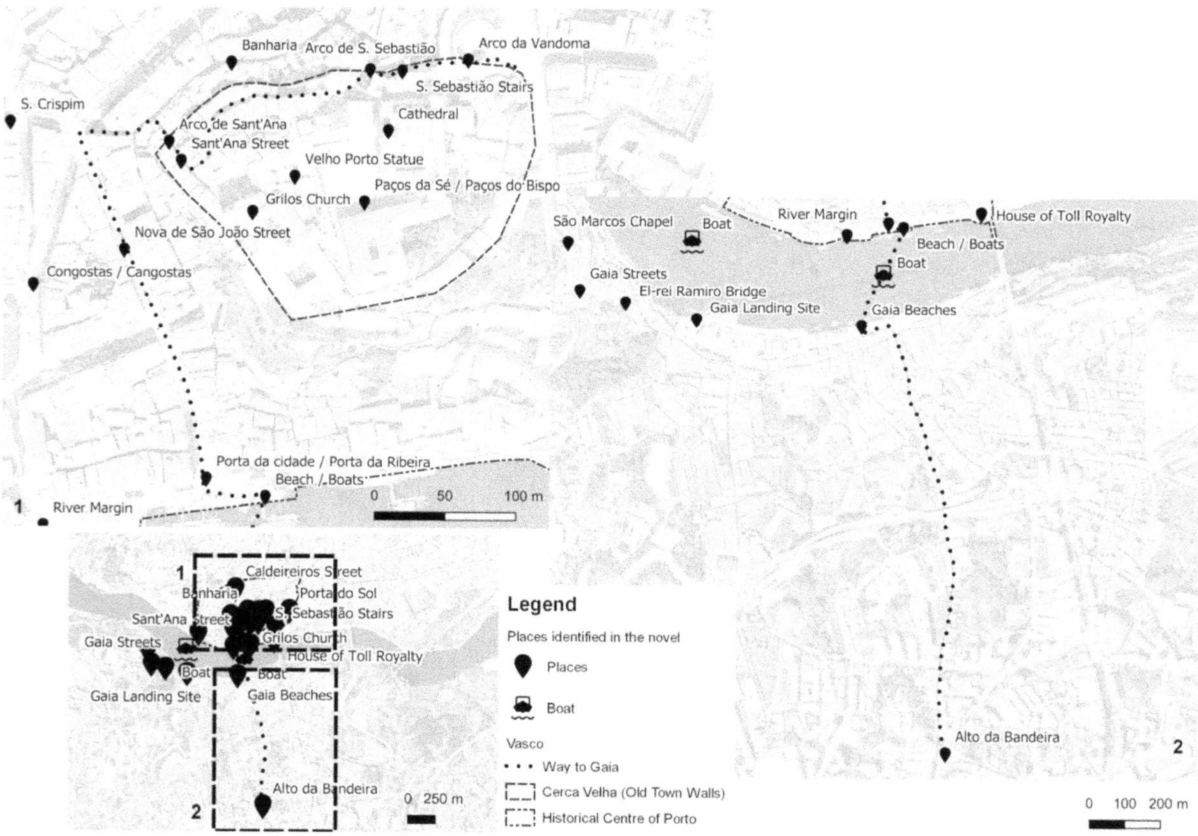

In order to meet D. Pedro I, Vasco requires his benefactor (the bishop) to get him the best sorrel he has. To do this, he invents a story that he is going to meet some friends "beyond the Douro to wait for the turtledoves early in the pine forests" (Garrett, 1845-1850, p. 32).

After getting his horse, Vasco leaves the cathedral, passes by the Arch of Vandoma and goes to the Arch of Sant'Ana to say goodbye to his beloved. The Arco de Vandoma, or Vandoma Gate, was considered of vital importance for the flow of people and products, as it provided a connection, via the Roman road, to the lands of Paredes and Penafiel (Oliveira, 2013, p. 29).

Then, he heads for the *Porta da Cidade* (or *Porta da Ribeira*), where he waits for the boatmen to arrive and take him to the beaches of Vila Nova. This gate also had another name. It was called *Porta da Senhora do Ó* because it passed under the chapel of *Senhora do Ó*, which adjoined the wall. The door gave access to the entrance from the river Douro to Ribeira square (Coutinho, 1962).

From there he departs in the direction of Grijó Monastery, passing through *Alto da Bandeira*. Garrett describes this "*Alto da Bandeira*" as "for where much blood was shed in our civil wars" (Garrett, 1845-1850, p. 38).

The Presence of the Narrator in the Novel

The author narrates the story, with the route described passing by several emblematic places of the city (figure 7). When describing Vasco's departure in search of D. Pedro, Garrett leaves the river banks, passes by the Cangostas (now Rua Mouzinho da Silveira), then sees S. Crispim, which was located next to the current *Largo de S. Domingos* (https://www.explorebonfim.com/capeladescrispim). After that, he passes by the statue of *Velho Porto*, which was located on Cathedral hill, saluting "the respectable emblem of our illustrious city" (Garrett, 1845-1850, p. 20), until he reaches the *Paços da Sé (Cathedral).* This enormous statue represented a warrior, a *lusitano* or Roman warrior carrying a spear sculptured in high relief on the wall of a house in the *Largo do Açougue*, a square that disappeared in the middle of the 20th century with the massive development and opening up of the *Sé* square (Basto, 1962). Unfortunately, with this urban growth, in this part of the city the old streets and their heritage have been lost.

Figure 7. The presence of the narrator in the novel.
Source: authors, based on Map data ©2015 Google (https://www.google.at/permissions/geoguidelines/attr-guide.html)

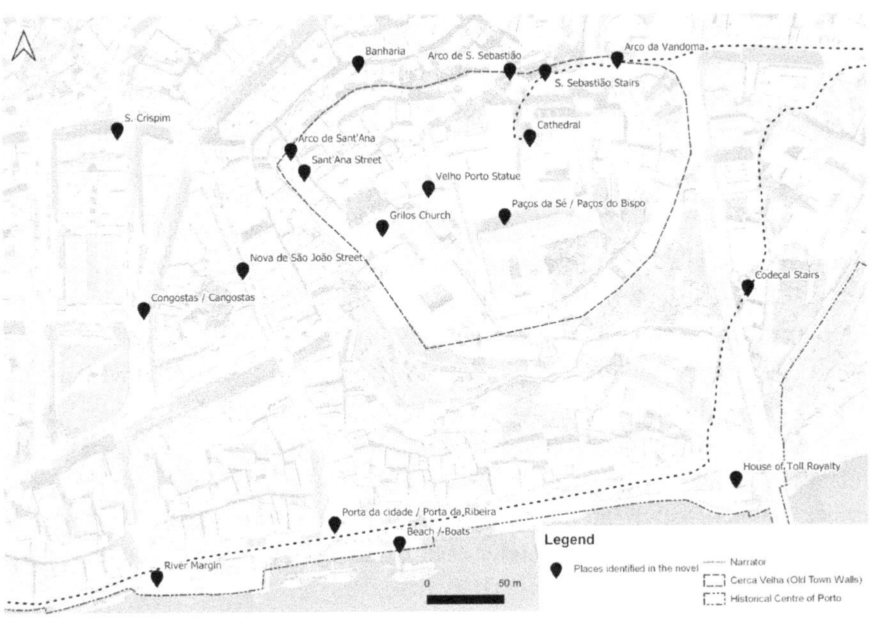

The Mutiny

The rebellion starts and the population begins to gather near the *Arco de Sant'Ana.* They leave from *Cangostas, Bainharia and Rua dos Caldeireiros,* main streets in medieval Porto (Sousa, 1994), and continue their rebellion until they reach the Cathedral (figure 8).

Figure 8. The mutiny itinerary.
Source: authors, based on Map data ©2015 Google (https://www.google.at/permissions/geoguidelines/attr-guide.html)

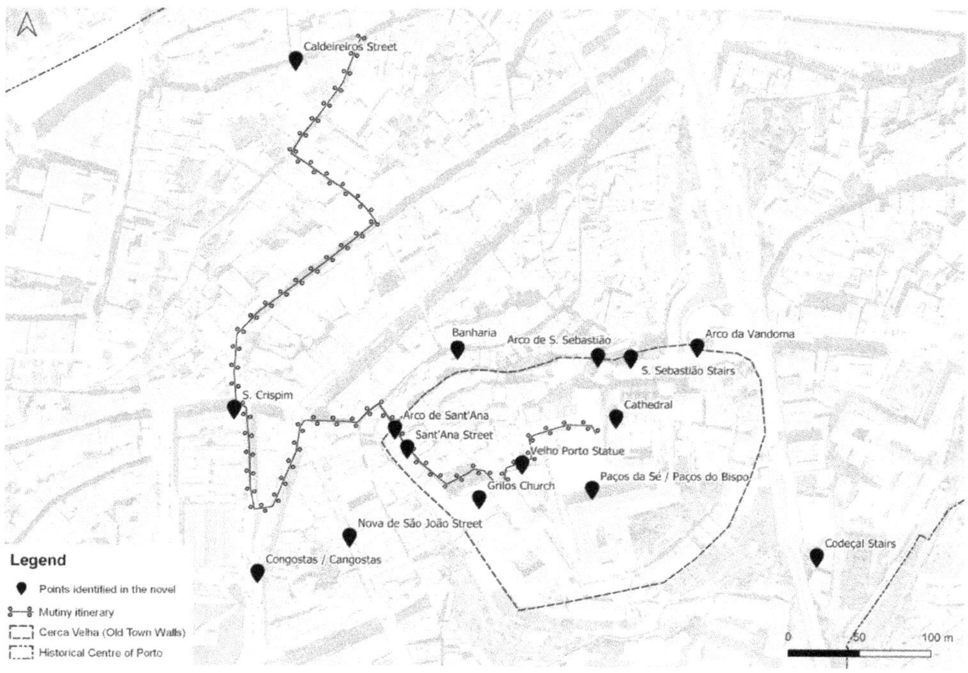

The Procession

At the same time, the clergy is preparing for a procession, which heads towards *S. Marcos*, in Gaia (figure 9). It is not known for sure where this chapel was located, with some references indicating that there was a chapel next to Gaia Castle and that this was the first Cathedral (https://www.castelosdeportugal.pt/castelos/Castelos(pre)SECXII/gaia.html).

Garrett also mentions this when he states "*Miragaia* was a sad little town, when Gaia was already a Roman city, and there was our first cathedral". (Garrett, 1845-1850, p. 58).

All the clergy gather at the Cathedral to begin the procession. They go through the Bishop's Palace, pass by the Saint Sebastian stairs and head towards the *Porta do Sol*. The *Porta do Sol* was originally a wicket, the Carvalho wicket, which later became known as the wicket of *Santo António do Penedo*, taking its current name, the wicket of the Sun (later door), due to its location to the east and, consequently, to an early morning ray of the sun (Oliveira, 2013).

They descend the *Codeçal* Stairs, which lead down to *Ribeira*. From there, they head towards *Porta Nova* or *Nobre*, an old wicket (*postigo da Praia*) that was elevated in importance by D. Manuel in 1522 to a door (*Porta*) (Oliveira, 2013). There they go on board "*Barca da Sé*", which takes them to the *Gaia Landing Site*. When they arrive in Gaia, they go up to the *Chapel of São Marcos*.

Figure 9. The Procession to São Marcos Chapel.
Source: authors, based on Map data ©2015 Google (https://www.google.at/permissions/geoguidelines/attr-guide.html)

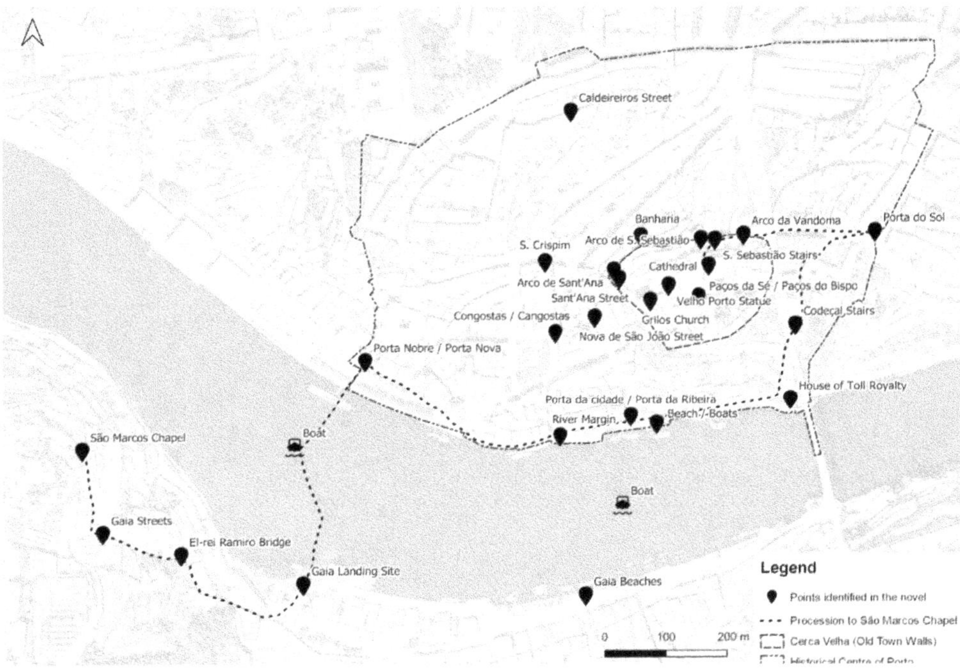

Vasco and His Mother

It is in this Chapel that Vasco appears again in the story. There he meets Rui and Garcia Vaz and they leave to plan how D. Pedro could sneak into the city of Oporto. Garrett imagines the path they take, passing by what is referred to as the "*romanesque bridge of El-rei Ramiro*", which the author identified in the pathway of Gaia Castle. On the map we have identified where the bridge is situated in *Rei Ramiro* street, although we don't know the exact location. In this street we can still find a fountain, but we don't know if it is the fountain that Garrett described in the novel. The author described *Paços de El-rei Ramiro* near the fountain, but there is no longer a way of establishing its current location. In this area, in the novel, the characters go into a *Fishermen's Tavern*, where they plan the entrance of D. Pedro into the city. After Rui and Garcia Vaz leave the tavern, Vasco stays and talks to the one known as the *Witch of Gaia*, also a spy of the king (Garrett, 1845-1850, p. 107). Around this time, he discovers what he already suspected: the Witch is his mother and the bishop is his father. At this point, he vows to avenge his mother, but his heart is always divided, since it was the bishop who raised him and always took care of him.

Vasco then sets off down the *Ladeiras de Gaia*, crossing the Douro and entering through the *Porta Nobre* (Noble Door) towards the Cathedral (figure 10).

Figure 10. Vasco in his way to Porto.
Source: authors, based on Map data ©2015 Google (https://www.google.at/permissions/geoguidelines/attr-guide.html)

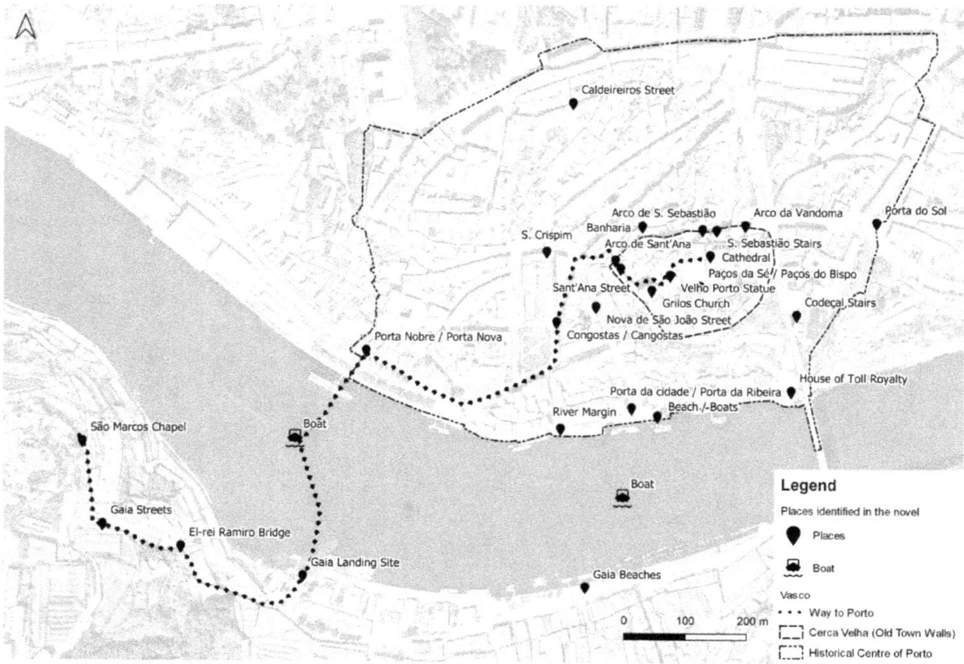

The End of the Novel

The masses in revolt had surrounded the Episcopal Cathedral (*Paços da Sé*) to demand Aninhas' release. Vasco had been leading the people, until King D. Pedro appeared in the middle of the crowd. Imposing his justice, he expelled the unworthy bishop, not before Vasco had knelt down and asked for mercy for his father. In the end, D. Pedro gave his blessing to the marriage of Vasco and Gertrudes.

The Literary Map Based on O Arco de Sant'Ana

The development of itineraries based on the novel by Almeida-Garrett meets the objectives of UNESCO (1994) that underline the importance of routes and itineraries as engines of development, as they have an enormous power in bringing communities together and facilitating cultural exchanges. Following this idea, ICOMOS (2008) refers to the importance of cultural itineraries as active instruments in intercultural dynamics and cultural encounters, promoting and valuing heritage and culture.

These itineraries, which have arisen from the pages of the novel, are embedded in a specific cultural context and can, in this way, contribute to the dissemination of knowledge and information.

There may be a growth in literary tourism, based on the most important and respected writers, as is the case of Almeida-Garrett, which may then stimulate the development of new products, namely cultural routes of literary interest.

The study of the literary content of a 19th century work is very important to access knowledge about the spaces and locations of that time, namely those that have already disappeared, and also to learn more about the scenarios of that time as seen by the author. It is possible through the reading of *O Arco de Sant'Ana* and its descriptions of landscape, to infer about the scenarios of the past and then to promote new itineraries based on its attractiveness to heritage tourists.

According to Shishmanova (2015), the integration of cultural routes in a network offers numerous possibilities for tourism and becomes a driver of culture, since an itinerary is a conveyor of cultural values, information, communication and history. Moreover, itineraries are excellent tools to help to maintain and preserve the territory and landscape and ensure their survival (Bambi et al., 2019).

CONCLUSION

Literary tourism, as Herbert (2001) argues, could attract heritage visitors to historical cities while promoting knowledge and providing historical recreation. Furthermore, literary tourist itineraries can offer a sustainable solution by reviving historical moments and promoting heritage education, where the emphasis is placed on a disappearing heritage and on the need to conserve the elements of this heritage so that future generations can also appreciate them.

In Porto and Gaia, which are both cities associated with the arts and writers, tourism could make use of both real life as well as emotional content of novels and capitalise on the new tendencies of international and national tourism asFerreira, Alén, Liberato and Liberato (2020) have pointed out.

In fact, Porto and Gaia have a rich historical heritage going back to medieval times and a significant number of well-known writers and artists that have created landscapes based around historical events. Novels, dramatic plays or poetry written in the past could be important documents to help us understand and fully appreciate and value local history and preserve our historical material and immaterial heritage.

The liberal conflict is a very important period in the city's history. Based on that, both cities could project and develop their liberal image as portrayed by certain writers. Garrett is one of the most important writers of the 19th century to explore these liberal places in his novels. Thus he has become an important contributor in the discovery of these places. In the same way, it is important to understand something about the real medieval city that has now disappeared, victim of the new urban architecture of city development, particularly in the last two centuries. In this novel, *O Arco de Sant'Ana*, it is also important to find out about the cultural artistic current, the new gothic vogue developed at the end of the 14th century and the criticism of the supporters of liberalism.

In this way, it is possible to create interesting new areas to visit in the cities of Porto and Gaia and to develop cultural and/or literary itineraries by joining both sides of the Douro River.

The use of methodological tools such as GIS is very helpful to tourism in several ways. With regard to this analysis, GIS are important in the creation of content for visitors. In this chapter, the use of GIS allowed the authors of the chapter to integrate literary tourism, historical information and urban analysis, through the identification and construction of itineraries, based on the *O Arco de Sant'Ana* novel. Nevertheless, there are some limitations in this study. It was possible to identify some of the monuments and places referred to by the author in the novel, but it was also found that some of them have disappeared with the growth in urbanisation. Therefore, a more detailed study about historical and archeological places would be very important to be able to understand and give due value to the places mentioned in this novel and also about Porto and Gaia in the 19th century.

This topic will be developed at a future date using other writers as sources, but still concentrating on the cities Porto and Gaia to try to re-create the urban landscape of the time.

In future research, it is expected to use combined methodologies, such as "oral history" and "oral memory" to complement the analysis already carried out. This cross information is a key to discovering more historical places in historical cities like Porto and Gaia and improving our knowledge about local heritage. In the near future, this study methodology could be used as an approach to explore other places, other novels and other writers in the history of Porto, and to develop literary tourism in both cities.

REFERENCES

Albuquerque, H., Costa, C., & Martins, F. (2018). The use of Geographical Information Systems for Tourism Marketing purposes in Aveiro region (Portugal). *Tourism Management Perspectives*, *26*, 172-178. doi:10.1016/j.tmp.2017.10.009

Alves, M. T. A. (1999). Almeida Garrett: O texto, o intertexto e as margens do texto na representação do escritor [Almeida Garrett: the text, the intertext and the text margins in the writer representation]. *Scripta, Belo Horizonte*, *3*(5), 142–154.

Amorim. (1881-1884). *Garrett: Memórias Biographicas.* Lisboa: Imprensa Nacional.

Arcos-Pumarola, J., Osácar Marzal, E., & Llonch-Molin, N. (2018). Literary urban landscape in a sustainable tourism context. *Human Geographies – Journal of Studies and Research in Human Geography*, *12*(2), 175-189.

Bahaire, T., & Elliott-White, M. (1999). The Application of Geographical Information Systems in Sustainable Tourism Planning. *Journal of Sustainable Tourism*, *7*(2), 159–174. doi:10.1080/09669589908667333

Baleiro, R., & Quinteiro, S. (2019). A small place, by Jamaica Kincaid: Envisioning literary tourism in Antigua. *Journal of Tourism and Cultural Change*, *17*(6), 676–688. doi:10.1080/14766825.2018.1529772

Bambi, G., Iacobelli, S., Rossi, G., Pellegrini, P., & Barbari, M. (2019). Rural tourism to promote territories along the ancient roads of communication: Case study of the rediscovery of the St. Francis's ways between Florence and La Verna. *European Countryside*, *11*(3), 462–474. doi:10.2478/euco-2019-0025

Basto, A. M. (1940). O Porto Medieval [The Medieval Porto]. *Memórias de Comunicações apresentadas ao Congresso de História Medieval, II Congresso do Mundo Português, 2, 675-708*

Basto, A. M. (1962). Desenvolvimento topográfico da cidade do Porto [Topographical development in Porto]. *História da cidade do Porto*, *1*, 116–158.

Beigi, H., Hossein, M., & Azani, M. (2015). Literary tourism and place identity of the central plateau of Iran. *Journal of Biodiversity and Environmental Sciences*, *5*(6), 309–406.

Cacilhas, N. (2020). Roteiro literário Levantado do Chão [Raised from the Ground Literary Roadmap]. *Dos Algarves: A Multidisciplinary eJournal*, *37*, 106-122. . doi:10.18089/DAMeJ.2020.37.7

CAOP. (2018). *Carta Administrativa Oficial de Portugal* [Official Administrative Map of Portugal]. Available at https://www.dgterritorio.pt/cartografia_e_geodesia/cartografia/carta_administrativa_oficial_de_portugal_caop/

Coelho, M. H. C. (2003). Bispos e reis: oposições em torno de bens e jurisdições temporais [Bishops and kings: oppositions over temporal goods and jurisdictions]. Lusitania Sacra, 15, 279-287.

Coutinho, B. X. (1962). Arquitetura militar e religiosa [Military and religious architecture]. In História da cidade do Porto [Porto city History]. Academic Press.

Cruz, J. M. (2009). @-Turismo ativo no Parque Natural da Serra da Estrela. Criação de rotas temáticas [Active tourism in Serra da Estrela Natural Park. Development of tematic routes] [Master Dissertation]. Universidade da Beira Interior.

David, S. N. (2007). *Cartas de amor à Viscondeza da Luz* [Love letters to the Viscountess of Luz]. Famalicão: Edições Quasi.

DMPU. (2018). Pessoas e Atividades. Dinâmicas Económicas. Relatório de Caracterização e Diagnóstico [People and activities. Economic Dynamics. Diagnosis and caracpterizazion report]. In *Revisão do Plano Diretor Municipal do Porto* [Revision of the Porto Municipal Master Plan]. Direção Municipal de Urbanismo, Departamento Municipal de Planeamento Urbano.

Donaldson, C., Gregory, I., & Murrieta-Flores, P. (2015). Mapping 'Wordsworthshire': A GIS Study of Literary Tourism in Victorian Lakeland. *Journal of Victorian Culture: JVC*, *20*(3), 287–307. doi:10.1080/13555502.2015.1058089

Engelmo-Moriche, A., Nieto-Masot, A., Horcajo-Romo, A., & Cárdenas-Alonso, G. (2021). Tourist geoportal of a rural region as a means of disseminating its offer and resources. Case study: Extremadura. *Revista de Estudios Andaluces*, *41*, 125–143. doi:10.12795/rea.2021.i41.07

Farsari, Y., & Prastacos, P. (2008). GIS Applications in the Planning and Management of Tourism. In A. Lew, M. Hall, & A. Williams (Eds.), *A Companion to Tourism* (pp. 596–607). Blackwell.

Ferreira, A., Alén, E., Liberato, P., & Liberato, D. (2020). Literary Tourism: A Cultural Trip? *Advances in Tourism. Technology and Smart Systems*, *171*, 505–515. doi:10.1007/978-981-15-2024-2_44

Ferreira, M. G. (2008). Teatro ao gosto de Almeida Garrett [Theatre in the style of Almeida Garrett]. *Teatro do Mundo*, *03*, 53–60.

Garrett, A. (2011). *O Arco de Sant'Ana* [The Arch of Sant'Ana]. Porto Editora.

Herbert, D. (2001). Literary Places, Tourism and the Heritage Experience. *Annals of Tourism Research*, *28*(2), 312–333. doi:10.1016/S0160-7383(00)00048-7

Hladnik, M., & Fridl, J. (2012). Space and its geographical presentation in Slovene historical narratives. *Slavistična Revija*, *60*(3), 443–455.

Hoppen, A., Brown, L., & Fyall, A. (2014). Literary tourism: Opportunities and challenges for the marketing and branding of destinations? *Journal of Destination Marketing & Management*, *3*(1), 37–47. doi:10.1016/j.jdmm.2013.12.009

ICOMOS. (2008). *Québec Declaration on the Preservation of the Spirit of Place*. Available at: https://whc.unesco.org/uploads/activities/documents/activity-646-2.pdf

INE. (2021). *Annual Provisional Resident Population Estimates, according to the administrative territorial organisation*. Available at: https://www.ine.pt/xportal/xmain?xpid=INE&xpgid=ine_indicadores&contecto=pi&indOcorrCod=0008273&selTab=tab0&xlang=en

Instituto Camões. (2021). *Dia Mundial da Língua Portuguesa, 5 de maio de 2021* [World day of the Portuguese language, May, 5th, 2021]. Available at https://www.instituto-camoes.pt/images/pdf_noticias/Dados_sobre_a_l%C3%ADngua_portuguesa-2021.pdf

Jiang, L., & Xu, H. (2016). Reading, tourism, and geography consumption in literary places. *Tourism Geographies*, *18*(5), 483–502. doi:10.1080/14616688.2016.1217033

Jiang, L., & Xu, H. (2017). The growth of literary places in ancient town tourism destinations: Based on the theories of Bourdieu, Danto and Dickie. *Journal of Tourism and Cultural Change*, *15*(3), 213–228. doi:10.1080/14766825.2016.1148153

Jiang, L., & Yu, L. (2020). Consumption of a literary tourism place: A perspective of embodiment. *Tourism Geographies*, *22*(1), 127–150. doi:10.1080/14616688.2019.1586985

Jurkus, E., Taminskas, J., Povilanskas, R., Kontautiene, V., Baltranaitė, E., Dailidė, R., & Urbis, A. (2021). Delivering Tourism Sustainability and Competitiveness in Seaside and Marine Resorts with GIS. *Journal of Marine Science and Engineering*, *9*(3), 312. doi:10.3390/jmse9030312

Kaczmarek, J. (2020). The aleatoric aspect of literary tourism: An example from Havana. *Tourism (Zagreb)*, *30*(1), 31–42.

Lopes, O. (1981). De «O Arco de Sant'ana» a «Uma Família Inglesa» [From "O Arco de Sant'ana" to "An English Family"]. Separata da «Revista de História», 4, 9.

Machado, A., Machado, C., Moreira, F. J., Reis, J., Aurindo, M. J., Almeida, M. M., & Faria, R. (2020). Story Map Caminhos d'O conspirador: Um percurso literário na vila de Marvão com Branquinho da Fonseca [Story Map Caminhos d'O conspirador: A literary journey through the town of Marvão with Branquinho da Fonseca]. *Dos Algarves: A Multidisciplinary e-Journal*, *37*, 88-105. . doi:10.18089/DAMeJ.2020.37.6

Marques, A. M. S. (2012). *O Anacronismo no Romance Histórico Português Oitocentista* [Anachronism in the 19th century Portuguese historical novel]. Edições Afrontamento.

Marques, L., & Cunha, C. (2013). Literary rural tourism entrepreneurship: Case study evidence from Northern Portugal. *Journal of Policy Research in Tourism, Leisure & Events*, *5*(3), 289–303. doi:10.1080/19407963.2013.801158

Mcadam, D. (1999). The Value and Scope of Geographical Information Systems in Tourism Management. *Journal of Sustainable Tourism*, *7*(1), 77–92. doi:10.1080/09669589908667327

Menezes, J. (2020). Casa do Rio Vermelho: From place of memory to literary place - a possibility of sustainable cultural tourism in Salvador, Bahia, Brazil. *Dos Algarves: A Multidisciplinary e-Journal*, *37*, 71-87.

Mukhametzyanova, L. R., Yao, M. K., Emanova, J. G., & Pozdnyakova, K. G. (2017). Aesthetic and semanticaccents of the neo-gothic interpretation in Russia. *Journal of Fundamental and Applied Sciences*, *9*, 7.

Murrieta-Flores, P., Donaldson, C., & Gregory, I. (2017). GIS and Literary History: Advancing Digital Humanities research through the Spatial Analysis of historical travel writing and topographical literature. *Digital Humanities Quarterly*, *11*(1). Available at https://www.digitalhumanities.org/dhq/vol/11/1/000283/000283.html

Murrieta-Flores, P., & Howell, N. (2017). Towards the Spatial Analysis of Vague and Imaginary Place and Space: Evolving the Spatial Humanities through Medieval Romance. *Journal of Map & Geography Libraries*, *13*(1), 29–57. doi:10.1080/15420353.2017.1307302

Oliveira, R. (2013). *Duas muralhas, duas cidades. A História Militar do Porto Medieval* [Two walls, two cities. The militar history of Medieval Porto] [Master Dissertation]. University of Porto.

Oliveira, S. (2017). *Um Porto de encontro entre Turismo e Literatura* [A meeting point between Tourism and Literature] [Master Dissertation]. University of Porto.

Ribeiro, J. M. (2020). Liberalismo e Revolução Liberal de 1820. Novas interpretações [Liberalism and the Liberal Revolution of 1820. New interpretations]. *História. Revista da FLUP*, *10*(2), 3–8. doi:10.21747/0871164X/hist10_2a1

Santana, M. H. (2004). *Edição crítica das Obras de Almeida Garrett* [Critical edition of the Works of Almeida Garrett]. Imprensa Nacional – Casa da Moeda.

Shishmanova, M. V. (2015). Cultural tourism in cultural corridors, itineraries, areas and cores networked. *Procedia: Social and Behavioral Sciences*, *188*, 246–254. doi:10.1016/j.sbspro.2015.03.382

Smith, K. (2003). Literary enthusiasts as visitors and volunteers. *International Journal of Tourism Research*, *5*(2), 83–95. doi:10.1002/jtr.419

Sousa, A. (1994). Tempos medievais [Medieval Times]. In L. A. Oliveira Ramos (Ed.), *História do Porto*. Porto Editora.

Sousa, P. R., & Fernandes, S. B. (2007). Aplicação dos sistemas de informação geográfica no turismo [Application of geographical information systems in tourism]. *Finisterra*, *42*(84), 105–118.

Taylor, J., & Pouliot, A. (2020). Introduction: Placing the author in ecologies of literary tourism. *Nineteenth-Century Contexts*, *42*(4), 381–389. doi:10.1080/08905495.2020.1782014

UNESCO. (1994). *Routes as part of our Cultural Heritage*. Convention concerning the protection of the world cultural and natural heritage, World Heritage Comittee, 18th session. WHC-94/CONF.003/INF.13. Available at: https://whc.unesco.org/archive/routes94.htm

UNESCO. (2003). *Convention for the Safeguarding of the Intangible Cultural Heritage*. Available at: https://ich.unesco.org/en/convention

Vitić-Ćetković, A., Jovanović, I., & Potočnik Topler, J. P. (2020). Literary tourism: the role of Russian 19th century travel literature in the positioning of the smallest European royal capital – Cetinje. *Annales. Series historia et sociologia*, *30*(1).

Wang, H., & Zhang, D. (2017). Comparing literary tourism in Mainland China and Taiwan: The Lu Xun Native Place and the Lin Yutang House. *Tourism Management*, *59*, 234–253. doi:10.1016/j.tourman.2016.08.008

Zhang, Q., & Xu, H. (2020). Understanding aesthetic experiences in nature-based tourism: The important role of tourists' literary associations. *Journal of Destination Marketing & Management*, *10*, 100429. doi:10.1016/j.jdmm.2020.100429

Chapter 9
Evaluating the Use of Digital Cartography to Showcase the Intangible Heritage:
The Case of Literary Tourism

Jordi Arcos-Pumarola
CETT-UB, Spain

Daniel Imbert-Bouchard Ribera
CETT-UB, Spain

Núria Guitart Casalderrey
CETT-UB, Spain

ABSTRACT

New narratives are essential for ensuring the sustainability of tourist destinations and improving visitor experience. One key resource destination that can be drawn on is intangible heritage, which digital cartography can help visitors to interpret. The overall objective of this chapter is to analyze—from a multidisciplinary perspective—the opportunities digital cartography offers for the exploitation of literary heritage. The authors present an evaluation tool (validated by experts), whose aim is to analyze the different dimensions and elements that should be incorporated in digital maps. The intention is to enable the analysis of existing digital literary tourism maps and to encourage the future use of the many options offered by digital cartography in maps of this nature.

INTRODUCTION

New narratives and attractions are essential for encouraging the redistribution of tourists and thus ensuring competitive and sustainable tourism. Following the official recognition of intangible cultural heritage and cultural expressions by UNESCO (2005; 2018), many tourism promoters have taken advantage

DOI: 10.4018/978-1-7998-8262-6.ch009

of intangible culture to create new products and attractions. Moreover, expressions of popular culture that are clearly linked to a tourist destination play an important role in shaping its image (Marine-Roig, 2015). This is because fictional narratives transmitted through television, literature, cinema, or online audiovisual platforms reach a large percentage of the population. The shared consumption of these cultural expressions generates a shared imaginary that affects the desires, conceptions, and expectations of potential visitors who have the book or film in mind when they visit (Iwashita, 2006).

This phenomenon is crucial for promoting destinations while it raises both challenges and opportunities for destination management and visitor experience since the perspective prompted by literature or cinema superimposes a new layer of significance on the landscape and the tourist resources of the destination. New foci are generated that are both embraced by local residents (Parramon & Medina, 2017) and lead to the creation of new cultural tourism products. The promotion of fiction-based tourism is thus essential for diversifying existing opportunities for tourism and adding new features to the image of destinations (Arcos-Pumarola et al., 2018). New cultural itineraries that provide a response to new tourist motivations and are communicated via maps may thus appear that provide alternatives to those contained in traditional tourist guides. Media tourism (McClinchey, 2015; van Es & Reijnders, 2018), which includes literary tourism, cinematographic tourism or, even emerging types such as video game tourism (Dubois & Gibbs, 2018), is a type of tourism that is undertaken in order to visit spaces linked to fictional narratives, their authors, or the characters that appear in them.

However, in order to exploit attractions connected to popular culture it is necessary to convert spaces into tourist attractions, spaces that, in many cases, are not immediately visible to the tourist, either because they have no obvious marker (MacCannell, 2003) or because they have been camouflaged in the original story and may therefore have a different appearance. This difficulty means that, in many cases, potential tourist attractions derived from fiction are absent from standard processes of iconization and spectacularization in urban destinations (Quaglieri Domínguez & Paolo Russo, 2010). Nevertheless, this also means that tourism products based on cultural expressions such as literature, cinema or television series may offer heritage-friendly experiences as they are based on the proactive interest of visitors seeking to visit spaces linked to their cultural interests. As Ashworth (2009) has asserted, tourists' previous interest in and knowledge of a specific heritage site produces tourism that is unlike some criticized and more stereotypical forms of cultural tourism.

This chapter focuses on literary tourism, a specific type of tourism linked to cultural expressions that can be defined as a visit to a location prompted by its links to a literary author or a literary work (Hoppen et al., 2014). This kind of tourism presents a difficulty that does not exist in other media tourism because while those who travel to see film locations, for example, can visit places originally seen on the big (and small) screen, literary tourists start with the idea of the locations they have gained from the written word and, therefore, have a greater need for tools to identify and interpret the places they wish to visit. Beeton (2005) has asserted that cinematographic tourism is an amplifier of literary tourism, since it makes it possible to reach a wider public, but not all literary works benefit from an audiovisual version that situates the works in a recognizable location, while it should not be forgotten that the role of the author in literary tourism is also crucial. In order to exploit the full potential for literary tourism, it is thus necessary to develop strategies and tools to facilitate literary tourists' visits to key sites.

One of the ways of doing so is with the help of maps. Until recently, traditional maps on paper have been the most typical means of locating and communicating the various tangible features in an area. Maps are a key component of tourist exploration and provide graphic and intuitive support to tourists throughout their trips. Advances in the digitalization of these information and visitor service tools and the advent of "neogeography" have transformed the way in which tourist destinations can be interpreted by both visitors and promoters (Konecny, 2011). In fact, the potential of digital maps is much greater than that of traditional maps. They are particularly valuable for their versatility and interactivity in disseminating intangible tourism products, while the opportunities offered by real-time geolocation, and their capacity to influence the decision-making capacity and behavior of visitors by providing information on locations and environments depending of the needs of the destination. These smart features make digital maps an excellent means of contributing to tourism management and movement around the destination. Not only a perfect platform for disseminating and locating new narratives and attractions linked to fiction, they are also—in the field of management—a tool for territorial rebalancing, as they can track the routes most used by visitors and collect data on tourist preferences in real time.

All of this makes it necessary to define what is needed by literary tourism maps to enable them to incorporate destinations' needs to disseminate and manage their heritage and to respond to the new concerns of tourists, who are increasingly accustomed to using digital media when traveling.

The objective of this study is therefore to develop a tool that includes the main parameters that should be included in digital literary maps. A weighting and rating system is employed both to evaluate existing digital tourist maps and to increase the range of options available to today's literary tourists. Lastly, on the basis of attributes that the academic literature considers suitable for the use of cartography in the field, we set out a series of recommendations for the tourism and culture sectors on optimizing the potential of this tool to exploit intangible heritage.

BACKGROUND

Literary Geography in a Tourism Context

Literary tourism is based on the existence of locations that possess a certain significance by virtue of their links with literary works or their authors. By connecting the symbolic world of literature to specific places, tourist promoters can identify a series of locations that one might call literary tourism resources. The creation of a range of literary tourism products involves identifying, understanding, and interpreting the resources that are contained within the literary geography of a territory.

Literary tourism resources are often related to visitors' initial knowledge of the author or the literary work (Hoppen et al., 2014; Squire, 1996). For this reason, locations can be split between those connected with the author (e.g., places in which the author was born, lived, dwelt, or those which affected his personal world) and those linked to his literary work (places in which the literary work takes place or which inspired the narrative).

These two aspects enable the creation of literary tourism products that enhance and transmit the literary geography of a territory; literary tourism, however, is not limited to just these two. There are also events such as literary festivals, second-hand book markets, festivities like Bloomsday, and Sant Jordi's Day, etc. (Arcos-Pumarola, 2019; Patricio Mulero & Rius-Ulldemolins, 2017). Literary tourism strategies should therefore not only encompass locations with links to writers and famous novels but also other types of tourism such as creative tourism and the different activities and events included in these festivals (Molina, 2018).

Spaces of memory, such as tombs and monuments, are also significant tangible resources in a literary geography of a territory. These places usually serve to commemorate a specific literary figure;

but, sometimes they can refer to literary works or to the conceptual or symbolic world of these works. For example, there are a number of monuments linked to fictional characters figures, such as Mercè Rodoreda's Colometa in Plaça del Diamant and Little Red Riding Hood in Passeig de Sant Joan, both in Barcelona (Uccella, 2008). Lastly, there are other spaces that are also of interest to literary tourists. Bookstores, public libraries, and cultural centers can all play a role in literary tourism, both for their contents and for their role in interacting and communicating with visitors.

Literary geography can thus be understood as an associative cultural landscape—an area on which culture has left a strong imprint—but it should be taken into account that the few tangible traces may not be very explicit (Donaire, 2008). This is because literary texts generate an imaginary that relates both to very specific locations and to much more general areas. Furthermore, literary tourism resources often lack distinctive physical characteristics, so they may well go unnoticed.

Tools for Discovering Tourist Destinations

Cartography is particularly important for exploiting literary geography. Visitor welcome packs and information are essential aids for tourism in many destinations (Planells Costa & Crespi Vallbona, 2014) and they are decisive in introducing visitors to the potential of the territory. Whether portraying local culture or natural features, they are transmitted by means of clear discourse and content that are represented in different formats (Ram & Hall, 2018). Often these resources are an ideal way to facilitate products such as itineraries or routes that are expressly designed to exploit a tourist attraction and make a major contribution to improving overall satisfaction with the tourist experience (Brent Ritchie et al., 2011; Wearing & Foley, 2017).

As a key source of information, maps act as ambassadors to the destination itself. As in other areas, cartography is a universal language for solving problems associated with the immediate environment (Rosselló, 2008).Of the two very different types of map—the topographic map with its inventory of all topographical elements located within a given territory and the thematic map with its items of interest in the area, tourist maps belong to the second category (Jancewicz & Borowicz, 2017). This applies equally to cultural maps. Cultural cartography is a very broad concept, as there may be a wide range of ways in which culture and maps can be linked, but the concept is based on the numerous uses of the map for representing cultural information (Arcila Garrido & López Sanchez, 2011).

Many destinations have recently become more interested in creating tourist maps with cultural information since they are fundamental for enabling tourists to find their way around destinations and to locate cultural items (Medynska-Gulij, 2003). Maps can be adapted to any tourist location or scale: cities, municipalities, regions or even countries. However, despite their versatility, they are more effective in places where the visitors are on foot as this makes it easier for them to interact and adapt more effectively to attractions (Forsyth, 2015).

In practice, tourism is unthinkable today without a map to accompany one both throughout one's stay and even after the visit (as a souvenir). It not only serves to describe the geographical area, since there are a wide range of options and variants that can be used to promote tourism. The tourist map—or city plan for urban environments—has two clear objectives: the first is to help tourists locate attractions and resources, routes, itineraries and complementary information; the second, as with other information in the destination, is to direct the tourist's attention to alternative points, redistributing flows to less frequented areas and so condition mobility and accessibility to the various resources and attractions present (Corral, 2009; García-Hernández et al., 2017; Simancas Cruz, 2016). Furthermore, some authors have drawn attention to the map's potential as a mechanism for promoting and reinforcing the identity and image of the destination (Domínguez-Mujica, 2007; Kałamucki & Czerny, 2015). Maps are thus instruments that serve to disseminate culture and tourism, to promote the image of the destination, and reinforce citizens' identity. As Fontal Merillas (2007) has pointed out, tourism-heritage maps do much to raise awareness of heritage.

It is not always easy to strike the right balance between an attractive design and an effective management tool. Apart from the fact that those who make maps have always had to make decisions about the territorial scale required (due to the different morphology of each destination and variations in the concentration of attractions), maps must be reviewed by experts from other disciplines who understand the perspective and needs of the tourist (Gorrini & Bertini, 2018). This is very important, as differing opinions frequently result in a failure to transmit the main ideas required in maps. Literary tourist maps require an interdisciplinary approach to ensure they transmit more than just points in space and help the visitor to establish a close connection with the places they wish to visit. The task of creating tourist heritage maps that do not simply provide information but seek interaction with the user brings benefits in a range of areas (Gamír, 2019). Research on methodologies and procedures, for example, has made it possible to embody the symbolic and cultural dimension of the territory on a map, involving the collaboration of various disciplines in the social sciences, including history, documentation, geography, and lastly, audiovisual communication, which has sought to incorporate narratives, illustrations, and components of a more subjective nature to describe landmarks (Arcila Garrido & López Sanchez, 2011; Gamír, 2019; Stadler, 2015).

To date, the most common format for information maps has been paper. There are normally a wide range of options available at tourist destinations, with numerous tourist maps seeking to stand out from other by including different elements that might be useful to the tourist. These can be as varied as complementary services and thematic routes, the exact location of chains of large stores, and large volumes of advertising that can sometimes make it difficult to read the map itself. Some maps are free and available to tourists (in tourist information offices and hotel receptions) while others cost money (in kiosks and specialized bookstores). Lastly, many are incorporated in typical tourist guides.

A last, more recent type of map that has increased the range of options available is the digital map. So-called collaborative cartography or neogeography (Turner, 2006) emerged at the beginning of this century as an alternative to more conventional techniques. On-screen maps are more interactive and up to date and transform the relationship between users and destination (Cartwright, 2012). One aspect in particular makes them stand out from conventional paper maps: so-called geolocation, the function that enables users to locate themselves in real time (Beltrán, 2016). Geolocation has radically changed the world of tourist maps, forcing tourism managers to reflect on how this new means of disseminating tourist information can be exploited in the future. In the case of literary tourism, digital cartography makes it possible to geolocate literary resources and use different types of information (audiovisual, textual, augmented reality, etc.) to enable visitors to interpret the place they are visiting (Pérez Isasi, 2018). The function of digital cartography is thus to mark a specific attraction and increase the amount of information available about tourist sites during their visit, a function that is of special interest to those managing intangible cultural expressions such as literary tourism.

MAIN FOCUS OF THE CHAPTER

The Importance of Identifying the Most Suitable Attributes to Be Included on a Digital Map

As mentioned above, the content in thematic maps must include the variables and attributes needed to convey the intended message. A number of authors have identified a range of parameters that maps on paper need to fulfill their function (Cabrera et al., 2009; Medynska-Gulij, 2003). While a characteristic range of aspects are important for tourist maps, standard features, such as the legend, north symbols, and scale, which are essential for understanding and using the map must not be forgotten.

The aim of tourist maps is to provide high quality information for tourists. This should be transmitted by certain elements, such as specific symbols to highlight items that may be of interest to tourists. The main way of doing this is by means of pictograms. (Beata Medynska-Gulij, 2008; Sarikanonp & Sahachaisaereep, 2010). Tourist maps also often display specific themed itineraries that provide itineraries linking key locations. Tourist maps have thus been defined by some authors (Medynska-Gulij, 2003) as maps that combine cartographic symbols with basic topographical information, the users' final impression depending on the balance struck between the two aspects.

Variables can be broken down into those to do with content and those to do with form. Authors such as Cabrera et al. (2009) have created a list of formal elements to be taken into account when drawing up a tourist map. These include design and ease of use, suitability and usefulness of information, helpfulness in orienting users, and quality of materials, all of which play a significant role in creating a traditional tourist map.

Digitalization, however, has prompted a complete rethink on what needs to be incorporated in tourist maps. Despite the paradigm change, conventional cartographical criteria such as title, scale, and font still apply, but there are now other specific criteria of a more technical nature. These include the capacity to customize maps with information and symbols that can be activated according to planning needs, the inclusion of multimedia and simulation features to expand content, and language accessibility and other functions (Pérez & Nel·lo, 2008; Tujaka, 2010)

The need to identify indicators of aspects that should be incorporated in tourist maps has some new academic studies. Pérez & Nel·lo (2008), for example, have compiled a list of 35 digital tourist map characteristics (see Table 1), which are broken down into sections on general characteristics (9 points), the information and the quality of the information in the map (15), and interactivity and technical connectivity (11). The indicators are based on the authors' literature review on the use of online geographic information by tourists.

Tujaka's (2010) interpretation also brings out the variables needed in an online map aimed at disseminating the values of a territory (see Table 2). The indicators are grouped by technical characteristics, design characteristics, information organization, map interactivity, quality and usability.

Table 1. Indicators used to analyze wine tourism maps

General Characteristics		Organization
		URL
		Is there a specific tourist website? (Y/N)
		Is it linked to other tourist portals? (Y/N)
		Is there a map? (Y/N)
		Own resource? (Y/N)
		Language
		Type of use (tourist, street, geolocation, others)
		Map accessibility (high, medium, or low)
Quality	**What Information Does Map Offer?**	Facilities in general
		Accommodation
		Bars and restaurants
		Shops
		Museums
		Places of interest
		Urban itineraries
		Routes
		Leisure activities
		Typical crafts and products
		Popular festivities
		Detail of information (high / low)
	Quality	Legibility (high, medium, low)
		Suitability of symbols (high/medium/low)
		Are there any features to enhance understanding (Y/N)
Interactivity and Functionality	**Interactivity**	Is it interactive? (Y/N)
	Functions	Display options (e.g. zoom)
		Printing
		Save image
		Send image
		Search function
		Advanced search (includes parameters)
		Distance calculator
		Edit (add, delete text, etc.)
		Create own route
		Download database
		Virtual visits

Source: Pérez & Nel·lo (2008)

Table 2. Analysis and evaluation indicators for a tourist map.

Indicator			Rating	Optimal Map
Usability	Access		1: direct 0: not direct	1
	Languages		From 0 to 5	5
	Interactivity types	Active symbols	1: yes 0: no	1
		Activable levels	1: yes 0: no	1
		Zoom	1: yes 0: no	1
	Format information	Window	1: yes 0: no	1
		File	1: yes 0: no	1
	Range of information	Nº levels	De 0 a 30	30
		Specific	1: yes 0: no	1
		Integrated	1: yes 0: no	1
	Functions	Print	1: yes 0: no	1
		Save	1: yes 0: no	1
		Send	1: yes 0: no	1
		Download	1: yes 0: no	1
		Measure distance	1: yes 0: no	1
		Other	1: yes 0: no	1
Design	Pictographic symbols		1: yes 0: no	1
	Multimedia		1: yes 0: no	1
	Unconventional layers		1: yes 0: no	1
	Title		1: yes 0: no	1
Max rating.			**53**	

Source: Tujaka (2010)

A specific methodology designed to review literary cartography based on different variables to be included as functions on an online map—a digital tool for tourism management—makes it possible to evaluate the limitations and advantages of future proposals for digital maps aimed at visitors. To this end, Tujaka (2010) proposes an evaluation methodology consisting of 20 indicators classified in just two categories: usability and design.

Given the potential of online maps for improving tourist experience of intangible cultural expressions, our intention is to explore this evaluation methodology in greater depth. The capacity of online maps' to enhance visits to certain areas of interest makes it necessary for us to reflect on how this capacity can be used to mediate between literary sites and the tourist.

METHODOLOGY

Given the existing lack of research into tools for evaluating digital literary tourism maps, our research was of an exploratory nature.

We created a literary map evaluation matrix to be used as a tool for digital tourism management in two stages. First, we undertook an extensive review of literature on a range of aspects: the usability and design of digital cartography (Tujaka, 2010), the evaluation of digital tools in heritage environments for educational purposes (Martínez Gil et al., 2018), the use of technology in the creation of cultural tourism experiences (Imbert-Bouchard Ribera et al., 2013; Marasco, 2020), and lastly, the variables needed to enable geographic information to be used as a tool for cultural planning and management (Arcila Garrido & López Sanchez, 2011). These sources, presented in the theoretical framework set out in this chapter enabled us to identify four dimensions and 30 indicators for evaluating the use of digital maps for literary tourism. The dimensions derived related to usability, content, interactivity and education, and design.

Second, a focus group that included two specialists in cartography, two in education, and one in the creation of experiences reviewed, validated, and amended the dimensions and indicators obtained in the first step. Although the group was small, it was nevertheless appropriate since there are still few studies on the subject and it was necessary to gain views from several complementary disciplines and fields. One of the experts came from a tourism experience management company, three were academics (two of them specializing in heritage education and one in cartography) and there was a technician qualified in innovation and digital strategy belonging to the Catalan map agency. The focus group in April 2021 was held by videoconference and lasted 75 minutes.

The data obtained from the literature review, together with the proposals for improvement and adjustments suggested by the focus group, resulted in a matrix of 35 indicators, which is set out in the next section (Tables 3, 4, 5 and 6). Furthermore, the panel of experts not only helped to redefine the list of indicators, but also to make suggestions on how they should obtained, considered, and interpreted. Moreover, the focus group also allowed us to detect a series of key issues that are essential in the use of this type of digital map. These reflections are also set out in this chapter.

RESULTS

As mentioned in the methodology section, the focus group was presented with an initial evaluation grid made up of thirty indicators divided into four dimensions, with 10 indicators relating to content, 8 to usability, 9 to interaction and education, and, lastly, 3 to design. Each indicator referred to an evaluable characteristic that could be applied to any literary tourism map. All characteristics included in a product to be evaluated counted one point, producing a final score that rated the degree of suitability and optimality of maps, both overall and dimension by dimension. There were two indicators that incorporated more than one element and therefore accounted for more than one points. The tables relating to each dimension that follow include the criteria that are needed to gain the scores.

The focus group discussion validated the proposed dimensions and the great majority of indicators. The result was a table that kept the same four initial dimensions: content, interaction and education, usability, and design, with 10, 9, 12 and 4 indicators respectively. On the suggestion of the focus group, some of the original indicators were modified and new ones were added. The evaluation table, divided into the four dimensions, is set out below.

Content

The final version of the evaluation table consists of 10 indicators for content (see Table 3). The experts suggested the following indicators: providing sources of information of proven quality to enhance content; structuring information to generate a range of different narratives; using images, audiovisual resources and technology to present the content in an attractive manner; geolocating the different literary resources of a territory in a well-organized and visual manner; and, finally, including other cultural events and attractions of the territory in the literary tourism map.

Table 3. Proposed indicators to evaluate the functionality of an online literary tourism map from the point of view of its content

<table>
<tr><th>Dimension</th><th colspan="2">Indicator</th><th>Rating</th><th>Explanation</th></tr>
<tr><td rowspan="11">Content</td><td colspan="2">Veracity of information</td><td>0/1</td><td>Details provided on sources of information used in map</td></tr>
<tr><td colspan="2">Volume of information</td><td>0/1</td><td>Provides online references and resources in order to increase the amount of information</td></tr>
<tr><td rowspan="2">Theming</td><td>Route suggestions</td><td>0/1</td><td>Starts by offering users thematic routes</td></tr>
<tr><td>Storytelling</td><td>0/1</td><td>Constructs itineraries around a literary figure or character and aims to involve the user in a related story</td></tr>
<tr><td colspan="2">Use of images and video</td><td>0/1</td><td>Provides information in visual and audiovisual format to contextualize or enrich text</td></tr>
<tr><td rowspan="2">Use of technology to create fictional content</td><td>Reproduction of fictional scenarios</td><td>0/1</td><td>Uses augmented or virtual reality to recreate spaces linked to the original narrative</td></tr>
<tr><td>Reproduction of characters in real scenarios</td><td>0/1</td><td>Superimposes fictional characters or literary authors on the location to generate parallel narratives</td></tr>
<tr><td colspan="2">Locations of literary tourism resources</td><td>0/2</td><td>Identifies exact locations and presents literary information to users (1). Also differentiates between locations linked to authors, books, and memory (1).</td></tr>
<tr><td colspan="2">Literary events</td><td>0/1</td><td>Gives information about literary events in the area</td></tr>
<tr><td colspan="2">Provides links to other thematic narratives linked to the area</td><td>0/1</td><td>Encourages users to follow other itineraries or visit other locations in the same area</td></tr>
</table>

Source: Own creation (2021)

A case that illustrates the importance of providing an extensive and rigorous database to build a digital literary tourism map is the one developed by Espais Escrits in Catalonia (Espais Escrits, 2018). This project, created by a literary heritage network that gathers together various Catalan literary heritage equipments, uses references to illustrate the importance of each location, which are correctly cited and permit the users to delve into the literary work according to their interests. Browsing into the map, we also see that the content of some locations is broadened through real pictures, audiovisuals or audio to expand the content of the literary map.

The map also suggests diverse literary routes to the users based on different topics related to Catalan literature. For instance, we can find a route through the spaces in Barcelona related to Mercè Rodoreda, the Catalan author of The time of the doves or Death in spring, among others. This route, composed of 22 locations, highlights the places of Barcelona more relevant to her biography and work, combining pictures, audio documents, quotes of her work, and other information. Another example of suggested routes within this map is the one dedicated to the modernist group in Reus, which gathers quotations of different authors to explain the relevance of this group at the end of the XIXth century. This project is an example of the priority to use quality information to build the digital map. Nevertheless, there is the need to find a balance between this and strategies to attract different non-expert publics. This could be done, for instance, with the use of technological resources to reproduce historical scenarios or characters, as the table of the present section shows, as well as through other strategies linked to the interactions with users that will be addressed in the next section.

Interaction and Education

This second dimension (see Table 4) is composed of 9 indicators and deals with the capacity of the map to transmit the content and to generate interactive dynamics that encourage visitors to explore the destination. Tools to evaluate visitor knowledge and set them challenges are rated here. We have also included questions on options to customize the map to suit user's interests, share content on social media, generate one's own content, and communicate directly with the literary groups in the area. Lastly, the experts also thought it was important to include basic interaction options, such as options to enable users to report possible errors, rate the map and download the content of the map for later offline use.

Table 4. Proposed indicators to evaluate the functionality of an online literary tourism map from the point of view of its interaction and education.

Dimension	Indicator	Rating	Explanation
Interaction and education	Tools to evaluate knowledge (tests, questions, etc.)	0/1	Includes tools for users to check their knowledge during the experience
	Strategies to motivate users (gamification, challenges, etc.)	0/1	Incorporates gamification and challenges in order to enrich user experience
	Links to social media	0/1	Enables users to share information on social media
	Customization of experience (tool to generate and plan itineraries)	0/1	Possesses tool to generate and plan own literary itineraries
	Interaction with local literary groups	0/1	Facilitates contact with literary groups in the destination by means of map.
	Function to report errors or suggest additional information	0/1	Reports errors and suggests improvements and additional information in the map.
	Download function	0/1	Users can download the map to use offline.
	Map rating	0/1	Users can rate the map.
	Interaction and creation of content by users	0/1	Users can leave comments about the literary locations on the map in order to enrich others' experience.

Source: Own creation (2021)

An interesting example that illustrates some of the elements considered by this dimension is the literary app Literapolis, set in Barcelona and developed by Iternatura and the Barcelona City Council (Iternatura, 2021). This app was created as a literary gymkhana, which means that the app connects the different literary locations through challenges. This app offers different routes based on particular literary works. For each book, there are various options considering whether the user wants to discover a book, answer questions about the plot, or solve mysteries regarding the story and its context. Therefore, it permits the user to customize their experience considering what they already know about the book and its author. This app also connects the user with the city libraries or other cultural equipments through the prizes that the user receives after completing the routes. In this sense, this app gives us some ideas to foster the interaction between the user and the literary atmosphere of the city through digital tools.

Related to the possibility of generating routes considering the user needs and motivations, Edinburgh's literary map *Lit Long: Edinburgh* (LitLong, 2021) is an excellent example, which provides the user with an extensive database of literary locations in Edinburgh and allows them to generate and save their paths, as well as to explore the ones created by other users. Furthermore, the Catalan literary map by Espais Escrits (2018) may also be used as a database to develop and build literary routes using the accurate information included in them. Thus, along with the precissenes of the content, it is also crucial that the digital map's platform enables and eases the creation and sharing of those user-created routes.

Usability

High quality usability is essential for online tourism-literary maps. This dimension includes 12 indicators (see Table 5). These include the various kinds of accessibility and the dissemination of contents by means of different resources and concepts (thematic layers, audiences, information hierarchies). There are also technical aspects such as the use of zoom, the incorporation of information in the map and the opportunity to create a printout. And when maps are used on mobile devices, geolocation may be used both for locating users and notifying users of content of interest. Lastly, when user data is used in user profiles or for geolocation, users must be informed about how their data will be used

When analyzing the practical connotations of the identified usability indicators, radically different cases stand out. On the one extreme, Edinburgh's literary interactive map and database of literature is versatile, offering many possibilities for user interaction such as the application of content filters based on interests (filter by work, author, genre or date) or the option to plan itineraries (LitLong, 2021). On the other extreme of ergonomics, Granada's Lorquian Route, which was developed by Granada City Council (Turismo Ciudad de Granada, n.d.), is limited in terms of language accessibility and interaction, as there are no options to activate different thematic layers. In addition, the content presentation is static, and the reference of the complete route is lost when accessing each of the points of interest.

Regardless of their differences, while both examples cover the basic features, none of them addresses the importance of stating physical or cognitive accessibility nor offer content adapted to different target groups. In this sense, considering the relevance of adapting literary digital maps in terms of accessibility should be addressed to promote universal access to the literary landscape of destinations.

Table 5. Proposed indicators to evaluate the functionality of an online literary tourism map from the point of view of its usability.

Dimension	Indicator		Rating	Explanation
Usability	Language accessibility		0/1	Availability in various languages
	Physical accessibility		0/1	Provides information on general accessibility and that of each point on route
	Cognitive, visual and/or auditory accessibility		0/1	Has tools to enable communication with persons who have learning difficulties or sight (e.g. color blindness) and learning impairments (via audio-guides)
	Suitability of content to optimize user experience	Use of thematic layers	0/1	Ability to activate and deactivate layers depending on interests of user
		Suitability to audience	0/1	Offers different content available to different groups (children, youth, adults)
		Information hierarchies	0/1	Different levels of information.
	Zoom and scale		0/1	Scale can be regulated by means of the zoom function
	Presentation of information		0/1	The information is presented on files or windows without the need to go to other websites and risk disconnection from other features
	Geolocation	Device positioning	0/1	The position of the user and points of interest are included on mobile devices.
		Notifications	0/1	Users are notified when they are close to a place of interest on which the map has information
	Available in print		0/1	The map can be printed or there is a paper alternative
	Data protection		0/1	If the map has a profile section or uses geolocation, data protection policies are communicated to users.

Source: Own creation (2021)

Design

The design dimension only has four indicators (see Table 6), but these are essential for assessing the quality of user experience and ensuring the optimization of map as an instrument. The experts suggested the customization of cartographic symbols to match different resources, the display of typical map features, and the strategic use of colors, interchangeable map bases, and fonts to suit the characteristics and themes of the map.

Table 6. Proposed indicators to evaluate the functionality of an online literary tourism map from the point of view of its design

Dimension	Indicator	Rating	Explanation
Design	Associated map symbols	0/1	Includes customized map symbols linked to the overall theme and the type of resources they refer to
	Basic map design elements	0/1	Has all essential map design features: scale, north symbols, reference systems.
	Suitable colors in thematic information	0/1	Suitable colors are used in symbols to help users to distinguish between the various qualitative and quantitative variables on the map.
	Base and typography	0/2	Base and typography style suit the themes presented (1). Base is in line with the needs of the map.

Source: Own creation (2021)

In terms of practical design analysis, an interesting case is the literary map promoted by the city of Barcelona as a UNESCO City of Literature (Barcelona Ciutat de la Literatura UNESCO, n.d.). This literary map stands out for using different cartographic symbols that refer to the wide range of resources it collects, such as theatres, literary murals, tombs, statues or specialized bookstores, among others. This permits the user to identify particular places for its interest linked to the literary heritage. Related to its content, this literary digital map also suggests literary paths to discover the city, broadening the information of each location and explaining, for instance, that Hans Christian Andersen stayed at the Orient Hotel or that George Orwell did the same at the Continental Hotel.

Nevertheless, although it could ease the map's creation process, the use of a generic Google Maps base and standard use of colors, does not serve to distinguish between categories or historical moments. This feature is shared with most of the examples used in the chapter, identifying fields for improvement in the use of suitable colors and, especially, the adaptation of the base and typography style to the topics and historical context.

FUTURE RESEARCH DIRECTIONS

The evaluation tool makes it possible to rate, optimize, and expand digital maps used for literary tourism and it can be used as a guide for the development of new applications.

While the methodology used for validating the dimensions and indicators to be used in the evaluation table served to identify a series of aspects in the use of digital cartography and technology for exploiting literary heritage, the table can either to be used for developing high quality tourist-cultural products or for stimulating fresh research in the field. Below we set out three areas that may prompt further interest.

Different Layers of Meaning

The representation of content in digital literary tourism maps is essential both for enhancing tourists' experience and enabling them to explore the destination. Literary tourism draws on narratives linked to literary works, authors, and their connection to other cultural elements in the destination. Visitors' own prior knowledge that also plays a role in the way they interpret the literary landscape.

The specialists consulted stressed that it is important for literary tourism products to take into account three different narrative levels: the original narrative, linked to the original literary work, the narrative deployed by the tourism manager or promoter, and, lastly, the narrative generated by users themselves. In this respect, the managers start with the original story, but based on the narrative linked to the authors or other cultural elements of the territory, they are able to decide on the content in the map and to develop a new narrative that incorporates several different elements. The managers thus play an important role in disseminating a narrative that matches the cultural resources of the territory to the interests of visitors and is able to prompt them to follow literary routes and put the story, author, or literary movement into context by means of links to explanatory resources. It is therefore necessary to develop an overall vision and skills linked to storytelling.

Dynamic Relation With Users

Users themselves play an increasingly important role in enhancing digital products. Their contribution to content is essential for enriching the product and making users feel more involved in the final result. In this sense, tourism literary products and the resources they offer should not be viewed simply as products directed at tourists but should benefit from interaction between the user and managers. The specialists pointed out that users can communicate with product managers with feedback on content and ratings, and with other users by sharing comments on locations on the map, their motivations and feelings, and by creating routes, etc. This enables users to customize their experience and to feel that they are playing a role in optimizing the tool. This does much to improve the competitiveness of the destination itself.

Moreover, this offers a deliberate and structured means of redistributing demand in the destination. Content can be modular and adapted to meet both the needs of the destination and the interests and curiosity of the visitor. The specialists agree that digital maps have the potential to be versatile and accessible tools that incorporate the various functions needed to enhance local management, visitor experience, and the image of the destination.

Usability and Design: Key Factors for Enhancing Content

Managers are responsible for deciding on the audiovisual resources that affect the perception of their location. Despite the opportunities generated by digital tools, the decision to use them must always be accompanied by an approach that prioritizes simplicity and ease of use, rather than technology for technology's sake. In addition to care over the content and narratives deployed, digital literary tourism maps must ensure that usability and design enhance tourist experience.

To this end, the specialists suggested that digital tools should not necessarily result in scrapping the conventional paper formats altogether; on the contrary, paper and digital cartography should complement each other. This is especially important in rural and ecotourism areas where mobile coverage may be poor.

They also highlighted the importance of information design—the presentation of information in a hierarchical order, with a first overview with titles, and the opportunity to increase the amount of information when deemed necessary by users. This kind of presentation enables the product to cater for and attract a range of users who may be less familiar with digital environments.

The design of the map itself is also important. The Google Maps platform is used as a basemap for the great majority of digital thematic maps, since it is an accessible, agile, and intuitive platform for those responsible for tourist maps. However, this option involves a series of limitations that may affect the final result. Errors in toponymy, outdated information, and low graphic resolution are common problems, while the American product's lack of coverage and detail has repercussions for tourist products. The platform offers more detail in urban and more densely populated areas where there is plenty of topological information; in rural or mountain environments, however, the basemap offers little help in creating maps based on linear itineraries or networks.

In this respect, it is worth highlighting the fact an increasing number of state and regional spatial and geographical institutions have developed specific programs in recent years that make data available to local communities. These may offer a worthwhile alternative. There are also private collaborative platforms that also offer excellent facilities for creating reference maps. These platforms often have better performance, more customization capabilities, and can also be integrated in third party websites. Knowledge of and use of these alternatives can help to increase the potential of digital literary maps.

CONCLUSION

The main objective of this chapter was to create a tool to evaluate literary maps. An evaluation grid consisting of 4 dimensions and 35 items was then created following an analysis of the concept of literary tourism and its associated challenges, a study of existing tourist map evaluation tools, and a discussion of a provisional model with a focus group of experts

Evaluation of the form and content of digital maps for literary tourism requires an analysis of a multidisciplinary nature and was broken down into the four dimensions raised in research by various writers. Dimensions 1 (usability) and 4 (design) were derived from previous work in cartography of a technical nature (Arcila Garrido & López Sanchez, 2011; Tujaka, 2010), whilst dimensions 2 (content) and 3 (interaction and education) drew on studies by researchers in several other areas, including literary tourism (Arcos-Pumarola, 2019; Hoppen et al., 2014; Patricio Mulero & Rius-Ulldemolins, 2017; Squire, 1996; Uccella, 2008), the use of digital aids in educational settings (Martínez Gil et al., 2018), and the use of technology and storytelling in cultural tourism (Imbert-Bouchard Ribera et al., 2013; Marasco, 2020). This work enabled us to provide a sound foundation on which to construct an up-to-date tool to evaluate digital maps to guide and inform tourists in current destinations.

Interdisciplinary research was needed to ensure that the maps are capable of doing more than simply inform tourists (Gamír, 2019) and are well suited to meet the needs of visitors and enhance territorial management (Planells Costa & Crespi Vallbona, 2014). This interdisciplinary work made it possible to address the specific challenges of the specific tourist typology for which the map is created and adapted. This is especially important for literary tourism, a type of tourism linked to an associative cultural landscape with tangible elements that may not be very evident (Donaire, 2008).

As mentioned in the introduction, the aim was then to evaluate existing literary tourism maps in order to identify opportunities for more functions as well as to provide support in developing new products. The indicators contained in the evaluation grid thus related to a series of essential features needed in a digital tourist map.

Nevertheless, those indicators cannot be considered without considering the context where the digital literary tourism maps are developed. That is so because some of the indicators included in the present research require stable and fast connectivity, for instance, to provide access to audiovisual resources or to develop gamified interactions with the user. This could become difficult in territories where access to adequate electronic devices or fast and stable internet connectivity is not widespread.

This also involves rural destinations where internet access is more complex than in urban areas. In those cases, the present indicators should be adapted to create the best experience for the visitor avoiding excessive loading waiting times, inaccessible links, or unsatisfactory interaction proposals. In this sense, a potential future research line could be structured around adapting the current technology to territories with technical restrictions and how to generate satisfactory and memorable experiences with shrank resources.

This chapter provides a multidisciplinary contribution to the use of digital tools in literary tourism and a sound foundation for creating and structuring similar products in the future. In addition to analyzing maps by means of the tool created, future research should also incorporate the views of tourists in order to identify the key elements needed in such maps. Furthermore, future research could also adopt a more propositional character and evaluate the suitability of specific products.

ACKNOWLEDGMENT

This work was supported by the project *Patrimonio Inmaterial y Políticas Culturales: desafíos sociales, políticos y museológicos* (PGC2018-096190-B-I00), funded by the Spanish Ministry of Science, Innovation and Universities / Spanish State Research Agency / European Regional Development Fund.

REFERENCES

Arcila Garrido, M., & López Sanchez, J. A. (2011). La cartografía cultural como instrumento para la planificación y gestión cultural. Una perspectiva geográfica. *Periférica Internacional. Revista Para El Análisis de La Cultura y El Territorio, 1*(12), 15–36. https://revistas.uca.es/index.php/periferica/article/view/1705

Arcos-Pumarola, J. (2019). Assessing Literary Heritage Policies in the Context of Creative Cities. *Journal of Spatial and Organizational Dynamics, 7*(4), 275–290.

Arcos-Pumarola, J., Marzal, E. O., & Llonch-Molina, N. (2018). Literary urban landscape in a sustainable tourism context. *Human Geographies, 12*(2), 175–189. doi:10.5719/hgeo.2018.122.3

Ashworth, G. J. (2009). Do Tourists Destroy the Heritage They Have Come to Experience? *Tourism Recreation Research, 34*(1), 79–83. doi:10.1080/02508281.2009.11081577

Beeton, S. (2005). *Film-induced tourism*. Channel View Publication. doi:10.21832/9781845410162

Beltrán, gersón. (2016). *Geolocalización online: la importancia del dónde*. Editorial UOC. https://www.researchgate.net/publication/293824412_Geolocalizacion_online_la_importancia_del_donde

Brent Ritchie, J. R., Wing Sun Tung, V., & Ritchie, R. J. (2011). Tourism experience management research: Emergence, evolution and future directions. *International Journal of Contemporary Hospitality Management, 23*(4), 419–438. doi:10.1108/09596111111129968

Cabrera, M. J., Arteaga, D. R., Carla, M. S., & Quijano, R. (2009). *Manual para hacer mapas turísticos urbanos, a partir del diagnóstico de áreas gravitacionales y focos urbanos*. Cicyt.

Cartwright, W. (2012). Neocartography: Opportunities, Issues and Prospects. *South African Journal of Geomatics, 1*(1), 77–91. doi:10.4314/SAJG.V1I1

Corral, C. (2009). La ciudad para el peatón: las ciudades históricas en la movilidad urbana. *Ingeniería y Territorio, 86*, 58–67.

Domínguez-Mujica, J. (2007). La cartografía en la promoción turística de Canarias (1880-1970). *Boletín de La Asociación de Geógrafos Españoles*. https://www.researchgate.net/publication/28202137_La_cartografia_en_la_promocion_turistica_de_Canarias_1880-1970

Donaire, J. A. (2008). *Turisme cultural. Entre l'experiència i el ritual*. Edicions Vitel·la.

Dubois, L. E., & Gibbs, C. (2018). Video game–induced tourism: A new frontier for destination marketers. *Tourism Review, 73*(2), 186–198. doi:10.1108/TR-07-2017-0115

Espais Escrits. (2018). *Mapa Literari Català.* Retrieved from: http://www.mapaliterari.cat

Fontal Merillas, O. (2007). El patrimonio cultural del entorno próximo : Un diseño de sensibilización para secundaria. *Enseñanza de Las Ciencias Sociales*, *6*(31), 31–47.

Forsyth, A. (2015). What is a walkable place? The walkability debate in urban design. *URBAN DESIGN International*, *20*(4), 274–292. doi:10.1057/udi.2015.22

Gamír, A. (2019). El giro espacial en las humanidades digitales y sus productos cartográficos. *Biblio 3w. Revista Bibliográfica de Geografía y Ciencias Sociales*, *0*(1), 1–27. doi:10.1344/b3w.0.2019.28482

García-Hernández, M., de la Calle-Vaquero, M., & Yubero, C. (2017). Cultural heritage and urban tourism: Historic city centres under pressure. *Sustainability (Switzerland)*, *9*(8), 1346. doi:10.3390u9081346

Gorrini, A., & Bertini, V. (2018). Walkability assessment and tourism cities: The case of Venice. *International Journal of Tourism Cities*, *4*(3), 355–368. doi:10.1108/IJTC-11-2017-0072

Hoppen, A., Brown, L., & Fyall, A. (2014). Literary tourism: Opportunities and challenges for the marketing and branding of destinations? *Journal of Destination Marketing & Management*, *3*(1), 37–47. doi:10.1016/j.jdmm.2013.12.009

Imbert-Bouchard Ribera, D., Llonch Molina, N., Martín Piñol, C., & Osácar Marzal, E. (2013). Turismo cultural y apps. Un breve panorama de la situación actual. *Heritage & Museography, 0*(13), 44-54–54.

Iternatura. (2021). *Literapolis BCN.* Retrieved from: https://www.iternatura.eu/proyectos/literapolis/

Iwashita, C. (2006). Media representation of the UK as a destination for Japanese tourists: Popular culture and tourism. *Tourist Studies*, *6*(1), 59–77. doi:10.1177/1468797606071477

Jancewicz, K., & Borowicz, D. (2017). Tourist maps – definition, types and contents. *Polish Cartographical Review*, *49*(1), 27–41. doi:10.1515/pcr-2017-0003

Kałamucki, K., & Czerny, A. (2015). Role of Cartographic Products in the Promotion of Tourism Based on the Example of Roztocze. *Barometr Regionalny. Analizy i Prognozy*, *1*(39), 145–151.

Konecny, M. (2011). Review: Cartography: challenges and potential in the virtual geographic environments era. Annals of GIS. *Annals of GIS*, *17*(3), 135–146. doi:10.1080/19475683.2011.602027

Litlong. (2021). *Lit Long: Edinburgh.* Retrieved from: https://litlong.org

MacCannell, D. (2003). *El turista. Una nueva teoría de la clase ociosa.* Melusina.

Marasco, A. (2020). Beyond virtual cultural tourism: History-living experiences with cinematic virtual reality. *Tourism and Heritage Journal*, *2*, 1–16. doi:10.1344/THJ.2020.2.1

Marine-Roig, E. (2015). Identity and authenticity in destination image construction. *Anatolia*, *26*(4), 574–587. doi:10.1080/13032917.2015.1040814

Martínez Gil, T., López Benito, V., Asensio Brouard, M., & Santacana Mestre, J. (2018). ¿Entretener o aprender? La evaluación de las tecnologías móviles en la Didáctica del Patrimonio [Entertainment or learning? Mobile technologies evaluation for heritage education]. *ENSAYOS. Revista de La Facultad de Educación de Albacete*, *33*, 39–52.

McClinchey, K. A. (2015). Travel books, place experience and the 'Je ne sais quoi' of Paris. *Tourism Geographies*, *17*(5), 701–718. doi:10.1080/14616688.2015.1066843

Medynska-Gulij, B. (2003). The effect of cartographic content on tourist map users. *Cartography*, *32*(2), 49–54. doi:10.1080/00690805.2003.9714252

Medynska-Gulij, B. (2008). Point symbols: Investigating principles and originality in cartographic design. *The Cartographic Journal*, *45*(1), 62–67. doi:10.1179/000870408X276602

Molina, G. (2018). Cities literary factory: The mobilization of literature in urban planning in Europe. *Cities (London, England)*, *77*, 4–12. Advance online publication. doi:10.1016/j.cities.2017.08.016

Parramon, P., & Medina, F. X. (2017). ¿Son Las Grandes Producciones Televisivas Un Atractivo Turístico De Base Cultural? Percepciones En Relación Con El Caso De Juego De Tronos En Girona. *International Journal of Scientific Management and Tourism*, *3*, 3–403.

Patricio Mulero, M., & Rius-Ulldemolins, J. (2017). From creative city to generative governance of the cultural policy system? The case of Barcelona's candidature as UNESCO City of Literature. *City. Cultura e Scuola*, *10*(January), 1–10. doi:10.1016/j.ccs.2017.05.001

Pérez Isasi, S. (2018). Hacia un mapa digital de las relaciones literarias ibéricas (1870-1930): Algunas reflexiones teóricas y metodológicas. *Artnodes. Revista de Arte. Ciencia y Tecnología*, *22*, 93–101.

Pérez, Y., & Nello, M. (2008). Los mapas turísticos disponibles a través de Internet. El caso de la Costa Dorada, Tarragona. *Ar@cne: Revista Electrónica de Recursos En Internet Sobre Geografía y Ciencias Sociales*. https://www.raco.cat/index.php/Aracne/article/view/114766/143621

Planells Costa, M., & Crespi Vallbona, M. (2014). *Sistemas y servicios de información turística*. Sintesis.

Quaglieri Domínguez, A., & Paolo Russo, A. (2010). Paisajes urbanos en la época post-turística. Propuesta de un marco analítico. *Scripta Nova, 14*(323).

Ram, Y., & Hall, C. M. (2018). Walking tourism in cities: Introducing the special issue. *International Journal of Tourism Cities*, *4*(3), 281–284. doi:10.1108/IJTC-09-2018-098

Rosselló, V. (2008). *Cartografia, paisatge i territori. Treballs de La Societat Catalana de Geografia*, 35-49.

Sarikanonp, C., & Sahachaisaereep, N. (2010). Graphical design features responding to tourist mapping need: A case of Bangkok's maps for foreign tourists. *Procedia: Social and Behavioral Sciences*, *5*, 1226–1231. doi:10.1016/j.sbspro.2010.07.265

Simancas Cruz, M. (2016). *La planificación y gestión territorial del turismo*. Síntesis. https://dialnet.unirioja.es/servlet/libro?codigo=674345

Squire, S. J. (1996). Literary tourism and sustainable tourism: Promoting "anne of green gables" in prince edward island. *Journal of Sustainable Tourism*, *4*(3), 119–134. doi:10.1080/09669589608667263

Stadler, J. (2015). Conceptualizing and Mapping Geocultural Space. *International Journal of Humanities and Arts Computing*, *9*(2), 133–141. doi:10.3366/ijhac.2015.0145

Tujaka, D. (2010). La e-cartografía turística, una TIC infrautilizada. In J. L. C. H. A. Jesús Guevara Plaza & A. Luis Aguayo Maldonado (Eds.), *Turitec 2010: VIII Congreso Nacional Turismo y Tecnologías de la Información y las Comunicaciones* (pp. 301–316). https://dialnet-unirioja-es.sire.ub.edu/servlet/articulo?codigo=3814712

Turismo Ciudad de Granada. (n.d.) *Ruta de Federico García Lorca*. Retrieved from: http://www.granadatur.com/rutas-tematicas/10-ruta-de-federico-garcia-lorca

Turner, A. J. (2006). Introduction to Neogeography. In A. Turner (Ed.), *OReilly Short Cuts*. OReilly. https://www.oreilly.com/library/view/introduction-to-neogeography/0596529953/index.html

Uccella, F. R. (2008). La patrimonialització literària a Catalunya: Una aproximació. *Mnemosine (Rio de Janeiro)*, *5*, 61–72.

UNESCO. (2005). Convention on the Protection and Promotion of the Diversity of Cultural Expressions. UNESCO. doi:10.1163/ej.9789004164543.1-0.38

UNESCO. (2018). *Basic Texts of the 2003 Convention for the Safeguarding of the Intangible Cultural Heritage*. https://ich.unesco.org

UNESCO Barcelona Ciutat de la Literatura. (n.d.). *Mapa Literari*. Retrieved from: https://ajuntament.barcelona.cat/ciutatdelaliteratura/ca/mapa-literari

van Es, N., & Reijnders, S. (2018). Making sense of capital crime cities: Getting underneath the urban facade on crime-detective fiction tours. *European Journal of Cultural Studies*, *21*(4), 502–520. doi:10.1177/1367549416656855

Wearing, S. L., & Foley, C. (2017). Understanding the tourist experience of cities. *Annals of Tourism Research*, *65*, 97–107. doi:10.1016/j.annals.2017.05.007

ADDITIONAL READING

Baleiro, R., & Quinteiro, S. (2018). *Key concepts in literature and tourism studies*. Universidade de Lisboa. Faculdade de Letras. Centro de Estudos Comparatistas.

Dodds, R., & Joppe, M. (2001). Promoting urban green tourism: The development of the other map of Toronto. *Journal of Vacation Marketing*, *7*(3), 261–267. doi:10.1177/135676670100700306

Leung, L. F., & Li, Z. (2002). Experimental evaluation of the effectiveness of graphic symbols on tourist maps. *Cartography*, *31*(1), 11–20. doi:10.1080/00690805.2002.9714176

Medyńska-Gulij, B. (2010). Map compiling, map reading, and cartographic design in pragmatic pyramid of thematic mapping. *Quaestiones Geographicae*, *29*(1), 57–63. doi:10.2478/v10117-010-0006-5

Uccella, F. R. (2013). *Manual de Patrimonio Literario. Espacios, casas-museo y rutas*. Trea.

KEY TERMS AND DEFINITIONS

Digital Cartography or Mapping: Online cartography that enables interaction with users.

Gamification: Use of techniques associated with games in other contexts and processes.

Geolocation: Capacity of a device connected to a communication network to determine its geographical position.

Intangible Cultural Heritage: Intangible cultural expressions and creations that have emerged from and are recognized by a cultural community.

Literary Heritage: Cultural heritage prompted by a literary work that includes tangible or intangible elements and literary spaces.

Literary Tourism: Tourism mainly motivated by a desire to visit to places linked to literary works or their authors.

Storytelling: Technique employed in the field of education to present a fact or experience by means of a real or fictitious story.

Chapter 10
Perspectives on the Women Literary Travelers:
City and Gender in Mrs. Dalloway and Quarenta Dias

Ana Carolina Mendonça Oliveira
Universidade Federal Fluminense, Brazil

Maria Jaqueline Elicher
Universidade Federal do Estado do Rio de Janeiro, Brazil

Márcia C. Moreira
Universidade Federal do Estado do Rio de Janeiro, Brazil

ABSTRACT

This chapter aims to analyze the novels Mrs, Dalloway (1925) and Quarenta Dias (2014) in the perspective of elucidating the view of the woman writer-character-traveller on the city, showing continuities and ruptures between the modern city and the contemporary city. Therefore, three paths of analysis are proposed: (1) the understanding of urban territories as a way of elaborating subjectivities and experiences; (2) the link between city and memory, place and identity; (3) the link between city and memory, place, identity, and gender. It was possible to verify that both in Mrs. Dalloway and in Forty Days, women have a central role in the construction of narratives about the city and that this is placed in a centrality-character.

INTRODUCTION

In recent decades, there has been a renewed interest in the possible connections between city history and literature, and between city representations and literature. Given that the city encompasses collective memory and the formation of identities, this chapter is interested in establishing a discussion about

DOI: 10.4018/978-1-7998-8262-6.ch010

urban literary narratives, gender, and memory. As stated by Dalcastagnè (2003), city space, today more than ever, is constitutive of characters, whether or not they're nomadic.

The reflection that instigates the authors of the chapter is towards the potential of using literature as a source for urban history, considering the city as its character-genre. This will be undertaken through the presentation of two literary works written by women (a classic and a contemporary) that touch the theme, although temporally and spatially distinct: Mrs. Dalloway, by Virginia Woolf and "Quarenta dias", by Maria Valéria Rezende.

The novel, *Mrs. Dalloway*, by Virginia Woolf (2021 [1925]), was described by Alan Pauls (2017, para. 1) as "a perfect example of modernist fiction". The story takes place on a single summer day in London (England), in 1923. Narrated through the characters' streams of consciousness, the novel invites us to wander through the city streets and the memories and thoughts awakened by observations of urban daily life. The idea of investigating the author's view of the city in this novel comes from the search for a perspective of "travel" and "traveler" from female characters during their brief journeys through the narrated territory.

The novel Quarenta Dias, by Maria Valéria Rezende (2014) is structured in a diary format written by the character Alice, who is the protagonist and narrator. The story begins with Alice, who is a retired teacher, moving the northwest to the south of Brazil, leaving her hometown, João Pessoa to the "strange" Porto Alegre, at her daughter's request. The diary registers, however, only start when she returns home after wandering for forty days in the streets of Porto Alegre searching for information about a missing fellow townsman. Throughout the text, the main interlocution of the character is with a Barbie doll, which is present on the cover of the diary.

Maria Valéria Rezende, born in 1942, is a contemporary brazilian writer, considered an "author of the *in between*, the people in transit, which allows her work to exceed frontiers" (Piaceski, 2019, p. 252). Aside from being a writer, Maria Valéria is also a nun and has dedicated many years of her life, since the 60's, to popular education and to social works that allowed her to travel the world. According to Piaceski (2019, p. 251), Maria Valéria "is a writer without frontiers also in her approach, since she writes for adults and children, poetry as much as prose, addressing themes such as fear, loyalty, social relations, gender violence, illiteracy, slavery and forced labour, money shortage and hungers, between many others".

The literary piece "Quarenta Dias", not yet translated to other languages, was created by Rezende in a mix of fiction and personal experiences from her past and from the 15 days that she lived in Porto Alegre, in 2011, as an invitation of a literary project in which Brazilian novelists would spend a specific amount of time in a different state than the one they lived in, in order to create novels aimed at rediscovering Brazil. The author, then a resident of João Pessoa in the state of Paraíba, travelled to Porto Alegre, in the south of Brazil, the same steps taken by her character, Alice. Other resemblances between author and character's life include the writing process of "Quarenta Dias" and the character's diary in the novel. Maria Valéria compares it:

The very way of composing the text, after getting back from the south, in lapses of written word in notebooks, booklets, emails to myself, fractions of readings and pieces of collected items in these wanderings, created the structure of the novel: a notebook of unburden, composed of scraps of winding and careless writing, in which the character tries to self-explain what has happen to her. (Own translation from Rezende, 2014a, p. 3)

The second writer, Virginia Woolf, was born in London in 1882, lived until 59 years old and is considered by many an 'icon of feminism'. Among other factors, she was worried about women's experiences as writers and as characters in her books. Besides her representativeness to the feminist movement, the author also stands out for her literary experimentation during English modernism. According to Mesquita (2020, p. 374): "Virginia Woolf was born in a conservative Victorian society and helped shape movements that, after the two great wars, modified decisively ways of life and ways of making art, being responsible for ruptures that shaped the world as we know it today."

Her work, "Mrs. Dalloway", alike Maria Valéria Rezende's "Quarenta Dias", is also inspired by various personal experiences, mixing reality and fiction, where "the real is not so much in what happens, but in the intersection between what happens and what each one apprehends, perceives." (Mesquita, 2020, p. 387). Clarissa Dalloway, the leading character, had already appeared in other novels by the author but in this work she becomes one of the leading and primary *flâneuse* and host of London and its urban life. Along the book, it is possible to walk through the internal processes of consciousness and perception of reality in each character. Yet, as pointed out by Mesquita (2020, p. 386-387), in Mrs. Dalloway "memories fight for attention, thoughts come and disappear haphazardly, and what is profoundly significant and what is absolutely trivial are treated with the same importance."

According to Ferraz (2020, p. 11), the work by Virginia Woolf presents a "vastness of interpretative possibilities [...] it is a book that observes subjectivity in its most primary form". While "facts and things in Mrs. Dalloway are nothing. They don't mean to say anything (apart from what the consensus dictates them to say) " (Pauls, 2017, para. 15).

In this duality, in harmonization to Campagnon's (1999) theories for the understanding of literature, and through the fenomenological approach (Gil, 2008) as methodology of choice, this chapter temporarily move away from the authors' intentions and dive into the special qualities of the literary text. Considering the characters (by themselves) as readers of the world (in the presented cases, London and Porto Alegre). And linking city descriptions to the awakening of memories and remnants of the past, and the *being* in the city to its identity and gender.

In both cases, the path undertaken by the characters becomes more important than the start or the end, and the character's perception becomes more important than the descriptions of places. It is important to note that in the two novels, even if they are separated in time for almost a hundred years, there is an effort to justify the presence of the characters in the city, seen as a public space that is still very masculin. In these bricolage cities of many and multiple experiences, the proposal is to analyze the character's paths, for whom walking and being in the city are inseparable parts of their sociability. The women in these novels write and create unique readings of the city, which are analyzed and spatialized throughout the research.

Thus, this chapter aims to analyze the novels Mrs. Dalloway [1925] (2021) and Quarenta Dias (2014), in order to elucidate the view of the woman writer-character-traveler about the city space, showing continuities and ruptures between the modern city and the contemporary city. The analysis starts in three directions: (a) understanding of the city as a character; (b) understanding of urban territories as a way of elaborating subjectivities and experiences; (c) the link between city and memory, place, identity, and gender.

This chapter is organized as following: in the first section, the authors contextualize the discussion with a brief theoretical justification about cities and urban territories as space-characters in literature. After, this chapter present a thematic analysis of the two novels, in three directions: (a) understanding of the city as a character; (b) understanding of urban territories as a way of elaborating subjectivities and experiences; (c) the link between city and memory, place, identity, and gender. The authors conclude the chapter by comparing the two corpus works of reference and then problematizing the before raised questions in the first thematic analysis about the interpretative relations to city, gender, identities, and women's memories whilst writers, characters and travelers. In the end, the authors present their final considerations.

CITIES AND URBAN TERRITORIES AS CHARACTER-SPACE

The city as an entity was understood, throughout the historical process, sometimes as a space of progress, sometimes as a space of disorder, sometimes as a space in opposition to the rural, this carrying the meaning of backward and traditional. Urban life received a differentiated evaluation and also began to be identified as the field for rationality and planning and, simultaneously, as a source of fragmentation and debasement of the individual (Oliveira, 2002).

Secchi (2009) in dealing with "La nueva cuestión urbana" starts from the observation of a double characteristic of the city throughout Western history: meeting place, integration, security, at the same time that it is a place of radical separation, differentiation, marginalization, and exclusion. In contemporary cities, especially in Europe and the United States, the emergence of what Secchi calls a more contrasted social topography, one that reveals inequalities, has been noticeable. For him, the metropolises, despite different temporalities, faced similar problems: spatial injustice, climate change, mobility, and rights to citizenship; and when the structure of the economy and society change, the urban question returns to the foreground. At every moment when the urban issue comes to the front, new themes, subjects, and ideas emerge, a fact that occurs nowadays.

Another of the author's thesis understands space as a social product, modeled by time. This conception of space has an inertia that refers to the economic, social and political dimensions; and changes in such dimensions do not signify a parallel transformation in space. It is this spatial inertia, the result of its modeling by time, that builds the trajectories through which the very changes in economy, politics and society can occur. This thesis aims precisely at rising the need to think that spatial policies and the management of the city and territory have repercussions on the very relations between the rich and the poor (and why not, on differences in gender and ways of occupying and taking care of the city), whether they are relations of integration or exclusion. For Secchi (2009), the city is also a space for the sedimentation of signs over time, and space, therefore, constitutes the indelible mark in the territory.

In regard to the city as a space-character in literature, in proximity to the literary geography thought out by Collot (2012), in a way, it shies away from its referential context in the concrete world and assumes a rather imaginative and subjective dimension. Literature transforms real territories "to build its own space, which is the imaginary and the written, that doesn't exist beyond the text, and that can't be transferred to any map of the known world." (Collot, 2012, p. 22). As much as there are recurrent tries to approach more closely and as real as literary representations can be, the space-character city is an image of the world, symbolically influenced by the author's point of view, inseparable from the characters that live in it, and by the reader's interpretations. According to Caldeira (2007, p. 152), "places gain identities through who is in them", and, henceforth, the city-character shapes its own identity, profoundly related to the values, imagination and perceptions of its characters, authors and readers.

In the impossibility of being the subject, the speechless character spokes through other's discourse, though the developments of the narrator's discourse. [...] the city's truth is a discursive construction made by the multiplicity of voices that circulate though space, including those of the anonymous spectators [readers] that are amused by the display of routine city life. (Own translation from Caldeira, 2007, p. 148)

Another important element concerns the issue of gender in urban literature. Literary criticism of the relationship between gender and urban space in the contemporary narrative of female authorship, nowadays, implicitly exalts and questions the assumptions of a still predominantly white, male and Eurocentric canon. The authors aim to investigate the critical content of female characters' representations in urban spaces, in the examples here treated, in which protagonists move through cities while negotiating their presence on the streets and in public spaces.

In this sense, Miranda (2013) points out that the global context of displacements, mobilities and territorial conflicts suggests the problematization of the notion of place, where gender intersects with other axes of difference (Kaplan, 1996) and produces the critical content of women characters' representations in urban spaces, in the path for deconstruction or maintenance of patriarchal oppression.

The narratives proposed here also contribute to the development of a new concept of space, more dynamic, heterogeneous and inclusive, in light of Massey's (2009) ideas, who defends the importance of problematizing the spatial dimension for the understanding of inequalities, which are also constituted spatially. The city as a social space presupposes inequalities that are, at the same time, intrinsic to it; like other social markers, gender and city are closely linked.

According to Morettoni (2020) the 19th and the 20th centuries represent a moment of great contributions to social theory and also pose as an intense stage for cultural, social, political, and economical changes, marked by progress, facination and fear, most of all in regard to the comprehension of the transition from urban classic to contemporary.

The definition of modernity or of that which is ancient/classic and modern have been discussed throughout decades. Modernity "is an exclusively western concept and doesn't appear in any other civilization" (Paz, 1984, p. 43). In the author's words: "no society or time has ever called itself modern – except ours" (1984, p. 39). According to Paz (1984), the one who first proposed a critic to modernity was Rousseau, as was him who found the adequate social space to reflect upon it, living in a time of constant transformation and, above all, of exponential urban growth. "In his work, the age which begins-the age of progress, of inventions, and of the development of urban economy – finds not only one of its basis, but its most furious denial as well" (Paz, 1984, p. 53).

Calling an age "modern" is something contradictory, because how will then be called the modern society of the future? Therefore, through all times, the classic division between antiquity and modernity is recovered (Pagotto, 2010). What Paz calls modernity, hence, only has ground because it is a critic of Cristian eternity.

The history of modernity – essential to the comprehension of current modernity and to the construction of the future one – can be divided in three phases (Berman, 1986 apud Pagotto, 2010). The initial phase concerns the period between the 16th and the end of the 18th centuries and the one who at first represents modernity is, as also understands Paz, Rousseau: the first to ever use the word *moderniste* with the meaning that it came to have in the 19th and 20th centuries. Rousseau is the foundation of modernity not only because he first used the word, but also because he manifested sensibility to the social transformations that began to affect the life of thousands of people; was he who visualized the social 'whirlwind' in which cities had transformed (Pagotto, 2010).

Among the classic thinkers that saw the rise and consolidation of modernity is Simmel (2005) [1903], who though about the development of consciousness and the appearance of the individual in big cities. In terms of the investigations of the modern way of life and the imperatives of life in big cities, Simmel (1973) [1903] when analysing the "metropolitan mental life" proposes a path in which he points out the particularities of routine life in major cities and its impacts in human life, from the analysis of how personalities self-adapt to the externalities of big cities, besides the psychological conditions that these generate. For Simmel, the focal point of modern life comes from the tension between interior and exterior, and from the existing conflict between the individual and society. In contrast to rural life, which presents a slow rhythm of sensorial stimuli and psychological life based on the emotional aspect, metropolises demand that its habitants have a sensibility and a psychological life that is capable of adjusting to the vicissitudes of speed and to the heterogeneity of stimuli that these present (Simmel, 1973 [1903], pp. 11-12).

Other of the author's analytic perspective regards the implications of the fact that big cities are home to monetary and economic production. Simmel argues that an intellectualized person becomes indifferent to all individuality. In the same amount, monetary economy and money make all individuals into an element in itself indifferent, reducing all qualitative aspects to a number, which can be measured. Form this analysis, Simmel (1973 [1903], p. 14) tries to demonstrate that modern life became more and more calculating, primarily due to the predominance of money, with its numeric determinations, its reduction of qualitative and quantitative values, its measures and calculus. Simmel argues that life conditions in big cities are simultaneously cause and effect of that calculating preciseness, which's necessity is created by the clustering of a great number of people with different interests, but that must integrate their intentions in a single highly complex organism that gathers all relations and activities in a unique detached and stable calendar (Simmel, 1973 [1903], p. 15). Time and space's rational management would be, therefore, a fundamental factor to the flow of life in metropolis and this process has constituted a central element in Simmel's theory of modernity.

The metropolis extracts from the man, as a creature who proceeds discriminations, a specific quantity of consciousness different from that which is extracted by rural life. In this, life's rhythm and the gathering of sensorial mental images flow more slowly, in a more habitual and cohesive way. It is precisely in this connection where the sophisticated character of metropolitan psychological life becomes comprehensible – in opposition to life in small towns, which rests over more profound and heartfelt relationships. (Own translation from Simmel, 1973 [1903], p.12)

Although the entreprising character of the big city (the classic one - London) and the contemporary one (Porto Alegre) is reinforced, in urban spaces they can be reinforced as a space for the production of identities and literary narratives under the eyes of the feminine, such as they are in the works analyzed in this chapter, as will be presented in the following.

A WOMAN AT MRS. DALLOWAY'S LONDON

In Mrs. Dalloway, novel written by Virginia Woolf and published in 1925, it is possible to experience, through the streams of consciousness belonging to different characters presented in the narrative, a summer day in the 1920s London. More specifically: a Wednesday in June 1923; walking through London's centre urban and modern life; represented/translated by the multiple and subjective looks of Clarissa, Lucrezia, Elizabeth and many other female characters in the novel (and also male characters, who this chapter will, however, not address).

The book's plot has two main cores that intertwine throughout the narrative: Mrs. Dalloway (Clarissa), her family and friends; and the couple Lucrezia (Rezia) and Septimus Warren-Smith. The story begins with Clarissa leaving in the morning to buy flowers for her party that will take place in that same night. On the way, she meets an old friend in the park and continues her path wandering through memories as she crosses Bond Street. Meanwhile, the Warren-Smith couple heads to Regent's Park for a walk until it's time for a doctor's appointment. The couple also wanders through memories of the past while experiencing the morning routine and common life in the park. Other secondary characters, antagonists and mere extras are introduced and intertwine the narrative of an apparent common day in the English capital that culminates, at the end of the day, in a tragedy for the Warren-Smith couple and for Clarissa Dalloway's party as well, both in their respective residences.

The city of London is both landscape and protagonist. "[...] a changing, dizzying ecosystem, which acquires a certain hallucinatory quality due to the force of its modernity, and which Woolf presents with the precision and detailing of a cartography maniac" (Pauls, 2017, para. 13). Each of the characters in the novel interacts and perceives the city according to their social, cultural, and emotional contexts. An interweaving of poetic and symbolic relationships through streets, parks, shops, traffic, passersby, events, and the beats of Big Ben ("The leaden circles dissolve in the air").

It is not an abstract or mythical city; no stylization serves to make it beautiful. The topography is meticulous and is designated by names and surnames, as if Woolf, as she unfolds the city before us, dreamed that someday, who knows, someone would walk through the pages of her novel. (Own translation from Pauls, 2017, para. 13).

Clarissa Dalloway is a 52-year-old woman from London's upper class, resident of Westminster and wife of a conservative politician, regarded as the perfect hostess, sometimes charming and vivacious, sometimes superficial and too preoccupied with what others think. Right in the first pages of the novel, already manifests enjoying walking around London, loves London life as she is captivated by the climate and small observations of everyday urban life.

In people's eyes, in the swing, tramp, and trudge; in the bellow and the uproar; the carriages, motor cars, omnibuses, vans, sandwich men shuffling and swinging; brass bands; barrel organs; in the triumph and the jingle and the strange high singing of some airplane overhead was what she loved; life; London; this moment of June. (Woolf, 2021 [1925], p. 2)

There is a strong relationship between time (months and seasons) and the landscape, between people, especially women, and their territories. Places in Mrs. Dalloway are perceived as active subjects transforming themselves as they are observed:

June had drawn out every leaf on the trees. The mothers of Pimlico gave suck to their young. Messages were passing from the Fleet to the Admiralty. Arlington Street and Piccadilly seemed to chafe the very air in the Park and lift its leaves hotly, brilliantly, on waves of that divine vitality which Clarissa loved. To dance, to ride, she had adored all that. (Woolf, 2021 [1925], p. 4).

Crossing Victoria Street, St. James's and Green Park, watching the buses on Piccadily, walking, gazing in the Hatchard's (bookstore) window, going up Bond Street to the flower shop. Fresh morning air; the bustle of the streets; the silence, mist and murmur of animals in the park; the sweet aroma of the earthy garden in the flower shop. Common pathways that rescue memories of the past and multiple sensations of belonging, individuality, and self-knowledge.

She felt very young; at the same time unspeakably aged. She sliced like a knife through everything; at the same time was outside, looking on. She had a perpetual sense, as she watched the taxi cabs, of being out, out, far out to sea and alone; she always had the feeling that it was very, very dangerous to live even one day. Not that she thought herself clever, or much out of the ordinary. (Woolf, 2021 [1925], p. 5)

It is also important to highlight that Mrs. Dalloway felt part of the city as if she were everywhere. She felt an affinity with beings, things, and people with whom she had never exchanged words or touches: "[...] some woman in the street, some man behind a counter – even trees, or barns" (Woolf, 2021 [1925], p. 130). Clarissa even had a theory that to know someone deeply it was necessary to know the places and people that complete them and to which they feel they belong. A transcendental theory of life and death, appearances and ephemerality.

Did it matter then, she asked herself, walking towards Bond Street, did it matter that she must cease completely; all this must go on without her; did she resent it; or did she not become consoling to believe that death ended absolutely? but that somehow in the streets of London, on the ebb and flow of things, here, there, she survived, Peter survived, lived in each other, she being part, she was positive, of the trees at home; there, ugly, rambling all to bits and pieces as it was; part of people she had never met; being laid out like a mist between the people she knew best, who lifted her on their branches as she had seen the trees lift the mist, but it spread ever so far, her life, herself. (Woolf, 2021 [1925], p. 6).

On the other hand, Lucrezia Warren Smith, - "a little woman, with large eyes in a shallow pointed face; an Italian girl" (Woolf, 2021 [1925], p. 11), young, simple, and impulsive who left Italy at twenty-four years old to live in England with her husband, an English soldier, who was then experiencing a "post-traumatic stress disorder" - has a completely different perception of the city of London. Longing for her homeland, and feelings of loneliness and sadness overwhelm her throughout the narrative, making her feel strange about the new place.

Far was Italy and the white houses and the room where her sister sat making hats, and the crowded streets every evening with people walking, laughing out loud, not half alive like people here, huddled up in Bath chairs, looking at a few ugly flowers stuck in pots!. (Woolf, 2021 [1925], p. 18)

Elizabeth, daughter of Mrs. Dalloway, also felt a similar strangeness for the city, due to her preference for being alone in the countryside or outdoors without being noticed, rather than at parties, amidst the agitation and adulation of people, as prefers her mother:

And Elizabeth waited in Victoria Street for an omnibus. It was so nice to be out of doors. She thought perhaps she need not go home just yet. It was so nice to be out in the air. So she would get on to an omnibus. And already, even as she stood there, in her very well-cut clothes, it was beginning ... People were beginning to compare her to poplar trees, early dawn, hyacinths, fawns, running water, and garden lilies; and it made her life a burden to her, for she so much preferred being left alone to do what she liked in the country, but they would compare her to lilies, and she had to go to parties, and London was so dreary compared with being alone in the country with her father and the dogs. (Woolf, 2021 [1925], p. 114)

Unlike the walk taken by her mother at the beginning of the novel, along a familiar path, one of those in which the feet already follow on their own and there is a mixture of freedom and security through the familiarity of everyday life; Elizabeth, on a passage, adventures herself on a bus trip through the city, aimlessly, resembling an inconsequential pirate and an innocent child at the same time. "Buses swooped, settled, were off – garish caravans, glistening with red and yellow varnish. But which should she get on to? She had no preferences. Of course, she would not push her way. She inclined to be passive" (Woolf, 2021 [1925], p. 114).

Being a woman in the city, and walking alone, aimlessly through unknown streets, is dealing with a mix of sensations such as insecurity and courage. As can be seen in the following excerpt:

She [Elizabeth] looked up Fleet Street. She Walked Just A Little Way Towards St. Paul's, Shyly, Like Some One Penetration On Tiptoe, Exploring Strange House By Night With A Candle, On The East Edge, The Owner Should Suddenly Fling Wide His Bedroom Door And Ask Her Business, Nor Did She dare wander off into queer alleys, tempting bystreets, any more than in a strange house open doors which might be bedroom doors, or sitting-room doors, or lead straight to the larder. For on Dalloway's came down the Strand daily; she was a pioneer, a stray, venturing, trusting. [...] But it was later than she thought. Her mother would not like her to be wandering off alone like this. She turned back down the Strand. (Woolf, 2021 [1925], pp. 116-117)

From these passages of "Mrs. Dalloway" by Virginia Woolf, it is possible to understand that the city of London is one of the main characters in the novel, constant element in the narrative, sometimes mergeing with the other characters, sometimes linking them in time and space by a ringing of bells, a corner that they cross, or the same plane that cruises the city sky and everyone watches it at different times and places. Urban territories in Mrs. Dalloway, such as squares, parks and street corners, express meeting, passing and leisure spaces. The city, even with its explicit and intrinsic social inequalities, promotes the meeting between women both rich and poor, young and old in the same spaces, shared by them even within different social and cultural contexts.

The female look over the city evokes Secchi's (2009) city, as a place of encounter, integration and belonging (similar to Clarissa Dalloway's point of view), but also of strangeness, solitude, separation and differentiation (similar to Lucrezia Warren-Smith's point of view). And more, it evokes the modern city comprehension as understood by Simmel (1973) [1903] when highlighting (through Elizabeth Dalloway's point of view) the tension between the individual and society, between rural and urban life, and between pretended freedom and invisibility when strolling freely in the city in comparison to an overbooked agenda of social events and appointments encouraged by London's urban society.

London city is henceforth perceived as a space and a character in "Mrs. Dalloway", with its explicit and intrinsic social inequalities, which at the same time promote encounters between rich and poor, old and young women, in various social and cultural contexts.

Even if the authors refer to the modern city of London, a big city already resembling a metropolis, under the female eyes it presents all traces of contemporaneity described by Simmel (1973) [1903]. The metropolis that demands sensibility and psychological conditions capable of rapidly adjusting to the vicissitudes and heterogeneity of routine stimuli. In this sense, comparatively, next session will henceforth meet the reality of Alice (protagonist from Quarenta Dias), habitant of a great contemporary city, in another time, in another social reality.

A WOMAN AT QUARENTA DIAS'S PORTO ALEGRE

The novel Quarenta Dias, by Maria Valéria Rezende (2014) is structured in a diary format belonging to the character Alice, the story's protagonist and narrator. The writtings registration occur when she returns home after wandering for forty days on the streets of Porto Alegre. Throughout the text, the main interlocution of the character is made with a Barbie doll, which is present on the cover of the diary. There is also an obvious reference to Lewis Carroll's Alice in Wonderland, the story of a girl who shrinks in size and enters an unknown world. It is an excellent metaphor for Alice from Porto Alegre, who sees her life turn upside down upon moving to the city.

It is noticeable how the author, Maria Valéria Rezende, slept on the streets of Porto Alegre for fifteen days in the process of writing the book, which is part of a project where several writers were invited to learn and write about different capitals in Brazil. Maria Valéria's experience was decisive for the series of references to urban landmarks in the city of Porto Alegre, with which the character interacts throughout the story.

The plot begins with Alice moving from João Pessoa to Porto Alegre as a request from her daughter who wanted to get pregnant and have the help of her mother. However, right from the start, there is a disagreement between the two, as the daughter decides to continue her studies in Europe. This is the trigger that drives Alice to the streets, wandering aimlessly for forty days, under the pretext of finding Cícero Araújo - the son of a friend and her countryman - who is missing. Nonetheless, this is almost an alibi for leaving the house and going out into the street, as it is obvious that there is no chance of meeting Cicero. During this time, she is confronted with the city itself and various characters marked by social and urban exclusion. More than that, Alice is crossed by questions about her own life and her feminine condition.

Other researchers have already pointed out that Alice's behavior would be that of a *flâneusis*, a resignification of the meaning of *flânerie*, since the *flâneur* has always been a typically masculine character in modernity (Wood, 2012; Stoll, 2020). During her time on the streets of Porto Alegre, Alice startes to collect little objects to tell her story, which she keeps in her diary. Given the above, the question is: would she be a contemporary *flâneusis*? Daniela Stoll's article (2020) is dedicated to trying to answer this question, but the clues are in Alice's account:

Fingia que dormia, as pálpebras entreabertas, mas espiava tudo e todos os que cabiam no meu raio de visão, rasteiro, até sentir que era hora de me mexer de novo, andar, continuar para qualquer direção, achar ou inventar novas pegadas de Cícero Araújo. (Rezende, 2014b, p. 165).

Translation by the authors of the chapter:

I pretended to be asleep, my eyelids half-open, but I watched everything and everyone that could be in my sight, crawling, until I felt it was time to move again, walk, continue in any direction, find or invent new footprints of Cicero Araújo (Rezende, 2014b, p. 165).

In another fragment she describes her path in suburban Porto Alegre as follows:

Já então eu sabia que andava por andar, por vilas, ruelas, becos e acampamentos de operários, ao léu, a qualquer lugar a que me conduzisse o ônibus que calhasse, dormitando embalada pelo sacolejo do veículo ou mirando vagamente a paisagem incógnita, as mesmas casinhas com grades e jardinzinhos, as mesmas, sempre as mesmas avenidas exageradamente largas pra meu olhar paraibano, me perguntando, aqui e ali, o que era 'brique', o que era isso ou aquilo que eu via escrito em placas por toda parte sem descobrir o significado, por que será que tem tanta casa funerária nesta cidade?, andava só pra não voltar, eu, rebelde peão de xadrez a correr atrás de um peão de obra imaginário, a ouvir histórias de gente quase reduzida a corpo e dor, quase. (Rezende, 2014b, p. 218)

Translation by the authors of the chapter:

I knew then that I was walking just for the feel of it, through villages, alleys, and workers' camps, at random, to any place where the bus would take me, sleeping rocked by the jolt of the vehicle or vaguely looking at the unknown landscape, the same little houses with bars and little gardens, the same, always the same exaggeratedly wide avenues for my eyes from Paraíba, wondering, here and there, what was 'brique', what was this or what I saw written on signs everywhere without discovering the meaning, why is there so many funeral homes in this city? I walked just so I wouldn't return, me, a rebellious chessman running after an imaginary workman, listening to stories from people almost reduced to body and pain, almost (Rezende, 2014b, p. 218)

Like a *flâneur*, Alice's diary records the character's experience of anonymously observing the city of Porto Alegre, as the fragment records:

Ninguém reparava em mim, talvez efeito dos meus cabelos que teimo em deixar grisalhos apesar da incansável insistência da Elizete. Credo, Alice, que desleixo!, nem parece que você é uma mulher inteligente e estudada, acha certo parecer uma velha bem antes mesmo de entrar nos sessenta? (Rezende, 2014b, p. 99)

Translation by the authors of the chapter:

Nobody noticed me, maybe the effect of my hair that I insist on leaving gray despite Elizete's tireless insistence. Jeez, Alice, how sloppy! It doesn't even seem like you're an intelligent and studied woman, do you think it's right to look like an old woman well before you're even in your sixties? (Rezende, 2014b, p. 99).

It is important to appreciate that the invisibility is initially due to her advanced age and, also in a secondary moment, when Alice starts to dress as Lola, a homeless woman with whom Alice builds a friendship. The following fragment highlights this moment:

Vesti as três camisas, uma por cima da outra, e amarrei na cintura o casaco de lã ruço, cheio de bolinhas, acabando de conquistar a total invisibilidade que eu desejava. Segui, assim, sentindo-me então mais confortável pra meter-me nas pequenas brechas da cidade. (Rezende, 2014b, p. 219)

Translation by the authors of the chapter:

I put on the three shirts, one over the other, and tied my brown wool coat, full of polka dots, around my waist, having just achieved the total invisibility that I craved. As so I went on, feeling more comfortable to get into the small loopholes in the city (Rezende, 2014b, p. 219)

On the streets, women are scrutinized by their clothes to determine the social class to which they belong. In this sense, women are objects of analysis and observation as much as the city. However, in Quarenta Dias (2014), the protagonist joins a huge contingent of people who are invisible to the city because of their marginalized condition.

Pra onde ir?, por enquanto, pra lugar nenhum, continuar escondida ali, invisível entre os invisíveis com suas garrafas térmicas e suas cuias de chimarrão, espiando, por todo o tempo que eu quisesse, aquele pedaço de mundo no qual tudo que a cidade quer esconder abre-se como um abscesso supurado. (Rezende, 2014b, p. 150)

Translation by the authors of the chapter:

Where to go? For now, nowhere, to remain hidden there, invisible among the invisible with their thermos bottles and their chimarrão gourds, spying, for as long as I wanted, that piece of the world in which everything the city wants to hide opens up like a suppurating abscess (Rezende, 2014b, p. 150)

This is exactly why Alice feels that she has penetrated the "back of the city", where marginalized identities of Porto Alegre live:

Continuei por semanas minha romaria pelo avesso da cidade, explorando livremente todas as brechas, quase invisíveis pra quem vive na superfície, pra cá e pra lá, às vezes à tona e de novo pro fundo, rodoviária, vilas, sebos e briques, alojamentos, pronto-socorro, portas de igrejas, de terreiros de candomblé, procurando meus iguais, por baixo dos viadutos, das pontes do arroio Dilúvio, nas madrugadas, sobrevivente, sesteando nas praças e jardins, debaixo dos arcos e marquises, sob as cobertas das paradas de ônibus desertas, vendo o mundo de baixo pra cima, dos passantes apenas os pés. (Rezende, 2014b, p. 235)

Translation by the authors of the chapter:

I continued my pilgrimage for weeks to the back of the city, freely exploring all the gaps, almost invisible to those who live on the surface, back and forth, sometimes to the surface and again in the background, bus stations, villages, secondhand bookstores and briquets, lodgings, emergency rooms, doors of churches, candomblé terreiros, looking for my equals, under the viaducts, the bridges of the Dilúvio stream, at dawn, survivor, slumbering in the squares and gardens, under the arches and marquises, un-

der the roofs of the deserted bus stops, seeing the world from the bottom up, only the feet of passers-by (Rezende, 2014b, p. 235)

Inés Madariaga (2004) performs an urban analysis from a gender perspective that echoes Alice's observations that the streets are named only after men: "Nem sei mais quantas vezes levei ao Borges, ao Bento, ao Protásio, ao Nilo, ao Osvaldo a minha desaparência" (Rezende, 2014b, p. 235). Translation by the authors of the chapter: "I don't even know how many times I took my disappearance to the neighbourhoods of *Borges, Bento, Protásio, Nilo, Osvaldo*." (Rezende, 2014b, p. 235). Since, even though women have also made history, politics and literature, they are not honored on plaques and monuments in the same proportion. A similar observation is made by Massey (2009), author of *Space, Place, and Gender*. The author highlights that:

...the intricacy and profundity of the connection of space and place with gender and the construction of gender relations. Some of this connection works through the actual construction of, on the one hand, real-world geographies and, on the other, the cultural specificity of definitions of gender. Geography matters to the construction of gender, and the fact of geographical variation in gender relations, for instance, is a significant element in the production and reproduction of both imaginative geographies and uneven development. (Massey, 2009, p. 2)

Thus, invisibility allows Alice to subversively penetrate the gaps and reveal the invisible histories and humanities of Porto Alegre. The character's transformation is evident when, at the end of the novel, she is no longer looking for Cícero Araújo, nor does she refuse the identity of a person living on the streets. In her words: "andar com Lola dava-me direitos de cidadania pelas ruas, assimilavam-me como uma a mais entre eles" (Rezende, 2014b, p. 237). Translation by the authors of the chapter: "Walking with Lola gave me citizenship rights in the streets, they assimilated me as one more among them." (Rezende, 2014b, p. 237).

Back at the apartment, Alice describes the feeling as follows:

"Saí andando, pensando em tudo o que ainda preciso escrever pra não sentir mais aquele frio na barriga, aquele aperreio que me dá quando me vejo de novo na rua, como se ela me agarrasse e não me quisesse mais largar, arrastando-me, rua-rio de novo" (Rezende, 2014b, p. 65)

Translation by the authors of the chapter:

I walked away, thinking about everything I still need to write so that I don't feel that butterflies in my stomach anymore, that distress I get when I find myself on the street again, as if it grabbed me and didn't want to let go, dragging me, street-river again (Rezende, 2014b, p. 65)

The experience resembles Walter Benjamin's (1994 [1989]) evaluation when pointing out that the *flâneur* is attracted by the urban masses. After all, as she describes it, it is on the streets where Alice identifies and recognizes herself, even though she is so far from her homeland.

Estranhezas, eu disse? Engraçado é que eu tinha a impressão de, afinal, quase nada ver de tão estranho assim, neste Sul tão longe de casa, o povo misturado de todas as cores, os petiscos de pobre, aquele tanto de negros gaúchos que eu nunca soube que existiam, violência e solidariedade, pobreza e necessidades, iguais às da minha terra, a pedir milagres. (Rezende, 2014b, p. 120).

Translation by the authors of the chapter:

Strangeness, did I say? It's funny that I had the impression that, after all, I saw almost nothing that was so strange, in this south so far from home, the mixed people of all colors, the poor people's snacks, that many number of black gauchos that I never knew existed, violence and solidarity, poverty and needs, equal to those of my land, asking for miracles (Rezende, 2014b, p. 120).

After all, the distant city of Porto Alegre was also a welcoming place for so many and diverse Alices who left their marks on the city. In fact, the metropolis is not only the setting for the novel, but it is a prominent and inseparable character from the other characters.

It is worth remembering once again that Simmel (1973) [1903], who affirms that the metropolis extracts from men (and also demands of him), as a creature who proceeds discriminations, a different quantity of consciousness, accelerated, fast, which leads the character to identify between their new acquaintances (characters from the streets of Porto Alegre), characteristics close to those of their fellow townsman, characters from other cities, of their hometown. These are urban characters, similar, homogenized, simply because they are habitants of cities.

THE CITY, GENDER, IDENTITIES AND MEMORY OF THE WRITER-CHARACTER-TRAVELER

The potential of a reading of a city based on her novels resides not only as a reporting force but rather as a gap in the homogeneous cementation of life and the city. Both Mrs. Dalloway [1925] (2021) and Quarenta Dias (2014) fulfill this role, with the addition of a focus on their female character's movement through the city.

Although the two novels take place in distant cities and temporalities, the challenge of being a woman moving through the city and the unknown is full of similarities. In this sense, it is useful to resort to the male figure of the *flâneur* and its female equivalent the *flâneusis*, which by definition is a "determined, resourceful individual keenly attuned to the creative potential of the city and the liberating possibilities of a good walk." Elkin (2016, p.8). Virginia Woolf called it "street haunting".

It is important to remember that the figure of the *flâneur* is recurrent in Baudelaire's work and Walter Benjamin's analyses. He is a traditionally masculine character of modernity, he is the figure of the guy who wanders the streets unhurriedly, looking, seeing, observing. Thus, the emergence of the *flâneur* was linked to the emergence of urbanism, with the recent agglomerations of people and their respective diversities.

In this sense, the ideals of modernity in search for freedom, equality and fraternity did not apply to everyone. The differences of class, race and gender were also forms of regulation in the urban space. In this particular research, the authors are interested in observing the regulation of women in the city in both Virginia Woolf's and Maria Valéria's novels.

Virginia Woolf herself had already called attention to the domestic experience of nineteenth-century writers, very different from the experience of men, who were always able to circulate freely in public spaces. For her, women, in addition to being discouraged from writing and not having their own money to support themselves when they dared to write, their stories were reduced because they were fundamentally domestic perspectives (Stoll, 2020). In Baudelaire's work, for example, Wolff (1985) highlights that the women who appear are a prostitute, a widow, an old woman, a lesbian, a murder victim and an unknown woman passing by who are target to the *flâneur*'s gaze.

According to Elkin (2016), the argument against the existence of *flâneusis* precisely concerns visibility, as the *flâneur* must be invisible in the crowd, to have its freedom to come and go, as well as to observe anonymously. This is a recurrent point in the work of Maria Valéria (2014), however, as observed by Pollock (1988), nineteenth-century women did not enjoy the freedom of being anonymous in the crowd.

For Elkin (2016), narratives focused on walking along the city's sidewalks repeatedly leave women out of the experience, so she assesses that:

The female flânerie - the flâneuserie - not only changes the way we move through space but intervenes in the organization of space itself. We claim our right to disturb the peace, to observe, to occupy (or not occupy) and organize (or disorganize) space on our own terms. (p. 196).

Elkin (2016) assesses that *flâneusis* can be:

She may be a writer, or she may be an artist, or she may be a secretary or an au pair. She may be unemployable. She may be a wife or a mother, or she may be totally free. She may take the bus or the train when she's tired. But mostly, she goes on foot. She gets to know the city by wandering its streets, investigating its dark corners, peering behind facades, penetrating into secret courtyards. [...]. She voyages out and goes where she's not supposed to; she forces us to confront the ways in which words like home and belonging are used against women. She is a determined, resourceful individual keenly attuned to the creative potential of the city and the liberating possibilities of a good walk" (p. 18)

In reality, it means moving women from the place of object to the subjects of gaze. There is yet another clear opposition between the flânuer's practically unrestricted mobility potential, versus the *flâneuse's* controlled and/or restricted displacement. If, on the one hand, *flâneuses* don't get the *flâneurs'* privilege of being and anonymous and alert, on the other, they are part of years demanding visibility to the female body in space, in the same towns where historically *flâneurs* circulated as their observation stages.

FINAL CONSIDERATIONS

In both cases, the path becomes more important than the starting or ending points and perception becomes more important than the description of places. It is important to note that in both novels there is an effort to justify the characters' presences in the city, seen as a public space that is still very much masculine, even with thought more than ninety years separate the two books.

The women in the analyzed novels, as they walked through the cities they write and create unique views that reinforce their role as subjects, showing a look over the cities that exists not only with recording strength, but also with the narrative that intertwines and solidifies the character's everyday lifes and city life. Both Mrs. Dalloway [1925] (2021) and Quarenta Dias (2014) fulfill this role, with the addition of bringing an outline that highlights the city as a space-receptacle of contrasts, aches, pains and also of encounter, integration, security (Secchi, 2009), at the same time that it is a place of separation, differentiation, marginalization and exclusion, all this from the perspective of the female gaze. It is precisely through female characters' looks, readings and perceptions, during their walks through town, that these place's roles as city-characters is reinforced.

Implications and particularities of women's displacement through urban public spaces in the two novels shows how pivotal are gender issues to literary analysis. It's imperative that you don't lose sight of the historically built standard of representation and perpetuation of the dominat discourse that reaffirms preconceived notions of women's role in society, as well as in terms of women's literary production.

Therefore, feminine's representation in both novels allows us to comprehend social space's pressure over women and their identity construction. This per passes, as demonstrated in the novels fragments, from their dressing codes, to behavior, displacements through town, indicative of sociocultural contexts subjected to the influence of big cities' vicissitudes.

Gender issues in literature appears as a perspective, henceforth, for new studies, through thematic analysis of other author's literary pieces form different times and nationalities. As well as interdisciplinary analysis between women traveler's literature and literary tourism and geography. Between the most significant limitations the authors highlight a stronger possibility for further development of literary analysis as comparative literary research and research methods in the field of literary studies. Methodological aspects that can also be included in studies that approach this theme.

In general, before the presented issues, it is possible to observe that the here appointed spaces under the female look, show in a significative manner that the configuration of cities and social relations there stablished can contribute para to the definition of female's roles and the characterization of subjects in general. Hence, the city and the displacements are concrete elements for the fictional being's construction and of what means to be a woman in a city. Common paths, reported in the novels by their protagonists, retrieve memories and sensations of belonging, individuality and self-knowledge.

Another perspective achieved with this chapter points towards the fact that there is still resistance to gender studies in literature as well as in academic environments in general, and that this is still considered a "specific study", perhaps to designate it as smaller knowledge, smaller literature. However, the novels here analyzed have shown how powerful women's strength is in producing eloquent and categorically feminine narratives of city space that they cohabit and record.

REFERENCES

Benjamin, W. (1994). O Flâneur [The Flaneur] (J. C. M. Barbosa & H. A. Baptista, Trans.). In Obras escolhidas III (pp. 33-65). Brasiliense.

Caldeira, S. P. (2007). A cidade-personagem: Pina Bausch e Ítalo Calvino [The city-character: Pina Bausch and Italo Calvino]. *Revista de Ciências Humanas*, *7*(2), 147-162. https://periodicos.ufv.br/RCH/article/view/3536

Collot, M., & Alves, I. (2012). Rumo a uma geografia literária [Pour une géographie littéraire] (I. Alves, Trans.). *Gragoatá*, *17*(33), 17–31. doi:10.22409/gragoata.v17i33.33006

Compagnon, A. (1999). *O demônio da teoria: literatura e senso comum* [Le démon de la Théorie: littérature et sens commun] (C. P. B. Mourão, Trans.). Ed. UFMG.

Dalcastagnè, R. (2003). Sombras da cidade: O espaço na narrativa brasileira contemporânea [City shadows: space in contemporary Brazilian narrative]. *Estudos de Literatura Brasileira Contemporânea*, (21), 33–53. https://periodicos.unb.br/index.php/estudos/article/view/8928

de Miranda, A. C. (2013). *Pensar o local: gênero e espaço urbano na narrativa brasileira contemporânea* [Doctoral Thesis, Universidade de Brasília]. https://repositorio.unb.br/bitstream/10482/12962/1/2013_AdelaideCalhmanMiranda.pdf

Elkin, L. (2016). *Flâneuse: women walk the city in Paris*. Farrar, Straus and Giroux.

Ferraz, M. (2020). Apresentação. In Mrs. Dalloway (pp. 11-13). Antofágica.

Gil, A. C. (2008). *Métodos e Técnicas de Pesquisa Social* [Methods and techniques of social research] (6th ed.). Editora Atlas.

Kaplan, C. (1996). *Questions of Travel: Postmodern Discourses of Displacement*. Duke University press.

Madariaga, I. (2004). *Urbanismo con perspectiva de género* [Urbanism with a gender perspective]. Instituto Andaluz de La Mujer. https://www.juntadeandalucia.es/institutodelamujer/institutodelamujer/ugen/sites/default/files/documentos/98.pdf

Massey, D. (2009). *Space, place and gender*. University of Minnesota Press.

Mesquita, A. C. (2020). Toda a vida em um só dia [All life in one day]. In Mrs. Dalloway (pp. 373-387). Antofágica.

Morettoni, M. M. (2020). A modernidade como destino e como questão: A cidade em Max Weber e Georg Simmel [Modernity as a destination and as a question: The city in Max Weber and Georg Simmel]. *Revista Ensaios*, *16*, 156–175. doi:10.22409/revistaensaios.v16.47766

Oliveira, L. L. (2002). Apresentação. In L. L. Oliveira (Org.), *Cidade: história e desafios* (pp 8-13). Editora FGV. http://cpdoc.fgv.br/producao_intelectual/arq/1264.pdf

Pagotto, A. M. de C. (2010). *A modernidade no ensaísmo político de Octávio Paz* [Masters Dissertation, Universidade Estadual Paulista]. http://hdl.handle.net/11449/93187

Pauls, A. (2017). Prefácio: O fio tênue da ficção [Preface: The fine thread of fiction]. In Mrs. Dalloway. Companhia das Letras.

Paz, O. (1984). *Os Filhos do Barro: do Romantismo à Vanguarda* (O. Savary, Trans.). Nova Fronteira.

Piaceski, D. P. F. (2019). Maria Valéria Rezende: Colorindo invisíveis por meio da literatura [Maria Valéria Rezende: coloring the invisible through literature]. *Revista Crioula*, *24*(24), 250–267. doi:10.11606/issn.1981-7169.crioula.2019.160624

Pollock, G. (1988). *Vision and difference: femininity and the histories of art*. Routledge.

Rezende, M. V. (2014a). Bastidores: Há 10 dias uma conhecida ligou no meio da noite [Backstage: 10 days ago an acquaintance called in the middle of the night]. *Pernambuco*, (102), 3. http://www.suplementopernambuco.com.br/images/pdf/PE_102_web.pdf

Rezende, M. V. (2014b). *Quarenta dias* [Forty days]. Objetiva.

Rodrigues, M. F. (Interviewer) & Rezende, M. V. (Interviewer). (2014, May 2). *Maria Valéria Rezende viveu na rua para escrever romance* [Maria Valéria Rezende lived on the street to write a novel]. Estadão. https://cultura.estadao.com.br/noticias/geral,maria-valeria-rezende-viveu-na-rua-para-escrever-romance,1161541

Secchi, B. (2009). *A cidade do século vinte* [The twentieth century city]. Perspectiva.

Simmell, G. (1973). [1903]. A metrópole e a vida mental. In *O fenômeno urbano* (pp. 10–24). Zahar Editores.

Simmell, G. (2005). A grande cidade e a vida do espírito (L. Waizbort, Trans.). *Mana*, *11*(2), 577–591. doi:10.1590/S0104-93132005000200010

Stoll, D. S. (2020). A flânerie de uma andarilha urbana [The Flânerie of an Urban Walker]. *Revista Estudos Feministas*, *28*(2), e57230. doi:10.1590/1806-9584-2020v28n157230

Wolff, J. (1985). The invisible flâneuse: Women and the literature of modernity. *Theory, Culture & Society*, *2*(3), 34–50. doi:10.1177/0263276485002003005

Wood, J. (2012). *Como funciona a ficção* [How Fiction Works] (D. Bottmann, Trans.). Cosac Naify.

Woolf, V. (2021). Mrs. Dalloway. Penguin Books. (Original publication 1925)

Chapter 11
Tourist Literature and the Architecture of Travel in Olga Tokarczuk and Patti Smith

Rita Baleiro
https://orcid.org/0000-0002-3188-5150
University of the Algarve, Portugal & CiTUR, Portugal

ABSTRACT

This chapter analyses two travel narratives within the scope of literature and tourism studies, aiming to explore the motivations to undertake journeying and the experience of (literary) pilgrims. The first is the novel Flights, by Olga Tokarczuk (2007), and the second is "How the Mind Works," by Patti Smith (2017). This chapter defines the umbrella concept of "tourist literature" and takes a cross-disciplinary perspective combining the hermeneutics process with findings from the literature review on tourism studies. The analysis of Flights reveals the touring identity and experience of a pilgrim and reflections about airports, travel guides, tourists, and their syndromes. The analysis of Patti Smith's short story uncovers the touring identity and experience of a literary pilgrim who is strongly motivated to undertake literary-inspired trips towards the authors' places.

INTRODUCTION

Over the last two decades, travel texts (*i.e.*, magazines and guidebooks) and travel stories (*e.g.*, Paul Theroux's and Sara Wheeler's) have been subjects of tourism analysis (Dann, 1999; McWha et al., 2015; Thompson, 2011), under the belief they may strongly influence those who travel and play a valuable role for tourism researchers, providing relevant insights into the touring experience (Robinson, 2004, p. 312). Trusting that the analysis and interpretation of literary texts can also have a profound influence on travellers (Onfray, 2007) and contribute with information about the travelling experience and its motivations (Hendrix, 2014), this chapter focuses on two narratives – tourist literature texts – with two main objectives. The first is to examine the narrators' touring identities and their construction of the travel experience. The second is to provide insights about (literary) pilgrims' motivations and experience,

DOI: 10.4018/978-1-7998-8262-6.ch011

the act of (literary) travel and some elements of the travelling experience (*e.g.*, airports, travel guides, tourists' syndromes, literary places). As this chapter is written within the cross-disciplinary study area of literature and tourism, the narrative hermeneutics is complemented with findings from a literature review on tourism studies.

Literature and tourism studies focus on the intersections between literature and tourism, and their starting point is considered to be the mid-1980s after the publication of Richard Butler's "Literature as an influence in shaping the image of tourist destinations" (1986). Since then, several researchers have explored the various links between literature and tourism by investigating tourist literature texts (Fonseca, 2020; Rodrigues, 2020), combining literary analysis and the concept of literary tourism (Alves, 2016; Oliveira, 2016) and designing literary trails (Almeida, 2016; Machado et al., 2020). Researchers have also studied literary places (Jiang & Xu, 2017), the construct of authenticity (Earl, 2008), the potential of literary tourism for local development (Hoppen et al., 2014) and literary tourism products and experiences (MacLeod, 2020). However, recent studies confirm that there are still scarcely researched topics (Çevik, 2020), namely the experience and motivation to go on literary travels (Macleod, 2020).

Thus, this chapter seeks to address this gap in the research by analysing and interpreting two travel narratives. In other words, two tourist literature texts. *Tourist literature* is an umbrella concept that refers to the *corpus* of literature and tourism studies. As such, the first section of this chapter focuses on presenting its definition. The two following sections showcase the analysis and interpretation of the selected literary texts, that share distinctive attributes of tourist literature. The first is *Flights*, by Olga Tokarczuk (2007), the winner of the Nobel Prize for Literature in 2018. Originally written in Polish, the book bears the title *Bieguni*, a noun that refers to an obscure and possibly fictional itinerant Slavic sect, who survived on the kindness of strangers and shared the belief that being in constant motion was the strategy to avoid evil (Kassabova, 2017). This novel merges a collection of voices and stories across time and space, and it portrays the experience of a pilgrim, her motivations and those she encounters, in a constant praise for "the crossing of boundaries as a form of life." (Charles, 2019). The second text, "How the mind works", by Patti Smith (2017), is a narrative which depicts a literary pilgrim's travel experience in France and England, investigating the minds, workspaces and gravesites of writers, searching for the places where she may emotionally encounter her favourite authors. The final section summarises the analysis of the two literary texts and highlights similarities and differences in the narrators' touring identities, experiences and motivations.

THE CORPUS OF LITERATURE AND TOURISM STUDIES: TOURIST LITERATURE

Within the process of identifying and selecting the *corpus* of literature and tourism studies, one main concept stands out: tourist literature. This refers to the literary texts involved in the "production of tourist space through travel narratives" (Robinson, 2004, p. 304), in the sense that they enclose spatial representations that can add value to places as tourist attractions. This definition reflected the perspective of cultural geographers. In 2014, the concept was further expanded by Harald Hendrix (pp. 19-29), who coined the term "tourist literature". According to this researcher's view, tourist literature describes the literary texts that can resignify space and foster literary tourism, motivating readers to visit places and landscapes associated with either the presence or memory of literary authors or with literary fiction representations (Hendrix, 2014, p. 22). Well-known examples of such literary fiction are the Brontë

Sisters' novels, which have created tourist spaces and have strongly influenced tourism patterns (Robinson & Andersen, 2002).

In 2018, Rita Baleiro and Sílvia Quinteiro elaborated on this concept by adding to the category of *tourist literature* the literary texts in which (literary) tourism practices, manifested in pilgrimage, travel and sightseeing, are the object of literary representation and, as such, their analysis may offer relevant insights into understanding the tourism experience, touring identities and the tourism industry. In their view, tourist literature may include travel guides authored by literary writers, literary fiction, memoirs, autobiographies, biographies, poems, and travel literature texts. Departing from a less formalist stance, these researchers follow a historical and communicative view of literary genres and subgenres, which declines the existence of frontiers that imprison texts in genres/subgenres, and, in line with Cohen (1986, p. 207), they believe that the genre of a text is defined by its interaction with other texts, that is, by difference and contrast as well as by the historical context of its reception, rather than by the existence of shared constellations of textual features. Thus, the authors propose the existence of *tourist literature* as a virtual literary genre, in the sense that it is up to the reader to identify the features that make some texts fitting for an analysis within the scope of literature and tourism studies (Baleiro & Quinteiro, 2018, pp. 19-26). In this type of research, one of the main challenges is to find literary texts whose semantic effects bear an analysis in the scope of tourism and literature studies. To that end, researchers must constantly browse the editorial market for literary texts, whose characters and plot promote a reflection about the act of (literary) travelling and the tourism industry.

In sum, in this chapter, *tourist literature* is defined as an umbrella concept that refers to a heterogeneous cluster of texts suitable for cross-disciplinary hermeneutics in the scope of literature and tourism studies. As such, tourist literature texts (i) can produce literary and cultural landscapes that motivate journeys to the in-between real and imagined locations and, hence, the creation of tourist trails; (ii) portray touring identities (*e.g.*, tourists, travellers and pilgrims), travel practices and experiences, and (iii) stimulate reflecting on those identities, practices and experiences in general and in the specific context of literary tourism.

The two texts under analysis in this chapter are *tourist literature*, but also travel literature texts. Contrary to tourist literature, *travel literature* is a literary genre with a long tradition: writing about the experience of travel is as old as the act of travelling itself, as there has always been a deep and persistent wish to communicate impressions, encounters and emotions (Robinson, 2004, p. 305). As such, Tim Youngs defines travel literature as "the most socially important of all literary genres. It records our temporal and spatial progress. It throws light on how we define ourselves and on how we identify others" (2013, p. 1).

The core of travel literature is a (real and unreal) journey, in which there is a "negotiation between self and other that is brought about by movement in space" (Thompson, 2011, p. 9). Due to its multiple motivations, travel experiences and encounters that result in a variety of literary styles and techniques, travel literature aggregates several subgenres (*e.g.*, autobiography, memoir, ethnographic commentary, scientific report). Consequently, Mike Robinson (2004, p. 307), one of the pioneers in literature and tourism studies, states that, rather than identifying and describing the features of travel literature, it is more beneficial to see it as a metaphor of travel itself. In other words, an autonomous group of texts whose ultimate force and energy results from the "disjuncture between 'being' home and 'being' away", and from the "experience of the unfamiliar or, often, the familiar in other locations." (Robinson, 2004, p. 311).

Therefore, instead of detailing taxonomy challenges in Olga Tokarczuk's and Patti Smith's texts, this study approaches the texts as "metaphors of travel", in line with Mike Robinson (2004). Metaphors of travel in which the act of travelling is much more than moving from place A to place B; metaphors of travel that reveal diverse touring identities, practices, motivations and experiences through the voices of two travelling narrators.

The following two sections focus on analysing the literary texts: first, Olga Tokarczuk's *Flights* and then Patti Smith's "How the mind works".

"Real Life Takes Places in Movement": Flights, by Olga Tokarczuk

Flights is a "constellation novel", in the author's own words (*in* Croft, 2019), a fictional fragmented travel narrative consisting of 116 texts. Some are two sentences long, and others are thirty-odd pages. Some are essayistic reflections about Wikipedia or maps; others are short-fictionalised histories, as they include real-life historical figures, such as the seventeenth-century Dutch anatomist Peter Verheyen, who kept his own amputated and preserved leg to which he writes letters; and still, others are fictional stories, like the one about a man on a Croatian island seeking out his wife and son, who have disappeared without a trace. The narrator portrays several centuries and locations, and "it is not always clear what is fiction and what is fact." (Bye, 2018). But more relevant to the analysis is the fact that the novel is a composition of multiple interweaved texts, in which journey, mobility, wandering and time are the main elements, and the act of travelling constitutes the core motif and the pilgrim's object of praise.

Pilgrims and Tourists: Motivations and Experience

In *Flights,* the narrator is a female solo traveller who defines herself as a "pilgrim" "in pieces, broken down", "invisible, see-through" who is "able to move around like a ghost, look over people's shoulders, listen in on their arguments and watch them sleep with their heads on their backpacks or talking to themselves, unaware of [her] presence, moving just their lips, forming words that [she] will soon pronounce for them." (2007/2017, p. 25).[1] These other "pilgrims", as she defines them, are the aim of her journeys (pp. 129, 404), with the intent of "trailing the errors and blunders of creation." (p. 24). She is interested in "anything that deviates from the norm" as "only what is different will survive", and she declines "the patterns so scrutinised by statistics that everyone celebrates with a familiar, satisfied smile on their faces." (pp. 23, 24, 23). The parallelism between religious pilgrims and tourists was initially drawn by MacCannell, in the belief that the pilgrims' desire to visit a religiously meaningful destination is similar to the tourists' attraction to destinations embedded with social, cultural and historical meaning (1976, pp. 42-43).

Tokarczuk's pilgrim-narrator does not long for visiting a specific religious site. Instead, she longs for a kind of spiritual experience, a desire to be part of a community of pilgrims such as her, of people who share her devotion and reverence for the very own act of travelling, as the only possible condition of life and the only chance of achieving salvation: "Real life takes place in movement", and "[...] a thing in motion will always be better than a thing at rest; [...] change will always be a nobler thing than permanence; [...] that which is static will degenerate and decay, turn to ash, while that which is in motion is able to last for all eternity." (pp. 61, 9). Her motivation is to permanently move, meet other pilgrims, learn from them, observe and celebrate the experience of fluidity, mobility and illusion (pp. 25, 58, 129).

The trigger for this impulse seems to be the motivation the narrator also finds in other pilgrims: the wish to reach the coordinates where the "right time and place" perfectly meet. Getting to the right place at the right time "would mean the code to the safe had been cracked, the combination revealed, the truth exposed." (p. 85). At that point, one will find "great love, happiness, a winning lottery ticket or the revelation of the mystery everyone's been killing themselves over in vain for all these years, or death." (p. 85). Only such findings can justify why people keep "moving around in a chaotic fashion" (p. 85): they hope to increase the chances of achieving the meaning of life.

The motivation for achieving a meaningful life propels pilgrims to undertake ongoing journeys and opposes them to "barbarians": "Barbarians don't travel. They simply go to destinations or conduct raids." (p. 58). In the narrator's view, tourists are "barbarians" and she repeatedly distances herself from that status. Contrary to tourists, her purpose is to experience the journey and the other pilgrims, her aim is neither the destination nor the voracity to photograph, buy and collect places and landscapes, as most tourists do (see Urbain, 1991). She wishes to be "the perfect observer" (p. 184), her "gaze" not constrained nor biased as that of tourists, who are often considerably limited by their imperfect knowledge and their preconceptions to comprehend and interpret what they see.

The notion of "gaze" was developed by Urry (1990, 2002) to conceptualise the experience of tourists and their relationships with people and places they encounter. The constraints shaping the gaze are multiple and derive from several attributes (socio-cultural, economic, environmental, class, gender, religion and age, see Urry, 2002; Urry & Larsen, 2011), but also by tourists' motivations and expectations. This is a complex concept, and most scholars, including Urry (1990, 2002), acknowledge the impossibility of a unified tourist gaze. In this chapter, this concept serves to interpret the narrator's gaze and motivation to embrace the new and strange phenomena she encounters whilst travelling, something that, in her view, most tourists do not do. According to Tokarczuk, tourists tend to develop a sense of superiority or repudiation when they encounter strange phenomena at their destination. That is why they are at their most comfortable when they return home. Adopting an "anti-tourist" identity, the narrator states, "they [the tourists] looked without seeing, their gazes sliding over everything, alighting only on whatever their mass-produced guidebooks pointed out to them specifically – the print equivalent of a McDonald's." (p. 381).

Jean Didier Urbain, the French sociologist and tourism specialist, also draws upon the metaphor of tourists as hordes of barbarians. In *L'Idiot du voyage*, Urbain describes tourists as cannibalistic tides devouring, colonising and murdering spaces and traditions (1991, p. 15). Although somehow toned down, Urbain's binary opposition between tourists and travellers is likewise present in Tokarczuk's narrative, namely in what refers to motivations and experience. In contrast to the narrator's touring identity, tourists prefer the pleasures of circular time and undertake circular journeys motivated by a wish to leave "in order to return [home]" (p.12, see also page 58). They are neither "real travellers" nor "pilgrims", the narrator states, as they are "relieved when they [get] back, with a sense of having fulfilled an obligation" (p. 12), eager to show others the photographs "safely tucked in plastic folders, like evidence in a detective's cabinet – evidence that they had been there." (p. 28). Just like those who prefer to stay home, tourists are dormant, willing to lead a settled life without the "energy [that] derives from movement – from the shuddering of buses, the rumble of planes, trains' and ferries' rocking." (p. 13). In contrast, as stated above, the narrator – as well as other pilgrims (and travellers) – thrive on movement and on the effects of conducting a linear journey, in a linear time, in which "every moment is unique", and life is lived "to the fullest" even if that means taking risks (p. 59). Pilgrims and travellers share this motivation and belief, but there is a distinctive trait between them: pilgrims embrace a sense of devotion to the act of travelling, an almost religious feeling towards the act of being constantly on the move.

The textual fragments that compose the novel create multiple effects. On the one hand, they hinder the reader's work, who initially struggles (at times with a sense of frustration) to perceive the unity among all the fragments and construct meaning. The obstacles the reader encounters can be interpreted as a metaphor for what travellers go through, as travelling can be like walking in the labyrinths of a strange, unknown language, where one needs to be willing to change places. On the other hand, the collection of short texts in *Flights* is also responsible for creating a reading rhythm similar to the rhythm of travel, as the reader moves from fragment to fragment (sometimes a short story, an essay, an ethnographic text, an aphorism, a dictionary entry), each setting its own particular pace. Just like it happens on a journey when "every traveller's time is a lot of times in one, quite a wide array. It is island time [...]; it is the time produced by the clocks in train stations. [...]" (p. 59).

The themes of fragmentation, journeying, wandering and time are the threads creating unity in *Flights*, from the narrator's liminal position between home and away, which places her in a privileged stance to assess critical elements of the tourism industry: airports, tourists' syndromes and travel guidebooks.

Airports, Tourists' Syndromes and Travel Guidebooks

The narrator describes airports as plenty more than travel hubs; they are "city-states" and "airport-republics" "with a stable location [and] citizens in flux." (p. 62). An airport is "an example of an extroverted system" where internal politics matter less than the ties with other airport republics, and in them, the "constitution is spelled out on every ticket" and "one's boarding pass is one's only identification as a citizen." (p. 62). Airports are portrayed as focal points of experience-oriented, individualised and experimental forms of service:

They hold conference centres, interesting art exhibits, festivals and product launches. They have gardens and promenades; they instruct: at Amsterdam's Schiphol you can see excellent copies of Rembrandt, and there is an airport in Asia that has a museum of religion – a fabulous idea. We have access to good hotels and wide variety of restaurants and bars from inside airports. There are little shops and supermarkets and shopping malls where you can not only stock up on provisions for the road, but also on souvenirs, in advance, so as not to waste any time once you get where you are going. There are gyms, places that offer both traditional and Eastern massage, hair stylists and customer service representatives from banks and mobile phone companies. And after satisfying the needs of our bodies, we can move on to spiritual succour at the numerous chapels and meditation spaces at airports. Sometimes they host readings and book signings for travellers. (pp. 61-62)

In Tokarczuk's novel, airports constitute a critical place for individuals, where they can learn from communicating and interacting with others.

To address the narrator's conceptualisation of airport and representation of passengers' spatial experience, the author of the chapter draws on Marc Augé's "opposed polarities" of place and non-place (1995, p. 64). According to the French anthropologist, in "supermodern" contemporaneity, the world has changed, and its best expression is constituted by non-places (1995, p. 69). He specifically refers to the human-made structures built for transport and consumption (*e.g.*, railways, shopping centres and airports), stating they impact the way individuals experience time, place and self. In his view, such structures are non-places, meaning they are anonymous, antiseptic, functional, transit environments in which a sense of group and society is lost under the objectives of functionality (1995, p. 86). This idea of loss is not portrayed in Tokarczuk's narrative, where airports – as examples of structures built for transport and consumption – are represented as places where individuals see their needs being met, and, albeit temporarily, they become citizens, share a constitution and a sense of belonging, as the narrator states. In them, the narrator acknowledges the heterogeneity of characters who experience and produce such environment: the "vociferous" men who travel in a group, wearing bright scarves, celebrating their sports team, and a "very young couple […] standing locked in an embrace […]." (pp. 80-81), among many other characters. Contrary to Augé, and in line with other tourism researchers, in *Flights* airports are depicted as places where "new forms of 'meetingness' and 'dwelling' [and] various sets of social relations are afforded possibilities for development" (Elliot & Radford, 2015, p. 1065). They are structures where passengers can do more than kill time.

However, "not all passengers in terminal are mobile in quite the same way" as Tim Cresswell states (2006, p. 222), and the narrator in *Flights* acknowledges this, describing the passengers who show "exhaustion and disorientation […] while wandering around the airport shops" or in "between gates" (p. 81) and those who do not belong to the privileged flows of passengers, such as "a man, poorly dressed, all rumpled […]." (p. 84).

Apart from meeting other pilgrims aiming to identify and understand the odd and the different, in an effort to fight the "terribly dangerous" assumption that humans are constant beings whose reactions can be predicted (p. 21), the narrator's motivation to travel is also to heal: "The chronicles of my travel would in fact be chronicles of an ailment." (p. 21). As the narrator acknowledges, for her, travelling is also the symptom of a syndrome, the "Recurrent Detoxification Syndrome", which describes the need to compulsively search "all things spoiled, flawed, defective, broken." (p. 22). Thus, the narrator appears to be as psychologically fragmented as the narrative structure.

While detailing her thoughts about guidebooks, the narrator refers to another tourist syndrome: the Paris Syndrome, which "is characterised by shock and by a number of physical symptoms like shortness of breath, heart palpitations, sweating, and arousal." (pp. 75-76). It often affects Japanese tourists, and it derives from a discrepancy between the tourists' expectations and the reality of the destination, which often bears no resemblance to the description in guidebooks. This might explain why the narrator despises guidebooks and chooses to travel with books: two books "[...] written with real passion, and a genuine desire to portray the world." (p. 76): Chmielowski's *New Athens* and Melville's *Moby Dick*. They are her pilgrimage companions.

In her opinion, guidebooks not only age the quickest but, most importantly, they are "a veritable scourge, an epidemic" responsible for ruining "the greater part of the planet" (p. 75). Guidebooks should also be accountable for damaging tourists' expectations, as they encourage them to travel only to discover that what they find does not meet what they expect. Additionally, in her view, "describing something is like using it – it destroys; the colours wear off, the corners lose their definition, and in the end what's been described begins to fade, to disappear." (p. 75). Travel guidebooks do exactly that: they destroy by describing and consequently "they have debilitated places, pinning them down and naming them, blurring their contours." (p. 75).

Towards the end of the narrative, the narrator-pilgrim shares an accomplishment: she has achieved what she considers to be the optimal condition of travelling: she does not know where she is anymore, which, according to her, is one of the most critical accomplishments in mobility. She is just there ("I'm here", p. 403). She has "progressed" – her actual word – and achieved what "travel psychology refers to as the key phase, the crowning phase […], whatever your destination might be, you are always heading in that direction." (pp. 402-403): life is happening in motion.

The very last sentence of the novel binds together the themes of travel, wandering, fragmentation and time: the narrator has boarded an aeroplane, once more, alone, since a "good pilgrim" always travels solo

(p. 171), and she is watching the flight attendants whose smiles are "a kind of promise that perhaps we will be born anew now, *this time in the right time and the right place*" (p. 410, our italics).

In *Flights*, the architecture of travel uncovers the ongoing journey as the only possible life condition. The novel unveils the touring identity of a pilgrim travelling on her own, moving fluidly, with the motivation to fight decay and stagnation. In her view, nothing should be set in stone, not even opinions; they are "excess baggage" that convey the "terribly dangerous" assumption "that we are constant, and that our reactions can be predicted." (p. 21).

In the following narrative, the narrator is also a pilgrim, but, contrary to *Flights*' narrator, she cherishes her "baggage", namely her "literary baggage", which is paramount to her travelling experience, whose main motivation is to visit the places of her favourite authors. In Patty Smith's short story, the reader finds a literary pilgrim touring identity and literature itself is the structural motif of the journey. Her "baggage" defines her literary knowledgeable gaze, which allows the narrator to identify literary places, even those which are not signalled with tangible icons.

"I Linger, Content to Be With the Ghosts of Writers Who Have Passed Into This Same Perimeter" – "How the Mind Works", by Patti Smith

"How the mind works" is the first part of a book entitled *Devotion* by Patti Smith, published in 2017. Like Tokarczuk's narrative, it is a tourist literature text, unlike it however, it is not fictionalised. It is, instead, a literary travel narrative that portrays the experience of a narrator on a journey across France and England seeking the places of authors, to pay homage to their creative minds and to understand how they how their mind work, what compels them to create literary art, where inspiration comes from and how "the burning imagination – at once unholy and divine" lifts and guides the pen (Smith, 2017).[2] The text is a combination of diary, inner reflections and observations, and it portrays the travel experience of a literary pilgrim. In fact, the hermeneutic process of the narrative reveals significant insights into the author's travel experience and motivations.

Literary Pilgrims and the Ability to Read the Destination Without a Map

Within the scope of tourism and literature studies, the touring identity of Patti Smith's narrator is that of a literary pilgrim, as she travels to France and England in search for the places of her favourite authors, primarily literary writers, but also other artists, such as painters and film directors. Artistic or literary pilgrimages are not a recent phenomenon and "they can be understood as a form of homage, a paying of tribute to works of fiction and writers within landscapes or settings they have made famous." (Stiebel, 2007, p. 93). Literary pilgrims are the first category of literary travellers (Herbert, 1996, p. 84); individuals who travel with the motivation to "celebrate the creativity of authors in places where their genius is believed to be located, first at their graves, but then also in the places where they were born and lived." (Hendrix, 2014, p. 27). In comparison to other travellers, pilgrims share a stronger sense of devotion towards the authors they admire. Besides, as David Herbert describes, literary pilgrims are often "well-educated tourists [...] with the cultural capital to appreciate and understand this form of heritage" (2001, p. 313; see also MacLeod et al. 2018, p. 388, and Watson, 2009, p. 3). Contrary to most tourists, the pilgrims' knowledge of authors' texts and biographies enables them to "read" the space (read is the actual verb Smith's narrator uses). The following quote illustrates that ability, as the narrator describes how she wondered through Paris literary places "without a map" (Smith, 2017), an instrument that would be most helpful for those, whose knowledge of literature and authors' biography would not be enough to read the city's literary landscape:

Walking down the narrow Rue du Dragon, old Sépulcre Street, which once boasted an imposing stone dragon. No. 30 a plaque in memory of Victor Hugo. Rue de l'Abbaye. Rue Christine. No. 7, Rue des Grands Augustins, where Picasso painted Guernica. These streets are a poem waiting to be hatched—suddenly it's Easter; eggs everywhere. I walk aimlessly, finding myself in the Latin Quarter, then cut over to Boulevard Saint-Michel, searching for number 37, where Simone was raised and the Weil family lived for decades. I have a flash of Patrick Modiano tracing one address after another, crisscrossing the entire city in search of a certain stairway. I think of Albert Camus, about to receive the Nobel Prize, making this same pilgrimage to the Weil residence, but for graver reasons—not mere curiosity but contemplation. (Smith, 2017)

In this quote, we learn the narrator's pilgrimage to Paris was twofold: on the one hand, she was travelling to visit Simone Weil's home; on the other hand, she was completing a pilgrimage previously undertaken by another author she also admires – Albert Camus's pilgrimage to Simone Weil's house. In other words, she was travelling towards the venerated author, following in the footsteps of a second author she greatly admires. It is clear that her journey map is no standard tourist map; her map is her literary baggage that carries information about the writers' biographies, the places they lived in, where they worked, and even where they are buried. Literary pilgrims' literary and cultural capital is often proportional to their devotion to authors or texts, and that affects the travelling experience.

Just like religious pilgrimages, secular literary pilgrimages "are tied to place in the same way as religious ones are — the abstract reasons for making the journey find concrete expression in reaching a specific site; which is imbued with semiotic and symbolic significance" (Stiebel, 2007, p. 94). The aim is to reach a literary place: a literary meaningful fraction of space.

Smith's narrator is able to identify the literary markers and hence find the literary places. Contrary to most tourists who need the tourism industry to signal places of interest for them, literary pilgrims are capable of "recognising" them (again, this is the actual verb the narrator uses) without the tourism industry's input. Their act of producing and recognising literary places stems from their literary knowledge. This is also to say that meaning is not inherent to the places: it is the outcome of a complex hermeneutics and perception process (Zhao et al., 2013) in which the tourists' literary repertoire, imagination and emotions actively intervene.

Urry stated that the tourists' gaze is "constructed through signs" (1990, p. 3), and, in the case of literary pilgrims, their gaze is assembled by their literary repertoire: the invisible baggage they carry on their journey. Along with her invisible baggage, she carries the books of the authors she admires, as these act as "a docent of sorts, setting a tone or even altering the course of a journey" (Smith, 2017). Indeed, the narrator's knowledge of literature and writers enables her to identify literary places, such as the Café Deux Magots, the Hôtel des Étrangers and the Hôtel de Lauzun:

The Deux Magots of the existentialists. The Hôtel des Étrangers, where Rimbaud and Verlaine presided over the Circle Zutique. The Hôtel de Lauzun with its chimeras and gilded halls where Baudelaire smoked hashish and penned the opening poems in Les fleurs du mal. The interior of our imaginations glowed, as we walked back and forth before these places synonymous with poets. Just to be near where they had written, sparred, and slept. (Smith, 2017).

Her interpretation of these places is inevitably tainted by her knowledge of literary references, which enables her to *produce* these literary places by outlining their boundaries, assigning them meaning and (re)organising the landscape of the city of Paris.

In the previous quote, another important trait of the literary pilgrim stands out: imagination (Magadán Díaz & Rivas García, 2012, p. 177). Tourism experience "involves not only tangible objects and tangible sights, but also emotion triggered by the senses or the mind" (Jiang & Yu, 2020, p.130). In the experience of literary tourism, imagination occupies centre stage in the co-creation of literary places (Brown, 2016; MacLeod et al., 2018). In collaboration with the senses, when literary pilgrims travel to literary places, they create a kind of symbioses between imagination and reality, often paying less attention to details and focusing on experiencing the site. At these places, the experience sets the work of imagination in motion, in a process in which literary travellers fill in the 'blanks', so that they construct meaning (Watson & Saunders, 2004). In the following quotes, the narrator articulates that experience of summoning the senses, emotion, intuition, memory and imagination (Jiang & Yu, 2020, p. 132):

I linger, content to be with the ghosts of writers who have passed into this same perimeter. Camus leaning against the wall smoking cigarettes. Nabokov reflecting on the curve of the Nautilus. (Smith, 2017).

I walk aimlessly, finding myself in the Latin Quarter, then cut over to Boulevard Saint-Michel, searching for number 37, where Simone was raised and the Weil family lived for decades. I have a flash of Patrick Modiano tracing one address after another, crisscrossing the entire city in search of a certain stairway. I think of Albert Camus, about to receive the Nobel Prize, making this same pilgrimage to the Weil residence, but for graver reasons—not mere curiosity but contemplation. (Smith, 2017)

The Literary Pilgrim's Homage to Literary Heroes

In tourism studies, the concept of place is paramount as travelling is "place-dependent, it happens in places […]." (Varley, Schilar & Rickly, 2020). For the experience of literary pilgrims, place is vital as they develop place attachment on the basis of the emotional engagement with authors, books, characters or places depicted in the texts. Therefore, in such literary-inspired journeys, where geography, knowledge of literature, the experience of reading and imagination converge (Tuan, 1977/2001), literary places are central components in fostering a process of meaning formation initiated in the act of literary reading. Hence, as far as the literary pilgrims' experience and motivations are concerned, places are the tangible elements where they can encounter the authors, and they are vital in achieving that sense of closeness and connectedness (Brown, 2016).

Writers' gravesites are one of such places. The experience of visiting such sites coincides with one of the oldest practices in this established tradition of travelling: "A practice that originates both in admiration for the author's work and in a desire to go beyond it […]." (Hendrix, 2009, p. 14). While visiting Sète in the south of France, the narrator does precisely that; she visits the cemetery "in search of Paul Valéry" to "find him and pay respects." (Smith, 2017). Significantly, the narrator does not state she wanted to visit Valéry's tomb; she affirms she wished to pay her respects, underlying yet again another of Harald Hendrix's conclusions regarding the literary pilgrim's motivations: to "worship" and have a "conversation with the dead." (2009, p.15). This pilgrim is author-oriented, much more than drawn to the geography of fictional texts. After visiting France, the narrator proceeds with her pilgrimage to another author's place in England. As such, after getting to St. Pancras railway station, she takes another train to

Ashford "to find Simone Weil's grave." (Smith, 2017). In Ashford, at the Bybrook Cemetery, she "walked up and down paths", searching for the grave, but it was raining, and she could not find it. After another hour had passed, she was almost "succumbing to the notion that it was an impossible task [...]. The sky grew darker" [and she] "sat on a bench somewhat demoralised." (Smith, 2017). At that exact moment, she began to wonder if Simone Weil would approve of such "pilgrimage" (again, the exact noun she chooses to write) and she admits Weil would not. At this point in the narrative, the reader accesses the narrator's emotional upheavals: her internal struggle as she doubts if she is actually doing something the author she admires would approve. Despite questioning her own actions, she persists, as she had brought "lavender from Sète in an old handkerchief to leave her, small bits of France, and recalling her love of her homeland, her longing to return, I pressed on." (Smith, 2017). Along with a desire for closeness and communion, a wish to pay homage, the desire to leave an offering is one of the common practices in literary pilgrims (Brown, 2016, p.171), also borrowed from religious pilgrims.

In "How the mind works", the reader learns that the narrator, besides keeping a diary with random observations, fragments of poems and notes, has also started writing a text entitled "Devotion". Although she also questioned her creative decisions in this text after returning home to New York, she decided to leave it as it was. The text was a "crime" where she "had inadvertently produced evidence" (Smith, 2017) of how the mind works. And even if she did not reach a definite answer as her conclusion was "absolutely meaningless", she realised that "most often the alchemy that produces a poem or a work of fiction is hidden within the work itself", in which writers leave shreds of evidence of their experience (Smith, 2017). As such, in "Devotion", she could "track a plethora of enticements, a forest of firs, Simone Weil's haircut, white bootlaces, a pouch of screws, Camus's existential gun." (Smith, 2017). These authors and their lives outline the architecture of the narrator's literary-inspired pilgrimage.

FINAL CONSIDERATIONS

This chapter contributes with insights into the travelling experience of (literary) pilgrims, and reflections about these touring identities, experiences and motivations, after the analysis and interpretation of two travel fictional narratives in the scope of literature and tourism studies. This process unveiled two literary texts in which the act of travelling is the object of literary representation. In both texts, the act of travelling is tantamount to a pilgrimage; a quest fulfilled in every stage of the journey and at every encounter with other pilgrims or at the authors' places. Both texts provide rich material that can inform on the experiential dimensions of touring, namely regarding the narrators' journeying identities. In them, the reader accesses two perspectives of the touring experience of pilgrims and their motivations, and reaches the conclusion that "Travel is the traveller. What we see isn't what we see but what we are." (Pessoa, 1982, p. 451).

As such, in *Flights*, the narrator understands travelling as an immersive and extensive physical displacement and an existential strategy to avoid decay and stagnation. A belief that is significant to the structure of the narrative and to the construction of this pilgrim's identity whose gaze is mediated by an unflinching belief that life happens in endless movement. Indeed, the narrator feeds on the very own act of travelling in a complicit relationship of intimacy with movement and other pilgrims. Her plea is for the acceptance and praise of mobility as the only possible way to find salvation.

In "How the mind works", travelling is a circular literary pilgrimage motivated by the cult of the authors. The quest is for literary meaningful sites, where a stronger connection to the revered authors is possible. Although the narrator is aware that the authors themselves might disapprove of such pilgrimage, her strong sense of mission, curiosity and devotion is the driving force underlying the pilgrim's experience.

Finally, because this study aimed at analysing two tourist literature texts and explore the narrators' touring identities and their construction of the travel experience, focusing on the motivations and experience of two pilgrims who happen to be women, future research should dive deeper into the female perspective.

REFERENCES

Almeida, M. M. (2016). "Viagens na minha serra": Percorrer a região do Caramulo na companhia de Branquinho da Fonseca ["Viagens na minha serra": Travel in the Caramulo region in the company of Branquinho da Fonseca]. In R. Baleiro, S. Quinteiro & I. D. Santos (Eds.), Viagens, relatos e itinerários [Travels, narratives and itineraries] (pp. 61-78). Universidade do Algarve.

Alves, I. M. F. (2016). A interioridade ou o itinerário transmontano de A.M. Pires Cabral [A.M. Pires Cabral's interiority or the Trás-os-Montes itinerary]. In R. Baleiro, S. Quinteiro & I. D. Santos (Eds.), Viagens, relatos e itinerários [Travels, narratives and itineraries] (pp. 47-60). Universidade do Algarve.

Augé, M. (1995). *Non-places: Introduction to an anthropology of Supermodernity*. Verso.

Baleiro, R., & Quinteiro, S. (2018). *Key concepts in literature and tourism studies*. Centro de Estudos Comparatistas, Faculdade de Letras da Universidade de Lisboa.

Brown, L. (2016). Tourism and pilgrimage: Paying homage to literary heroes. *International Journal of Tourism Research*, *18*(2), 167–175. doi:10.1002/jtr.2043

Butler, R. (1986). Literature as an influence in shaping the image of tourist destinations. In J. S. Marsh (Ed.), *Canadian studies of parks, recreation and tourism in foreign lands* (pp. 111–132). Trent University.

Bye, S. G. (2018). Flights by Olga Tokarczuk [Review of the book *Flights*]. *World Literature Today*. https://www.worldliteraturetoday.org/2018/july/flights-olga-tokarczuk

Çevik, S. (2020). Literary tourism as a field of research over the period 1997-2016. *European Journal of Tourism Research*, *24*, 2407.

Charles, R. (2019, 10 October). The Swedish Academy took a year off to fix the Nobel Prize in literature. It's still broken. *The Washington Post*. https://www.washingtonpost.com/entertainment/books/the-swedish-academy-took-a-year-off-to-fix-the-nobel-prize-in-literature-its-still-broken/2019/10/10/23f1b6da-eb7d-11e9-9306-47cb0324fd44_story.html

Cohen, R. (1986). History and genre. *New Literary History*, *17*(2), 203–218. doi:10.2307/468885

Cresswell, T. (2006). *On the move*. Routledge.

Croft, J. (2019, 10 October). The Nobel Prize was made for Olga Tokarczuk. *The Paris Review*. https://www.theparisreview.org/blog/2019/10/10/the-nobel-prize-was-made-for-olga-tokarczuk/

Dann, G. (1999). Writing out the tourist in space and time. *Annals of Tourism Research, 26*(1), 159–187. doi:10.1016/S0160-7383(98)00076-0

Earl, B. (2008). Literary tourism: Constructions of value, celebrity and distinction. *International Journal of Cultural Studies, 11*(4), 401–417. doi:10.1177/1367877908096003

Elliott, A., & Radford, D. (2015). Terminal experimentation: The transformation of experiences, events and escapes at global airports. *Environment & Planning, 33*(6), 1063–1079. doi:10.1177/0263775815595407

Fonseca, A. M. (2020). Dentro do Segredo: Representações da Coreia da Norte na narrativa de viagens de José Luís Peixoto [Inside the Secret: Representations of North Korea in José Luís Peixoto's travel narrative]. *Dos Algarves: A Multidisciplinary e-Journal, 37*, 22-35. doi:10.18089/DAMeJ.2020.37.2

Hendrix, H. (2009). From Early Modern to Romantic Literary Tourism: A Diachronical Perspective. In N. J. Watson (Ed.), *Literary tourism and nineteenth-century culture* (pp. 13–24). Palgrave. doi:10.1057/9780230234109_2

Hendrix, H. (2014). Literature and tourism: Explorations, reflections, and challenges. In S. Quinteiro & R. Baleiro (Eds.), Lit & Tour: Ensaios sobre literatura e turismo [Lit & Tour: Essays on literature and tourism] (pp. 19-29). Húmus.

Herbert, D. (2001). Literary places, tourism and the heritage experience. *Annals of Tourism Research, 28*(2), 312–333. doi:10.1016/S0160-7383(00)00048-7

Herbert, D. T. (1996). Artistic and literary places in France as tourist attractions. *Tourism Management, 17*(2), 77–85. doi:10.1016/0261-5177(95)00110-7

Hoppen, A., Brown, L., & Fyall, A. (2014). Literary tourism: Opportunities and challenges for the marketing and branding of destinations? *Journal of Destination Marketing & Management, 3*(1), 37–47. doi:10.1016/j.jdmm.2013.12.009

Jiang, L., & Xu, H. (2017). The growth of literary places in ancient town tourism destinations based on the theories of Bourdieu, Danto and Dickie. *Journal of Tourism and Cultural Change, 15*(3), 213–228. doi:10.1080/14766825.2016.1148153

Jiang, L., & Yu, L. (2020). Consumption of a literary tourism place: A perspective of embodiment. *Tourism Geographies, 22*(1), 127–150. doi:10.1080/14616688.2019.1586985

Kassabova, K. (2017, 3 June). Flights by Olga Tokarczuk review: The ways of wanderers. *The Guardian*. https://www.theguardian.com/books/2017/jun/03/flights-by-olga-tokarczuk-review

MacCannell, D. (1976). *The tourist: A new theory of the leisure class*. Schocken Books.

Machado, C., Moreira, F. J., Reis, J., Aurindo, M. J., Almeida, M. M. & Faria, R. (2020). *Story Map* Caminhos d'O conspirador: Um percurso literário na vila de Marvão com Branquinho da Fonseca [Story Map the paths of the conspirator: A literary journey in the village of Marvão with Branquinho da Fonseca]. *Dos Algarves: A Multidisciplinary e-Journal, 37*, 88-105. doi:10.18089/DAMeJ.2020.37.6

MacLeod, N. (2020). 'A faint whiff of cigar': The literary tourist's experience of visiting writers' homes. *Current Issues in Tourism*. Advance online publication. doi:10.1080/13683500.2020.1765996

MacLeod, N., Shelley, J., & Morrison, A. M. (2018). The touring reader: Understanding the bibliophile's experience of literary tourism. *Tourism Management, 67*, 388–398. doi:10.1016/j.tourman.2018.02.006

Magadán Díaz, M. & Rivas García, J. (2012). *Estructura, economía y política turística* [Structure, economy and tourism policy]. Septem Ediciones.

McWha, M. R., Frost, W., Laing, J., & Best, G. (2015). Writing for the anti-tourist? Imagining the contemporary travel magazine reader as an authentic experience seeker. *Current Issues in Tourism, 19*(1), 85–99. doi:10.1080/13683500.2015.1013525

Oliveira, M. (2016). Lisboa, Livro de Bordo, de José Cardoso Pires: A construção de um guia cultural [Lisboa, Livro de Bordo, by José Cardoso Pires: The construction of a cultural guide]. In S. Quinteiro, R. Baleiro & I. D. Santos (Eds.), Turistas, viajantes e lugares literários [Tourists, travellers and literary places] (pp. 85-102). Universidade do Algarve.

Onfray, M. (2007). *La theorie du voyage: Poétique de la géographie* [The theory of travel: Poetics of geography]. Livre de Poche.

Pessoa, F. (1982/2002). *The book of disquiet* (R. Zenith, Trans.). Penguin Classics.

Robinson, M. (2004). Narratives of being elsewhere: Tourism and travel writing. In A. A. Lew, C. M. Hall, & A. M. Williams (Eds.), *A Companion to tourism* (pp. 303–315). Blackwell Publishing. doi:10.1002/9780470752272.ch24

Robinson, M., & Andersen, H.-C. (2002). Literature and tourism: Reading and writing tourism texts. *Continuum.*

Rodrigues, J. (2020). Literatura de turismo sobre o Algarve: O caso dos romances de Rolf Osang [Tourist literature about the Algarve: The case of Rolf Osang's novels]. *Dos Algarves: A Multidisciplinary e-Journal, 37*, 36-50. doi:10.18089/DAMeJ.2020.37.3

Smith, P. (2017). How the mind works. In *Devotion*. Yale University Books. doi:10.2307/j.ctv1bvnfgr.3

Stiebel, L. (2007). Going on (literary) pilgrimage: Constructing literary trails with particular reference to KwaZulu-Natal. *Scrutiny, 12*(1), 93–106. doi:10.1080/18125440701398596

Thompson, C. (2011). *Travel writing: The new critical idiom*. Routledge. doi:10.4324/9780203816240

Tokarczuk, O. (2007/2017). *Flights* (J. Croft, Trans.). Fitzcarraldo Editions.

Tuan, Y.-F. (1977/2001). *Space and place: The perspective of experience* (8th ed.). University of Minnesota Press.

Urbain, J.-D. (1991). *l'idiot du voyage, histoires de touristes*. Petite Bibliothèque Payot.

Urry, J. (1990). The tourist gaze: Leisure and travel in contemporary societies. *Sage (Atlanta, Ga.).*

Urry, J. (2002). The Tourist Gaze 2.0. *Sage (Atlanta, Ga.).*

Urry, J., & Larsen, J. (2011). The tourist gaze 3.0. *Sage (Atlanta, Ga.).*

Varley, P., Schilar, H., & Rickly, J. M. (2020). Tourism non-places: Bending airports and wildscapes. *Annals of Tourism Research*, *80*, 102791. Advance online publication. doi:10.1016/j.annals.2019.102791

Watson, C., & Saunders, R. (2004). The production of literary landscapes. In M. Robinson & D. Picard (Eds.), *Conference proceedings tourism and literature: travel, imagination and myth.* Centre for Tourism and Cultural Change.

Watson, N. J. (2009). *Literary tourism and nineteenth-century culture*. Palgrave. doi:10.1057/9780230234109

Youngs, T. (2013). *The Cambridge introduction to travel writing*. Cambridge University Press.

Zhao, L., Chen, Q., & Zhou, W. (2013). The ontology description and intention construct of tourism experience under phenomenology perspective. *Luyou Xuekan*, *28*(10), 97–106.

KEY TERMS AND DEFINITIONS

Literary Pilgrim: Literary pilgrims are the first type of literary travellers, and they are devoted and well-educated visitors with the cultural capital to appreciate and produce literary landscapes and sites.

Literary Tourism: Literary tourism is a niche of cultural and heritage tourism, and it refers to the act of travelling to visit places and landscapes associated with the presence or memory of literary authors or literary fiction representations.

Tourist Literature: An umbrella concept that serves to describe a cluster of literary texts whose attributes make them suitable for cross-disciplinary hermeneutics in the scope of literature and tourism studies.

Travel Literature: A literary (sub)genre that refers to a cluster of textual typologies that portray a fictional or real journey by a narrator-traveller. It is challenging to present an ultimate definition of travel literature because its textual typologies are multiple and diverse, and the nature of the literary phenomenon is complex.

ENDNOTES

1 From now on the quotes from the primary text (*i.e.*, the English translation of *Bieguni*) will be integrated in the continuous text marked with the page number.

2 All quotes are from the book's kindle edition, in which pages are not identified.

Chapter 12
Could Television Streaming Sites Lead the Charge for Film-Induced Tourism Post COVID?
Seeking the "Sites/Sights" Seen on the Small Screen!

Jennifer Stewart
https://orcid.org/0000-0001-9366-5279
Technological University Dublin, Ireland

ABSTRACT

By focusing on television film-induced tourists, this chapter will contribute to a better understanding of tourist behaviour in relation to motivations for travel to filming locations. The chapter combines, analyses, and critiques the main debates raised by key authors in relation to identifying the motivational factors that prompt site-specific film tourism as well as providing contributions from this author's 2016 research on television film-induced tourism in Ireland. The chapter is divided into the following sections: a brief history of film and television and a review and discussion on film tourism, a breakdown of the different categories of film tourist and a summation of the various motivations for television induced film tourism, followed by an insight into the concept of authenticity in film tourism studies and the use of technologies such as virtual reality and augmented reality as a means to provide a more immersive experience post COVID-19.

INTRODUCTION

This chapter comes at a time when managers in tourism and hospitality industries need to look to the future in a dynamic and competitive market and utilise film-induced tourism development to gain competitive advantage. As a result of the world lockdown due to the global Coronavirus pandemic, television viewing and media streaming site subscriptions boomed. Thus, film tourism will possibly be

DOI: 10.4018/978-1-7998-8262-6.ch012

a niche form of cultural tourism that could be a key to rebuilding the tourism industry post COVID-19. Film induced tourism is still a relatively new field of study, and the appeal is not yet fully understood or explored. There have been many studies completed investigating the relationship between film and tourism, but they have primarily dealt with: 1. destination image planning and management; 2. investigating the impacts of film tourism on host communities; 3. authenticity. Subsequently, after identifying gaps in the current literature around the topics of tourist motivation and the experience of being a film tourist, further research was undertaken to identify, characterise and understand both the possible and actual motivations of the film tourist; specifically, from the consumer perspective to examine the effects of film on tourist behaviour.

It is well-recognised that visits to film and television series locations is a growing global phenomenon in the tourism field, and with that comes a growing amount of tourism research. Such research examines the relationships between tourism and popular media (e.g., Beeton, 2016; Busby & Klug, 2001; Connell, 2005, 2005a, 2012; Connell & Meyer, 2009; Kim, 2010; McWha & Beeton, 2018; Reijnders, 2011; Riley & van Doren, 1992; Roesch, 2009; Santos & Kim, 2017; Yen & Teng, 2013). However, consumer demands and behaviour will be permanently altered by the pandemic, and all stakeholders in the tourism industry will need to adapt. As travellers strive for meaningful short breaks and touristic experiences, one way in which destinations can bounce back would be using immersive technologies and the redesigning of tourism offerings. Thus, this chapter will begin by providing a brief history of film and television and a review and discussion on film tourism. The chapter also offers a breakdown of the different categories of film tourist and a summation of the various motivations for film tourism, followed by an insight into the concept of authenticity in film tourism studies and the use of technologies such as Virtual Reality and Augmented Reality as a means to provide a more immersive experience. The objective is to provide a critical overview of film tourism research to date and present the motivational factors that prompt site-specific television film-induced tourism, with contributions from the author's 2016 research on tourist motivation in the context of television film-induced tourism from the television series *Game of Thrones*.

BACKGROUND

History of Film and Television

The origins of cinema began in the late 19th century, when an employee of the American inventor Thomas Alva Edison created a machine called the Kinetoscope in 1894. People could watch short films through a peephole viewer window at the top of the machine; however, the major disadvantage of the Kinetoscope was the fact that the moving pictures could only be seen by one person at a time. In 1895 the Lumière brothers from Paris, France, came up with an improved version of the Kinetoscope, which they called the Cinématographe. This device projected images onto a screen to give the illusion of movement allowing several people to watch a film together.

The first major movie studio was opened in Hollywood, Los Angeles in 1909 (Hellman, 2005). Roughly a decade later cinema was to become a mass leisure time activity, creating popular culture on a scale that had never been seen before. Further developments in 1925 saw John Logie Baird, a Scottish inventor; succeed in transmitting moving silhouette images. In 1926, he demonstrated the transmission of an image of a face in motion via radio; this landmark event is regarded as the world's first true demonstration of television (Baird, 2020).

This meant that film which was once only viewed in the cinema could now be viewed at home through television. Fast-forward to the 21st Century and a sitting room without a television is almost unthinkable because it has become a natural part of our environment and culture. Television has evolved into the most significant tool for mass media communication and changed into our primary source for news, weather, and entertainment. Nowadays, almost all households have ownership of multiple televisions, satellite systems, DVD players and computers. The transmission of televised films, drama and other programmes into the home continues to be a major facet of modern life, through the likes of Netflix, Disney+, Amazon Prime and SKY TV.

Focusing on television series, today's offerings have a much more complex storyline than they used to have, which implies a longer story length. This chapter will be focusing on television-based film tourism, so at this point it is important to note the differences between cinema and television. Roesch (2009) talks about how the cinema movie focuses on an issue that must be resolved by the end of the film. However, television is based on several episodes in a series, or multiple series with longer drawn out storylines. Therefore, the viewer has more time to become engaged in the plot and develop empathy with the characters, which is important in determining the motivations for film tourism.

2020 was a year like no other, and a year in which the world tourism industry experienced unprecedented losses. Film tourism could be a niche form of cultural tourism that is key to rebuilding this ailing tourism industry post COVID-19. As a result of the global Coronavirus pandemic "by mid-April 2020, a third of the global population was under full or partial lockdown" (Hudson, 2020, p.28). These lockdowns ranged from legally enforced "mandatory geographic quarantines" by Governments to "non-mandatory recommendations to stay at home, closures of certain types of businesses, or bans on events and gatherings" (ibid, p.28). As a result of the world lockdown, television viewing and media streaming site subscriptions boomed. Netflix surpassed 203 million subscribers in 2020 with a net increase of 37 million new subscribers (Netflix, 2021). As of the first quarter of 2021 Amazon Prime Video had surpassed 150 million subscribers (Keyes, 2020), and Disney's direct-to-consumer portfolio including Disney+, ESPN+, and Hulu accounted for more than 146 million global paid subscribers (Walt Disney Company, 2021). A study by Ofcom in the United Kingdom found that adults spent nearly 6 and a half hours a day watching TV and online video and viewing figures for video streaming services were up by 71% on 2019 (Ofcom, 2020). It is clear that the time spent per viewer watching television and online entertainment increased as the spread of COVID-19 worsened.

Film Tourism

Film tourism is a niche element of tourism activity and is simply defined as a genre within tourism that provides a link with the location of a film or a television series – during or after shooting. Other definitions include "Film tourism is a specific pattern of tourism that drives visitors to see screened places during or after the production of a feature film or a television production" (Roesch, 2009, p.6) and "film tourism is the effect of both cinema film and of television film on the numbers of visitors coming to the place where the filming is believed to have taken place" (Tooke & Baker, 1996, p.87).

A number of early researchers on this topic highlight the potentially lucrative effects of film tourism for a destination and speculated that the influence of film and television on consumer choice would increase (e.g., Butler, 1990; Cohen, 1979; Happer & Philo, 2013; Urry, 1990). In 1996 Tooke and Baker carried out a study in the United Kingdom, which illustrated that films or television series have a big impact on the choice of tourism destination. According to the study there can be an increase of up to 93% in visitors within a year of a film/television series being released, in comparison to recorded tourism figures pre-filming at the study location. Television film-induced tourism proves itself as a very powerful marketing tool and with long-term effects, for example, as a means of sustainability and an answer to seasonality since visiting filming locations can be undertaken the whole year through (Busby & Haines, 2013; Garrison & Wallace, 2021; Kim *et al.*, 2019; Radomskaya, 2018).

There are several aspects that give TV shows some advantages over feature films in terms of their tourism-inducing properties. Destination marketers can assess the tourism-inducing potential of a TV show during the release of the first season, before investing in a marketing campaign accompanying the following seasons. According to Roesch (2018), official imaging media are actively controlled and implemented by tourism marketers. A successful example being the *Game of Thrones* social media campaign carried out by Tourism Ireland in 2014 and 2015, which sold Northern Ireland as the fictional land of Westeros.

Film-induced tourism can manifest itself in many different forms. For some people, visiting the filming location may be the only goal of their trip, for others it may be part of an organised tour, or the tourist may have happened upon the site incidentally. Some of these locations may have already been tourist attractions, while other locations are turned into attractions only because of their appearance in a film or television series. This is especially the case where film crews leave sets behind after filming finishes.

Connell (2012) devised a clear model as shown in Figure 1, which comprises the scope of film tourism to include:

- Visits to real town and countryside locations that have been employed as film shooting locations, for example, Tollymore Forrest, Downpatrick (NI) – *Game of Thrones.*
- Tours of studio sets, for example, Warner Bros. Studio Tour, California – *Friends.*
- Visits to theme parks and attractions with a link to film and TV, such as *Universal Studios* or *Disneyland*, California.
- Tourists visiting locations where filming is taking place, for example, the flood of tourists that descended on Wicklow, Ireland during the recent filming of *Vikings.*
- Visits to a location marketed as a filmic location. Tourists may not have experienced the television series but interest in the location is brought about by marketing imagery, such as, the *Glass of Thrones* campaign, Tourism Ireland– promoting *Game of Thrones* tours to Northern Ireland.
- Participation in an organised tour of a portrayed location, for example, *Game of Thrones* Tours which are commercially operated tours with tour guides around film locations in Northern Ireland.
- Participation in organised tours of film celebrity homes, for example, Southern California is renowned for its many celebrity estates, and bus tours driving through Beverly Hills looking for the stars' homes is now a time-honoured tradition and a favourite pastime of tourists.
- Visits to film festivals like the *Cannes Film Festival* in France.
- Visits to cities for film premieres, to watch the film or to star spot celebrities on arrival.

Figure 1. The scope of film tourism
Source: Connell, 2012, p1010

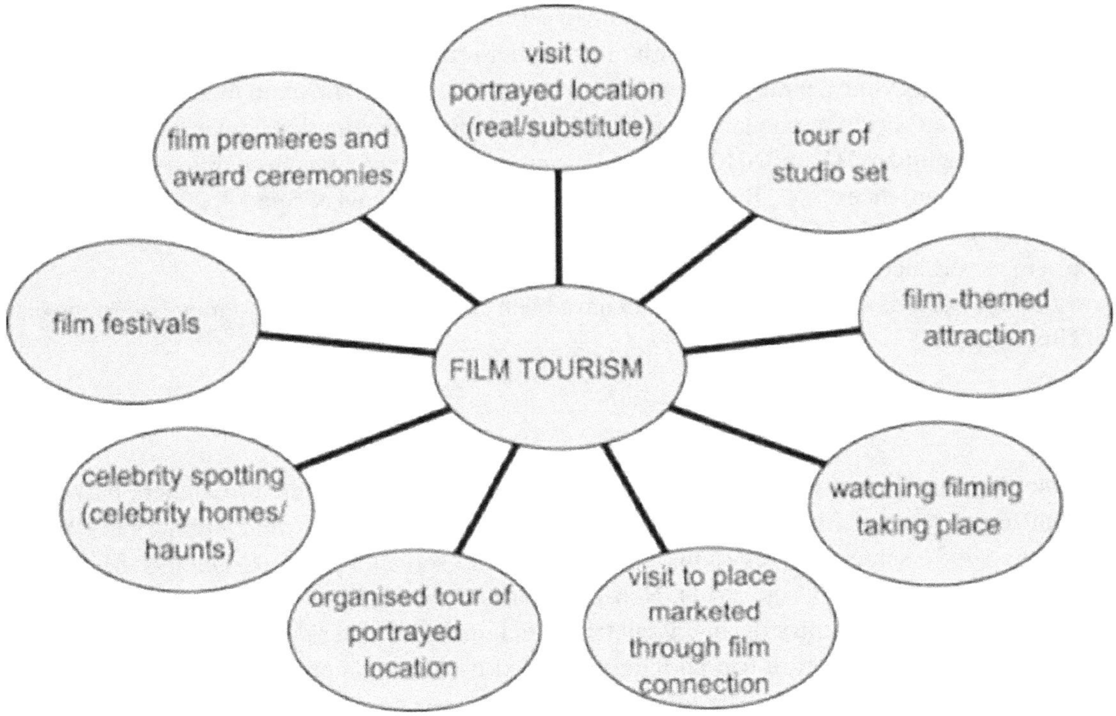

Within the last two decades film-induced tourism has gained increasing attention from academics and the industry alike. Connell (2012, p.1007) claims that several "high-profile tourism destinations utilise film-related aspects in inbound marketing campaigns", for example, the UK, USA, Korea, Australia, and New Zealand. Current worldwide television film induced tourism examples includes: *Outlander* and *Downton Abbey* in the United Kingdom; *Neighbours* and *Home and Away* in Australia; *Parks and Recreation* and *Lost* in the United States etc. Such locations and others have been documented in previous studies examining the effects of film tourism on destinations (e.g., Bakiewicz *et al.*, 2016; Butler, 2011; Carl *et al.*, 2007; Connell, 2012; Edensor, 2005; Hanefors & Mossberg, 2002; Jones & Smith, 2005; Kim *et al.*, 2007; Pearce, 2005). Every day, the film universe captivates people. Films are not commonly made with the purpose of attracting tourists to a destination. However, they do influence viewers indirectly as a secondary element of movie viewing through the use of composition, sets, props, actors, costumes, and lighting (Butler, 1990). The very best films can take on a life of their own in the viewer's imagination. A good film or television series can instil the feeling that we have expanded our circle of friends. Therefore, film tourism has the means and the scope to reach everyone. A visit to a location will take the tourist into the very core of the film's universe where, through their own senses, they will experience a new dimension of the television or film production. Film tourism represents a gateway to new and more intense ways of experiencing destinations. This is where there is vast potential for television film tourism because television series have always played a uniting role within the family. Television is something that interests everyone – across generations (Hudson & Ritchie, 2006; Signorielli & Morgan, 2001; Spigel, 2013; Wheeler, 2015).

Television series are becoming increasingly interesting from a film-tourism perspective, particularly as tourism reopens after COVID-19. Most citizens around the world have been in lockdown of varying intensities for nearly eighteen months at the time of writing and because television series have a longer life cycle than feature film productions, with most of them running for several seasons, the level of engagement with the storylines, plots and destinations is much stronger. The main focus of current research remains primarily on "destination marketing, branding and image, destination choice and motivation, and impacts and management" (Kim, 2012, p.387). However, some academics have undertaken research on film site tourist experiences (e.g., Beeton, 2005, 2010, 2016; Macionis & Sparks, 2009; Roesch, 2009). Notwithstanding this exploratory work, there is little research that has investigated the understanding of film tourism experiences and motivations associated with a television series.

Now that the various forms of film tourism have been examined, it is time to move on to the categories of film tourists.

Film Tourists

In the same way that there are various definitions of film tourism, academic researchers have also categorised different types of film tourists - such as, film pilgrims (Riley & van Doren, 1992), film tourists (Couldry, 1998), screen tourists (Kim *et al.*, 2009), elite screen tourists (Connell & Meyer, 2009), and purposeful film tourists (Croy, 2011). Non-film tourists are usually tourists who are travelling to a film site as part of their itinerary during a day trip which includes other destinations. Film tourists are those who take a pilgrimage film trip and are motivated by nostalgia and place identity because of a film (Jewell & McKinnon, 2008). The television film tourist is attracted by the first-hand experience of the locality captured on the small screen. It also must be said that the degree of fanaticism of the television film location tourists can change over time. In the first years after release of the series, the film locations are more likely to be visited by tourists with a very high degree of fanaticism. As time goes by, this degree of fanaticism decreases and film or television series as a main goal for travelling moves slowly to the background.

Research shows that most tourism marketing focuses on the destination and lacks focus on the consumer. Consumers can be greatly influenced by images and may make purchase choices based on what they have seen in the movies and on television. Therefore, the medium of film can have an impact on its audience. Bolan and Williams (2008, p.382) also note "an expanding body of literature suggests that film can influence people's travel decisions and entice them to visit particular destinations they have seen on the cinema screen" or television. (e.g., Araújo Vila et al., 2021; Butler, 2011; Chang, 2015; Hudson & Ritchie, 2006; Ismail et al., 2017; Rittichainuwat & Rattanaphinanchai, 2015)

Macionis (2004) devised an alternative set of categories of film tourists: The Serendipitous Film Tourist; the General Film Tourist; and the Specific Film Tourist. Figure 2 details the distinctions between each type of Film Tourist

Figure 2. Continuum of film-induced motivation
Source: Macionis, 2004, p94

Increasing interest in film →		→
Serendipitous Film Tourist	General Film Tourist	Specific Film Tourist
• those who just happen to be in a destination portrayed in a film	• those who are not specifically drawn to a film location but who participate in film tourism activities while at a destination	• those who actively seek out places that they have seen in film
Motivations include: • social interaction • novelty	Motivations include: • escape • novelty • education • nostalgia	Motivations include: • ego-enhancement • self-actualisation • pilgrimage • self-identity • vicarious experience • fantasy • status/prestige • romance • nostalgia
Increasing self-actualisation motivations →		→
Decreasing importance of authenticity →		→
Increasing importance of Push Factors →		→

There is a need to better understand film/screen tourist profiles as there is a general lack of research in understanding these travellers, including their socio-demographic characteristics, travel patterns and screen product preference. One question which we must ask is whether understanding the profile of film location tourists helps to understand their motivations for travel? Regarding the primary travel motivations of film tourists, these seem to be quite like those of other tourists.

TOURIST MOTIVATION

Generally, motivation represents the reasons behind peoples' actions, desires, and needs. Tourist motivation is centred on the questions - Why do people travel? and, Why do they choose the holidays and destinations they do? It is closely related to people's needs and wants. This section will utilise previous motivation theories to understand the travellers' desire to visit film locations.

There were early attempts to understand tourism motivation in the 1970's, 1980's and 1990's, the most popular theories adapting Maslow's Hierarchy of Needs concept (Maslow, 1970) by putting forward the notion of a motivational career – with more experienced tourists showing higher needs than less experienced travellers (McIntosh, Goeldner & Ritchie, 1995; Pearce & Caltabiano, 1983). Attempts to categorise motivations related to tourists' choice of holiday destination suggest that these motives can be classified into *push* and *pull* factors (Crompton, 1979; Dann, 1977; McIntosh & Goeldner, 1990; Trimurti & Utama, 2020; Yoo *et al.*, 2018;). The Push-Pull framework provides a simple intuitive approach for explaining the motivations underlying tourist behaviour. Push relates to the socio-psychological factors that *push* a person towards travel and have been described as motivational factors, or needs, that arise due to a state of instability or tension in the motivational system. Whereas *pull* factors on the other hand, have been described as those factors related to the features, attractions, or attributes of the destination itself, that pull a person towards a certain location, such as beaches, recreation facilities and tourist sites, and these differ from one destination to another (Dann, 1981; Dunne, 2009; Iso-Ahola, 1982; Kim *et al.*, 2003; Klenosky, 2002). Destination choice therefore comes from tourists' perceptions of destination attributes and their apparent values.

Push-Pull Framework

For this chapter, the author employs the conceptual framework adopted by Crompton in 1979 which can encompass all the motives for travel. Crompton (1979) identified seven socio-psychological push factors and two cultural pull factors guiding visitors' choice of a holiday destination.

The Socio-Psychological Motives are the *Push* Factors. *Escape from a Perceived Mundane Environment* refers to what Crompton calls "a temporary change of environment" which is "physically and socially different from the environment in which one normally lives" (ibid, p.416). *Exploration and Evaluation of Self* relates to the desire to develop as a person; it may involve the need to challenge oneself, to learn new skills, attain new knowledge or find out about oneself. *Relaxation* refers to a "mental state rather than a physical relaxation" and means "taking the time to pursue activities of interest" (ibid, p.417). *Prestige* refers to when the trip represents a combination of both symbolism and self-expression - often associated with more expensive and exotic holidays. *Regression* relates to the need to "regress to a less complex, less changeable, less technologically advanced environment", for instance, looking for the simple life (ibid, p.418). *Enhancement of Kinship Relationships* relates to the desire to enhance or enrich family relationships. And *Facilitation of Social Interaction* represents a need to "meet new people in different locations" (ibid, p.420), which is about a desire to mix with the locals and other tourists in the area.

Cultural Motives are the *Pull* Factors. *Novelty* represents the things that are different from the traveller's everyday life and environment, such as beaches, recreation facilities, and cultural attractions. *Education* relates to the pull of studying abroad to attain a better or higher level of education or to learn new skills/gain knowledge. If a trip is perceived as the one opportunity in a lifetime to see a cultural phenomenon, and the opportunity is not grasped, then the educational benefits will be lost.

Explanation of underlying motives offers useful insights into understanding the television film location tourists' decision process. Although the push/pull framework is simplistic and dichotomous in design, this author believes that the socio-psychological motives and the cultural motives are not polar opposites, but instead fit together like the pieces of a puzzle to entice the tourist to travel and subsequently form an overall motivation to take a trip. The author proposes that the desire for education is something inherent within the individual and is something you either seek or you do not. Therefore, it is not necessarily dependent on the location and as such should not be classified as a pull attribute of the destination. Similarly, if the pull factor of novelty is "substituted with *curiosity*, then the motive is more like a push factor" (Hsu & Huang, 2008, p.18). Novelty refers to the desire to see new places or do things in a different environment, so changing novelty to curiosity (the desire to know and to understand) makes more sense as a push factor.

Hence, for that reason, the author recommends that to develop the tourist profile, the pull attributes of destinations should be disregarded, and focus should be placed solely on push factors as primary motives for travel. The pull attributes of a destination are secondary to the motive and desire to travel and can be interchangeable between many destinations but the primary motives for travel remain unchanged. Thus, are site-specific motivators *push* or *pull* factors? The term site-specific would lead one to believe that the motivations are linked to *pull* attributes of place; however, is suggested that this is not necessarily the case. In relation to television film induced tourists, the question remains as to what motivates tourists to engage in site-specific tourism and does it align with Crompton's motivation framework.

DISCUSSION

To contextualise the discussion, the author conducted a quantitative study of 125 tourists participating in *Game of Thrones* bus tours from Dublin and Belfast over the summer of 2016 at the height of the series' popularity. The tourists were surveyed at 5 of the most popular filming locations – Tollymore Forest Park, Inch Abbey, Castle Ward, Dark Hedges Estate, and Ballintoy Harbour. The author adopted a positivist worldview to conducting the research and the resulting empirical study led to conclusions based on data and specifically collected information using a deductive process as the research is based on prior logical reasoning – Push/Pull Framework. The participants were asked multiple questions about their motivations for visiting the film locations. The survey instrument yielded categorical data, with data analysis building themes and relationships among variables. This non-experimental form of research is of correlational design and the results are incorporated into the following discussion.

Kork, (2016, p.154) suggests that the "core element of a film that affects whether the film will influence the viewer to travel is the portrayal of the destination", and that the plot and characters are less important. However, the author's research findings indicate that most respondents are motivated to take a film location tour because of the performance-oriented storylines of the show. In this instance, the viewers are more motivated to visit a location because of its association with the television series, characters etc., than for its natural visual beauty. The intrinsic place qualities were a lesser stimulator with respondents being marginally motivated by attributes of place such as the scenery and landscape. Yet, it is doubtful if the natural beauty of the location would be enough to draw the tourists to visit if the location were not associated with a recognised television programme. An interesting finding was that respondents were seeking education and knowledge, declaring that they desired to learn more about television programming, for example, how the television series was filmed. The author believes that character association also creates a strong emotional tie to cast members, with respondents specifically travelling in the hopes of meeting the actors on site. The author's research also highlighted visiting somewhere new as a major motivation to travel.

Figure 3 shows a conceptual diagram designed to highlight how the site-specific motivations align with Crompton's motivation theory. This contribution is developed by the author specifically for television film-induced tourism.

Figure 3. Site specific film tourist motivation
Source: Author

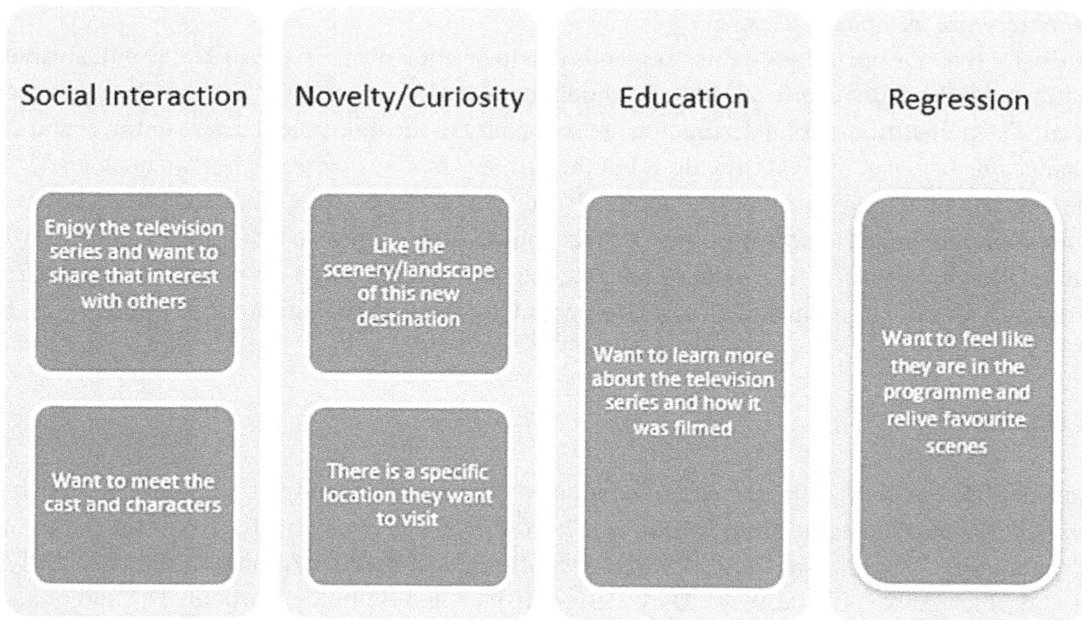

Film Tourist Profile

Connell (2012, p.1016) reports that, "until the mid-2000s, few studies focused on the characteristics, motivations and behaviour of film tourists". Roesch (2009) conducted research about the on-site experiences of film tourists at three different movie locations, specifically Salzburg (Sound of Music), Tunisia (Star Wars) and New Zealand (The Lord of the Rings). Not surprisingly, the profile of visitors varied depending on the locations and the themes of the movies. The age range of the film tourists was diverse and led Roesch to identify three major age groups:

- Gappies are backpackers who have taken a year off before or after college. They are in their early 20s, travel on a tight budget and have more time available at the destination than the other groups;
- Revivals, are people aged between 40-50 years who often travel with their partner or children to see the movie locations; they have less time, but more money available;
- Initials, describe people aged 60 and over

This following section will build on this conceptualisation and concentrate on film tourist characteristics as identified by the author and the *Game of Thrones* tour participants.

Socio-Demographic Profile

It has been observed that film tourists tend to be highly educated, wealthy, middle-aged individuals (Busby & Klug, 2001; Kim & O'Connor, 2011). The author's findings disagree with this assumption because almost half of all respondents were under 30 years of age. Another observation from previous studies is that the majority of film tourists are female (Hudson *et al.*, 2011), the author's research mirrors this, with 35% of the respondents being male and 65% female. With regards to nationality, almost half of the respondents were Irish (45%). British citizens were also well represented at 20%. The remaining 35% of respondents were from various origins including North America (and Canada), South America, Europe, Asia, and Australia/Oceania. This gave the study a cross-cultural aspect, which is supported by the work of previous researchers. The author's research also corroborates the work of Buchmann *et al.* (2010), confirming that most respondents took the tour as a means of enhancing social interaction with friends and kinship relationships with family. The Irish participants were more inclined to travel with family and spouses/partners while the remaining nationalities were mainly travelling with friends. The younger respondents (18-30) showed the highest propensity for social interaction by travelling with friends, whereas, those classed as middle aged (31-45) were travelling as couples with their spouse or partner, thereby enhancing close relationships by experiencing the location most likely with the person with whom they had watched the series.

Types of Film Tourist

Since most bus tours to film-shooting locations use film tourism as the main theme of the tour, the target of such tours is primarily what could be called *fanatical film tourists.* It is the author's opinion that it is highly unlikely that serendipitous or incidental film tourists would be on an organised film location trip. Quite possibly, the tour may catch some general film tourists, but the percentage is too insignificant to record. As expected, *fanatical film tourists* are more likely to take a pilgrimage film trip and are motivated by nostalgia and place identity because of television. Furthermore, for *fanatical film tourists*, the re-enactment and photographing of performances like those acted out by characters seen on the show reinforces their emotional attachment and creates personalised memories by providing tourists with a first-hand experience of the locality. The author's research confirms the evocation that *fanatical film tourists* must physically be at the sites of their favourite films, to stand in the footsteps of their movie heroes and relive special moments from the film.

Figure 4 presents the author's interpretation of existing theories, combined with new information gained from the author's research study and creates a film tourist motivational profile. The primary motivations as identified by Crompton (1979) reveal that the respondents recognise their need for social interaction, curiosity, education, and regression. These motivations manifest themselves as site specific motivators by appearing in the following manner:

- Social Interaction – respondents wish to share their interest with others because they all enjoy the show, and they want to meet the cast/actors because they have formed an emotional attachment to certain characters.
- Curiosity – respondents like the scenery/landscape of the destination or there is a specific location they want to visit.
- Education – respondents want to learn more about the television series and how it was filmed.
- Regression – respondents want to feel like they are in the programme and relive favourite scenes by dressing up in character costumes and performing re-enactments.

The profile also details the main socio-demographic data of gender, age, and travel behaviour, as recorded by the author. The motives recognised were social interaction, novelty/curiosity, education, and regression. The study respondents were primarily Generation Y (the generation born in the 1980s and 1990s, comprising primarily the children of baby boomers and typically perceived as increasingly familiar with digital and electronic technology) and were motivated to take a film location tour because of the performance-oriented storylines of the show, rather than by attributes of the place.

Figure 4. Conceptual fanatical film tourist profile
Source: Author

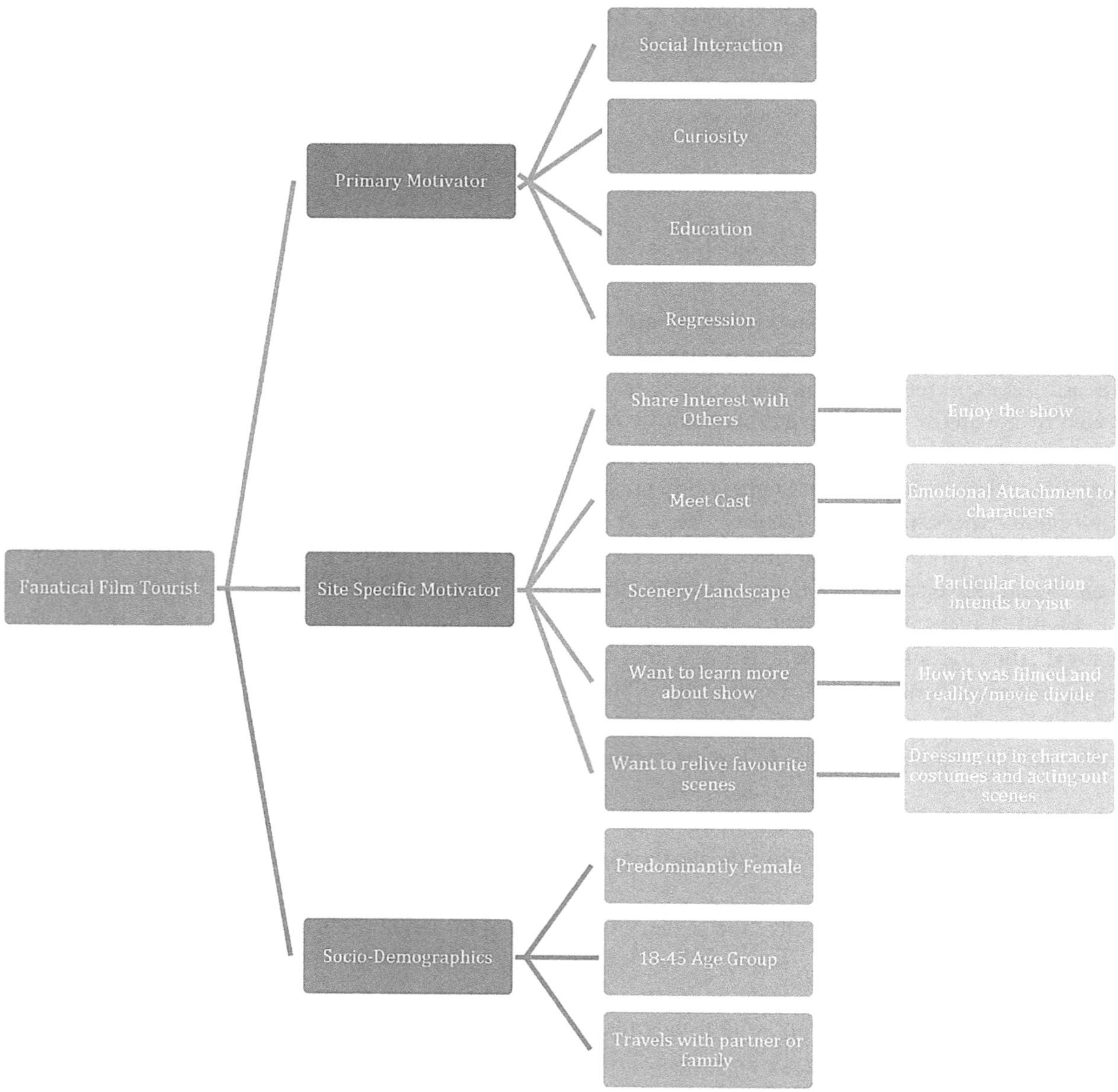

As part of this exploration, tourist motivation research must consider the issue of whether it matters if the location seen in a film is real, artificial or a substitute location to the one portrayed.

Authenticity and the Tourist Gaze

A theme in evaluating and understanding tourist motivations and their on-site experiences has arisen around the concept of authenticity. It is argued that the original form of the authenticity notion has been only partially employed in the work of tourist researchers (e.g., Cohen, 1979; MacCannell, 1999; Pearce, 2005; Rickly & McCabe, 2017; Tiberghien *et al.*, 2020; Torchin, 2002). The major development of the authenticity concept in the tourist literature is that "all tourists seek authenticity but are frustrated in their attempts because of the creation of front stage, inauthentic environments" (Pearce & Moscardo, 1986, p.122).

Accurate and inaccurate perceptions of authenticity need to be considered, together with tourists' preference levels for authenticity to explain tourist satisfaction. Bolan *et al.*, (2011) ascertained that displacement in relation to film-induced tourism refers to when a movie is shot in one location but is representing somewhere else entirely. For example, the Dark Hedges in County Antrim portraying the King's Road in *Game of Thrones,* or various scenes from the movie blockbuster *Braveheart* representing the Scottish fight for independence, set in Scotland but filmed in Ireland. MacCannell (1999, p.14) confirms that the "rhetoric of tourism is full of manifestations of the importance of the authenticity of the relationship between tourists and what they see", however little is known about whether film-tourists have greater, if any, expectations of visiting the real locations of a film when they visit a particular destination.

One way in which tourists perceive the authenticity of the on-site experience is through what Urry calls the Tourist Gaze; he suggests that a "variety of non-tourist practices such as film, TV, literature, magazines, records and videos, construct and reinforce the gaze" (Urry, 1990, p. 3). Urry and Larson (2011, p.2) state that "people gaze upon the world through a particular filter of ideas, skills, desires and expectations, framed by social class, gender, nationality, age and education". These filters are what allow tourists to decide whether what they see is beautiful or interesting, and whether the nature of the scene is real or staged. The impact of film on people's image construction is not a new notion and has been previously documented in the literature (e.g., Happer & Philo; 2013; Riley & van Doren; 1992). Major films often cause a tourist influx to destinations previously unvisited before the location was shown on the silver screen. This is also true for TV series as they can present millions of viewers with substantial information about a destination over a relatively short period of time. As the viewers become more involved and emotionally attached to the characters and plots, an awareness and interest in filming locations and settings is generated, resulting in actual visits. The author's research findings support this argument and show that the images portrayed of locations in films and on television were important to respondents when selecting holiday destinations. The respondents confirm that they have been exposed to various images of destinations from television which led to them being enticed to visit. Thus, the author concludes that television has a substantial role in encouraging tourists to visit a destination and that the main driving force for tourists to travel to the film location may be the complex process of association construction that is caused by watching the television series. The respondents place value on images derived from film as important in choosing a holiday destination.

Fictional versus authentic film locations is one of the more complicated debates around film tourism. Bolan *et al.* (2011) discuss the displacement aspect or in-authenticity of experience for the film tourist and suggest they may become disappointed when they do not see exactly what they saw on screen. Nonetheless, a visit to a television film location can also result in great satisfaction if the viewer understands that, by nature, a film or television production is only make-believe, and is therefore content to be at the location without expecting an exact replica of what they saw on screen. So, this begs the question, does authenticity actually matter?

The author's research asked the respondents if their trip had met their expectations regarding the image they had of the location. Most stated yes, they were satisfied that what they saw was the same as what they had seen on the small screen. However, a small percentage of respondents were not satisfied and claimed that the location was not the same as was shown on television. The imagined destination was not the same as the real destination, thereby creating disappointment in the in-authentic experience. The authenticity of experience depends on the motivations of the tourist, the type of film tourist and their expectations, and whether they feel they have been duped.

The Future of Film Tourism

Consumer demands and behaviour will be permanently altered by the global COVID-19 pandemic, and all stakeholders in the tourism industry need to adapt to this and subsequent changes. As travellers strive for meaningful short breaks and touristic experiences, one way for destinations to bounce back would be using immersive technologies and the redesigning of tourism offerings. As has already been established, the largest market segment for film tourism is Gen Y, a generation well adapted to the use of digital and electronic technologies. Research has also shown that the new generation of television film induced tourists are interested in the performance-oriented storylines of the show and want to feel like they are part of the action.

Alternative reality or immersive technology – which encompasses virtual reality (VR), augmented reality (AR), and mixed reality (MR) – could become the next big phenomenon in tourism (Bec *et al.*, 2021; Beck & Egger, 2018; Beck *et al.*, 2019; Yung & Khoo-Lattimore, 2017). These technologies replace or expand the physical world by the creation of 360-degree space, allowing users to look in any direction and see content. It can take a variety of forms, from virtual reality which shuts out real life spaces to augmented reality, which superimposes a layer of digital content over the physical world. In VR, the user is immersed in an entirely artificial world. In AR, the user sees the real world overlaid with digital data. Virtual Reality enables travel to places that would otherwise be difficult to reach in the current climate. Destination Management Organisations could develop apps or programmes for use with devices such as HTC Vive, Oculus Quest, PlayStation VR, Samsung Gear, Google Daydream Viewer or even Google Cardboard. Enthusiasts then have the option of experiencing film locations virtually in their own home environment rather than travelling to the original site. On the other hand, Augmented Reality allows the user to see the real world overlaid with a layer of digital content. This digital content layer can include sensor-based data, sound, video, graphics, GPS information, or other datasets. For the fanatical film tourists that are more motivated to be present physically in the locations and walk in the footsteps of their favourite characters, to feel part of the programme, AR can transport them into the show at the actual site where it was filmed.

The tourism industry is highly competitive, so companies or destinations wishing to differentiate themselves from the competition and attract a larger number of customers (particularly younger, tech-savvy ones) should invest in AR or VR tools to bring novelty to the decision-making process. The fast pace of technology advancements, and numerous ICTs used in tourism destinations is having the effect of providing tourists with more real and immersive virtual environments, which convey strong social presence. The less which technology users perceive artificial or mediated experience, the more strongly social presence occurs. With the use of this technology DMOs and tour companies can create a whole new environment for a tour. Immersive technology could transport tourists to destinations across the world using virtual reality, and immersive environments can be generated with augmented reality overlays. Today's society revolves around technology, even more so during the Covid dictated lockdowns over the past year and a half; therefore, it is vital for the tourism industry to keep up. Immersive Technology could have a major impact on the future of destination marketing. It can enhance the experience for tourists and aid in the selling of the overall product. Numerous destinations are now marketing what they have to offer through the means of 360 virtual reality videos. VR shows huge promise for destination promotion due to the immersive sensations it can present, and its ability to communicate the way a distant destination or experience feels. As a promotional tool it has the potential to reduce perceived risks, helping travellers make more informed decisions with more realistic expectations (Griffin *et al.*, 2017; Klein, 2003). Particularly during this time of global crisis and the COVID-19 pandemic, with cultural activities and tourism restricted by stay-at-home quarantine orders, film tourism has the potential to rise to this challenge by exploiting AR and VR technologies.

IMPLICATIONS FOR INDUSTRY

The information presented in this chapter is important at a practical level for DMOs and tourism suppliers. Film locations are more likely to be visited by tourists with a very high degree of fanaticism, immediately following the film's release or the television series airing. As time goes by, this degree of fanaticism can decrease, and the film or television series is no longer a main motivation for travelling. Thus, there needs to be a greater recognition of film induced tourism, and by using the *Fanatical Film Tourist* profile, DMOs can target the relevant market segments via their socio-demographic or behavioural/motivational characteristics, thereby keeping their product placement firmly in the interest of consumers. It is recommended that film locations should be equipped with the tools and knowledge to develop a strategy to manage the visitor influx caused by film tourism in a sustainable, but financially rewarding way, and this will most likely include investing in technology such as VR and AR to create more meaningful and immersive experiences.

FUTURE RESEARCH DIRECTIONS

This chapter has contributed to the existing knowledge of film tourism from the perspective of a television series filmed in Ireland via the author's empirical research, and it would certainly be advantageous to build on this through further research. Specifically, research that would study whether the main findings identified here are more widely representative would be of value. For instance, are the site-specific motivations transferable to other mediums and genres within film tourism? Would a study of *Star Wars* film tourists in Ireland illicit similar site-specific motivations? It would have been interesting to carry out a comparative study exploring contrasts between the medium of film and that of television series induced travel, examining the motivations of tourists, in an Irish and global context. Further study would involve in-depth interviews, increased numbers of questionnaire participants and focus groups. Longitudinal research could examine whether the socio-demographic characteristics, or site-specific motivations of the *fanatical film tourist* change, and measure lasting on-location tourism effects. Outlying external factors that can facilitate destination awareness and motivation to visit a location seen in a TV series have also not been addressed here. These factors would include destination marketing activities or other *pull* attributes of a destination. Other influences on the consumer decision making process could be word of mouth and reference groups. Future studies should develop the theoretical and practical contributions by testing the role of these influences and other possible mediators.

CONCLUSION

This chapter has explored current debates around film tourism. Film tourism is a niche product and offers potential for destination marketers and tour operators via special interest tours. There is a great awareness of the phenomenon of film tourism within the area of Tourism Studies as can be demonstrated by the literature. It has been shown that film tourism has a motivational role in the tourism destination decision making process and several researchers have attempted to identify the range of impacts arising from film tourism via destination and film case studies. A thorough review of secondary information sources revealed that necessary academic attention has not been given to researching and understanding what specific aspects motivate a viewer to participate in television film tourism, such as, the scenery or scenic landscape in the movie, the narrative storyline, the music / film score, identification with the characters, or the actors themselves.

Therefore, this chapter has aimed to further develop a range of themes that identify elements of interest from the film tourism consumer, and has addressed gaps in the current literature around the topics of tourist motivation and the experience of being a film tourist - particularly the understanding of film tourism experiences and motivations associated with a TV series. A film represents a one-off opportunity, whereas a television series creates and encourages interest in the destination over its longevity, and can act as an advert for the location, which in turn, serves to entice the tourist to the destination. People travel because they are pushed by their own internal forces. Characters somehow establish a connection with the viewer, and when this is heightened further by the effect of an emotional or gripping storyline, beautiful scenery and music; films help to market destinations to the tourism consumer in an open and engaged way. Television film induced tourists are more highly motivated to take a film location tour because of the performance-oriented storylines of the show, than by the attributes of place. Research shows that these motivations can be aligned with the underlying primary motivations for travel, as identified by Crompton (1979). The main findings of the author's research combined with theories from literature led to development of a new film tourist profile, the *fanatical film tourist* as shown in figure 4. The profile details the tourists' primary (push) motivations culminating in the need to take a trip. Next the secondary motivations or site-specific television induced motivations for the film tourist were identified, and these were amalgamated with the primary motivations to provide a more in-depth look at film tourist motivation. Furthermore, the pull attributes of a destination are secondary to the motive, and desire to travel and can be interchangeable among many destinations, but the primary motives for travel remain unchanged. It is clear from a range of viewpoints that film-induced tourism has something novel to offer. It has been proven that locations instilled with the magnetism of some of the lead characters make attractive places to visit because of this association. Throughout filming, the destination will have the opportunity to attract attention and therefore tourists to the region. The long-term affiliation between the viewer and the television series creates an emotional attachment that draws the viewer back to the filming location for many years.

A major difficulty when dealing with motivation theory is that the principal classification frameworks are very broad and general. Unfortunately, every film tourist is different, and every film trip is different. However, this chapter recommends that the pull attributes of destinations should be disregarded and focus should be placed on push factors as primary motives for travel in motivation theory research; by eliminating the extrinsic factors and concentrating on intrinsic motivations the *Fanatical Film Tourist* profile should be transferable to other locations and studies, and particularly developed with regard to television film tourism.

The notion that the whole audience for any film or television series will react by deciding to visit the location is unlikely. However, if some viewers are strongly influenced by the image and character association revealed through films or television series, then there must be more focus on the consumer and the impact of such media on their travel decisions. This chapter has revealed the influence of film and television media as a means of igniting people's curiosity for former filmed locations and stimulating tourism motivation and demand for filmed destinations. As previously noted, immersive technology, such as VR and AR, could have a major impact on the future of destination marketing for this product segment - it can enhance the experience for tourists and aid in selling the overall product.

The main concern for the future is to guarantee that each and every tourist that visits a filming location has a positive overall impression of what they have seen and leave happy and satisfied.

REFERENCES

Araújo Vila, N., Fraiz Brea, J. A., & de Carlos, P. (2021). Film tourism in Spain: Destination awareness and visit motivation as determinants to visit places seen in TV series. *European Research on Management and Business Economics*, *27*(1), 100135. Advance online publication. doi:10.1016/j.iedeen.2020.100135

Baird, M. (2020). *Television and Me: The Memoirs of John Logie Baird*. Berlin Ltd.

Bakiewicz, J., Leask, A., Barron, P., & Rakic, T. (2016). Exploring film-induced tourism. In M. Kozak & N. Kozak (Eds.), *Destination Marketing: An international perspective* (pp. 97–109). Routledge.

Bec, A., Moyle, B., Schaffer, V., & Timms, K. (2021). Virtual reality and mixed reality for second chance tourism. *Tourism Management*, *83*, 104256. Advance online publication. doi:10.1016/j.tourman.2020.104256

Beck, J., & Egger, R. (2018). Emotionalise Me: Self-reporting and Arousal Measurements in Virtual Tourism Environments. In B. Stangl & J. Pesonen (Eds.), *Information and Communication Technologies in Tourism 2018*. Springer. doi:10.1007/978-3-319-72923-7_1

Beck, J., Rainoldi, M., & Egger, R. (2019). Virtual reality in tourism: A state-of-the-art review. *Tourism Review*, *74*(3), 586–612. doi:10.1108/TR-03-2017-0049

Beeton, S. (2005). *Film-Induced Tourism*. Channel View Publications. doi:10.21832/9781845410162

Beeton, S. (2010). The Advance of Film Tourism. *Tourism and Hospitality Planning and Development*, *7*(1), 1–6. doi:10.1080/14790530903522572

Beeton, S. (2016). *Film-Induced Tourism* (2nd ed.). Channel View Publications. doi:10.21832/9781845415853

Bolan, P., Boy, S., & Bell, J. (2011). We've seen it in the movies, let's see if it's true. *Worldwide Hospitality and Tourism Themes, 3*(2), 102–116. doi:10.1108/17554211111122970

Bolan, P., & Williams, L. (2008). The role of image in service promotion: Focusing on the influence of film on consumer choice within tourism. *International Journal of Consumer Studies, 32*(4), 382–390. doi:10.1111/j.1470-6431.2008.00672.x

Buchmann, A., Moore, K., & Fisher, D. (2010). Experiencing Film Tourism: Authenticity & Fellowship. *Annals of Tourism Research, 37*(1), 229–248. doi:10.1016/j.annals.2009.09.005

Busby, G., & Haines, C. (2013). Doc Martin and film tourism: The creation of destination image. *Tourism (Zagreb), 61*(2), 105–120.

Busby, G., & Klug, J. (2001). Movie-Induced Tourism: The Challenge of Measurement and Other Issues. *Journal of Vacation Marketing, 7*(4), 316–332. doi:10.1177/135676670100700403

Butler, R. (1990). The Influence of the Media in Shaping International Tourist Patterns. *Tourism Recreation Research, 15*(2), 46–53. doi:10.1080/02508281.1990.11014584

Butler, R. (2011). It's only make believe: The implications of fictional and authentic locations in films. *Worldwide Hospitality and Tourism Themes, 3*(2), 91–101. doi:10.1108/17554211111122961

Carl, D., Kindon, S., & Smith, K. (2007). Tourists' experiences of film locations: New Zealand as 'Middle Earth. *Tourism Geographies, 9*(1), 49–63. doi:10.1080/14616680601092881

Chang, D. Y. (2015). A Study of TV Drama Series, Cultural Proximity and Travel Motivation: Moderation Effect of Enduring Involvement. *International Journal of Tourism Research, 18*(4), 399–408. doi:10.1002/jtr.2058

Cohen, E. (1979). A Phenomenology of Tourist Experiences. *Sociology, 13*(2), 179–201. doi:10.1177/003803857901300203

Connell, J. (2005). Toddlers, tourism and Tobermory: Destination marketing issues and television-induced tourism. *Tourism Management, 26*(5), 763–776. doi:10.1016/j.tourman.2004.04.010

Connell, J. (2005a). What's the story in Balamory: The impacts of a Children's TV Programme on small tourism enterprises on the Isle of Mull, Scotland. *Journal of Sustainable Tourism, 13*(3), 228–251. doi:10.1080/01434630508668555

Connell, J. (2012). Film Tourism – Evolution, progress and prospects. *Tourism Management, 33*(5), 1007–1029. doi:10.1016/j.tourman.2012.02.008

Connell, J., & Meyer, D. (2009). Balamory revisited: An evaluation of the screen tourism destination-tourist nexus. *Tourism Management, 30*(2), 194–207. doi:10.1016/j.tourman.2008.06.001

Couldry, N. (1998). The view from inside the 'simulacrum': Visitors' tales from the set of Coronation Street. *Leisure Studies, 17*(2), 94–107. doi:10.1080/026143698375178

Crompton, J. (1979). Motivations of pleasure vacations. *Annals of Tourism Research, 6*(4), 408–424. doi:10.1016/0160-7383(79)90004-5

Croy, G. W. (2011). Film Tourism: Sustained economic contributions to destinations. *Worldwide Hospitality and Tourism, 3*(2), 159–164. doi:10.1108/17554211111123014

Dann, G. M. S. (1977). Anomie, Ego-enhancement and Tourism. *Annals of Tourism Research, 4*(4), 184–194. doi:10.1016/0160-7383(77)90037-8

Dann, G. M. S. (1981). Tourist Motivation: An Appraisal. *Annals of Tourism Research, 8*(2), 187–219. doi:10.1016/0160-7383(81)90082-7

Dunne, G. (2009). *Motivation and Decision Making in City Break Travel: The Case of Dublin*. VDM Verlag Dr Muller Aktiengesellschaft & Co.

Edensor, T. (2005). Mediating William Wallace: audio-visual technologies in tourism. In D. Crouch, R. Jackson, & F. Thompson (Eds.), *The media and the tourist imagination. Converging cultures* (pp. 105–118). Routledge.

Garrison, S., & Wallace, C. (2021). Media Tourism and Its Role in Sustaining Scotland's Tourism Industry. *Sustainability, 13*(11), 6305. doi:10.3390/su13116305

Griffin, T., Giberson, J., Lee, S. H., Guttentag, D., Kandaurova, M., Sergueeva, K., & Dimanche, F. (2017). Virtual Reality and Implications for Destination Marketing. *Travel and Tourism Research Association: Advancing Tourism Research Globally*, 29.

Hanefors, M., & Mossberg, L. (2002). TV travel shows – A pre-taste of the destination. *Journal of Vacation Marketing, 8*(3), 235–246. doi:10.1177/135676670200800303

Happer, C., & Philo, G. (2013). The Role of the Media in the Construction of Public Belief and Social Change. *Journal of Social and Political Psychology, 1*(1), 321–336. doi:10.5964/jspp.v1i1.96

Hellman, P. T. (2005). *Historical Gazetteer of the United States*. Routledge.

Hsu, H. C. C., & Huang, S. (2008). Travel Motivation: A Critical Review of the Concept's Development. In A. G. Woodside & D. Martin (Eds.), *Tourism Management: Analysis, Behaviour and Strategy* (pp. 14–27). CABI. doi:10.1079/9781845933234.0014

Hudson, S. (2020). *COVID-19 and Travel: Impacts, Responses and Outcomes*. Goodfellow Publishers Limited. doi:10.23912/9781911635703-4387

Hudson, S., & Ritchie, J. R. B. (2006). Promoting Destinations via Film Tourism: An Empirical Identification of Supporting Marketing Initiatives. *Journal of Travel Research, 44*(4), 387–396. doi:10.1177/0047287506286720

Hudson, S., Wang, Y., & Gil, S. M. (2011). The Influence of a Film on Destination Image and the Desire to Travel: A Cross-Cultural Comparison. *International Journal of Tourism Research, 13*, 177–190.

Ismail, M. N. I., Sumarjan, N., Hadi Khan, N. F. A., & Hanafiah, M. H. (2017). Movie Touring: The Influence of Film on Viewers' Travel Intention. In A. Saufi, I. Andilolo, N. Othman, & A. Lew (Eds.), *Balancing Development and Sustainability in Tourism Destinations*. Springer. doi:10.1007/978-981-10-1718-6_11

Iso-Ahola, E. (1982). Towards a social psychology theory of tourism motivation: A rejoinder. *Annals of Tourism Research*, *9*(2), 256–262. doi:10.1016/0160-7383(82)90049-4

Jewell, B., & McKinnon, S. (2008). Movie Tourism—A New Form of Cultural Landscape? *Journal of Travel & Tourism Marketing*, *24*(2), 153–162. doi:10.1080/10548400802092650

Jones, D., & Smith, K. (2005). Middle-earth meets New Zealand: Authenticity and location in the making of the Lord of the Rings. *Journal of Management Studies*, *42*(5), 923–945. doi:10.1111/j.1467-6486.2005.00527.x

Keyes, D. (2020). Amazon has surpassed 150 million Prime subscribers globally. *Business Insider*. https://www.businessinsider.com/amazon-surpasses-150-million-prime-subscribers-2020-2?r=US&IR=T

Kim, S. (2010). Extraordinary Experience: Re-enacting and Photographing at Screen Tourism Locations. *Tourism and Hospitality Planning & Development*, *7*(1), 59–75. doi:10.1080/14790530903522630

Kim, S. (2012). Audience involvement and film tourism experiences: Emotional places, emotional experiences. *Tourism Management*, *33*, 387–396.

Kim, S., Long, P., & Robinson, M. (2009). Small Screen, Big Tourism: The role of popular Korean television dramas in South Korean Tourism. *Tourism Geographies*, *11*(3), 308–333. doi:10.1080/14616680903053334

Kim, S., & O'Connor, N. (2011). A cross-cultural study of screen-tourists' profiles. *Worldwide Hospitality and Tourism Themes*, *3*(2), 141–158. doi:10.1108/17554211111123005

Kim, S. S., Agrusa, J., Lee, H., & Chon, K. (2007). Effects of Korean TV dramas on the flow of Japanese tourists. *Tourism Management*, *28*(5), 1340–1353. doi:10.1016/j.tourman.2007.01.005

Kim, S. S., Kim, S. S., & Petrick, J. (2019). The Effect of Film Nostalgia on Involvement, Familiarity, and Behavioral Intentions. *Journal of Travel Research*, *58*(2), 283–297. doi:10.1177/0047287517746015

Kim, S. S., Lee, C., & Klenosky, D. (2003). The influence of push and pull factors at Korean national parks. *Tourism Management*, *24*(2), 169–180. doi:10.1016/S0261-5177(02)00059-6

Klein, L. (2003). Creating virtual product experiences: The role of telepresence. *Journal of Interactive Marketing*, *17*(1), 41–55. doi:10.1002/dir.10046

Klenosky, D. (2002). The "Pull" of Tourism Destinations: A Means-End Investigation. *Journal of Travel Research*, *40*(4), 396–403. doi:10.1177/004728750204000405

Kork, Y. (2016). How Film Tourists Experience Destinations. In V. Katsoni & A. Stratigea (Eds.), *Tourism and Culture in the Age of Innovation: Second International Conference IACuDiT, Athens 2015* (pp. 145-56). Springer International Publishing AG. 10.1007/978-3-319-27528-4_10

MacCannell, D. (1999). *The Tourist: A New Theory of the Leisure Class*. University of California Press.

Macionis, N. (2004). Understanding the Film-Induced Tourist. In W. Frost, G. Croy & S. Beeton (Eds.), *International Tourism and Media Conference Proceedings* (pp. 86-97). Tourism Research Unit, Monash University.

Macionis, N., & Sparks, B. (2009). Film-Induced Tourism: An Incidental Experience. *Tourism Review International*, *13*(2), 93–101. doi:10.3727/154427209789604598

Maslow, A. H. (1970). *Motivation and Personality* (2nd ed.). Harper & Row.

McIntosh, R. W., & Goeldner, C. R. (1990). *Tourism: Principles, practices, philosophies*. John Wiley & Sons.

McIntosh, R. W., Goeldner, C. R., & Ritchie, J. R. B. (1995). *Tourism: Principles, practices, philosophies*. John Wiley & Sons.

McWha, M., & Beeton, S. (2018). Engaging with popular media through tourism: A brief introduction. *Tourist Studies*, *18*(3), 257–260. doi:10.1177/1468797618782254

Netflix. (2021). *Q4 2020 Financial Statements*. https://ir.netflix.net/financials/financial-statements/default.aspx

Ofcom. (2020). *Media Nations 2020 UK Report*. https://www.ofcom.org.uk/__data/assets/pdf_file/0010/200503/media-nations-2020-uk-report.pdf

Pearce, P. (2005). *Tourist Behaviour. Themes and Conceptual Schemes*. Channel View Publications. doi:10.21832/9781845410247

Pearce, P., & Caltabiano, M. (1983). Inferring Travel Motivation from Travellers Experiences. *Journal of Travel Research*, *22*(2), 16–20. doi:10.1177/004728758302200203

Pearce, P., & Moscardo, G. (1986). The Concept of Authenticity in Tourist Experiences. *Journal of Sociology (Melbourne, Vic.)*, *22*(1), 121–132. doi:10.1177/144078338602200107

Radomskaya, V. (2018). *Popular culture as a powerful destination marketing tool: an Australian study* (Doctoral dissertation). James Cook University, https://researchonline.jcu.edu.au/58966

Reijnders, S. (2011). *Places of the imagination: media, tourism, culture*. Ashgate Publishing Limited.

Rickly, J. M., & McCabe, S. (2017). Authenticity for Tourism Design and Experience. In D. Fesenmaier & Z. Xiang (Eds.), *Design Science in Tourism. Tourism on the Verge*. Springer. doi:10.1007/978-3-319-42773-7_5

Riley, R., & van Doren, C. (1992). Movies as Tourism Promotion: A 'Pull' Factor' in a 'Push' location. *Tourism Management*, *13*(3), 267–274. doi:10.1016/0261-5177(92)90098-R

Rittichainuwat, B., & Rattanaphinanchai, S. (2015). Applying a mixed method of quantitative and qualitative design in explaining the travel motivation of film tourists in visiting a film-shooting destination. *Tourism Management*, *46*, 136–147. doi:10.1016/j.tourman.2014.06.005

Roesch, S. (2009). *The Experience of Film Location Tourists*. Channel View Publications. doi:10.21832/9781845411220

Roesch, S. (2018). Film Tourism in the Golden Age of Television. In *The Routledge Handbook of Popular Culture and Tourism*. Routledge. doi:10.4324/9781315559018-8

Santos, C., & Kim, S. (2017). Promoting Tourism, Projecting Power: The Role of Television Commercials. *Journal of Travel & Tourism Marketing, 34*(2), 192–208. doi:10.1080/10548408.2016.1156610

Signorielli, N., & Morgan, M. (2001). Television and the Family: The Cultivation Perspective. In J. Bryant & J. A. Bryant (Eds.), *Television and the American Family* (2nd ed.). Lawrence Erlbaum Associates, Inc.

Spigel, L. (2013). *Make Room for TV: Television and the Family Ideal in Postwar America*. The University of Chicago Press.

Stewart, J. (2016). *Seeking the 'Sites/Sights' Seen on the Small Screen! An investigation of tourist motivation in the context of television film-induced tourism* [Unpublished master thesis]. Dublin Institute of Technology, Dublin, Ireland.

Tiberghien, G., Bremner, H., & Milnec, S. (2020). Authenticity and disorientation in the tourism experience. *Journal of Outdoor Recreation and Tourism, 30*, 100283. Advance online publication. doi:10.1016/j.jort.2020.100283

Tooke, N., & Baker, M. (1996). Seeing is believing: The effect of film on visitor numbers to screened locations. *Tourism Management, 17*(2), 87–94. doi:10.1016/0261-5177(95)00111-5

Torchin, L. (2002). Location, location, location: The destination of the Manhattan TV tour. *Tourist Studies, 2*(3), 247–266. doi:10.1177/14687976020023002

Trimurti, C. P., & Utama, I. G. B. R. (2020). The examination of the Bali destination marketing chain model based on tourist motivation. *International Journal of Scientific and Technology Research, 9*(3), 4491–4495.

Urry, J. (1990). *The tourist gaze: leisure and travel in contemporary societies*. Sage Publications Ltd.

Urry, J., & Larson, J. (2011). *The Tourist Gaze 3.0*. SAGE Publications Ltd. doi:10.4135/9781446251904

Walt Disney Company. (2021). *The Walt Disney company reports first quarter earnings for fiscal 2021*. https://thewaltdisneycompany.com/app/uploads/2021/02/q1-fy21-earnings.pdf

Wheeler, K. S. (2015). The Relationships Between Television Viewing Behaviors, Attachment, Loneliness, Depression, and Psychological Well-Being. *University Honors Program Theses,* Paper 98. https://digitalcommons.georgiasouthern.edu/honors-theses/98

Yen, C., & Teng, H. (2013). Celebrity Involvement, Perceived Value, and Behavioural Intentions in Popular Media-Induced Tourism. *Journal of Hospitality & Tourism Research (Washington, D.C.), 39*(2), 225–244. doi:10.1177/1096348012471382

Yoo, C. K., Yoon, D., & Park, E. (2018). Tourist motivation: An integral approach to destination choices. *Tourism Review, 73*(2), 169–185. doi:10.1108/TR-04-2017-0085

Yung, R., & Khoo-Lattimore, C. (2017). New realities: A systematic literature review on virtual reality and augmented reality in tourism research. *Current Issues in Tourism, 22*(17), 2056–2081. doi:10.1080/13683500.2017.1417359

ADDITIONAL READING

Bolan, P., & Kearney, M. (2017). Exploring film tourism potential in Ireland: From Game of Thrones to Star Wars. *Revista Turismo & Desenvolvimento, 27/28*, 2149–2156.

Kim, S., & Kim, S. (2018). Perceived values of TV drama, audience involvement, and behavioural intention in film tourism. *Journal of Travel & Tourism Marketing, 35*(3), 259–272. doi:10.1080/10548408.2016.1245172

Kim, S., & Reijnders, S. (Eds.). (2018). *Film Tourism in Asia. Evolution, Transformation, and Trajectory*. Springer. doi:10.1007/978-981-10-5909-4

Potočnik-Topler, J., & Špenko, T. (2019). Film tourism as a tool of tourism development: The Representation of Scotland in the Outlander TV series. TIMS. *Acta, 13*(2), 79–88. doi:10.5937/timsacta1902079P

Sousa, B., Malheiro, A., Liberato, D., & Liberato, P. (2021). Movie Tourism and Attracting New Tourists in the Post-pandemic Period: A Niche Marketing Perspective. In J. V. de Carvalho, A. Rocha, P. Liberato, & A. Peña (Eds.), *Advances in Tourism, Technology and Systems. ICOTTS 2020. Smart Innovation, Systems and Technologies* (Vol. 208). Springer. doi:10.1007/978-981-33-4256-9_34

Teng, H., & Chen, C. (2020). Enhancing celebrity fan-destination relationship in film-induced tourism: The effect of authenticity. *Tourism Management Perspectives, 33*, 1–11. doi:10.1016/j.tmp.2019.100605

Tkalec, M., Zilic, I., & Recher, V. (2017). The effect of film industry on tourism: Game of thrones and Dubrovnik. *International Journal of Tourism Research, 19*(6), 705–714. doi:10.1002/jtr.2142

KEY TERMS AND DEFINITIONS

Fanatical Film Tourist: Specific dedicated film tourists or purposeful film tourists, as they are taking part in an organised tour exclusively for a specific television programme and its film locations.

Film Tourism: Tourist visits to a destination or attraction because of the destination being featured on television, video, or the cinema screen.

Primary Motivators: Initial socio-psychological motivations that create a need within people to travel – exploration and evaluation of self, escape from a perceived mundane environment, relaxation, health and fitness, enhancement of kinship relationships, facilitation of social interaction, novelty, cultural experiences, education, prestige.

Secondary (Site Specific) Motivators: An expansion of the primary motivators in the way that they relate to site-specific film tourism. A primary motivator may be facilitation of social interaction, whereas the secondary motivator (site specific motivator) may be enjoying the television series and want to share that interest with others or want to meet the cast and characters of the show.

Site-Specific Film Tourism: Tourism to one particular location because it featured on television.

Chapter 13
Experiencing the Story:
The Role of Destination Image in Film-Induced Tourism

Eleni Michopoulou
https://orcid.org/0000-0002-1857-4462
University of Derby, UK

Aleksandra Siurnicka
University of Derby, UK

Delia Gabriela Moisa
University of Derby, UK

ABSTRACT

The importance of destination image in film tourism has been recognized by scholars and practitioners. However, despite a large number of research papers related to the destination image within the field of film tourism, several issues remain unclear. This chapter provides insights into how movies influence the featured destination's image by focusing on specific film tourists' perceptions, their motivations, and emotional relation to the movies. The chapter begins by offering a film tourism definition followed by film tourist typology with the context of film fans. Then, factors influencing film tourism destination image are examined, in particular destination marketing activities, film-specific factors, and destination attributes. Two case studies will also be provided to better showcase the findings from the literature review. Theoretical and practical implications are also presented.

INTRODUCTION

An increase in international travel and the development of the entertainment sector translated into growth for film tourism, which within a relatively short time became a rapidly growing sector of the tourism industry (Hudson & Ritchie, 2006). As Cardoso et al. (2017) note, this phenomenon is included in the relatively new subject of 'film-induced versus destination branding image' research. When looking at the

DOI: 10.4018/978-1-7998-8262-6.ch013

contents of destination branding, destination image emerges as the crucial one, as it is the core variable that influences consumers' choices for destination selection and travel related decision making (Cardoso et al., 2017). Therefore, assessing the influence of a particular screen production on the featured destination's image becomes an important issue within film-induced tourism. Roberson & Grady (2015) suggest that one of the ways to understand the phenomenon of film tourism is to focus on tourists who are at the same time members of fan communities.

A wide scope of theoretical and practical knowledge within the field of film tourism literature has been developed, raising issues such as film tourism phenomenon in general, destination image, marketing and promotional activities, film tourists' motivations, activities and experiences, and others (Domínguez-Azcue et al., 2021, Gjorgievski & Trpkova, 2012 Hahm & Wang, 2011, Macionis & Sparks, 2009). However, the literature on fans as screen tourists remains scarce. Hence, this chapter focuses on this relatively small group of visitors, represented by fans of fantasy movies. The aim of the chapter is to identify which specific destination image components are more important to specific film tourists; in this case fans of fantasy films. The chapter also considers how particular destinations use films as an aid in promoting and developing their destination image, what location attributes play a significant role in inducing the core screen tourists, and what characteristics should movies possess to generate the potential to induce tourism.

Film Tourism

One of the most recognised definitions of film tourism is the one provided by Cardoso et al. (2017, p. 24), acknowledging that "*film-induced tourism takes place whenever tourists decide to visit a place after having been attracted by projected audiovisual images*". Thus, it can refer to visiting actual locations where the movies were set, as well as to places related to the movies, like theme parks or production studios (Beeton, 2005). Film tourism is however much wider phenomenon - including activities such as traveling to take part in film festivals, film-themed attractions, film premieres and awards ceremonies, or visiting places that have been marketed through film connection (Connell, 2012).

The positive influence of film tourism on destinations has been well established (Domínguez-Azcue et al., 2021). As a general rule, the higher the number of tourists, the greater impact on economy. A UK based research showed that 8 out of 10 individuals chose their holiday destination based on a recent film they watched and 1 in 5 "actually make the 'pilgrimage' (Beeton, 2005, p.34) to the location of their favourite film" (Hudson et al., 2011, p.178).

The global publisher Travelzoo also compiled research into screen tourism and its impact on the UK's economy and communities. The South Shropshire Tourism Board estimated 'Atonement' brought in £1.3 million to the area before the film was even released, Shibden Hall, as seen in 'Gentleman Jack', has seen visitor numbers increase by 700% in 2019, and East Cambridgeshire visitor numbers rose to nearly 4 million after the first series of 'The Crown' was broadcast (Morris, 2019). Northern Irland's 'Game of Thrones' tourists contributed in excess of £50M to the local economy in 2018 (Morris, 2019).

The most relevant aspect of film tourism is therefore its positive impact on economy, but also its capacity to endure, proving that films keep attracting visitors each year, while also reducing seasonality problems (Hudson & Ritchie, 2006).

According to Olsberg SPI (2015), the scope of values of the location can be significantly broadened when screen production turns out to be successful. The report 'Quantifying Film and Television Tourism in England' produced by Olsberg SPI in association with VisitEngland gathered data from eight movie locations, including Alnwick Castle (famously home to Harry Potter's Hogwarts), the village of Bampton (Downton Abbey), West Bay (Broadchurch), Castle Howard (Brideshead Revisited, Death Comes to Pemberley), Holkham (Shakespeare in Love, The Duchess), Lyme Park (Pride and Prejudice, The Awakening) Puzzlewood (Merlin, Doctor Who) and Wollaton Hall (The Dark Knight Rises). According to the numbers, the value from both overseas and national visitors can vary from £0.1 million to over £4 million per year (Olsberg SPI, 2015). DMOs often capitalise on this chance to tap into the additional promotion and see an increase in both visitors' numbers in film locations and in profits. Inbound marketing campaigns have been targeted especially in Australia, New Zealand, Korea, the USA, and the UK (Connell, 2012).

Film Tourists Typologies and Motivations

Several attempts to distinguish types of film tourists based on their motivations have been undertaken (Macionis, 2004; Rittichainuwat & Rattanaphinanchai, 2015; Kim & Kim, 2017). Macionis (2004) identified three types of screen tourists formed on the push and pulls motivation theory of Dann (1977), who described push factors as internal drivers of visitors like escape or nostalgia, and pull factors as external drivers, such as beautiful scenery or location attributes. The first category distinguished by Macionis (2004) relates to serendipitous tourists, who happen to be in the movie location by accident. The second category belongs to general film tourists, who may relate to the film-related activities, but screen production was not their main motivation for coming to the location. And the third one refers to specific film tourists, whose main purpose is visiting film locations and looking for a deep film experience (Macionis, 2004).

Figure 1. Typology of film tourists and Continuum of Film-Induced Motivations.
Adopted from Macionis (2004).

Increasing interest in film →		→
Serendipitous Film Tourist	General Film Tourist	Specific Film Tourist
• those who just happen to be in a destination portrayed in a film	• those who are not specifically drawn to a film location but who participate in film tourism activities while at a destination	• those who actively seek out places that they have seen in film
Motivations include: • social interaction • novelty	Motivations include: • escape • novelty • education • nostalgia	Motivations include: • ego-enhancement • self-actualisation • pilgrimage • self-identity • vicarious experience • fantasy • status/prestige • romance • nostalgia
Increasing self-actualisation motivations →		→
Decreasing importance of authenticity →		→
Increasing importance of Push Factors →		→

Visitors are inspired to visit film tourism destinations not only because of the attractive physical surroundings but also for motives of pilgrimage, nostalgia, and escape (Macionis & O'Connor, 2011). Macionis (2004) distinguishes fantasy as one of the specific film tourists' motivations. Increasing interest in film and self-actualization motivations goes in pair with decreasing importance of authenticity; which is especially important in the context of fantasy movies. In this rationale, it can be assumed that specific film tourists are fans of visited location's movies.

According to Connell (2012), the concept of fandom in movie-induced tourism has its source in the emotional relation of the viewer (fan) to the 'viewed' (fan's interest object). Taking this further, fantasy movies' fans can be classified as what Roesch (2009) calls '*romantic gazers*', the type of tourists who have more emotional expectations when visiting a location than the random tourist. Visitors of this type prefer visiting places alone or with a close friend and aim to appreciate the time spent on location. Roesch (2009) also claims that the film site is perceived by them as a kind of sacred place rather than as a mere tourist destination. Therefore, they strive for a deeper experience of the visited place. This sentiment was also maintained by Oviedo-García et al. (2014, p. 722) whose study showed that screen tourists to Seville were motivated by having "first-hand experiences related to the film" and appreciation of the moment spent on location. Important factors were also fantasy (imagining themselves in the role of actors in the film), novelty (referring to a unique experience when visiting a location, touring the film), a feeling of personal involvement with the film, the possibility of linking the movie's plot with places of visitor interest, and finally, personal film-location connection (having a positive personal feeling because of the undertaken trip) (Oviedo-García et al., 2014).

Several studies found core film tourists as a minority among film-induced visitors (Rittichainuwat & Rattanaphinanchai, 2015; Olsberg SPI, 2015). Such results have been confirmed by study of Araújo Vila et al. (2021) where results showed that for 10,4% of tourists to the TV series' locations in Spain, screen production was the main motivational factor to visit. However, to get a deeper insight into the phenomenon of film tourism, there is a need to examine other elements that play a role in inducing film tourists, such as destination image as discussed in the next section.

Destination Image

Destination image is generally defined as "a set of beliefs, ideas, and impressions that people have of a place or destination" (Baloglu & McCleary, 1999, p. 871). It entails a distinguished cognitive component which relates to a people's perceptions of the destination's physical attributes, and an affective component, which refers to people's emotional reception of a place (Prayag & Ryan, 2012). Pike & Ryan (2004) maintain that one more dimension needs to be examined, conative image, referring to the behavioral aspect of the place in a certain period of time. Interactions between those three components form an overall image of the destination.

Numerous studies regarding destination image issues can be found in tourism literature, in which diverse approaches have been applied, as it is a basic factor shaping visitors' attitudes toward tourist destinations (Yoon & Uysal, 2005; Hahm, & Wang, 2011; Afshardoost & Eshaghi, 2020). The formation of a destination image "stimulates information search behavior concerning that destination and this heightened interest level leads to potential travel behavior to visit the location" (Hahm & Wang, 2011, p.167). Zhang et al. (2014) indicate that perceiving the destination image in a positive way results in an increase of interest in visiting and revisiting among tourists, and well rated trip quality. Afshardoost and

Eshaghi's study (2020) suggested that the strongest influence on tourists' behavioral intentions, (intention to visit, re-visit and recommend), comes from overall and affective images, followed by cognitive image.

In terms of destination image formation however, Prayag & Ryan (2012) suggest three components, namely: organic, induced, and autonomous. The organic image is generated by non-formal sources of information like people's knowledge or actual experiences from visiting destinations. External and formal sources, such as destination marketing activities, the Internet, brochures, travel agencies, are responsible for creating the induced image. Finally, the autonomous image can be improved through media information sources and pop-culture which, as suggested later by has much more powerful influence on destination image than previously thought; making pop-culture also a strong agent of destination image formation (Lee & Bay, 2016)

Destination Image in Film Tourism

Numerous destinations use screen productions to assist in creating and promoting a positive image of their destination, noticing the immense influence they have on enticing potential tourists (Hahm & Wang, 2011). Hence, destination image becomes a key issue to explore within the context of film tourism; and as such it has been studied under various prisms, including: the general role of destination image in film tourism (Chen, 2018), movies' influence on changing of destination image' perceptions (Kim & Richardson, 2003), and the role of destination image in film tourists decision-making process (Whang et al., 2016).

Under the umbrella of cultural tourism, film-induced tourism represents the cultural heritage of a destination, having ''great potential to advance cultural exchange and understanding'' (Kim et al., 2007, pp. 1351), while being considered an expression of visual arts and local traditions (Gjorgievski & Trpkova, 2012), A key factor for tourists and film tourism is represented by authenticity, as people are given the chance to live out their fantasies of their facvorite actors or movies in sometimes mythical places (Hall, 2003).

A field of clear expansion in recent literature, video games based on real locations apply the same principle (Dubois et al., 2021; Dubois & Gibbs, 2018). Based on this developing nexus, Shaheer's (2021) paper conceptualized the nascent link between video gamed-induced tourism and heritage attractions, arguing that video games could potentially represent a driving force for restoration, preservation and conservation of heritage attractions on tourism destinations. For example, Mochocki (2021) presented a detailed analysis of the player's perception of heritage authenticity with the central example of Assassin's Creed Unity and its famous representation of the Notre-Dame cathedral.

In order to get more insight into factors influencing film tourism destination image, the subsequent discussion will be based on the model proposed by Hudson & Ritchie (2006), according to which a screen tourist is motivated to travel to the location by push or pull factors (motives) influenced by one or more of three agents: *destination marketing activities*, *film-specific factors*, and *destination attributes*. Hudson & Ritchie (2006, p.390) mention that the success of the movie in inducing film tourism depends also on two more factors which are film commission and government efforts, what includes issues such as "lobbying, tax breaks, scouting services, dedicated Web sites, active promotion", and location feasibility, including such elements like "resources, cost, taxes, labor, and expertise". This chapter will focuses on the first three components, *destination marketing activities*, *film-specific factors*, and *destination attributes*.

Destination Marketing Activities

Hudson & Ritchie (2006) separate destination marketing activities into these undertaken before and after the release of the movie. As one of the activities before the release, they indicate promoting actively locations to film producers by DMOs. An example of that can be the 3-year UD$7 million program introduced in 2004 by the Singapore Tourism Board with the aim to entice main international film-makers to set projects there (Hudson & Ritchie, 2006). Some DMOs hire public relations specialists to cooperate with TV studios. However, promotion after the release of movie creates more possibilities to entice visitors, as they have already had their first impressions of the destination depicted in the film, enriched with an emotional relation to it.

DMOs use the time after the movie's release to contribute to marketing activities regarding film tourism in order to increase the interest of additional visitors enticed by the movie. One of the crucial activities is the collaboration with the movie industry (Hudson & Ritchie, 2006). DMOs use relations with movie committees "to track productions and film releases so the organizations are in a position to act as soon as they see the signs of film tourism" (Hudson & Ritchie, 2006, p. 391). Another effective way to entice film tourists are movie maps, the example of what can be the VisitEngland movie map campaign started already in 1996, featuring 200 screen locations from 60 years of UK's cinema history, and becoming rapidly one of VisitBritain's most popular printed products (Hudson & Ritchie, 2006). A new series of movie maps show up together with new movies depicting the UK locations. Research undertaken by Pratt (2009) showed that The Pride & Prejudice 2005 movie map marketing campaign was the main driver in enticing potential visitors to the film locations.

Different authors including Giuliana Bruno (2002) in her *Atlas of Emotions*, and Tom Conley (2006) in *Cinematic Cartography* explored the various dimensions of maps in films, including the narrative, symbolic, emotional and political aspects, but also explored films as maps. These authors established cinema not only as a visual art, but also as a spatial art.

Among other important marketing activities, film walks and guided tours can also play a significant role in destination promotion. Examples of those include the California tour related to the movie Sideways, and National Treasure movie self-guided tours that follow Nicolas Cage footsteps (Hudson & Ritchie, 2006). Hotels, museums, and attractions frequently take advantage of the opportunity that featuring in movies gives to promote themselves to drive visitations; such as The Crown Hotel at Amersham in the UK advertising the room used by Andie MacDowell and Hugh Grant in 'Four Weddings and a Funeral' movie (Hudson & Ritchie, 2006).

Attributes featuring on TV productions and on advertising forms like travel brochures, postcards, guidebook photography, and the Internet can help in the process of creating the image of a destination (Busby & Haines, 2013). Dedicated film Websites that show locations linked to the movie, as well as provide information regarding tours and other film-related activities are therefore highly recommended (Croy & Walker, 2003).

The fan motivations to travel are frequently special to the individuals. For some of them, the leading motivating factor can be the desire for an authentic experience, while for others it can be closeness to a celebrity (Roberson & Grady, 2015). Thus, a few researchers placed their focus on investigating the influence of celebrity involvement in creating a destination image. Veasna et al. (2013) assume that when considering brand signal theory, impressions of visitors regarding related destination image can be significantly influenced when they trust a reliable destination source. Lee et al. (2008) indicate that fans of celebrities admire and trust them. Thus, celebrity involvement can induce fans to perceive locations represented by people they admire in a positive way and to find visiting those places as a form of pilgrimage (Chen, 2018).

Film-Specific Factors

Riley & Van Doren (1992, p. 269) indicated that when a movie possesses one or more characteristics out of “uniqueness, status, or timely significance” it can generate the potential to induce tourism. The uniqueness of the movie relies on distinguishing it from other productions and contributes to audience recognition. Status plays an important role as it draws “potential viewers to a socially condoned feature on the ‘big screen’’ (Riley & Van Doren, 1992, p. 269). Finally, timely significance reduces the possibility of lack of interest linked to anachronistic screen productions (Riley & Van Doren, 1992).

Tooke & Baker (1996) claim that if there is a close relationship between storyline and site in the movie and if viewers engage emotionally in the story and link those experiences with the destination, the film becomes an effective driver in inducing tourism. This becomes especially important in the case of core screen tourists/fans, who can develop place attachment to a particular unvisited location, on the basis of their emotional engagement in the movie, story, characters, and places depicted there (Hosany et al., 2019). Kim (2012) acknowledged the importance of the character when an individual feels close or becomes involved with it, while Beric et al. (2013) confirmed that the film experience is private and personal because the individual personally interprets the media.

Hudson & Ritchie (2006) extended those findings and listed more of the film-specific factors that can contribute to induce visitors to the locations. Those relate to both characteristics of the movies and linked locations, and include the success of the movie and its location'saccessibility and recognizability, together with the length of screen exposure., As movie settings may hold emotional attachment, tourists are also influenced by the image that they want to explore,, pollution free environments, and a place with a tangible icon tourists can identify with.

Nonetheless, and equally important role in film induced tourism is played by the film's destination attributes, and these are discussed in the following section.

Destination Attributes

In order to entice visitors, the destination should possess various attributes (Hudson et al, 2011). In tourism literature, a number of approaches to the classification of destination attributes can be found (Moon & Han, 2018; Eom et al., 2020). However, destination attributes important to film tourists cannot be fully understood by adopting traditional tourism classifications, as this specific group of visitors (and especially fans), significantly differs from generic tourists (Beeton, 2005; Connell, 2012). This argument is further supported by Güzel & Güzel's (2017) who showed that there is a significant difference between perceptions of destination attributes between film tourists and non-film tourists. Film tourists indicated shopping facilities, culture and climate, and activities as important attributes. Moreover, it was found that the activities and locality of destination attributes influence film tourists' satisfaction, whilst for non-film tourists only the physical environment had an impact on satisfaction (Güzel & Güzel, 2017).

In the matter of film tourism, Riley & Van Doren (1992) propose that the perfect model to attract individuals to a movie location includes stunning landscape features, a distinctive cultural and social outlook, and an image or icon that visitors can refer to and want to discover. Potočnik-Topler & Špenko

(2019) also suggested that for film-induced visitors, destination attributes such as distinguished scenery, sets and backdrop, icons, awareness, and brand are particularly important.

As Roberson & Grady (2015) maintain, for film fans an important factor related to destination is location authenticity, as for most film tourists the visit to a chosen location is extraordinary because of the link between scenery and movie. Tourists judge authenticity in a personal way and attribute value to those judgments. Therefore, film tourists search for authentic experiences build upon the place and their interpretations of the experience (Buchmann et al., 2010).

In terms of film tourist experiences, Reijnders (2011) in his research related to Dracula fandom tourism, discovered the existence of the 'inner experience of the Dracula tourist', which can be characterized by the relation between two opposite modes. The first mode is a tourist's wish to compare pictures he knows from the book/film and the landscape he is visiting. When traveling to movie locations, fans hope to discover some 'truth' behind the story. And the second mode says that when being on-site, tourists wish to create a kind of symbiosis between imagination and reality; and accomplish that by paying less attention to details but focusing more on experiencing the location in general; that is "to taste, see, hear, feel and smell it—in order to experience the story anew thanks to these sensory stimuli" (Reijnders, 2011, p. 246).

Researchers also emphasise the importance of recognizable scenery and iconic locations as factors influencing a successful movie tourism product (Beeton, 2010; Hudson & Ritchie, 2006). Jewell & McKinnon (2008) stress that the authenticity of a film lies in the link of facts, myths, and imagination. Hence, 'experience and a new reality based on tangible remains' are far more important than scientific historical evidence (Jewell & McKinnon, 2008, p. 155). This was also mentioned by Macionis (2004) in his statement that importance of authenticity decreases with increase of tourists push motivations; still, authenticity remains important in terms of the scenery that film tourists can relate to when engaging with their imagination and experiencing 'hyper-reality'.

Roberson & Grady (2015) stress the importance of a fans's willingness to re-enact the movie through traveling to the film connected sites, interaction, and consumption (buying souvenirs, taking photographs, meeting participants). Fans can replicate movie scenes or fan objects by taking photographs, or standing in the same places as their stars, offering them the chance to interact with the location in a much more personal way, rather than they would have by just watching the film(Roberson & Grady, 2015). To tourists, souvenirs represent a 'facilitator of memory' or 'tangible evidence to relive the experience' or just 'evidence' of their travel (Wilkins, 2011). However, fans are frequently collectors of memorabilia not only for the mentioned above reasons but perceiving it also as a "path to friendships and community" (Roberson & Grady, 2015, p. 54), what distinguishes them from random visitors.

Several studies found also guided tours (Carl et al., 2007) or film-related activities (Olsberg SPI, 2015) as factors which can enhance film tourists experience and attractiveness of film location. An example of such activity could involve the opportunity to dress up in Edwardian epoque style at Lyme Park, location of Pemberley in the Pride and Prejudice 1995 BBC series. Guided tours may enhance knowledge related to movies and location, particularly by providing interesting details about movies' making-offs around the location (Carl et al., 2007).

Consideringall the above, it is suggested that the focus should be placed on how tourists can experience alternative reality based on tangible remains when visiting a film tourism destination and what attributes can positively influence film destination image. The creation of the destination image directed to specific film tourists should take into account the authenticity of location [or tangible representation of location used in the movie (Beeton, 2010)], recognizable sceneries showcased in screen productions, tangible remains like movie props tourists can relate to, film-related activities such as guided tours, and film-related merchandise (e.g. souvenirs).

White (2017) reinforced the extraordinarily importance of a destination's highlight of the 'emotional hook of the film's story' and its ability to connect that emotion to the location where the movie was filmed. Furthermore, he confirms that there are many film fans who would happily pay large amounts of money to visit an authentic location (White, 2017). Therefore, DMOs should consider strategies for increasing emotional attachment to the movie and the associated site, as suggested by Teng and Chen (2020). These strategies could involve exclusive fan clubs or celebrity activities as a method of 'reinforcing the fans' sense of belonging toward film celebrities.

Case Studies: The Lord of The Rings (New Zealand) and Harry Potter (UK)

In this section, two locations of fantasy movies will be considered to better showcase the above-described factors: destination marketing activities, film-specific factors, and destination attributes, as applied in real life examples.

A common example of screen tourism destination image among researchers became New Zealand with its numerous locations used in the Lord of The Rings motion picture trilogy. Within the time of five years since the release of the first movie in 2001, the number of tourists increased by 40%. By 2018, the country has had 3.6 million visitors per year. Moreover, one out of five overseas visitors admit the LOTR series was the leading factor that motivated them to travel to New Zealand (Arnold, 2019).

Harry Potter has been filmed in numerous locations across the UK, of which the iconic one is Alnwick Castle, which appeared in the first two movies in the eight-part series: Harry Potter and the Philosopher's Stone (2001) and Harry Potter and the Chamber of Secrets (2002). After Alnwick Castle was exposed in movies, the number of tourists to the location rose from 60,000 to 300,000 in 2003, the number of visitors is still around 290,000 (Bąkiewicz et al., 2017). According to the Report for Creative England Olsberg SPI (2015), around 20% of visitors to the site could be described as core screen tourists, motivated particularly by seeing Harry Potter movies.

Destination Marketing Activities

New Zealand taking advantage of the wide possibilities coming from the settings of The Lord of the Rings series became the classic example that illustrates the use of the film in tourist promotion. The Government of New Zealand acknowledged the opportunity of tourism promotion given by movies and developed the image and perception of the country as a tourism destination by providing funding to encourage and estimate positive results of the trilogy (O'Connor et al., 2008). As Buchmann & Frost (2011, p. 57) comment, "the movies and the 'making of' videos promote New Zealand as a green, exotic faraway place possessing mythical qualities". Tourism New Zealand put emphasis on promoting LOTR and its film locations throughout the country by developing part of its website devoted to this topic. The website noted more than 1 billion views within a year after the 2002 Academy Awards and after New Zealand being announced by numerous ads as the best-supporting country (Hudson & Ritchie, 2006). The national airline Air New Zealand used LOTR trilogy to its promotional campaigns that linked the carrier and the country to the movies. Due to these numerous promotional activities, links between Middle-Earth and New Zealand have been reinforced (Carl et al., 2007). Moreover, various businesses across New Zealand take advantage of promotional opportunities given by LOTR trilogy. One example is the Middle-Earth winery, located in the Nelson region, the synonymous geographical centre/Middle-Earth of New Zealand (Arnold, 2019).

Alnwick Castle uses its website as a tool of promoting itself as a movie location. It provides there a wide range of information regarding filming in the location, including a section dedicated to the Harry Potter series (Alnwick Castle, n.d.). Promotion tools are also themed scheduled events such as the Broomstick Training sessions or Battleaxe to Broomsticks tour. Moreover, Alnwick is advertised as 'Harry Potter's castle' by Northumberland Tourism via the website and numerous promotions (Olsberg SPI, 2015). One of the hotels near Alnwick Castle took its name after the tavern in the Harry Potter films, what can additionally entice film-induced visitors to choose this particular accommodation over another (The Inn Collection, n.d.). Much more of such examples can be found across the UK, like Harry Potter themed apartment in Edinburgh or De Vere House in Lavenham in Suffolk (Delahaye, 2020).

Film-Specific Factors

Carl et al. (2007) claim that the Lord of The Rings series meets all three factors introduced by Riley and Van Doren. The uniqueness of those movies refers to the '80s when it has been believed yet that Tolkien's novels cannot be translated into a popular dramatic production such as movie. Researchers attributed the unusual status of the trilogy to being filmed simultaneously on at several sets over 1,5 years as one project and paid attention to the fact that the movie's motives regarding friendship and the issue of rescuing the world coincided in time with the important problem of global politics and the fight against terrorism in 2001 (Carl et al., 2007).

Harry Potter is also undeniably ingrained within popular culture. The wide range of merchandise associated with the novels by J.K. Rowling and the movies, including housewares, clothing, websites, Internet forums, and even music, led the Harry Potter icon to the cultural forefront (Dunn, 2006), and has consequently gained a large number of fans.

Lord of the Rings and Harry Potter are placed in the range of the first 30 top movies based on the cumulative worldwide box office (The Numbers, 2021). The film series can have stronger impacts than single productions, because of the greater screen time and extended viewing opportunities. Moreover, a particularly impactful factor for the success of both movies is that they are literary adaptations, thus they are productions having a wide pre-existing audience base (Olsberg SPI, 2015). Features that connect them, i.e. adventure, fantasy, action, alternate realities and worlds, magic, and the rivalry between good and evil, contribute to their being deeply embedded in the culture of the 20th century, which provides those movies with time significance and status that still draws potential viewers.

Destination Attributes

Screen tourists want to escape into the in-between real and imagined world and experience cultural landscapes presented on the screen. Thus, the landscape becomes attraction used to "increase potential tourists' awareness of a destination and familiarize them with it" (Carl et al., 2007, p. 51). In the movies based on The Lord of the Rings trilogy by Tolkien, an alternative image of New Zealand has been presented as it shows the landscape within a fantasy framework (Busby et al., 2013). In the study of Buchmann et al. (2010, p. 239), one of the interviewees admitted that "New Zealand is, indeed, Middle-Earth", what shows that visiting the real-world film' setting in connection with imagination and emotional relation to the movie provides specific film tourists with a deep experience of the location.

In Carl et al. (2007) research, some of the interviewees admitted that they enjoyed evidence of the film's sets remained, highlighting especially Hobbiton with hobbit holes or the party tree. In order to make the experience 'authentic', some of the core screen tourists are willing to re-enact scenes from films. In Carl et al.'s (2007) study, an interviewee described positively standing on the same hill where Eowyn has been standing in the movie and looking at the same view. Elsewhere, others enjoyed dancing like hobbits under the party tree. Hence, they could experience a real representation of the fantasy world featured and enhance the experience by re-acting movie scenes.

Nation-wide tours and site-specific operators offer a scope of products focused on the trilogy's locations, including coach tours, jet boat experiences, hiking tours, cave explorations, and even helicopter flights (New Zealand, n.d.). Merchandise is also widely available, for example see the Weta Cave shop offering a wide range of LOTR-related products (Tours Weta Workshop, n.d.). This is particularly important as tour operators and guides with deep knowledge and behind-scenes insights, appreciation of natural landscapes, and the possibility of immersion in the Middle-Earth world, are the key three factors influencing the satisfaction of movie tourists (Carl et al., 2007)

In the case of Harry Potter, when visiting Alnwick castle, tourists have the opportunity to be in the exact places where the Harry Potter movies were filmed, allowing them to put themselves in places of actors, re-act scenes or just immerse deeper into the film experience. One of the fans described her experience of visiting Alnwick Castle as follows: "I could barely contain my excitement when I walked in and saw the Outer Bailey, where Harry and his fellow students learned to fly broomsticks with Madam Hooch" (Sunsets and Suitcases, 2017). A broomstick activity is just one of the film-related activities developed to entice and satisfy tourists visiting the Alnwick Castle. In terms of film-related merchandise, an onsite and online shop that provides a wide range of Harry Potter products is available (Alnwick Castle, n.d.).

Both LOTR and Harry Potter locations offer attributes as discussed earlier, including recognizable scenery or tangible remains, film-related activities, and merchandise.

CONCLUSION

Some past studies have focused on specifying the importance and creation of the destination image in film tourism (Tkalec, Zilic, & Recher, 2017; Connell, 2012; Thelen, Kim, & Scherer, 2020). Specific film tourists base their intention to visit more on particular destination image attributes rather than on the general image of the destination (Hahm & Wang, 2011; Wen, et al., 2018). Yet, little is known about film fans' perceptions of the destination image and destination attributes in film tourism literature (Thelen, Kim, & Scherer, 2020; Teng, 2020). Thus, this study answers calls by authors such as Araújo Vila, et al. (2021) and Teng (2020) for further research on the topic. The chapter highlights such attributes to provides a more in-depth understanding of the fans' perspectives of the film destination image.

IMPLICATIONS FOR PRACTICE

Understanding the impact of films on film tourists' (file fans') perceptions of destination image can provide clues for (re)constructing the destination image formation process' model. Hence, destinations can use this kind of knowledge to attract and promote themselves as filming locations to movie companies and producers.

By better understanding how film tourists perceive and experience movie locations, DMOs and destination managers can enhance their destination image to better meet the expectations of this relatively small group; which, with still improving cinematography, development of pop culture and international travel, has the potential to develop into an increasingly significant market segment

FUTURE RESEARCH DIRECTIONS

Whilst this study sheds some light into the film tourists' perspectives of the film destination image, there are still a number of issues ripe for further investigation. For example, further research should focus on how a location can enhance its attractiveness for the core screen tourists and maximize their presence- in a way that will not alienate other types of tourists, reduce place authenticity, or affect other destination facets. Researchers could also investigate the perceptions of non-visitors to gain insights into a general understanding of destination image in terms of movie-induced tourism. Also, as different types of movies can influence people's interest, intention to travel, or perception of a destination image, exploring the impacts of distinct genres and types of motion pictures would be particularly interesting.

REFERENCES

Afshardoost, M., & Eshaghi, M. (2020). Destination image and tourist behavioural intentions: A meta-analysis. *Tourism Management*, *81*, 104154. doi:10.1016/j.tourman.2020.104154

Araújo Vila, N., Fraiz Brea, J., & de Carlos, P. (2021). Film tourism in Spain: Destination awareness and visit motivation as determinants to visit places seen in TV series. *European Research on Management and Business Economics*, *27*(1), 100135. doi:10.1016/j.iedeen.2020.100135

Arnold, N. (2019). *More than Middle-Earth: How Film Tourism Changed New Zealand*. Discovery. Retrieved 14 March 2021, from https://discovery.cathaypacific.com/middle-earth-film-tourism-changed-new-zealand/

Bąkiewicz, J., Leask, A., Barron, P., & Rakić, T. (2017). Management Challenges at Film-Induced Tourism Heritage Attractions. *Tourism Planning & Development*, *14*(4), 548–566. doi:10.1080/21568316.2017.1303540

Baloglu, S., & McCleary, K. (1999). A model of destination image formation. *Annals of Tourism Research*, *26*(4), 868–897. doi:10.1016/S0160-7383(99)00030-4

Beeton, S. (2005). *Film-induced tourism*. Channel View Publ. doi:10.21832/9781845410162

Beeton, S. (2010). The Advance of Film Tourism. *Tourism And Hospitality Planning & Development*, *7*(1), 1–6. doi:10.1080/14790530903522572

Beric, D., Kovacevic, M., Simat, K., & Bozic, S. (2013). Film tourism: A contemporary resource for promoting Serbia. *Turizam*, *17*(1), 18–28. doi:10.5937/Turizam1301018B

Bruno, G. (2002). *Atlas of Emotion – Journeys in Art, Architecture and Film*. Verso.

Buchmann, A., & Frost, W. (2011). Wizards everywhere? Film tourism and the imagining of national identity in New Zealand. In E. Frew & L. White (Eds.), *Tourism and national identities – An international perspective* (pp. 52–64). Routledge.

Buchmann, A., Moore, K., & Fisher, D. (2010). Experiencing film tourism. *Annals of Tourism Research, 37*(1), 229–248. doi:10.1016/j.annals.2009.09.005

Busby, G., Ergul, M., & Eng, J. (2013). Film tourism and the lead actor: An exploratory study of the influence on destination image and branding. *Anatolia*, *24*(3), 395–404. doi:10.1080/13032917.2013.783874

Busby, G., & Haines, C. (2013). Doc Martin and film tourism: The creation of destination image. *Tourism (Zagreb)*, *61*, 338–348.

Cardoso, L., Estevão, C., Fernandes, C., & Alves, H. (2017). Film induced tourism: A systematic literature review. *Tourism & Management Studies*, *13*(3), 23–30. doi:10.18089/tms.2017.13303

Carl, D., Kindon, S., & Smith, K. (2007). Tourists' Experiences of Film Locations: New Zealand as 'Middle-Earth'. *Tourism Geographies*, *9*(1), 49–63. doi:10.1080/14616680601092881

Castle, A. (n.d.). *Harry Potter.* Retrieved 13 March 2021, from https://www.alnwickcastle.com/explore/on-screen/harry-potter

Chen, C. (2018). Influence of celebrity involvement on place attachment: Role of destination image in film tourism. *Asia Pacific Journal of Tourism Research*, *23*(1), 1–14. doi:10.1080/10941665.2017.1394888

Conley, T. (2006). *Cartographic cinema*. University of Minnesota Press.

Connell, J. (2012). Film tourism – Evolution, progress and prospects. *Tourism Management*, *33*(5), 1007–1029. doi:10.1016/j.tourman.2012.02.008

Croy, W. G., & Walker, R. D. (2003). Rural tourism and film - Issues for strategic regional development. In D. Hall, L. Roberts, & M. Mitchell (Eds.), New Directions in Rural Tourism (1st ed., pp. 115-133). Ashgate Publishing Limited. doi:10.4324/9781315248097-22

Dann, G. (1977). Anomie, ego-enhancement and tourism. *Annals of Tourism Research*, *4*(4), 184–194. doi:10.1016/0160-7383(77)90037-8

Delahaye, J. (2020). *Magical Harry Potter themed breaks in the UK that all muggles need on their radars*. Mirror. Retrieved 14 March 2021, from https://www.mirror.co.uk/travel/uk-ireland/harry-potter-themed-breaks-uk-21658845

Domínguez-Azcue, J., Almeida-García, F., Pérez-Tapia, G., & Cestino-González, E. (2021). Films and Destinations—Towards a Film Destination: A Review. *Information (Basel)*, *12*(1), 39. doi:10.3390/info12010039

Dubois, L.-E., & Gibbs, C. (2018). Video game–induced tourism: A new frontier for destination marketers. *Tourism Review*, *73*(2), 186–198. doi:10.1108/TR-07-2017-0115

Dubois, L.-E., Griffin, T., Gibbs, C., & Guttentag, D. (2021). The impact of video games on destination image. *Current Issues in Tourism*, *24*(4), 554–566. doi:10.1080/13683500.2020.1724082

Dunn, C. (2006). *The cultural phenomenon of Harry Potter.* The Stanford Daily. Retrieved 13 March 2021, from https://www.stanforddaily.com/2016/10/24/the-cultural-phenomenon-of-harry-potter/

Eom, T., Han, H., & Song, H. (2020). Discovering the perceived attributes of CBT destination travelers in South Korea: A mixed method approach. *Tourism Management, 77*, 104013. doi:10.1016/j.tourman.2019.104013

Gjorgievski, M., & Trpkova, S. (2012). Movie Induced Tourism: A New Tourism Phenomenon. *UTMS Journal of Economics (Skopje)*, 3.

Güzel, Ö., & Güzel, B. (2017). The Influence of Films on Destination Attributes (DAs) for Arabian Tourists in Turkey. *International Journal of Business & Information, 12*(3), 209–242. doi:10.6702/ijbi.2017.12.3.1

Hahm, J., & Wang, Y. (2011). Film-Induced Tourism as a Vehicle For Destination Marketing: Is it Worth the Efforts? *Journal of Travel & Tourism Marketing, 28*(2), 165–179. doi:10.1080/10548408.2011.546209

Hall, C. M. (2003). *Tourism Issues, Agenda Setting and the Media, e-review of Tourism Research.* Academic Press.

Hosany, S., Buzova, D., & Sanz-Blas, S. (2019). The Influence of Place Attachment, Ad-Evoked Positive Affect, and Motivation on Intention to Visit: Imagination Proclivity as a Moderator. *Journal of Travel Research, 59*(3), 477–495. doi:10.1177/0047287519830789

Hudson, S., & Ritchie, J. (2006). Promoting Destinations via Film Tourism: An Empirical Identification of Supporting Marketing Initiatives. *Journal of Travel Research, 44*(4), 387–396. doi:10.1177/0047287506286720

Hudson, S., Wang, Y., & Gil, S. M. (2011). The influence of a film on destination image and the desire to travel: A cross-cultural comparison. *International Journal of Tourism Research.* Advance online publication. doi:10.1002/jtr.808

Jewell, B., & McKinnon, S. (2008). Movie Tourism—A New Form of Cultural Landscape? *Journal of Travel & Tourism Marketing, 24*(2-3), 153–162. doi:10.1080/10548400802092650

Kim, H., & Richardson, S. (2003). Motion picture impacts on destination images. *Annals of Tourism Research, 30*(1), 216–237. doi:10.1016/S0160-7383(02)00062-2

Kim, J. (2014). The antecedents of memorable tourism experiences: The development of a scale to measure the destination attributes associated with memorable experiences. *Tourism Management, 44*, 34–45. doi:10.1016/j.tourman.2014.02.007

Kim, S. (2012). Audience involvement and film tourism experiences: Emotional places, emotional experiences. *Tourism Management, 33*, 387–396. doi:10.1016/j.tourman.2011.04.008

Kim, S., & Kim, S. (2017). Segmentation of potential film tourists by film nostalgia and preferred film tourism program. *Journal of Travel & Tourism Marketing, 35*(3), 285–305. doi:10.1080/10548408.2017.1284705

Kim, S. S., Agrusa, J., Lee, H., & Chon, K. (2007). Effects of Korean television dramas on the flow of Japanese tourists. *Tourism Management, 28*(5), 1340–1353. doi:10.1016/j.tourman.2007.01.005

Lee, S., & Bai, B. (2016). Influence of popular culture on special interest tourists' destination image. *Tourism Management, 52*, 161–169. doi:10.1016/j.tourman.2015.06.019

Lee, S., Scott, D., & Kim, H. (2008). Celebrity fan involvement and destination perceptions. *Annals of Tourism Research, 35*(3), 809–832. doi:10.1016/j.annals.2008.06.003

Macionis, N. (2004). Understanding the film induced tourist. In *Proceedings of the international tourism and media conference*. Tourism Research Unit, Monash University.

Macionis, N., & O'Connor, N. (2011). How can the film-induced tourism phenomenon be sustainably managed? *Worldwide Hospitality and Tourism Themes, 3*(2), 87–90. doi:10.1108/17554211111123032

Macionis, N., & Sparks, B. (2009). Film-induced Tourism: An Incidental Experience. *Tourism Review International, 13*(2), 93-101.

Mochocki, M. (2021). Heritage sites and video games: Questions of authenticity and immersion. *Games and Culture*. Advance online publication. doi:10.1177/15554120211005369

Moon, H., & Han, H. (2018). Destination attributes influencing Chinese travelers' perceptions of experience quality and intentions for island tourism: A case of Jeju Island. *Tourism Management Perspectives, 28*, 71–82. doi:10.1016/j.tmp.2018.08.002

Morris, J. (2019). Screen tourism: how much money is this tourism trend making in the UK? *Tourism Review News.* Retrieved 10 July from https://www.tourism-review.com/film-related-inbound-tourism-in-uk-news11231

O'Connor, N., Flanagan, S., & Gilbert, D. (2008). The integration of film-induced tourism and destination branding in Yorkshire, UK. *International Journal of Tourism Research, 10*(5), 423–437. doi:10.1002/jtr.676

Olsberg SPI. (2015). *Quantifying Film and Television Tourism in England. Report for Creative England in association with VisitEngland*. Olsberg•SPI.

Oviedo-García, M., Castellanos-Verdugo, M., Trujillo-García, M., & Mallya, T. (2014). Film-induced tourist motivations. The case of Seville (Spain). *Current Issues in Tourism, 19*(7), 713–733. doi:10.1080/13683500.2013.872606

Pike, S., & Ryan, C. (2004). Destination Positioning Analysis through a Comparison of Cognitive, Affective, and Conative Perceptions. *Journal of Travel Research, 42*(4), 333–342. doi:10.1177/0047287504263029

Potočnik-Topler, J., & Špenko, T. (2019). Film tourism as a tool of tourism development: The Representation of Scotland in the Outlander TV series. *TIMS. Acta, 13*(2), 79–88. doi:10.5937/timsacta1902079P

Pratt, S. (2009). A movie map conversion study: A case study of Pride & Prejudice. *Advances In Tourism Destination Marketing: Managing Networks, 22*, 59–73.

Prayag, G., & Ryan, C. (2012). Antecedents of Tourists' Loyalty to Mauritius. *Journal of Travel Research, 51*(3), 342–356. doi:10.1177/0047287511410321

Reijnders, S. (2011). Stalking the count. *Annals of Tourism Research, 38*(1), 231–248. doi:10.1016/j.annals.2010.08.006

Rittichainuwat, B., & Rattanaphinanchai, S. (2015). Applying a mixed method of quantitative and qualitative design in explaining the travel motivation of film tourists in visiting a film-shooting destination. *Tourism Management, 46*, 136–147. doi:10.1016/j.tourman.2014.06.005

Roberson, R., & Grady, M. (2015). The "Shawshank Trail": A Cross Disciplinary Study in Film Induced Tourism and Fan Culture. *Almatourism, 6*(4), 47–66. doi:10.6092/issn.2036-5195/4953

Roesch, S. (2009). *The experiences of film location tourists.* Channel View Publications. doi:10.21832/9781845411220

Shaheer, I. (2021). The nexus of video games and heritage attractions. *Current Issues in Tourism*, 1–5. Advance online publication. doi:10.1080/13683500.2021.1915254

Sunsets and Suitcases. (2017). *A trip to Hogwarts (Alnwick Castle).* Retrieved 15 March 2021, from https://sunsetsandsuitcases.com/trip-hogwarts-alnwick-castle/

Teng, H.-Y. (2020). Can film tourism experience enhance tourist behavioural intentions? The role of tourist engagement. *Current Issues in Tourism.* Advance online publication. doi:10.1080/13683500.2020.1852196

Teng, H. Y., & Chen, C. Y. (2020). Enhancing celebrity fan-destination relationship in film-induced tourism: The effect of authenticity. *Tourism Management Perspectives, 33*(Jan), 1–11. doi:10.1016/j.tmp.2019.100605

The Inn Collection. (n.d.). *The Hog's Head Inn, Alnwick.* Retrieved 14 March 2021, from https://www.inncollectiongroup.com/hogs-head-inn/

The Numbers. (2021). *All Time Worldwide Box Office.* Retrieved 9 January 2021, from https://www.the-numbers.com/box-office-records/worldwide/all-movies/cumulative/all-time

Thelen, T., Kim, S., & Scherer, E. (2020). Film tourism impacts: A multi-stakeholder longitudinal approach. *Tourism Recreation Research, 45*(3), 291–306. doi:10.1080/02508281.2020.1718338

Tkalec, M., Zilic, I., & Recher, V. (2017). The effect of film industry on tourism: Game of Thrones and Dubrovnik. *International Journal of Tourism Research, 19*(6), 705–714. doi:10.1002/jtr.2142

Tours Weta Workshop. (n.d.). *The Weta Cave | Weta Workshop Tours.* Retrieved 15 March 2021, from https://tours.wetaworkshop.com/wellington/wetacave/

Veasna, S., Wu, W., & Huang, C. (2013). The impact of destination source credibility on destination satisfaction: The mediating effects of destination attachment and destination image. *Tourism Management, 36*, 511–526. doi:10.1016/j.tourman.2012.09.007

Wen, H., Josiam, B., Spears, D., & Yang, Y. (2018). Influence of movies and television on Chinese Tourists perception toward international tourism destinations. *Tourism Management Perspectives, 28*, 211–219. doi:10.1016/j.tmp.2018.09.006

Whang, H., Yong, S., & Ko, E. (2016). Pop culture, destination images, and visit intentions: Theory and research on travel motivations of Chinese and Russian tourists. *Journal of Business Research*, *69*(2), 631–641. doi:10.1016/j.jbusres.2015.06.020

White, S. (2017). Tourism guru off to 'Star Wars' site. *Otago Daily Times*. https://www.odt.co.nz/news/dunedin/tourism-guru-star-wars-site

Wilkins, H. (2010). Souvenirs: What and Why We Buy. *Journal of Travel Research*, *50*(3), 239–247. doi:10.1177/0047287510362782

Yoon, Y., & Uysal, M. (2005). An examination of the effects of motivation and satisfaction on destination loyalty: A structural model. *Tourism Management*, *26*(1), 45–56. doi:10.1016/j.tourman.2003.08.016

Zealand, N. (n.d.). *The Lord of the Rings Trilogy Experiences | 100% Pure NZ*. Retrieved 15 March 2021, from https://www.newzealand.com/int/feature/the-lord-of-the-rings-trilogy-experiences/

Zhang, H., Fu, X., Cai, L., & Lu, L. (2014). Destination image and tourist loyalty: A meta-analysis. *Tourism Management*, *40*, 213–223. doi:10.1016/j.tourman.2013.06.006

Chapter 14
Analysis of Budapest as a Film Tourism Destination

Árpád Ferenc Papp-Váry
John von Neumann University, Kecskemét, Hungary

Tímea Zsófia Tóth
Alexandre Lamfalussy Faculty of Economics, University of Sopron, Hungary

ABSTRACT

The Hungarian film industry has undergone a huge change recently, becoming a pull sector and re-positioning the country in Europe. The economic performance of Budapest's film industry is growing steadily, which has also been facilitated by the strategic steps of recent years, making the capital a key player on the film map of Europe. In the case of Budapest, the diversity of the city is clearly outlined, as it primarily serves as a backdrop for various film productions. Using primary and secondary sources, the research focuses on the supply elements of film tourism in Budapest and the changes in the economic environment affecting the film industry. In the analysis, different types of film tourism are explored through interviews with the actors of film tourism. The research seeks to highlight factors that can boost film tourism in Budapest and make the city even more prominent on the film map of Europe.

INTRODUCTION

Over the past decades, the Hungarian capital has appeared in many Hollywood movies – but mostly not as Budapest. Instead, there was a case when the movie "Munich" featured the city as the double of three European capitals, London, Rome and Paris at the same time. Budapest has already 'played' Berlin (Spy Game, Atomic Blonde), Moscow (Red Heat, A Good Day to Die Hard, Red Sparrow), Buenos Aires (Evita) and even Houston/Peking (Martian), along with several other examples. In recent years, the city has also appeared more and more frequently as Budapest – in this connection, and in a rather strange way, the city is very popular in spy films such as I Spy, Spy, and Spy Who Dumped Me. But what makes the Hungarian capital suitable to be the location for so many movies? And how can you build on this popularity in tourism?

DOI: 10.4018/978-1-7998-8262-6.ch014

The cultural and creative industries play a central role in boosting competitiveness, generating benefits that are unmatched in other areas. The segment includes activities that are based on creative skills and competences, having both cultural and economic impact. A typical example is the world of film, which can be defined as both film art and film industry, depending on whether it is approached from a creative or a business perspective. The film industry, as a broad segment of the creative economy, has the potential to stimulate the economy, which is clearly influenced by the regulatory, financing and economic environment, and it is therefore advisable to look at national specificities. In Hungary, the performance of the film sector has increased almost twenty-fold since 2004, with a quantifiable tax revenue generating and GDP increasing effect. In examining the Hungarian film industry, it is worthwhile to look at the changing economic environment, the changes in the performance of the film sector and the impact of legislation and legislative changes.

The boom of the film industry itself clearly facilitates the strengthening of the economic sectors linked to it. This theorem is also the basis of the research, as it attempts to identify the focal points that make Budapest a prominent player on the European film tourism map. To do this, however, it is essential to take into account the locations that can be associated with Budapest on the film screen, the actors on the supply side of film tourism, as well as the city's film tourism products.

Using primary and secondary sources, the research focuses on the supply elements of film tourism in Budapest and the changes in the economic environment affecting the film industry.

The focus is on films, that can generate travel in case of Budapest, being the capital itself or a backdrop venue. When examining the supply elements, different types of film tourism are explored, by visiting the given tourism actors: film tourism in a presumed location (Runaway Productions), film-themed tours, and visits to film studios. In addition, tourism products, that fit the consumer habits of the postmodern tourist, will be presented.

The case study shows on the one hand the incentives that have made the Hungarian capital attractive as a filming location, and on the other hand presents the initiatives currently linked to film tourism. In connection with the latter aspect, the walking tours by Moviewalking Budapest, the cycling tours by Tandem Budapest, and the Film Destination Budapest smartphone application will be discussed. Last but not least, the authors attempt to define how Budapest's marketing communication could be based on the fact that the city has "played so many capitals on the screen".

The research clearly seeks to draw attention to the factors, that can boost film tourism in Budapest and make the city even more prominent on the film map of Europe.

BACKGROUND: THE FILM INDUSTRY AS PART OF THE CREATIVE ECONOMY

The Creative Economy and the Film Industry

From the 2000s onwards, the creative industries, or the creative sector itself, became a central focus of research. The creative economy encompasses industries in which different forms of creativity, such as scientific, technological, economic or cultural creativity, are strongly dominant (Keresnyei & Egedy, 2015). The development of creative cultural service industry and the cultural economy has a positive impact on both national culture and economic growth. The development and expansion of cultural heritage and related industries is part of the national economic strategy to strengthen European competitiveness. In Europe, the cultural and creative industries have become one of the fastest growing sectors, with a direct impact on economic growth. In the last decade, the use of intellectual and cultural heritage within the creative industries has clearly become a value creator for the economy, as settlements with developed cultural industry have a higher workforce retention capacity and a better quality of life in these areas (Csizmadia, 2016).

There are several approaches to defining and classifying creative industries in the literature (Howkins, 2004; Throsby, 2001; Hesmondhalgh, 2007), but all of them consider the film industry to be an integral part of the creative economy. When grouping the creative industries, several models (Howkins model, WIPO copyright model, Concentric Circles model) clearly classify the film and video industries as core industries. The range of definitions and classifications used by researchers and professional organisations is very wide and varied, but there is still no uniform classification of the creative industries.

The database produced by the United Nations Conference on Trade and Development (UNCTAD) measures the value of global exports of creative goods and services, based on the output of crafts, graphic and interior design, books, fashion, film, music, online and print media and audiovisual materials. The global market for creative products has shown significant growth, rising from USD 208 billion in 2002 to USD 509 billion in 2015. Europe is the largest exporter of creative products among developed regions, with exports of creative products worth USD 171 billion in 2015, up from USD 85 billion in 2002. The growth rate of creative product exports in the European Union was 5.5% in the period 2002-2015, with employment in the cultural and creative industries affecting nearly 12 million people in the European region. In 2015, Hungary was among the top 20 developed countries in creative exports (20th), with the United States, France, Italy and the United Kingdom in the top positions (United Nations, 2018).

The development of the Hungarian creative industry and the potential of the sector has become more and more evident to the players of Hungarian economy every year. In 2016, the creative industry in Hungary employed 4% of the workforce and produced 6.5% of the GDP (Ministry of Innovation and Technology, 2019). Hungarian products – as is typical of creative products as a whole – have a high added value and play an important role in improving the Hungarians' quality of life.

THE ECONOMIC ENVIRONMENT OF THE HUNGARIAN FILM INDUSTRY

In Hungary, Act II of 2004 on Motion Picture marked the first steps towards the legal regulation of the film industry. The Act II of 2004 on Motion Picture defined and structured the role of the state in film production support, as well as the level of support and forms of control. Its primary objective was to create a tax policy that would create a balance between the economy and film production, and to regulate Hungarian film rights and asset management. Regarding the assessment of film subsidies, it was stressed that aspects such as cultural expectations and budget should be taken into account (Zachar, 2006). The Hungarian National Film Fund (from December 2019: National Film Institute, NFI) was established during the structural transformation of the sector, with the aim to develop an effective operational strategy to increase the competitiveness of the Hungarian film industry.

Its primary aim was to support Hungarian films and film events, film books, magazines and training courses. The support scheme was extended to include a deposit account, through which sponsoring companies could make a contribution to the production of a production. This had the great advantage of providing tax relief on the amount deposited in the account. The National Film Institute also had a clear objective of promoting Hungarian films, primarily through international film festivals.

In 2012, the amendment of the Act on Motion Picture was adopted, which determined the revenues needed for the organisation's operation, most of which came from the central budget and other revenues. It also included a provision that the organisation would receive 80% of the gambling tax on the Hatoslottó lottery game. This amounted to HUF 4.4 billion (EUR 12,503,216) based on the 2017 payment. The support scheme enabled the production of Hungarian films that achieved international success. Examples include Son of Saul (2015) or On Body and Soul (2017). With this form of funding, the Hungarian film industry is following the European trend. The National Lottery, established in the UK in 1994, provides direct support to the film industry, with a fixed percentage of the proceeds received by the British Film Institute. Similarly, the national film industry is supported through the proceeds of lotteries in Greece, Finland and Switzerland (www.figyelo.hu, 2018).

Screenplays play a decisive role in the awarding of grants, thus they are the basis for direct grants for Hungarian films. To this end, script development is also a priority in the call for proposals. The decision-making process consists of three steps: first, the script is submitted to pre-readers, then to the script development team and finally to the jury. Andy Vajna, who served as Government Commissioner for Film Industry in Hungary from 2011 to 2019, was a member of this committee. The clear aim of the "Vajna era" was to create a completely renewed film financing system and to attract more foreign productions to the country by the reforms of the tax system.

Figure 1. Film industry competitiveness ranking
Source: National Film Institute, www.nfi.hu, 2017

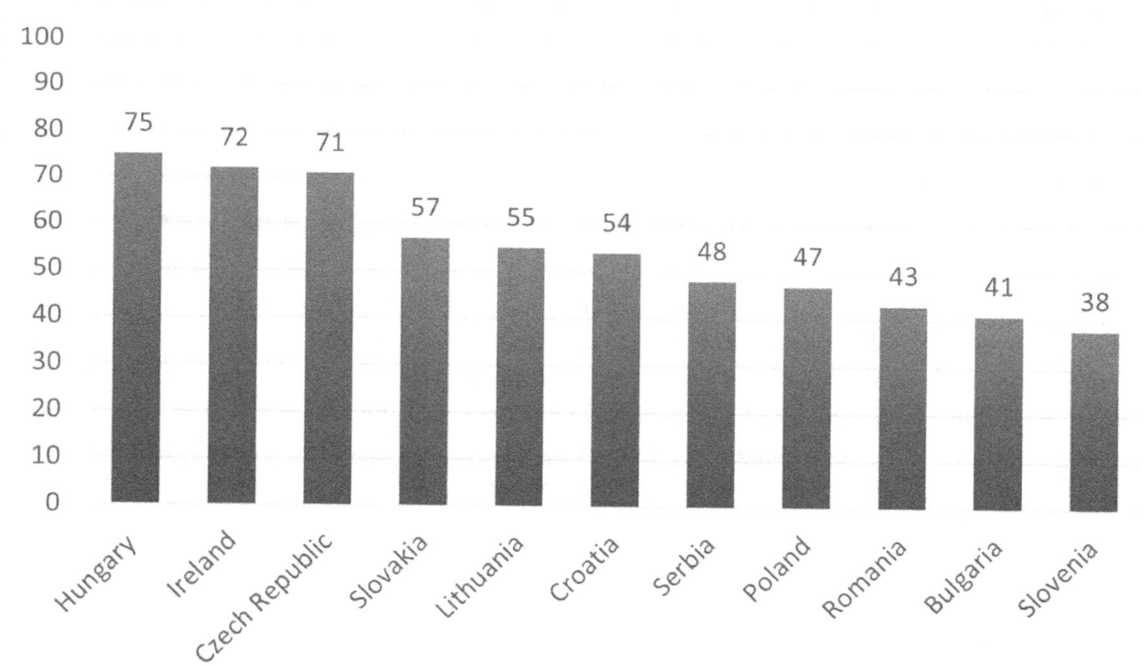

The year 2012 brought another change, as the European Commission authorised an increase (from 20% to 25%) in the amount of tax relief for film production costs in Hungary. From this turning point onwards, more and more national (that is, Hungarian) films were made, and their market share increased (from 3.7% in 2014 to 4.2% in 2015) (Zachar, 2013). However, this trend seemed to fall back in 2016. One of the main reasons for this could be the strengthening of the film industry in the region and the increased resources devoted to the production of foreign co-productions (foreign films shot in Hungary). The level of support has been even higher since 2018, when it was increased to 30% at the initiative of the Hungarian National Film Fund. This strategic move is also attributable to Andy Vajna, who paved the way for the creation of one of the region's most favourable film support schemes.

These moves have led to a clearly visible recovery in the film industry, as confirmed by a 2017 survey by tax consultancy OrienTax. The analysis of the competitiveness of the film industry looked at eleven countries with similar economic characteristics to Hungary. The criteria examined were currency stability, average price and wage levels, political environment, available film industry infrastructure and language skills. The level and effectiveness of subsidies and the development of film infrastructure also showed a strong competitive advantage (*Figure 1*). Countries could score a maximum of 100 points in the test.

The analysis has shown that the Czech Republic is a clear competitor in the region, as its film infrastructure and development is at a similar level to Hungary. In terms of price and wage levels, Hungary is in the middle of the range, being more competitive than the main rival, the Czech Republic. In terms of the policy and regulatory environment (based on World Bank data), Hungary lags slightly behind both the Czech Republic and Ireland. The country worst performed in terms of language skills, due to the high proportion of dubbed films. The overall first place is mainly due to Hungary's average or above average performance in all categories except language skills (National Film Institute, 2017).

The annual survey of the National Media and Infocommunications Authority also confirms the fact that the film industry is generating higher figures in terms of economic indicators every year. Since 2004, the performance of the film sector has increased almost twenty-fold. In 2013, total film expenditure amounted to only 54.2 billion HUF (155,825,000 EUR), but this figure has risen to 125.5 billion HUF (360,561,500 EUR) in 2016. In 2018 and 2019 its value remained above HUF 110 billion (EUR 316,360,000) (*Figure 2*).

Figure 2. Performance of the film industry in Hungary
Source: National Film Institute, www.nfi.hu, 2021

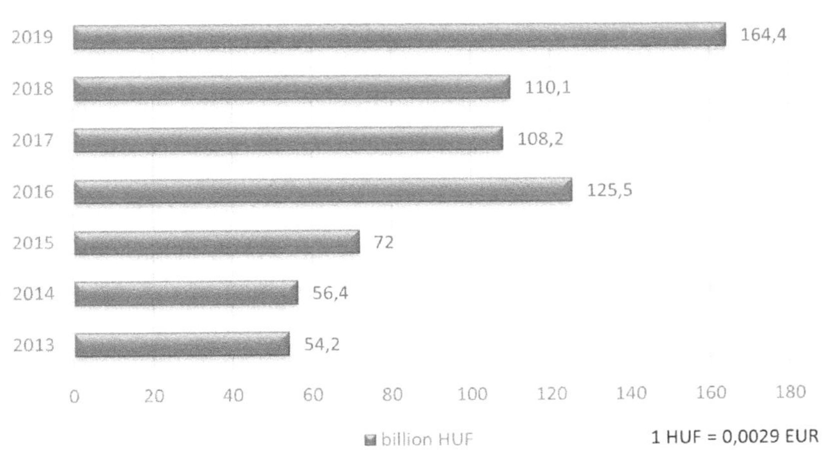

The surge in 2016 was most likely due to the fact that the international production Blade Runner 2049, starring Ryan Gosling and Harrison Ford, was made in Hungary. Looking at the production costs for that year, this production clearly stood out with its HUF 28.5 billion (EUR 82,023,000), making it the most expensive film ever made in Hungary on a work-for-hire basis. In 2018, the production of 333 films was registered in Hungary, with 30 more Hungarian films than in 2017. The share of work-for-hire has also increased, with 50 foreign productions shot in Hungary in 2015 and 2016, rising to 64 in 2017 (Nemzeti Filmintézet, 2019).

FILM INDUCED TOURISM

The previous chapters clearly demonstrate that the impact of the film industry on creative economy – and therefore on economy as a whole – is increasing year by year. However, the economic impacts that can be linked to films are not only found in the industries discussed earlier, but also in tourism. Film tourism is clearly a multidisciplinary field of science with marketing and business aspects at its core (Beeton, 2010). The essence of film tourism is best explained by Anna Irimiás: 'to see it and fall in love, to fall in love and go on a journey' (Irimiás, 2015). This perspective also reflects the clear role of film success in creating the motivation to travel and attracting visitors. A successful film can also trigger strong emotional attachment, so the success of the film also plays a key role in the success of film tourism (Sulyok, 2009). It is not an insignificant fact that films have the maximum potential to influence attitudes about a destination and thus have a maximum impact on the image of the country itself (Beeton, 2005; Gunn, 1972; Papp-Váry, 2019). Of course, the perception of a destination can be both negative and positive. A perfect example of negative perception and representation is the film Borat: Cultural Learnings of America for Make Benefit Glorious Nation of Kazakhstan. After its release, the comedy film clearly drew attention to the country, despite the Kazakh government's total opposition to the making of the film. The film's negative image of Kazakhstan led to a multi-million dollar campaign under the slogan "Kazakhstan – Heart of Eurasia". Despite all its negative connotations, the divisive comedy could spread the word about Kazakhstan even to people who had never heard of the country before (Papp-Váry, 2009).

Through this phenomenon, it becomes clear that motion pictures play an important role in the choice of destination and the formation of opinions about a specific place. Films have the maximum potential to convey relevant information about a destination in a short time (Kim & Richardson, 2003; Gartner, 1994), thus it is important for destinations to emphasise positive elements in this process. The strong correlation between the increase in the number of visitors to a destination and films has been confirmed by a large body of international literature (Riley & Van Doren, 1992; Riley et al., 1998; Grihault, 2003), despite the fact that film-induced trips are very difficult to measure, as their impact is very long term (Sulyok, 2009). Several years may elapse between the film seen and the trip to the destination, which is a fundamental difficulty in measurement processes.

The travel-generating effect of films can be reinforced by local destination marketing organisations by developing a marketing strategy that can create a stable and reliable image while preventing the phenomenon of overtourism (Hudson & Ritchie, 2006; Morgan & Pritchard, 1998; Cynthia & Beeton, 2009). Of course, this requires a coordinated effort between marketing organisations and film production companies. The literature has approached the study of film tourism from many angles, but for the purposes of this research, the following themes stand out: the impact on the local economy (Croy, 2003; Pratt, 2015; Stachowiak & Stryjakiewicz, 2018) and the relationship with the areas of country image and destination marketing (Beeton, 2005; Bolan & Williams, 2008; Riley et al., 1992; Irimiás, 2015; Hudson & Ritchie 2006; Baker, 2012, Papp-Váry, 2019).

The relationship between film tourism and the economy is central to the present research. The demand generating effect of films has been identified in many areas, with hotspots in American cities (Los Angeles, New York, Cleveland, Philadelphia), Asia (Maya Bay, Singapore, Hong Kong, Zhangjiajie National Forest Park) and New Zealand (Waikato region, Mount Taranaki), but European destinations are also represented in the statistics. The Hobbit Trilogy in 2013 is shown to have increased tourist arrivals by 104,745 from 9 major origins, which resulted in a US$374.62 million income increase for New Zealand in that year. In 2014, tourist arrivals increased by 110,066, along with an additional US$397.18 million tourist receipts. In total, the Hobbit Trilogy brought an additional US$771.80 million international tourist receipts to New Zealand (Li et al., 2017). On the continent, in addition to Barcelona (Vicky Christina Barcelona) and London (Harry Potter), Dubrovnik (Game of Thrones) is already a top destination for film tourists (Sulyok, 2009; Tóth, 2016). Through visitor surveys at eight filming locations in England, such as Alnwick Castle (Harry Potter) and Lyme Park (Pride and Prejudice), it was found that the on-screen induced tourism contributed between £100 million and £140 million to the national economy in 2014 (Creative England, 2015).

The consumer trends of film tourists can be paralleled with the characteristics of the postmodern tourist, i.e. mass tourism is rejected, and individual and experiential services are emphasised. The postmodern tourist no longer stares at things like a traveller of the industrial society, but has a need for active participation (Orva - Deli-Gray, 2011). Thanks to this process, tourism products have also changed completely, adapting to the needs of tourists.

Basically, a tourism product is defined as any service or product offered by tourism enterprises in competition with or complementing each other (Puczkó & Rátz, 2011). Film tourism products are also characterised by a maximum focus on the attraction. In this case, these attractions are related to the shooting, production or distribution of films or television programmes (Irimiás, 2015). Of course, successful film tourism products can only be created if the motivations and needs of film tourists are at the centre of the development of supply.

BUDAPEST AS A FILM DESTINATION

The research focuses on the factors that can consolidate Budapest, the capital of Hungary, on the continent's film tourism map. In addition to the changes in the economic environment of the Hungarian film industry discussed earlier, the study focuses on exploring the capital's film destinations, the actors of film tourism (including the professional relationships between these actors) and the film tourism products that can be associated with them. Having examined secondary sources, the literature review also summarises the current economic environment of the Hungarian national film industry, the concept of film tourism and its impact on the economy. The research also includes an enumeration of the films that were shot in the capital, as these have provided the guidelines for the development of film tourism products. The study will also highlight the fact that the city is often not displayed as Budapest itself on the screen. However, this trend seems to have changed in the last five years, which is not insignificant in terms of the country image. It should be noted, however, that a location can generate film tourism products even when it appears as a double or stand-in location in a specific film.

The primary research was based on a purely qualitative method, using in-depth professional interviews, which were conducted in person or electronically. In order for the film location to become a successful film tourism destination, close cooperation between all those involved in film tourism is necessary. In the film industry, cooperative partners that can make Budapest a fruitful film destination include destination management professionals, owners of tourism service providers, film professionals, local residents and, of course, film tourists (Irimiás, 2015). It was with this idea in mind that the interviewees were approached.

In the context of those in control of destination management, the marketing staff of the Hungarian Tourism Agency and the Budapest Festival and Tourism Centre assisted the research (Dániel Szabó and Márk Újvári). The owners of tourism service providers were represented by the marketing director of Korda Film Studio (Ilona Meruk), the director of ORIGO Film Studio (Márta Fekszi Dr.) and the managers of various themed walking tour companies, who all took part in the research. The biggest help came from Balázs Basa, one of Hungary's most respected film tourism experts, a film location researcher, according to his own opinion.

There are two roles that Budapest usually plays in films. The first – which is also important for the image of the country – is when the city depicts itself on the screen. The second category includes films where the city is a double location, masked as another city. The literature also provides a more detailed categorisation, whereby the roles of the capital city can be divided into even smaller categories, five in total. In the first, Budapest is presented as another city; in the second, the filming takes place in Hungary and the story is also set there. In the third, the country appears, but the film is not shot there; the fourth category includes films with a Hungarian character (Casablanca, The Whole Nine Yards), but there is no other territorial connection with Hungary, and in the fifth group includes films in which Hungarian appears as a gibberish language (Blade Runner, 1982) (Papp-Váry & Marthy, R., 2009). It is rare, but there are cases when a character in the movie is not Hungarian, and they suddenly start speaking the language. Perhaps the most famous example of this is "Blade Runner", in which Harrison Ford is arrested in Hungarian.

Budapest on the Screen

Budapest, City of a Thousand Faces

It is fair to say that in most cases, Budapest has appeared on the silver screen as another city, but the trend has started to reverse in recent years. Thanks to the region's outstanding film tax reliefs, the Hungarian capital has recently been home to super-productions such as Spy (2015), Blade Runner 2049 (2017), Spy Who Dumped Me (2018), Gemini Man (2019), Dune (2021) and Black Widow (2021).

First, it is worth taking stock of the works where Budapest takes on a different form, some of which will be presented in more detail. One such major production is Munich (2005), directed by Steven Spielberg. In the film Budapest is among other things Rome, London and Paris; the Opera House and its surroundings are mostly shown in the film, and the Puskás Stadium serves as the location of the Munich stadium. The atmosphere of the city of Munich is mostly embodied in streetscapes and doorways, provided by the houses along Andrássy Avenue (Irimiás et al., 2013). Another production of similarly high calibre was Evita (1996), in which the Hungarian capital was portrayed as Buenos Aires. Budapest was chosen for this project because Buenos Aires at the time was already too modern for the story set in the 1940s. The production was hallmarked by Madonna and Antonio Banderas, and its locations included Keleti Railway Station, Szabadság Square and Andrássy Avenue. The fifth part of Die Hard (2013) was also shot in the capital – the story is set in Moscow, with Üllői Street and Kálvin Square playing the role of the main locations, and the huge firefights took place in Heroes' Square (Hősök tere).

The construction sites of Üllői Street have also inspired other filmmakers. The film In the Land of Blood and Honey (2011) was shot here, set in Bosnia-Herzegovina in south-eastern Europe. Budapest has also appeared on the screen as Berlin, including in Spy Game (2001), despite the fact that Brad Pitt was walking along the bank of the Danube with the Liberty Bridge and Gellért Hill in the background. The other big production in which the Hungarian capital also appeared as Berlin was the 2017 film Atomic Blonde, starring Charlize Theron. The big sensation of 2015 was the American sci-fi film The Martian, shot in Hungary, in which the Bálna conference building on the Danube bank served as the NASA headquarters in the film, and the entrance to the Korda Film Studio was decorated as the futuristic entrance of NASA. In addition to these two locations, the Palace of Arts was also featured in the film, serving as the location of the China National Space Administration. In the background, the Rákóczi Bridge is very clearly visible, as well as a green and white train passing by.

A Bollywood (that is, Indian) film made in 1999 also fits perfectly into this bunch of films. The film is titled Hum Dil De Chuke Sanam – I have given my heart away, darling – and its promotional poster features the Széchenyi Chain Bridge, an iconic location of the Budapest panorama. The film is historically set in Rome, Italy, but also features scenes in which Budapest could not deny itself. There are aerial panoramic shots of the Citadel, the Parliament, the Széchenyi Bath, Matthias Church and Heroes' Square. The protagonists attend a luncheon at a gypsy party on board the Columbus ship, and also take a taxi ride on the Chain Bridge and along the Danube. All the while we can hear Hungarian background noise, as the man in the street speaks to them in Hungarian. The film was a huge success in its home country, so much so that a few years later the Hungarian capital could welcome another Indian production (Raabta). This will be discussed in more detail in a later chapter.

Budapest as Budapest

There has been a recent increase in the number of films in which Budapest featured as itself. One of the first movies from the early days was the action-comedy I Spy (2002), in which Eddie Murphy and Owen Wilson team up to escape an action-packed situation. As a result, we see a shootout at the Budapest Castle Hill Funicular and they hang together over Clark Adam Square, with a full view of the Széchenyi Chain Bridge and the Danube bank. The film's production team included Andy Vajna, who made it possible to feature the capital city as itself. Unfortunately, this turn of events did not work out for the fifth Die Hard movie, where the writers clearly considered Moscow to be a more authentic Mafia location. In the action comedy Spy (2015), the capital almost suffered the same fate as in the aforementioned Bruce Willis film, because the director's original plan was to masquerade Budapest as Paris or Venice. However, Paul Feig fell in love with the Hungarian capital and rewrote the story. He said that the city of Budapest was so exotic and beautiful that he did not want to display it in any other way.

The Four Seasons Hotel Gresham Palace, located behind the Széchenyi Chain Bridge, was one of the main locations for Spy, which was a unique occasion as the crew was the first who were given permission to film in this hotel. Both the crew and the hotel made the most of this opportunity. The film features the Lobby with its huge crystal chandelier, the Royal Suite and of course the hotel's iconic main entrance. During the action scenes we also see the Basilica, Váci Street and the Great Market Hall. Four Seasons also took advantage of the unique opportunity by creating a special package called Spy In Budapest.

Table 1. The most popular filming locations in Budapest

Locations	Movies	Cities on the Screen
Széchenyi Chain Bridge	Au nom de tous les miens (1983)	Warsaw
	Royce (1994)	Kiev
	Gloomy Sunday (1999)	Budapest
	I have given my heart away, darling (1999)	Rome
	An American Rapsody (2001)	Budapest
	I Spy (2002)	Budapest
	In Secret (2013)	Paris
Hungarian State Opera House	Mephisto (1981)	Berlin
	Evita (1996)	Buenos Aires
	Rasputin (1996)	St. Petersburg
	Il fantasma dell'opera (1998)	Paris
	Munich (2005)	Paris, London, Rome
	Monte Carlo (2011)	Monte Carlo
	Dracula (2013)	London
	Red Sparrow (2018)	Moscow
Heroes' Square	Red Heat (1988)	Moscow
	Evita (1996)	Buenos Aires
	The Luzhin Defense (2000)	St. Petersburg
	Chernobyl Diaries (2012)	Kiev/Chernobyl
	100 Degrees below Zero (2013)	Paris
	A Good Day to Die Hard (2013)	Moscow
Museum of Fine Arts	Mephisto (1981)	Berlin
	Evita (1996)	Buenos Aires
	Sunshine (1999)	Berlin
	The Company (2007)	Washington
	A Good Day to Die Hard (2013)	Moscow
City Park - Vajdahunyad Castle	Bluebeard (1972)	Vienna
	Étoile (1989)	Paris
	Crime and Punishment (1998)	St. Petersburg
	The Prince and the Pauper (2000)	London
	Good (2008)	Berlin
	The Borgias (2011)	Rome/Florence
Gozsdu Courtyard	Red King, White King (1989)	Moscow
	Citizen X (1995)	Moscow
	Last Run (2001)	Kiev
	Spy Game (2001)	East-Berlin
	Max (2002)	Berlin
	Underworld (2003)	Imaginary city

Continued on following page

Table 1. Continued

Locations	Movies	Cities on the Screen
Kossuth Square	Étoile (1989)	Paris
	An American Rhapsody (2001)	Budapest
	The Debt (2010)	Kiev
Museum of Ethnography	Mephisto (1981)	Berlin
	Evita (1996)	Buenos Aires
	Copying Beethoven (2006)	Vienna
	Einstein and Eddington (2008)	Cambridge
	The Rite (2011)	Vatican
	La banda Picasso (2012)	Paris
	The Raven (2012)	Baltimore
	Dracula (2013)	London
Margaret Island	Étoile (1989)	Paris
	Sunshine (1999)	Berlin
	Munich (2005)	Paris, London, Rome
	Good (2008)	Berlin
	Houdini (2013)	Atlantic City
Fisherman's Bastion	Crossed Swords (1977)	London
	Mephisto (1981)	Budapest
	The Phantom of the Opera (1983)	Budapest
	Gloomy Sunday (1999)	Budapest
	I Spy (2002)	Budapest
	Tinker Tailor Soldier Spy (2011)	Budapest

Source: Based on Basa & Názer (2014), self-edited (2021)

This programme includes a Behind the Scenes tour, where participants can walk through the locations featured in the film.

The table below (*Table 1*) is based on the major productions and their locations in Budapest, and shows not only the pairs of films and locations, but also the name of the double location (if there was any). It is also clear from this information that the Hungarian capital is much more often portrayed on screen as Rome, Paris, Berlin or Moscow.

The 2019 film Gemini Man, starring Will Smith, has caused a stir in the capital. Budapest is a key location in the story of the futuristic action film. It film was shot in the courtyard of the Budapest University of Technology and Economics, among other locations. The American actor/musician came to Budapest exclusively for the action film Gemini Man. However, he also took time out to see the sights of the capital. Of course, he did this in an unusual way. During his time here, he toured the roof of the Hungarian State Opera House, which is under renovation, but also climbed to the top of the Széchenyi Chain Bridge. On the bridge, he completed an Instagram challenge, a dance video of which has been viewed by almost 20 million people. He has been broadcasting it on his social platforms. The special content has been seen by millions of people, generating fantastic publicity for the Hungarian capital.

In 2018, he and his fellow musicians produced the music video for the 2018 FIFA World Cup Official Song at the famous entrance to the Sziget Festival, called Bridge K.

Budapest is still a popular filming location, as the big sensation of 2021, the Marvel movie Black Widow, was also shot in the Hungarian capital following Atomic Blonde, Red Sparrow and Spy Who Dumped Me.

In the film, Budapest appears as itself and with the help of the main character, everyone can learn to pronounce the name of the city correctly. The movie Black Widow shows many locations in Budapest, which are related to the main character's past; Scarlett Johansson and Florence Pugh are in action in Budapest for most of the plot. Of particular interest is the fact that both actresses have previously filmed on the streets of Budapest.

BUDAPEST AND FILM-INDUCED TOURISM

The Supply Elements of Film Tourism in Budapest

The analysis of the supply-side was supported by in-depth interviews with professionals. The selection of the interviewees was based on the assumption that film tourism in the capital and the tourism products available in the sector are intertwined in a central management system. Therefore, the Hungarian Tourism Agency and the Budapest Festival and Tourism Centre were contacted in the first phase of the investigation. Based on the publications and online presence of the Hungarian Tourism Agency, we can state that film tourism is currently not given much weight, and it does not play a prominent role. In the Nemzeti Turizmusfejlesztési Stratégia 2030 (National Tourism Development Strategy 2030), the sector is mentioned in passing as one of the pillars of cultural tourism. The Budapest tourism strategy prepared by Budapest Brand Zrt. (established on 1 November 2020 with the merger of Budapesti Fesztivál és Turisztikai Központ and Budapesti Városarculati Kft.) already includes film tourism. The strategy is based on three pillars: diversity, sustainability and safety. Its basic aim is to make Budapest's values more visible and to raise the level of tourism in the capital (Szerdahelyi, 2020).

The in-depth interview with the Budapest Festival and Tourism Centre also revealed that there is no clearly defined direction for film tourism in Budapest, as the Hungarian Tourism Agency would have to appoint a film tourism specialist. Special requests for film productions are very rare, but the Indian production Raabta, which was shot in Budapest in 2016, has been a highlight in the life of the Budapest Festival and Tourism Centre. The international production resulted in a 20% increase in Indian visits to the capital, clearly demonstrating the travel-generating character of films. The problem of the capital not being itself on the screen but being featured as other cities was also often raised. The Budapest Festival and Tourism Centre promotes Budapest as a film location (called Budapest on the silver screen) in a number of publications, such as the Brilliant Budapest brochure. They also highlight the opportunities offered by the Budapest Card, which allows tourists to take a discounted 3D historical virtual tour of Buda Castle (the set of several films). The publication also promotes the restaurant of Danubius Hotel Gellért, highlighting the fact that this is where Madonna sang her iconic song in the film Evita.

During the interview, VisitBritain's campaigns and approach to film tourism were mentioned several times. In the UK, it is common practice to develop complete regional tourism products and marketing campaigns based on film screenplays. The adoption of this good practice and the use of its templates would be of great help in boosting film tourism in the Hungarian capital.

In this context, it was noted that the only way to achieve a qualitative leap in the range of film tourism products in the capital would be to integrate them into a larger service provider. However, the convenience of the Hungarian private sector makes it difficult to move forward with the expansion of supply. In the context of film tourism products, various companies organising thematic tours were mentioned, as well as Korda Film Studio, as a lone giant in the sector.

Film Studios

Korda Studios is a state-of-art film studio complex located 30 km (18 miles) west of Budapest in the wine-making village of Etyek. The studio is named after Sir Alexander Korda, a Hungarian-born British film producer and director. Korda Studios were inaugurated in April 2007. Altogether six studios were built from a total budget of 90 million Euros (www.kordastudio.hu, 2021). The interview took place in the visitor centre on the premises of the film studio, with the facility's senior event manager assisting with the research. The interview revealed that, thanks to the favourable development of the film tax relief, there is a clear sense that film production in Hungary is enjoying a renaissance. The complex follows the trends of American film studios, which means that it has created a visitor centre for people interested in film by putting together the sets and other elements and props associated with film shoots. The tour also includes a close-up look at outdoor locations: a New York street block, a medieval village and a Renaissance set.

The studio consists of two parts, the visitor centre and the studio area, which is exclusively for professional staff.When it opened in 2007, the facility had only four studios, but thanks to continuous improvements, it now has seven studios for filming. Each year, around thirty productions work within its walls, including two or three movie films, around ten commercials, one or two series and around five to ten music videos. Among other things, the 2018 Robin Hood remake starring Jamie Foxx was made here. The studio also produced the aforementioned superproductions Blade Runner 2049, Terminator: Dark Fate, released in 2019, and The Martian, released in 2015. The Martian spent around 38% of its estimated budget of $108 million (more than HUF 11 billion in total) in Hungary. The cast list included many Hungarian names, not only as supporting actors but also as make-up artists and set and prop designers.

The interview revealed that the studio's nearly 84% occupancy in 2016 can be clearly attributed to the revised support system, making the country one of the most important film destinations after the UK. The key criteria that contributed to this were the affordable costs, significant tax relief, the development of hotels and hospitality, and a diverse climate and cityscape. Productions can arrive with a crew of up to 100-150 people and spend up to six to eight months in Hungary. During this time, they stay in five-star hotels, bring their families, get to know Hungarian gastronomy and, of course, spread the country's reputation around the world. This is the basis of business tourism – in this case, film MICE (Meetings, Incentives, Conferences, and Exhibitions) tourism. This process is also capable of boosting leisure tourism, in the event of a return visit. It is also thanks to this that there are further economic and country image opportunities for foreign film productions visiting Hungary, the film studio managers say.

It is also worth mentioning that commercials are starting to discover our country with a noticeable shift eastwards, as Korda Film Studio has recently worked with companies such as Shell, Nintendo and O2. The studio is also bolder in connection with its current expansion, as it has set up its own production company, which no longer only deals with commercials. Recently, there has also been a growing focus on film series, so that when it comes to film production, they no longer distinguish between feature films and series. The newly created in-house production company has already produced a $30 million Canadian series. Thanks to the boom, Korda Film Studios has been home to the History Channel's Houdini series starring Adrien Brody, Netflix's Marco Polo history series and National Geographic's series on Picasso's life, starring Antonio Banderas.

The primary product of the Visitor Centre is a film production experience tour, which attracts around 17-20,000 visitors a year. The Visitor Centre also offers a range of experience products for companies, families and schools, which can be booked individually. In line with European practice, the complex is trying to reach an increasingly wide range of people, so these interactive programmes form the backbone of the attraction. Visitors can take part in the shooting of their own film, become a voice actor, investigate with the 'Etyek CSI', take the whole family to a slumber party, or hold a wedding or a corporate event there. The Film Park also has a permanent exhibition where visitors can learn about the Korda brothers' adventurous career in international filmmaking as part of the Korda Memorial Exhibition.

Opened in 2007, the complex has everything needed for professional film production at the highest level. Korda Filmstudio is part of the Gránit-Pólus Zrt. Group, so any extra requirements (accommodation, catering) for major international productions are handled exclusively through partners within the group. In the last year or two, a joint initiative with the municipality of Etyek has been launched for the Etyek Piknik event, but it is still in its infancy. During the interview it became clear that there is a uniform differentiation in the market, i.e. those who want to work in seclusion and tranquility prefer Korda Film Studio, while those who want to shoot more in the capital prefer Origo Studio. The reason of this is that it is easier to get to the central locations in the capital. Origo is a Hollywood standard studio complex in Budapest.

When visiting Origo Studio, the director of the complex gave detailed information about its operation. During the interview it was revealed that Origo Film Group consists of several units, each segment providing a specific service. Origo Studios is in charge of studios, blacklots, warehouses and other workshops, Origo Rentals is responsible for the lighting park, trailers and toy cars, Origo Digital is in charge of film post-production, Origo Catering is responsible for catering and Origo Film is responsible for service and financing. They are not affiliated with any other film companies outside the group. The interview revealed that the occupancy rate of the units depends largely on the nature of the production. Each production has different needs when it comes to the Hungarian capital, as a production that requires a lot of studio shooting will make better use of the nine different studio units, as was the case with Blade Runner 2049. The large-scale production was shot both at Korda Film Studio and Origo's facilities. Other films are shot in more real-world locations, so they tend to use the facilities in post-production, which is what happened in the case of A Good Day to Die Hard.

Origo is basically a host studio, so its main aim is to increase the number of productions arriving there, and at the same time to sell as much of the service portfolio as possible. In recent years, the studio's occupancy rate has increased significantly in all areas. This is due to the fact that in the last three years two or three feature films have been shot in Hungary. In 2018, several feature films and a number of series were made there, and from August 2020, three major productions, Dune, UWOMT (The Unbearable Weight of Massive Talent) and Birds of Paradise, were in the works. Their clients also include Netflix and regional television productions.

The director of the film studio drew attention to some very important pillars in connection with Hungarian film production and the studio. Origo Film Studio was built to Hollywood standards from the start, and therefore meets American expectations. It is obvious that the crews want to work in an environment that supports their work and provides them with the appropriate and modern technology. In addition to this, Hungary is very attractive architecturally, and Hungarian landscapes can evoke many other locations. That is why scenes in Rome, Moscow, Florence and Berlin have been shot in Budapest, so it is very cost-effective to produce this kind of work here. It is important to mention that Hungarian filmmakers have excellent skills, so productions can find local professionals who can offer the best quality at a good price. Hungarian filmmaking has a long tradition anyway, which is also an important factor when choosing locations for foreign productions. The film industry in the United States, and the cinema industry in general, owes a lot to Hungarians, most notably Adolph Zukor and William Fox. Both of them founded film companies that later became the largest and most important film companies in the USA. Zukor was the founder of Paramount Pictures, and Fox was the founder of Fox (later 20th Century Fox), bearing his name. Together, these elements form a package of offerings capable of attracting high calibre film productions to Hungary.

However, it is worth mentioning an external professional relationship that supports the training of film professionals. The studio has a close connection with Budapest Metropolitan University, which has a small stage and studio at the disposal of students, who also have the opportunity to visit productions made in the studio as part of their work placement. However, private visitors, such as tourists or film lovers are not allowed to visit the facility.

Film-Themed Tours

In the context of the analysis of the supply side, a number of businesses in Budapest organising themed tours were contacted. However, the interviews revealed that they did not have any film-themed tourism products. The companies Budapesti Séták and Budapest Beyond Sightseeing organise walks with different themes related to cultural tourism. Sétaműhely and Tandem Budapest are the only companies that organise film-related thematic programmes in the capital. The programme "Éljen a Grund! – családi séta, Mese és valóság a Pál utcában" – family walk, is organised twice a month. The walk is based on the Hungarian author Ferenc Molnár's The Paul Street Boys, one of Hungary's best-known and most read youth novels. It reveals basic patterns of human behaviour and iconic life situations that have almost guaranteed its popularity. It has been translated into more than thirty languages. The walk lasts 2-2.5 hours, during which participants can try out a popular toy of the era (marbles), explore the location called Grund and recall the circumstances of the writing the novel through the memories of Ferenc Molnár. The tour is very popular, so it is extremely difficult to register for the event with a limited number of participants (www.setamuhely.hu, 2021).

The other company with a film-themed programme is Tandem Budapest, which offers a sightseeing bike tour for participants. The tour titled Világ körüli út ("Walk around the world") presents locations where famous and popular films have been filmed. At the stops, the guide plays film clips and tells interesting stories about the making and the stars of the films. The route is: Hősök tere, Kodály KöröND, Kálmán Imre utca, Kossuth tér, Duna part, Szent István tér, Madách tér, Gozsdu udvar, Rózsák tere, 56-osok tere. The following films are evoked during the tour: Evita (Buenos Aires - Hősök tere, Szabadság tér), A Good Day to Die Hard (Moscow - Szabadság tér, Városligeti fasor, Vörösmarty utca, Szépművészeti múzeum, 56-osok tere), Gloomy Sunday (Budapest), I Spy (Budapest), Hellboy II (Budapest), Red Cop (Moscow), Mortal Kombat (imaginary location), Spy Game (East Berlin), Underworld (imaginary - Gozsdu Court), Munich (Rome, Paris, Netherlands, London), Spy Game (Berlin) (www.tandembudapest.hu, 2021).

Of the companies visited, Moviewalking was the one that responded to the market vibrations to the maximum. The company, which originally organised themed walks, underwent a major transformation and a complete platform change in 2017. Under the leadership of Balázs Basa, a tourism product was created to meet the needs of the postmodern tourist, which was unique in the Hungarian film tourism market. In October 2017, the Hungarian film tourism app Guide@Hand Film Destination Budapest was launched. Jointly developed by Filmdestination.hu and the eLearning Department of the Institute for Computer Science and Control, Hungarian Academy of Sciences, it is available in Hungarian and English, and works on both iOS and Android systems. The app allows users to get acquainted with the Budapest film tourism guide with the widest range of information to date, including filming locations and extra content. It illustrates the filming locations with pictures and videos, and can also be operated offline, so users can follow their route on the app's map without an internet connection. In the future, an audioguide function will be added to the product to make the film walk experience even more complex (www.guideathand.com, 2021).

The lack of coordination of the film tourism sector and the lack of a centrally controlled film marketing activity was mentioned several times during the discussion with Balázs Basa. The film tourism expert repeatedly stressed a state-led and organised film tourism policy would be needed to successfully position the capital and the country, and the films shot here should be used in marketing activities. Many attempts have been made to link film-themed services with each other, but these efforts have not been successful in recent years.

The interviews clearly revealed the elements of the capital's film tourism offer. It can be said that very few of the elements highlighted in the literature can be found in Budapest. Among the supply elements of film tourism, we can distinguish on-location film tourism, for-profit commercial attractions, runaway productions, film tourism linked to film production, unique film events, armchair travel (television travel films) (Irimiás, 2015). The supply elements offered in Budapest are summarised in the table below *(Table 2)*.

Table 2. The supply elements of film tourism in Budapest

Type of Film Tourism	Attraction	Example
On-locations **Nostalgic film tourism**	Visit filming locations that evoke times/ ages gone by	*Szomszédok (Neighbours) series:* The Gazdagréti housing estate in Budapest The series was one of the most popular, best-known and most popular Hungarian television series, which was broadcast from 1987 to 1999.
Double locations– Runaway Productions **Film tourism at a supposed filming location**	Films or television series not shot on location where the plot is set	*Evita:* The filming location is partly Budapest, featured as Buenos Aires *The Martian:* The filming location is partly Budapest, featured as Peking and Houston *Spy Game:* The filming location is partly Budapest, featured as East Berlin
For-profit commercial attractions **Film themed tours**	Film themed trips	*Éljen a Grund!* – family walk, Mese és valóság a Pál utcában *Világ körüli út (A trip around the world)* – bike sightseeing with filming locations *GUIDE@HAND Film Destination Budapest*
Film tourism in connection with filmmaking **Film studio visits**	Guided tours of an operating film studio	*Korda Film Park and Visitor Centre*

Source: Based on Beeton (2005), Irimiás (2015), self-edited (2021)

As can be seen from the summary table, one of the most characteristic features is the supply element related to the double location, which is clearly due to the fact that the capital is very rarely featured as itself in film productions. The film themed tours and the programmes organised around the attractions of working film studios are the most popular elements of the offer, but more thematic tours should be added to the palette of film tourism in Budapest.

FUTURE RESEARCH DIRECTIONS

Due to the multidisciplinary nature of film tourism, it is possible to develop future research directions by linking it with a number of other disciplines. One of the most fundamental insights can be linked to film studies. In the context of Budapest, it is worth examining how films and television series of different genres influence the image of the capital and travel decisions. Another interesting perspective would be to look at why the Hungarian capital is most often chosen for spy and action films. What are the buildings and iconic locations appearing in films that essentially carry this atmosphere?

Another connection point would be to further examine the economic pillars mentioned in the previous chapters of this research. The 2017 competitiveness study carried out by tax consultancy Orien Tax provides an excellent basis for this. In the future, however, the study would not focus on countries, but on European cities that are at the heart of film production and film tourism. The focus of this research will be on factors that clearly contribute to the measurement of the direct and immediate economic impact of film productions. This study will provide an excellent picture of what further economic steps are needed to ensure Budapest's secure place on the European film map, and of what other locations in Hungary, apart from Budapest, could be more prominent in films shot here (Spy - Lake Balaton).

On the demand side, the study of the motivations of film tourists from a cognitive and neuropsychological perspective provides an interesting insight. This kind of feedback from film tourists specifically visiting Budapest would provide perfect guidance for the development and design of tourism products required in experience economy. One excellent methodology for this is cognitive mapping, which can be used in relation to films in a number of aspects.

The phenomenon of overtourism, which is also becoming more and more common in tourism as a whole, has also made its way into film tourism destinations, thus clearly highlighting the issue of sustainability and the study of film tourism. In the context of film-generated tourism and sustainability, pillars such as changes in visitor numbers, the economic and infrastructural environment and the situation of local residents in the destination can be analysed. In many cases, the local population in priority film tourism destinations is confronted with the negative impacts of this specific segment, and the issue of sustainable tourism in rural areas becomes even more acute. Barcelona, Dubrovnik, Paris and Venice are potential study locations in this respect.

In relation to sustainability, the impact of COVID-19 could also be examined, as the pandemic situation has led to dramatic changes in the tourism sector, including film tourism. The pandemic has completely transformed the tourism market, and film destinations facing the overtourism phenomenon have faced new challenges since spring 2020. The focus of the examination may be popular film tourism destinations that have already had to introduce regulations to prevent the phenomenon of overtourism. Such potential locations may include Venice, Dubrovnik, Paris, Barcelona, London, Matamata (New Zealand) and Koh Phi Phi Leh (Thailand).

CONCLUSION

The analysis of the Hungarian film sector has shown that the standardization of legislation and the support system, as well as the development of infrastructure, have brought about huge changes. The film industry has succeeded in generating economic indicators that were unprecedented in the past decades. This is clearly confirmed by the competitiveness study and by the total film industry expenditure of over HUF 100 billion year on year. The National Film Institute has clearly sought to successfully reform the Hungarian film industry in line with European trends. Following regional examples, it has also encouraged the organisation of professional events, training and education for occupations in short supply in film industry and expanded the opportunities for film funding. Professional workshops have also been organised to bring young people closer to the world of film, with the potential to build a generation of film-literate people in the future. These good practices have long been applied in France, Spain and the UK. Education and vocational training have a role to play in ensuring the continued high quality of the professional workforce, which has proved to be a key pillar in the competitiveness analysis. Favourable tax conditions and market prices, transparent legislation and high quality allow serious international productions to continue to choose Hungary. These essential elements are the factors behind the success of Hungary's film industry, which can be further enhanced in the future by increasing the popularity of national (Hungarian) films, expanding the range of professional trainings and international professional events, as well as broadening the spectrum of grant support opportunities. The development of the film industry in Hungary has not stopped either, as the demand for quality studios is increasing due to the delayed productions caused by the COVID situation. Currently, a film investment worth EUR 113,710,800 is underway in Mogyoród (25 km from Budapest), called Astra Filmland Project. The aim of the development is to create a film ecosystem that, in addition to its unique studio elements (e.g. large deep water pool), will offer a wide range of additional services that crew members may need during the production (architect's workshops, camera, lighting and other technical equipment rental, costume storage, post-production studio, English-speaking kindergarten and school for the family members of foreign crew members).

In addition, four studios will be added to the National Film Institute complex in Fót (21 km from Budapest), one of which will specialise in Hungarian films. The development will add a further 9600 m^2 of usable floor space to the current 2600 m^2 of studio capacity. An important quality change is that 2 out of the 4 new studios can be connected to each other, so that the future complex can meet the needs of large-scale productions even in parallel (www.telex.hu, 2021).

The positive changes in the economic environment have also resulted in a boom in film production in Hungary, which is the basis of film tourism. Film tourism can develop spontaneously and consciously in film locations, but it is more fruitful for the destination if it is developed in a controlled and planned way (Irimiás, 2015). It was clear from the interviews that there is a complete lack of coordinated management of film tourism services in the film tourism segment, which would require a thoughtful and long-term management. Such a move would increase the city's competitiveness and Budapest could be repositioned on the European film map. To achieve this, an open-minded management is needed, capable of change and development, alert and aware of new market trends. An important aspect would also be to feature Budapest as Budapest, which would be a major contribution to the successful adaptation of the practice applied at VisitBritain, the national tourism agency of Great Britain. The research on locations has also highlighted that there is a strong emphasis on the being featured as a double destination. Although this phenomenon is improving, it is still not sufficiently significant to form the basis of complex marketing activities. Nevertheless, even with a double role, the Hungarian capital is still able to make successful marketing moves. One outstanding example of this is the filming of Indian productions (I have given my heart away, darling, Raabta) here, which has attracted local tourists to the capital. The number of tourists from India has clearly increased in Budapest after the premiere of the films.

If Budapest was featured as itself more often, it would possibly increase the share of potential film tourists in leisure tourism. This process would also boost the expansion of the supply of film tourism products. Very few visitors arrive in the capital with a specific interest in film tourism products. Emphasizing Budapest and presenting it as itself, in its own right, would allow the destination to attract not only casual film tourists but also general and dedicated ones.

The in-depth interviews also revealed that there is an urgent need to expand the offer of film tourism products, which does not require a major financial investment. However, structured organisation and partnerships are essential, which are currently not sufficient in Budapest. A dynamic network of contacts would allow the Hungarian capital to develop a truly effective film community brand and, by doing so, increase the attractiveness of the destination, with a greater emphasis on the success of Budapest's film tourism services.

The research has highlighted the fact that Budapest can indeed show a thousand faces on the film screen, making the city increasingly popular with international film productions. The reasons for filming in the Hungarian capital can be summed up clearly: the favourable tax regime, the excellent infrastructure, the diverse climate, the multifaceted cityscape and, of course, the highly qualified film professionals all play a significant role. These pillars strengthen Budapest' positiong on Europe's film map and ensure that it remains one of the most popular film shooting destinations in the European market. In addition, the most popular filming locations table clearly highlights the filming locations that play an important role in the city's tourism marketing. Their use in the marketing of the capital would be essential, as this strategy would not only reach potential film tourists. A promotional film with the slogan "See it live as Budapest now!" would be an absolutely feasible tool in promotion. In the video, scenes cut from movie films would show Budapest as Paris, Buenos Aires, Rome or Berlin, but with a crossfade cut in the film, the film would represent the real location. The impact of all this would be further enhanced by the famous actors (Will Smith, Matt Damon, Scarlett Johansson, Ryan Gosling) who appear in the film. This campaign would be a clear motivation for the viewer to admire Budapest for what it is. The realisation of this promotional film could be a much-needed highlight in the marketing efforts to promote Budapest.

As Dániel Kresméry (one of the owners of Hero Squared, an international film production company) says, it is hard to imagine better country marketing than Will Smith talking about how much he loves Budapest, Matt Damon saying he would love to live here, or Brad Pitt bringing his family along to the Hungarian capital.

REFERENCES

Árva, L., & Deli-Gray, Zs. (2011). Posztmodern turizmus, a posztmodern marketing [Postmodern tourism, postmodern marketing]. In P. Róbert (Ed.), The socio-economic situation in Hungary in the first decades of the 21st century (pp. 1–11). Academic Press.

Baker, B. (2012). Branding and the opportunities of movies: Australia. In N. Morgan, A. Pritchard, & R. Pride (Eds.), *Destination Brands* (pp. 239–250). Routledge.

Basa, B., & Názer, Á. (2014). *Itt forgott - Főszerepben Budapest [Filmed here - Starring Budapest]*. Akadémiai Kiadó.

Beeton, S. (2005). Film-induced Tourism. In *Multilingual Matters*. Channel View Publications. doi:10.21832/9781845410162

Beeton, S. (2010). The Advance of Film Tourism. Tourism and Hospitality Planning & Development, 7(1).

Bolan, P., & Williams, L. (2008). The role of image in service promotion: Focusing on the influence of film on consumer choice within tourism. *International Journal of Consumer Studies*, *32*(4), 382–390. doi:10.1111/j.1470-6431.2008.00672.x

Creative England. (2015). *Quantifying film and television tourism in England*. Report for Creative England in association with VisitEngland by Olsberg, SPI.

Croy, W. G., & Walker, R. D. (2003). Rural tourism and film-issues for strategic regional development. In D. Hall, L. Roberts, & M. Mitchell (Eds.), *New directions in rural tourism* (pp. 115–133). Ashgate Publishing Ltd.

Csizmadia, N. (2016). *Geopillanat - A 21. század megismerésének térképe* [Geomoment - Map for understanding the 21st century]. Budapest: L'Harmattan Kiadó.

Cynthia, D., & Beeton, S. (2009). Supporting Independent Film Production Through Tourism Collaboration. *Tourism Review International*, *13*(2), 113–119. doi:10.3727/154427209789604624

Figyelő. (2018). *születésnapját ünnepli a Hatoslottó* [The Six Number Lottery celebrates its 30th birthday]. Retrieved from: https://figyelo.hu/hirek/30-szuletesnapjat-unnepli-a-hatoslotto-21927/

Gartner, W. C. (1993). Image Formation Process. *Journal of Travel & Tourism Marketing*, *2*(2-3), 191–216. doi:10.1300/J073v02n02_12

Grihault, N. (2003). Film tourism-the global picture. *Travel & Tourism Analyst*, *5*, 1–22.

Guide@Hand. (2021). *Film Destination Budapest*. Retrieved from: https://guideathand.com/hu/downloads/guide@hand

Gunn, C. A. (1972). *Vacationscape: Designing Tourist Regions*. University of Texas.

Hesmondhalgh, D. (2007). *The Cultural Industries*. SAGE Publications.

Howkins, J. (2004). *Az alkotás gazdagít: Bátorítás kreatív embereknek [The creative economy. How People Make Money from Ideas]*. HVG.

Hudson, S., & Ritchie, J. R. B. (2006). Promoting Destinations via Film Tourism: An Empirical Identification of Supporting Marketing Initiatives. *Journal of Travel Research*, *44*, 387–396.

Innovációs és Technológiai Minisztérium. (2019). *Együttműködési megállapodást kötött az ITM és a Moholy-Nagy Művészeti Egyetem* [Cooperation agreement between ITM and Moholy-Nagy University of the Arts]. Retrieved from: https://www.kormany.hu/hu/innovacios-es-technologiai-miniszterium/gazdasagstrategiaert-es-szabalyozasert-felelos-allamtitkar/hirek/egyuttmukodesi-megallapodast-kotott-az-itm-es-a-moholy-nagy-muveszeti-egyetem

Irimiás, A. (2015). *Filmturizmus. A filmek és televíziós sorozatok turisztikai szerepének és hatásainak geográfiai vizsgálata [Film tourism. Geographical analysis of the role and impact of films and television series on tourism]*. Akadémiai Kiadó.

Irimiás, A., Papp-Váry, Á., & Szűts, Z. (2013). Filmturizmus [Film tourism]. *Figyelő Hetilap*, *2013*(7).

Keresnyei, K., & Egedy, T. (2015). Adalékok a kreatív gazdaság elméletéhez [Contributions to the theory of the creative economy]. *Földrajzi Közlemények*, 30–42.

Kim, H., & Richardson, S. L. (2003). Motion picture impacts on destination images. *Annals of Tourism Research*, *30*, 216–237.

Korda Filmstúdió. (2021). *About us*. Retrieved from: https://kordastudio.hu/

Li, S., Li, H., Song, H., Lundberg, C., & Shen, S. (2017). The economic impact of on-screen tourism: The case of The Lord of the Rings and the Hobbit. *Tourism Management*, *60*, 177–187.

Morgan, N., & Pritchard, A. (1998). *Tourism promotion and power: Creating images, creating identities*. John Wiley & Sons Ltd.

Nemzeti Filmintézet. (2017). *Megéri hollywoodi produkciókat az országba csábítani* [It is worth attracting Hollywood productions to the country]. Retrieved from: https://nfi.hu/hu/filmipar/magyarorszagnak-megeri-hollywoodi-produkciokat-az-orszagba-csabitani

Nemzeti Filmintézet. (2019). *110,1 milliárd forintból készültek filmek Magyarországon 2018-ban* [HUF 110.1 billion spent on films in Hungary in 2018]. Retrieved from https://nfi.hu/hu/filmipar/1101-milliard-forintbol-keszultek-filmek-magyarorszagon-2018-ban

Nemzeti Filmintézet. (2021). *Rekordot döntött a magyar filmipar* [The Hungarian film industry sets a record]. Retrieved from: https://nfi.hu/hu/filmipar/rekordot-dontott-a-magyar-filmipar-1

Papp-Váry, Á. (2009). Országimázs a moziban [Country image in the cinema]. *Marketing & Media*, *4*(9), 13–26.

Papp-Váry, Á. (2019). *Országmárkázás. Versenyképes identitás és imázs teremtése [Country branding. Creating a competitive identity and image]*. Akadémiai Kiadó.

Papp-Váry, Á., & Marthy, R. (2009). Magyar "Borat"-ok: Budapest és Magyarország imázsa a hollywoodi filmekben [Hungarian "Borats": the image of Budapest and Hungary in Hollywood films]. In "Új marketing kihívások a 21. században – Fenntartható fogyasztás", a Magyar Marketing Szövetség Marketing Oktatók Klubja, Kaposvár. Academic Press.

Pratt, S. (2015). The Borat Effect: Film-Induced Tourism Gone Wrong. *Tourism Economics*, *21*, 977–993.

Puczkó, L., & Rátz, T. (2011). *Az attrakciótól az élményig [From attraction to experience]*. Akadémiai Kiadó.

Riley, R., Baker, D., & Doren, C. S. V. (1998). Movie induced tourism. *Annals of Tourism Research*, *25*, 919–935.

Sétaműhely. (2021). *Sétakínálat* [Offer of Walks]. Retrieved from: https://setamuhely.hu/setak/

Stachowiak, K., & Stryjakiewicz, T. (2018). The rise of film production locations and specialized film services in European semi-peripheries. *Hungarian Geographical Bulletin, 67*(3), 223–237.

Sulyok, J. (2009). Filmturizmus, avagy a filmekben rejlő lehetőségek a turisztikai desztinációk számára [Film tourism or the potential of films for tourist destinations]. *Turizmus Bulletin, 2009*(1).

Szerdahelyi, K. (2020). *Jön a Budapest Brand Zrt., új turisztikai stratégia a fővárosban* [Budapest Brand Zrt. is coming; new tourism strategy for the capital]. Retrieved from: https://turizmus.com/szabalyozas-orszagmarketing/jon-a-budapest-brand-zrt-uj-turisztikai-strategia-a-fovarosban-1171906

Tandem Budapest. (2021). *Budapesti túrák* [Budapest Walks]. Retrieved from: http://tandembudapest.hu/budapest/

Telex.hu. (2021) *Készüljünk fel Hollybudra? Filmes bérgyártásban Magyarország a globális verseny élmezőnyébe tart* [Shall we prepare for Hollybud? Hungary is heading to the top in the global competition in contract film production]. Retrieved from: https://telex.hu/gazdasag/2021/08/04/budapest-kornyeke-mar-ma-is-vezeto-filmgyartasi-kozpont-de-a-nagy-boom-miatt-ujabb-meretes-beruhazasok-indulnak

Throsby, D. (2001). *Economics and Culture*. Cambridge University Press.

Tóth, R. A. (2016). Széles vászon, nagy haszon [Wide screen, big profits]. *Turizmus Trend – Vendég & Hotel, 2016*(5).

United Nations. (2018). *United Nations Conference on Trade Development*. Retrieved from: Creative Economy Outlook: https://unctad.org/en/PublicationsLibrary/ditcted2018d3_en.pdf

Zachar, B. (2006). *Van egy magyar filmtörvény 1* [There is a Hungarian Film Act I]. Budapest: Elektronikus Periodika Archívum és Adatbázis. Retrieved from: http://www.epa.oszk.hu/00300/00373/00029/filmtorv.html

ADDITIONAL READING

Beeton, S. (2001). Smiling for the Camera: The Influence of Film Audiences on a Budget Tourism Destination. *Tourism, Culture & Communication, 3*(1), 15–25. doi:10.3727/109830401108750661

Busby, G., & Klug, J. (2001). Movie-induced tourism: The challenge of measurement and other issues. *Journal of Vacation Marketing, 7*(4), 316–332. doi:10.1177/135676670100700403

Croy, W. G., & Buchmann, A. (2009). Film-induced tourism in the high country: Recreation and tourism contest. *Tourism Review International, 13*(2), 147–155. doi:10.3727/154427209789604642

Iwashita, Ch. (2008). Roles of flms an television dramas in international tourism: The case of Japanese tourists to the UK. *Journal of Travel & Tourism Marketing, 24*(2-3), 139–151. doi:10.1080/10548400802092635

Riley, R., & Van Doren, C. (1992). Movies as Tourism Promotion: A Push Factor in a Pull Location. *Tourism Management, 13*(3), 267–274. doi:10.1016/0261-5177(92)90098-R

Roesch, S. (2009). *The Experiences of Film Location Tourists*. Channel View Publications. doi:10.21832/9781845411220

Squire, S. (1993). The Cultural Values of Literary Tourism. *Annals of Tourism Research*, *21*(1), 103–120. doi:10.1016/0160-7383(94)90007-8

Tooke, N., & Baker, M. (1996). Seeing is Believing: The Effect of Film on Visitor Numbers to Screened Locations. *Tourism Management*, *17*(2), 87–94. doi:10.1016/0261-5177(95)00111-5

KEY TERMS AND DEFINITIONS

Creative Industry: The creative industry is a set of autonomous sectors, whose activities are all geared towards the production of intellectual products and cultural values based on human creativity, also inducing an economic process. It forms a bridge between art, culture, business, and technology.

Film Tourism Product: There is an appeal or attraction at the heart of the product, which is a material or intellectual element linked to the shooting, production or distribution of films or television programmes.

Film Tourist: Film tourists are travellers who visit a film destination for work purposes and not just for leisure. The motivations of leisure tourists may include the location seen in the film, the plot, the film star and other intrinsic motivations.

Film-Induced Tourism: This sector is linked to cultural tourism, including trips where the primary motivation of travellers is to visit a film production location. However, it also includes film business trips, visits to film theme parks and film studios.

Postmodern Tourism: The individual is at the centre; mass tourism is rejected. The most important elements are the other motivations of visitors (learning, shopping). It is characterized by heterogeneous demand; visitors' consumption patterns are completely different. There are just a few return visits to familiar, established places.

Postmodern Tourist: The postmodern tourist is well-informed, does not accept the opinions of others or mass products, wants to organise his/her own travel programme and wants to actively participate in the activities. The post-industrial tourist "participates."

Chapter 15
From Nuclear Disaster to Film Tourism:
The Impact of the Chernobyl Mini-Series on the Exclusion Zone

Tímea Zsófia Tóth
Alexandre Lamfalussy Faculty of Economics, University of Sopron, Hungary

Árpád Ferenc Papp-Váry
John von Neumann University, Kecskemét, Hungary

ABSTRACT

The research explores how a well-known historical location associated with disaster tourism became a destination of film tourism. Thanks to the HBO miniseries, the nuclear accident zone around Chernobyl registered a record number of tourists in 2019. The study includes a complex tourism analysis of Chernobyl. The examination of the demand and supply sides of local film tourism is followed by an analysis of elements such as the number of visitors and the supply of thematic routes. The study also analyses interviews with local travel companies and information on their social media platforms. The clear aim of the research is to find out how a successful film may affect the tourism of a location. Using the results of the research, the study discusses how a site previously associated with disaster tourism utilizes the opportunities of film tourism, as well as the opportunities provided by a possible change of profile.

INTRODUCTION

"Pop culture is changing the tourism industry - the proof is in Chernobyl"; "Chernobyl is now an official tourist attraction"; "Disaster tourism in Chernobyl has been boosted by a new series on the site"; "Chernobyl to become official tourist attraction, Ukraine says". Since the summer of 2019, headlines like these have appeared on CNBC, Forbes, turizmus.com, CNN and many other media outlets. The industry is already familiar with the impact of film-related tourism, but this phenomenon has so far been studied mostly in relation to feature films and motion pictures on fictional events. The series has created a spe-

DOI: 10.4018/978-1-7998-8262-6.ch015

cific trend under the banner of disaster tourism, to which film tourism itself can be organically linked. In defining disaster tourism, however, it is essential to define dark tourism, since these two concepts are often used synonymously, despite the fact that dark tourism is used as a generic term in the literature.

In the 21st century, postmodernism has created a whole new focus in tourism, with a clear emphasis on experience and uniqueness. The area of the Chernobyl disaster has also been a case in point, with a new and interesting fusion of tourism offerings that makes the destination distinctly unique. The film series presented in 2019 (and its impact) has further promoted the development of the area, considering the needs of post-modern tourists.

The Chernobyl disaster has also become a part of common knowledge, a well-known event in human history. In a rather morbid way, it was even associated with the Ukrainian nation itself in South Korean television network MBC's coverage of the opening ceremony of the 2021 Tokyo Olympics.

HBO's five-part miniseries Chernobyl has fundamentally changed the way we understand and perceive location and disaster tourism. The impact of the HBO series surprised even the producers, who had not expected such acclaim from viewers and the industry alike. The production on the 1986 tragedy topped the charts thanks to its historical authenticity, a script that took nearly two and a half years of painstaking research and an excellent cast. The research aims to examine the process of how a historical disaster site became a popular tourist destination, the impact of the series on the area, the changes in infrastructure and the film tourism attractions and products available in the area.

BACKGROUND: POSTMODERN TOURISM

Film Tourism

In the literature review, film tourism and disaster tourism are defined, but after clarifying the concepts, it is also worth examining the characteristics and motivations of disaster tourists and the typology of disaster tourism itself. Film tourism is the attraction of visitors through films, where the success of the film is clearly a key factor. There is a strong link between story and location, so the success of the film plays a key role in the success of film tourism (Sulyok, 2009). As a multidisciplinary discipline, film tourism focuses on both marketing and business aspects (Beeton, 2010) – in 2012, nearly 40 million tourists worldwide chose their destinations according to what they saw in the movies. Destinations with an outstanding film tourism industry are the most likely to attract young people and city visitors. Brazil, Russia, India and China are considered as prominent sending markets (Tóth, 2016).

Motion pictures are extremely influential in the potential tourists' destination choices, as these films are perceived as unbiased compared to traditional advertising and convey relevant information about the destination in a short time (Kim & Richardson, 2003; Gartner, 1994). The various destination marketing organisations have a very important role in shaping and positively reinforcing the image of a destination in tourists. It is also an important fact that good films contribute to a positive perception of the filming location and the country, and to country image itself (Beeton, 2005). Films can have a strong influence on the perception of the destination, which can be both positive and negative, as the viewer can develop positive or negative attitude towards the destination. Considering all that, we can say that the perception of the country image can be greatly influenced by the film (Gunn, 1972; Papp-Váry, 2019). Close coordination of the work of destination marketing organisations, film agencies and production companies is necessary to develop a successful marketing strategy. Such a concept makes it possible to prevent the phenomenon of overtourism, strengthen the image of the destination and promote the development of travel motivation associated with the film location (Hudson & Ritchie, 2006; Morgan & Pritchard, 1998; Cynthia & Beeton, 2009).

While the concept of film tourism is easy to define, the impact of film on tourism is difficult to measure, and only a small proportion of locations that benefit from film tourism examine its effectiveness. The primary reason for the difficulty of measurement is that the impact of films on tourism is manifested in an extremely long term, with several years elapsing between the time of viewing and the time of travel (Sulyok, 2009). The decision to travel, or merely to plan a trip, is not made immediately after a film experience, and therefore up to ten or twenty years may elapse between the two events. However, cinematographic images clearly have the potential to make the tourist travel as a result of their spectacles, (famous, well-known) characters and real or fictional locations (Kim & Richardson, 2003; Hao & Ryan, 2013; Irimiás, 2015). A successful film is therefore clearly capable of inducing travel whether its location embodies itself or a completely different destination. This is also the case with Chernobyl, where the successful series boosted the number of tourists not only in the disaster zone, but also in the Lithuanian locations used as a double location. This phenomenon will be discussed in more detail in a later chapter.

The strong correlation between films and the increase in the number of visitors to a destination has been supported by a large body of international literature (Riley & Van Doren, 1992; Riley et al., 1998; Grihault, 2003). The academic literature has examined the field of film tourism from several perspectives, mostly focusing on the topics of travel motivations (Kim et al., 2003; Peters et al., 2011; Oviedo-Garcia et al., 2016), impacts on the local economy (Croy, 2003; Pratt, 2015; Stachowiak & Stryjakiewicz, 2018), and the role of films in destination marketing and country image (Beeton, 2005; Bolan & Williams, 2008; Riley & Dolan, 1992; Irimiás, 2015; Papp-Váry, 2009; Hudson & Ritchie 2006; Baker, 2012, Papp-Váry, 2019).

DISASTER TOURISM

The definition of disaster tourism should be based on the categories of alternative tourism. The latter umbrella term naturally includes many specific types of tourism, such as heritage tourism, ecotourism, or disaster tourism as discussed above (Puczkó & Rátz, 2005). Mass tourism and alternative tourism are opposites within the sector – while the number of participants is high and programmes are mostly implemented in large groups in the former category, operators pay more attention to satisfying needs and interests that are different from the average in alternative tourism, with a focus on individuality and small group programmes tailored to the individual (Michalkó, 2012). This particular branch of tourism is becoming increasingly popular with tourists (Lennon & Foley, 2000), as evidenced by the fact that in 2017 more than 2 million people visited the Auschwitz-Birkenau Memorial. However, this phenomenon can also be found in other areas, with tourism in Chernobyl, the known site of the 1986 nuclear disaster, continuing to grow despite a 48% drop in tourism in Ukraine in 2014 due to the tense political situation (The Globe Post, 2018).

Disaster tourism is a relatively new field in tourism, but the phenomenon itself can be traced back in history, with Thomas Cook's tours and Mark Twain's guided tours being good examples. The concept itself is nothing other than the visit of sites associated with natural disasters, tragedies, battle scenes, where the primary aim is not to help but to increase adrenaline levels and satisfy individual curiosity (Palancsa, 2012). The term itself evolved more than twenty years ago, due to tourists' love of travelling to dangerous places and visiting disaster areas. The basic thought that comes to mind is: "It is a good thing that it did not happen to me". But there can also be a more personal connection to the visit, one example being when family stories lead to visiting World War II sites or people follow tragedies involving friends (e.g. terrorist attack, disaster site). In this form, the traveller encounters these events in a way that he or she does not actually experience the tragedy, but still becomes part of the events. Disaster tourism and dark tourism are used synonymously in a number of works in the literature, but while the former expression refers to a specific area, the latter is an umbrella term. The importance of dark tourism comes from the fact that it, by its very nature, can trigger strong and complex reactions in tourists (Buda, 2015; Seaton, 2009). In the international literature, the term 'dark tourism' was first used in 1996 by the authors Foley and Lennon, but at that time they focused exclusively on the relationship between death and tourist attractions. Dark tourism is the circulation of people to places characterised by distress, atrocity, or sadness and pain. As a more specific component of dark tourism, "disaster tourism" denotes situations where the tourism product is generated within, and from, the aftermath of a major disaster or traumatic event (Gotham, 2007, 2015). A number of perspectives and researches on 'dark tourism' have been published, which have led to the emergence of various terms based on travel motivation, such as 'thanatourism' = defined as 'travel to a location wholly, or partially, motivated by the desire for actual or symbolic encounters with death' (Seaton, 1996), 'morbid tourism', and 'black-spot tourism' (Stone, 2006). However, dark tourism is clearly the most widely used term (Buda, 2015; Dunkley, Morgan, & Westwood, 2011). The common thread in defining disaster tourism and dark tourism is that a tragedy occurred in a given location, resulting in the death of many people and/or severe natural destruction. Although the two concepts have been conflated in much of the literature, the fundamental difference lies in the impact of the events. While dark tourism involves locations that have no direct impact on the present day (concentration camps, prisons, terrorist sites), the impact of disaster tourism is still felt today, even decades after the event (e.g. Chernobyl or the location of Hurricane Katrina's destruction). Of course, survivors of dark events or others personally involved are still deeply affected by the sight or the story itself to this day.

On the demand side, the focus is primarily on the motivations of tourists, as this also allows us to distinguish several typologies within the category. The largest group is made up of locations where individual or mass deaths have occurred. These include battlefields (Gallipoli), death camps (Auschwitz), genocide sites (Cambodia) and places where famous people have died (Michael Jackson – Holmby Hills). The other popular category is war memorials, cemeteries and crypt sites, whose visitors are motivated by aspects similar to those of a pilgrimage. Groups with questionable historical authenticity include categories where sites completely unrelated to the location of the event are visited. These include, for example, museums displaying weapons of death and attractions showing tragic events (Sharpley 2005). The latter category also has an interesting additional segment that functions as a theme park in dark tourism. These artificially created attractions are known as 'dark fun parks', such as the London Dungeon, which takes visitors back to England in the 1880s, or Dracula Park in Romania (Stone, 2006). The reasons for travel motivations of tourists in this specific segment are influenced by a number of factors, which, based on research by M. Pirveli (2008), are summarised in the following table by D. Rucińska and M. Lechowicz. (*Table 1*).

Table 1. Reasons for undertaking disaster tourism

Motivations for Disaster Tourism	Specific Motivations for Disaster Tourism
Need to change location	Space and time compression
Recreational needs	Individual preferences
Cognitive needs	Global interest in the world
Need for human contact	Contact with people harmed by natural disasters
Need for new experiences	Experiencing emotions
General social needs	Expansion of social awareness
Additional tourism conditions	Economic status, need to aid others

Source: Rucińska & Lechowicz (2014)

Based on the elements collected, it can be said that there is a very broad spectrum of motivations for the participants of disaster tourism. There are tourists who want to increase their social awareness, to meet people who have lived through traumatic experiences themselves. However, there are others who are specifically looking for authenticity and want to get a first-hand experience of the disaster-stricken zone in the destination.

It is man's fundamental curiosity and interest in the events of the past that has allowed disaster tourism to develop such a broad spectrum. This has given rise to different nuances, first summarised by Stone (2006). There are a number of categories between the dark and light end of the scale, but while at one end of the scale the focus is on location and historical authenticity, education, presentation and low infrastructure, the lighter end is characterised by entertainment, purposeful offerings, meeting the needs of tourists and advanced tourism infrastructure. An accurate understanding of nuances is aided by Sharpley's (2009) supply-demand model, which is based on the relationship between attractions and visitor motivations (Dallen, 2011). The model distinguishes four nuances: 1) pale tourism (limited visitor knowledge, spontaneous visits), 2) grey tourism demand (interested tourists, planned visits), 3) grey tourism supply (highlighting dark tourism attractions, museums, monuments), 4) black tourism (trips induced by attractions of disaster tourism). Of course, visiting disaster tourism locations gives tourists the opportunity to learn about and relive the events of the past, but it is important to underline that respect and ethical behaviour is one of the most important factors on the visitors' side, which has also generated a lot of media attention in the context of Chernobyl.

DISASTER TOURISM AND FILM TOURISM

It is worth examining how the visitor can be described in the case of the present study, where the concepts of disaster and film tourism are intertwined. The search for experiences and their consumption has become a lifestyle nowadays, which is a clear characteristic of the postmodern tourist. The visitor wants to participate and become an integral part of the tourist experience, thus tourism products with high quality, intense and multidimensional experiences are favoured (Árva & Deli-Gray, 2011). A tourism product is nothing else than a set of services and products offered by different complementary or competing tourism enterprises (Puczkó & Rátz, 2011). Attractiveness, also known as attraction, is central to both disaster tourism and film tourism, so it is possible to design thematic services after understanding the motivations of tourists. In the disaster tourism segment, these are tragedy-related tangible (material) or intangible (intellectual) elements linked to the shooting, production or distribution of films or television programmes (Irimiás, 2015). One international example of the merging of disaster and film tourism is the film series Narcos. Produced by Netflix, the series depicts the circumstances of drug wars in South America, highlighting the role of Pablo Escobar. The series has led to a surge in the number tourists exploring sites linked to the drug lord, including ones in Medellín, Colombia. Thanks to the series, themed tours and souvenir shops have been set up at Escobar-related locations, despite the fact that the drug war also claimed the lives of many civilians in the area.

Looking at the subject of this article, the sites (potential attractions) of the Chernobyl miniseries, a number of destinations featured prominently in the film have been opened to visitors over the years (e.g. the Pripyaty residence quarter, the control room of the nuclear power plant).

It is also interesting to note that today, research on dark tourism is not only focused on the psychological analysis of the tourist experience (Biran & Buda, 2018; Morten et al., 2018), but also on the consequences and emotional impact of disasters in a specific area (Amujo & Otubanjo, 2012; Tucker et al., 2017, Wright & Sharpley, 2018) and possible future directions (Wright, 2018).

CHERNOBYL ZONE AND TOURISM

From Disaster Tourism to Film Tourism

The research focuses on the miniseries and the Chernobyl-related tourist destination, approached from several perspectives. The basic methodology can be divided into two parts: the first part consists of a detailed review of the literature, focusing on the importance of the motivations of tourists and the typology of disaster tourism, the results of which are presented in the previous section. In the following secondary research, the series was analysed. Its analysis is closely linked to the third stage, the pillars of the primary research, thus the two are discussed together.

In the primary research, the supply side was examined in detail, focusing on the attractions and thematic trips that can be associated with film and disaster tourism. The analysis was based on in-depth interviews with local travel agencies – interviews with experts – and content analysis of their various social media platforms (Facebook, Instagram, TripAdvisor). Interviews with the three selected travel agencies (Gamma Travel, Chernobylwel.come, Chernobyl Tour) were conducted electronically. The research basically sought answers to the questions that lie at the border between film and disaster tourism, i.e.:

- Can a film about a disaster become a travel generating factor?
- Is such a film suitable for boosting tourism in a destination?
- If so, what cinematic elements are needed and how can the destination use them in its marketing activities?

The issues raised by the research are essentially very complex, which is why a comprehensive tourism analysis was needed from both the supply and demand side.

CHERNOBYL - THE PLACE OF THE DISASTER

Chernobyl is presented below as a unique tourist destination. However, before the analysis, it is worth clarifying the location of the studied area and what actually happened on 26 April 1986. The Ministry of Energetics of the USSR selected the site, 30 kilometres from the city of Chernobyl, and large-scale construction work began in 1970 on what was to be the world's largest power plant. Meanwhile, the town of Prypiat had become a regional centre near Chernobyl, thanks to its excellent transport links (railway station, port), hospital and marketplace. It was a modern, almost luxurious town in the Soviet Union when the first reactor was started up in 1977, having 14,000 inhabitants with an average age of 25. (www.chernobylwel.com, 2021).

The tragedy occurred precisely at 1:23 AM (local time) on 26 April 1986, in Block 4 of the nuclear power plant, located only 120 kilometres from Kiev, the capital of Ukraine, close to the border with Belarus, both of which were parts of the Soviet Union at the time. The largest nuclear disaster in human history may have been caused by a confluence of various factors. The nuclear power plant, a major centre of the Soviet military's strategic military programme, had an outdated safety system and a very low level of automation. The actual explosion happened due to a coincidence of several factors: there was an experiment going on (test of the inertial range of the turbo-generator unit), the operators of the plant lowered the energy output (almost until the stoppage of the nuclear chain reaction) which was followed by an immediate increase in the energy output. The stoppage of the cooling water to the reactor core combined with the previously described issue led to the overheating of the fuel and the destruction of the reactor core. A series of professional errors and the unprepared experiment led the facility straight to disaster: after the first explosion, the combination of radioactive steam and hydrogen blew the 1,200 ton cover off the reactor and destroyed the roof; the second explosion had a light blue blaze, after which a mushroom cloud rose above the reactor (www.chernobylwel.com, 2021; www.gamma-travel.com, 2021). The explosion caused the reactor to burn for another 10 days and released a lot of radioactive contamination into the air. The explosion at Chernobyl blew up radioactive substances to the altitude of 1.5 kilometres in the air. Due to the wind, high levels of radioactive contamination were also measured in southeastern Scandinavia, Poland, (then) Czechoslovakia, Austria, Hungary and Romania. The most affected territories were Ukraine and Belarus (part of the former Union of Soviet Socialist Republics a.k.a. USSR), which decided to evacuate parts of their countries forever because of the contamination of the environment. The Soviet government was aware of the events on the day of the tragedy, but did not inform the public of the damage until two days later. Evacuations from the area around the plant began some 36 hours later, and the 30 kilometre area around the plant was declared a no-go zone. The other three blocks of the reactor were completely closed, and a concrete sarcophagus was erected over the damaged fourth block in November 1986. This unit was reinforced with a 36,000-tonne protective casing in November 2016. There is a variety of information regarding the number of victims. There is no uniform documentation including the exact number of casualties.

Most of the people exposed to lethal doses of radiation were nuclear plant workers and firefighters carrying out rescue operations. According to official reports, nearly 600,000 people were affected, and altogether over 10 million people are considered to be ill owing to the Chernobyl catastrophe, with 3.2 million just in Ukraine (the rest is reported in Belarus and Russia) (chernobylwel.com, 2021).

The name of the site has become a catch-all term, but Pripyat and Chernobyl are also located within the 30 km zone. The zone is no longer an uninhabited area, being re-populated mainly by retired people and workers involved in the recovery work. The military protected destination has a single entry point at the settlement of Dityatki, where the first visitors, mainly officials and journalists, were allowed to enter in 2006. In 2011, it became a "fully" public area, open to tourists, but strict rules must be observed during the programmes. The 30 km zone itself was declared a nature reserve in 2016.

CHERNOBYL AS A MINISERIES

In a statement, Ukraine's current president Volodymyr Zelensky said that Chernobyl has so far cast Ukraine in a negative light, but it is time to reposition it and show the world, scientists, ecologists, historians and tourists that the area is a unique place on the planet where nature has been able to reborn after a man-made disaster (tourism.com, 2019). In the light of this, Zelensky had a clear goal to boost tourism in the Chernobyl zone, which was greatly helped by the HBO series (Helmeczi, 2020). Thanks to its increasing popularity, the President of Ukraine officially declared the site of the nuclear disaster a tourist attraction in July 2019.

The series was co-produced by HBO and Sky UK, and premiered in May 2019. The series kicks off exactly two years and one day after the explosion in April 1986, at the home of Soviet inorganic chemist Valery Legasov (played by Jarred Harris). The Russian language is also deliberately used in all episodes, with native speakers, without subtitles or commentary. The actors in Chernobyl do not speak with a Russian accent – a conscious choice made by the show's creator Craig Mazin and director Johan Renck. Mazin explained, "If we make it in a different language, we are narrowing it to people that speak that language and we don't, so right off the bat there is a bit of a disconnect there and we felt that not only did we not want to get performances and performers we could connect with and so move them through the story together." (Mitchell, 2019).

The script was written after nearly two and a half years of research, which is one of the reasons for the historical authenticity of the work (Schwartz, 2019). Many sites online have compared historical reality with the film, but no elements have been found to have distorted the events of the disaster. They were true to reality with the portrayal of the story of Lyudmilla Ignatenko and her firefighter husband Vasily, and the recruitment of local people for the cleanup work. The only part of the storyline that was changed was the role of the scientists who helped with the response works. In real life, many scientists and specialists assisted Legasov, but in the series, a fictional character, Ulana Khomyuk, represents the scientific society.

Thanks to its authentic storytelling, a great cast and the perfection of its visuals, the series won 10 Emmy Awards in 2019, including Outstanding Cinematography for a Limited Series or Movie. It also won two BAFTA Awards (Best Leading Actor, Best Miniseries) in 2020, and was voted one of the top five series of all time by viewers. The most famous and well-known source for movie, TV and celebrity content is provided by IMDB The list is ranked by a formula which includes the number of ratings each show received from users, and value of ratings received from regular users. To be included on the list, a series or miniseries must receive ratings from at least 5000 users, and a TV series must also have aired at least 5 episodes (www.imdb.com, 2021).

Filming took place in the spring and summer of 2018, but the creation of an authentic visual world was a big challenge before filming started. The filmmaking team had to lay special emphasis on creating an authentic Soviet lifestyle and a contemporary representation of the nuclear power plant itself and the city of Prypiat. The work was also made more difficult by the fact that the symbols and scenery of a state that no longer existed had to be conjured up on screen. Documentary films and photo series from the period helped a lot in the preparation, but creative solutions were also used, with items such as ashtrays, plates and period Soviet uniforms being sourced from Ukrainian eBay and marketplaces in Kiev. Finding a location was not an easy task for the creators of the series either, as they were not allowed to shoot inside the objects located in the zone; therefore they decided to use double locations. (Turp-Balazs, 2018).

Two double locations were used while filming, both in Lithuania, which has a very favourable film tax system and, as a post-Soviet state, parts of it have perfectly preserved the atmosphere of the time. The Lithuanian capital Vilnius is increasingly popular with film crews. Apart from Chernobyl, the city recently hosted film crews working on Netflix's yet-to-be-released The Last Czars, BBC's War and Peace and HBO/Sky's Catherine the Great. Although all four shows were set in different historical periods, Vilnius was able to successfully host all of them, among many others (Turp-Balazs, 2018). Film professionals working in the city also consistently highlight the city's fantastic infrastructure, easy accessibility and high-quality film workers. In addition, the Lithuanian economic environment is also favourable for film crews shooting here, as since 2014 they can save up to 30% of their costs due to tax rebates. Statistics from the Lithuanian Film Centre also support the boom in the Lithuanian film industry: during the period between 2014 and 2018, a total of 115 films used the tax incentive. As a result, the amount of non-Lithuanian capital spent on filming in Lithuania rose from 1.25 million euros in 2014 to 45.5 million euros in 2018 (Turp-Balazs, 2018).

The Ignalina nuclear power plant in Lithuania – which has been out of operation since 2009 – is actually the "twin sister" of the Chernobyl site, making it the perfect location for the interior shots in the plant. The powerplant is currently in the stages of decommissioning as part of the country's condition to join the European Union.

The other double location was Fabijoniškės, near the Lithuanian capital Vilnius, which represented the residential area of Pripyat in the film. All in all, the show took almost 1,000 hours to make, was shot in 40 locations, with nearly 5,000 background actors participating in the making of the film. (Turp-Balazs, 2018).

The film series on the tragedy has had a clear impact on Ukrainian tourism, with hundreds of tourists visiting the actual site of the disaster after the screening. Curiosity and the sites seen on the screen encouraged viewers to experience the real-life aftermath of a historical disaster of this scale. It is an interesting factor, however, that the series has had an impact not only on the original site but also on the tourism of the double sites. In the residential neighbourhood of Fabijoniškės, next to the Lithuanian capital, many apartments have been converted into Airbnb accommodation, reflecting the atmosphere of the Soviet Union perfectly. These accommodations are extremely popular, as all the furnishings exude the unique Soviet feel. In addition, themed tours are organised to the Ignalina nuclear power plant, where visitors can get to know the plant in a three-and-a-half hour tour and climb to the top of the building. According to Natalija Survila-Glebova, the plant's Lithuanian spokeswoman, most of the tourists arriving here are Lithuanians, but there are also tourists from Poland, Latvia and the UK.

The locations shown on the film became a potential target for those who wanted to learn more about the disaster site as a result of the series. Indeed, crowds of people started to visit the zone after the series was shown, with local travel agents also experiencing the consequences. The series on the 1986 tragedy had a profound effect on tourists with a particular interest in the area, making it clear that the concepts of disaster tourist and film tourist were one and the same in this case.

CHERNOBYL AS A TOURIST DESTINATION

In order to carry out a complex analysis of the area, local travel agencies were contacted to examine the supply side. During the expert interviews, the managers of the travel agencies Gamma Travel, Chernobylwel.come and Chernobyl Tour answered the questions. Gamma Travel has been organising thematic trips since 2013, and its website is informative and easy to use. The interface includes an instant online booking system, which displays the type and date of the next tour as a calendar entry. Chernobylwel.come has been at the disposal of tourists since 2008, with a special offer including online tours (e.g. Chernobyl live: The control room #4 online experience) and a website with a blog interface with diary-like entries. The third bureau examined is Chernobyl Tour, which has also been organising thematic tours since 2008, but is the only one of the agencies examined to offer a special offer based on the series. On their website, they define their values with the following description: "CHORNOBYL TOUR® unites companies and self-employed professionals proficient in the fields of Chornobyl accident, ecology and tourism. The project goal is to liquidate an "information contamination" in the society as a result of the ChNPP accident; to organise lectures and trainings for both novices and professionals right on the spot of the accident; to teach radiation and survival skills in condition of anthropogenic catastrophes. Since its foundation in 2008, CHORNOBYL TOUR® has become a brand which embodies quality and professionalism, attentive client service, and easy-to-get information presentation." (https://www.chernobyl-tour.com/about_us.html).

In the summary of the results, the authors of the chapter will present information from the interviews and the travel agencies' websites, followed by the content of their social media platforms. It became clear from the interviews with marketing managers that the most popular sending countries are the UK, Poland, Germany, the USA and the Netherlands. The trend in visitor numbers to the zone has undoubtedly shown an increase from 2017 with 50,000 visitors, rising to 75,000 in 2018 and 125,000 in 2019. In October 2019, a record monthly number of 17,038 visitors was recorded in the area. Visitors were found to cover a very wide age spectrum, and to be both male (approximately 60%) and female (40%).

Information received electronically revealed that most people had planned to travel in spring 2020 as a result of the series, but this was cancelled due to the COVID pandemic. The travel agencies' offers are very diverse, with some offers available at all three agencies surveyed, but there are also very specific programme packages. All programmes depart from Kiev, the capital of Ukraine, and shuttle services are available to take tourists to the zone. The most popular programmes are one-day tours, which can be booked at between €82 and €99 (Gamma Travel €89, Chernobylwel.come €99, Chernobyl Tour €82). A comparison of the travel agencies' offers is shown in the table below *(Table 2)*.

All three travel bureaus offer a 2-3 day tour, a special private tour and a nuclear power plant tour. Chernobylwel.come offers a special online tour, while Chernobyl Tour offers kayak and air tours (*Table 3*). Chernobyl Tour also offers a service specifically based on the series called "HBO Chernobyl series TOUR". The programme is based on the series and includes visits to sites within the zone, also using tablets and photos during the tour to show how particular locations appear on the screen.

Locations include the basement of the Chernobyl nuclear power plant, the fire department from which the first firefighters left for the scene, the medical unit of the city of Pripyat, which received the first victims, the legendary bridge of Pripyat town, from which, according to the series, citizens of Pripyat watched the fire at the station, and the canteen of the Chernobyl Nuclear Power Plant. The constantly evolving infrastructure has made it possible to offer more and more specialised tours and site visits within the zone. With the involvement of the local population, more and more information offices have been set up on the site, which facilitate the coordination of tourists and the sale of services (tours and souvenirs) linked to the programmes. In terms of souvenirs, a very wide range of offers can be found at all three offices. In addition to classic souvenirs (T-shirts, fridge magnets, pens, mugs, key rings), dosimeters are also available. A dosimeter is a device that continuously records the total intensity of ionising radiation. This device is also used during tours to confirm that the visitor is really safe, as it can record radiation data over a longer period of time. The travel agency Chernobylwel.come has created its own souvenir site, Chernowear, where you can find special Chernobyl-inspired items (https://www.chernowear.com/, 2021).

Table 2. Offers of the surveyed travel agencies

	Gamma Travel	Chernobylwel.come	Chernobyl Tour
1 Day Tour to Chernobyl and Prypiat	• Safety first! It is safer than you thought it would be • Small groups (up to 20 people) • 12 hours of the unforgettable experience • More than 35 very best locations including Chernobyl, Prypiat, Power Plant and Duga Total: 89€	• It is the only tour you can find, where you meet and talk to the real heroes of Chernobyl – liquidators and witnesses. • Small groups (up to 18), no big buses, no shopping tour • Super safe: 3 Dosimetric controls/ day, Free Dosimeter, Proven Routes • Full service (incl. permits, lunch - traditional Ukrainian lunch, dosimeters, insurance) Total: 99€	• All formal permissions • Comfortable auto transportation Kiev-Zone-Kiev • Professional English-speaking guide • Teaching skills of radiation survival • Personal certificate about visit to the Chernobyl zone Total: 82€
2 Days Tour to Chernobyl	• All-inclusive (permits, accommodation, transportation, insurance, dining and maximum safety) • Over 50 stunning locations • Friendly, professional guides with perfect English • Small groups (5-12 people) • Skilled drivers and comfortable vehicles (AC, wifi, TV) • One night in Chernobyl • 36 hours of unique experience Total: 279€	• It is the only tour you can find, where you meet and talk to the real heroes of Chernobyl – liquidators and witnesses. • Small groups (up to 15), no big buses, no shopping tour • Tour starts in Kyiv, continuing with 2-hour drive to the Zone • Super safe: 3 Dosimetric controls/ day, Free Dosimeter, Proven Routes • The price is final (incl. dining, accommodation etc.), no hidden fees Total: 279€	• Chernobyl Tour has a program of 2-, 3-, 4, 5, 6, and 7-days trips • The Chernobyl NPP, towns of Pripyat and Chernobyl • In 3- and 4-day trips you also get the on-site basic radiation survival training • Accommodation in Chernobyl Hotel • Personal certificate about visit to the Chernobyl zone Total: 205€
Private Tours	• Private 1-2 days tour to Chernobyl and Prypiat • 3-5 days Private/photo tour to Chernobyl and Prypiat	• Private 1-5 days tour to Chernobyl and Prypiat	• Private 1-4+ days tour to Chernobyl and Prypiat • Different routes: Basic, Basic+, Mock-up, Protective
Chernobyl Power Plant and Prypiat Tour	• 1 rector that changed the history of humankind • 3 ways to provide your radiation safety (personal dosimeter, dosimetric radiation check, human radiation spectrometry) • Safety guaranteed! • 5 hours of an eye-opening experience Total: 150/200€	• 2 days tour • Super safe • Secret places of Prypiat • Full service (dining, accommodation) Total: 449€	• 4,5 hours inside the Chornobyl nuclear power plant • Walk around the ghost- town of Prypiat • Lunch at the canteen # 19 of the Chornobyl nuclear power plant Total: 449€

Source: Based on the websites of the travel agencies, self-edited (2021)

Table 3. Special offers of the surveyed travel agencies

<table>
<tr><th colspan="3">Special Tours</th></tr>
<tr><th>Gamma Travel</th><th>Chernobylwel.come</th><th>Chernobyl Tour</th></tr>
<tr>
<td rowspan="5">Chernobyl Air Tour
• 2 hours of unforgettable emotions and breathtaking views
• Locations: The town of Prypriat, Nuclear Power Plant, Chernobyl river port, OTH Radar – „Duga"
• Possibility to show the Chernobyl zone to children and people under 18 years
<u>Total: 240€</u></td>
<td rowspan="5">Chernobyl Online Tours
• Chernobyl Live: The Control Room #4 Online Experience
<u>Total: 29€</u>
• Chernobyl Live: Explore Chernobyl City
<u>Total: 29€</u>
• Chernobyl Live: At The Epicenter in Chernobyl Power Plant
<u>Total: 29 €</u></td>
<td>Quest in Chernobyl Zone
• Team Game
• The quest is conducted in Ukrainian
• The game takes place with the participation of 10 players
• A whole day of participation in an exciting quest on the territory of the radar DUGA-1 and the military town of Chernobyl-2
• Visiting the observation deck of the Chernobyl nuclear power plant
• Lunch in the Chernobyl power plant canteen
<u>Total: 68€</u></td>
</tr>
<tr><td>"HBO Chernobyl" series TOUR
• The real locations of the Chernobyl exclusion zone
• Learn the real stories of people whose characters were reflected at the series
<u>Total: 152€</u></td></tr>
<tr><td>Kayaking tour
• Absolutely new type of tourism in the Chernobyl zone
• A unique combination of active tourism, the observation of nature and an informative guided tour
• The route goes through the territory of the Chernobyl biosphere radioecological reserve
• High possibility to meet wild animals and a variety of birds
<u>Total: 125€</u></td></tr>
<tr><td>Sherp ATV Extreme Tour
• Sherp all-terrain vehicles
• The duration of the ride is 15-20 minutes
• Off-road and steep slopes in the Zone
<u>Total: 23€</u></td></tr>
<tr><td>Chernobyl Air Tour – fly above the radiation
• See legendary places from a bird's-eye view (~ 200-500 m)
• Duration of the tour: 3-6 hours
• Fly up to the giant antennas DUGA-1 and town CHERNOBYL-2, flying over the ghost town Pripyat, see ChNPP from new angles
• Discover the secrets of the abandoned Exclusion Zone territory
<u>Total: from 197€</u></td></tr>
</table>

Source: Based on the websites of the travel agencies, self-edited (2021)

The interviews also revealed that new social initiatives have emerged in relation to the zone. The Chernobylwel.come travel agency website specifically highlights that 1% of the tour price goes directly to support Chernobyl Babushkas, for every Tripadvisor review they plant a tree in the Chernobyl zone ("Together we are helping the zone recover") and with the help of visitors they support children with cancer via the charity Dobry Anjel.

In addition, they have provided free access to the site for workers cleaning up environmental damage, started waste selection and recycling programmes through voluntary projects, established the Chernobyl

University in 2020 and applied for UNESCO World Heritage status last year. It has become clear that the promotion of the destination will be done in close cooperation with travel agencies, and they may not rely purely on state support. They only consult with governmental organisations and ministries on professional matters concerning the zone.

In addition to the websites, the analysis of the travel agencies' marketing activities also included their social media pages. Nowadays, people spend most of their free time on social networking sites, so it is useful to reach potential tourists through these channels. Of course, these platforms have a clear aim: to attract new customers, retain existing ones and provide a customer service system. Besides the two most popular social platforms, the analysis also includes a dedicated travel platform that helps nearly 470 million travellers to navigate the wide range of offers. Tripadvisor is a platform with user-generated content and comparison and shopping functions. Available in 28 languages, the platform allows users to browse reviews and ratings of a wide range of accommodation, restaurants, programmes, airlines and cruises. Tripadvisor thus provides content that is basically created by tourists who have already experienced the service, ensuring maximum feedback from users that can be used in a comparative analysis.

There is not much variation between the travel agency websites: all three have a Chernobyl story, gallery, blog interface, detailed safety instructions and a shop or souvenir sub-page. Of course, the tours are also detailed and the social networking sites are directly accessible from the website.

Considering social media sites, Facebook, Instagram and TripAdvisor were used as the basis for the analysis (*Table 4*). The research did not only compare the number of followers and reviews, but also highlighted the activity, the type of content and the use of the film series.

Table 4. Analysis of the social sites of the surveyed travel agencies

	Facebook	Instagram	Tripadvisor
Gamma Travel	• Followers: 14.697 • 2-3 posts/week • 360˙ panoramic images • There is no reference to the series	• Followers: 4285 • 200-300 interactions/post • There is no reference to the series	• #6 of 331 Tours in Kiev • 538 reviews (510 – excellent), most for a full day tour • Travellers' Choice 2020
Chernobylwel.come	• Followers: 66.172 • 2-3 posts/week • They use the series on their site.	• Followers: 46.700 • They use the series on their site • Archive videos vs. reality	• #10 of 331 Tours in Kyiv (Kiev) • 2543 reviews (2309- excellent), most for a full day tour • Quality guarantee since 2015
Chernobyl Tour	• Followers: 23.735 • 2-3 posts/week • 100-200 interactions • They use the series on their site	• Followers: 58.900 • Total 2702 posts • 2500-3000 interactions/post • Fresh content every 2-3 days • #chernobylhbo	• #2 of 185 Outdoor Activities in Kiev • 3318 reviews (3119 – excellent), most for a full day tour • Travellers' Choice 2020 • 2014-2019 Certificate Excellence

Source: Based on Facebook, Instagram, TripAdvisor, self-edited (2021)

As can be seen from the table, the two oldest of the travel agencies surveyed (Chernobylwel.come and Chernobyl Tour, both operating since 2008) have the largest base, but both focus on different platforms. The study clearly shows that Chernobylwel.come reaches more potential visitors through its Facebook platform, while Chernobyl Tour reaches more potential visitors through its Instagram platform.

It is interesting to note that all three travel agencies have an outstanding performance on TripAdvisor, which is based on travellers' reviews. In terms of reviews, one-day tours generate the most reviews and positive feedback, which support the information provided in the interviews. The two oldest businesses are using the potential of the series in their marketing activities, with references to the series appearing on both their Facebook and Instagram platforms. All posts must include the tags #hbo, #hbogo #hbo-chernobyl or #chernobylhbo, clearly referring to the creators of the series.

FUTURE RESEARCH DIRECTIONS

Several future researches can be formulated on this topic, since – as the literature review has shown – the special fusion of disaster and film tourism is a very new trend in tourism. A further (even additional) basis for this may be the fact that more and more news and research results are appearing in all press platforms, suggesting that fission in the plant is not slowing down. The phenomenon is probably caused by rainwater infiltration, which makes it difficult to take the planned preventive measures. The plan is to dismantle the damaged reactor and clean up the hazardous radioactive waste that has been locked there. However, this process must be complemented by the management of the new problem. This may also be another adrenaline factor for disaster 'fans', as they could be faced with a weakened mini-disaster again when they travel to the Chernobyl site. This special measure could generate new investigation factors and perspectives for visitors. The phenomenon of overtourism in relation to disaster and film tourism may be a further research topic. This specialised area is becoming increasingly popular with tourists, so the management of overtourism, its prevention and even its regulation by law is a growing problem. In the current situation, the pandemic has curbed this phenomenon, but as soon as we gradually return to pre-pandemic life, we may be faced with this problem again. Potential areas: Chernobyl, Cambodia, the Golan Heights (Israeli side).

A case study and analysis of similar film content could also be an interesting research topic. The Narcos series, cited as an example in this study, is absolutely suitable for such an investigation. The series – which depicts a drug war that claimed many civilian lives – has resulted in a flood of tourists to Medellín, Colombia, causing the establishment of new tourist businesses in the area. A complex tourism analysis of this phenomenon could provide new answers about the motivation of tourists and the tourism potential of film series. Opinions on the legacy of the notorious drug lord and the related tourism boom are divided among local residents, so much so that in 2019, the local government demolished one of the most popular properties, in front of which many tourists had taken their iconic selfies. Regarding this problem, the change in the local residents' attitudes towards the phenomenon of increased visitor numbers generated by disaster/movie tourism could also be an interesting subject of study.

This is entirely relevant to the case of Chernobyl, where the reception of tourists has not been uniformly positive. Some tourism companies explicitly did not support the creation of thematic offers. "This is not a tourist business, this is not a show. This is a particular side of a man-made disaster, and there are hundreds of scientists working to clean up the aftermath." (Vilkos, 2006).

In the field of disaster tourism and film tourism, particular attention should be paid to identifying and defining travel motivations. One way of investigating this could be to conduct questionnaires or in-depth interviews with tourists who have already visited potential areas. In this case, more in-depth information on the travel-inducing effects of disaster and film tourism and on the offer of the tourism destination can be obtained. In the context of the Chernobyl series and travel motivation, a specific segment could be described by the question of visiting the real location or the double site. As the miniseries shows the tragedy on Ukrainian territory, it is clear that it will boost tourism primarily in the real location. However, the series also used double locations in Lithuania, where it was also able to attract visitors. It would therefore be worth investigating which of these destinations would be visited by tourists having watched the series, and whether they would be interested in both destinations.

The present study also deals with social media platforms and the role of these platforms in marketing. An interesting research can be built up to further investigate social networking sites, where not only travel agency sites but also influencers involved in film tourism and content posted by visitors can be analysed. Here, the popularity of the destination itself (the most popular locations within the zone) could be examined, as well as the management of the negative effects of the pandemic and the subsequent recovery process. Of course, future research may also face difficulties; in this case these may include contacting tourists who have already visited the zone, contacting social networking site operators, and contacting official organisations with close links to the examined area in connection with the phenomenon of overtourism.

CONCLUSION

The study clearly showed that the Chernobyl miniseries has created a unique segment of film and disaster tourism. The authenticity of the story and the precision of the execution guaranteed the success of the series, which is a prerequisite for film tourism as well. Curiosity and adrenalin boost proved to be very important in the study of travel motivations, reinforced by the willingness to travel generated by the film. These were also maximised by the special tours on offer (flying, kayaking and canoeing) and the programmes linked to the series.

The main pillars of the analysis of the supply side were the creation of a sense of security, the wide range of programmes, accessibility and the professionalism of the tours. The series has led to an increase in the number of visitors to the zone, thanks to the introduction of edutainment (or "learning while being entertained"), in addition to continuous improvement. The development process has included the restoration of many of the sites seen in the series and the adaptation of all the offerings to the needs of the postmodern tourist. This continuous development process has enabled the destination to move from the category of dark tourism to a lighter, grey category. However, in order for this process to continue in the future, there is a need for closer coordination of professional relations and well-thought-out and long-term tourism management.

Factors that could facilitate repositioning in the future include the expansion of the range of film tourism products (more interactive solutions, applications), strengthening alternative tourism services and a greater use of the series in marketing activities. When developing the offer of film tourism products, it would be useful to create an application that would be based on the series, guiding visitors around the zone. It could be used not only to present archive footage but also to include commentary from the crew who made the series. In addition, backstage information can also be shared, such as the difficulties of filming a particular scene, or the similarities and differences of the double location and the original site.

Chernobyl Tour is the absolute leader in the provision of alternative tourism services among the three tourism enterprises studied, and this approach should also be reflected in the case of other operators. The post-modern tourist is looking for an experience, being part of a programme where he or she is an integral part of the action, experiencing the events seen on the screen from a completely unique perspective.

An excellent example of this is the 2 to 3-hour flight tour, which allows under-18s to explore the zone, as well. In its marketing activities, Gamma Travel does not make any use of the opportunities offered by the series. The other two travel agencies, Chernobyl Tour and Chernobylwel.come, use the imagery of the series on their Facebook and Instagram pages. In addition to comparative archive footages, they use the characters' images and iconic quotes to promote their thematic tours and the development work underway in the destination.

Taking the results into account, the research has shown that a successful series can induce travel motivation despite its focus on a disaster area. This increases the number of visitors to the area, which has an obvious positive impact on the destination up to a certain point. The fundamental question is whether all visitors really come to learn about the destination, or whether they are just driven by the 'crowd phenomenon'. There is also a question whether there is a point when profit maximisation becomes the focus rather than the credible communication of the history of the area and the background to the disaster. We hope that in the future Chernobyl will maintain a healthy balance of sustainability through continuous improvement and will remain open for many years to come for visitors who want to learn about the site of the tragedy with the appropriate respect.

REFERENCES

Amujo, O. C., & Otubanjo, O. (2012). Leveraging rebranding of 'unattractive' nation brands to stimulate post-disaster tourism. *Tourist Studies*, *12*(1), 87–105. doi:10.1177/1468797612444196

Árva, L., & Deli-Gray, Zs. (2011). Posztmodern turizmus, a posztmodern marketing [Postmodern tourism, postmodern marketing]. *Magyarország társadalmi-gazdasági helyzete a 21. század első évtizedeiben*, 1-11.

Baker, B. (2012). Branding and the opportunities of movies: Australia. In N. Morgan, A. Pritchard, & R. Pride (Eds.), *Destination Brands* (pp. 239–250). Routledge.

Beeton, S. (2005). Film-induced Tourism. In *Multilingual Matters*. Channel View Publications. doi:10.21832/9781845410162

Beeton, S. (2010). The Advance of Film Tourism. Tourism and Hospitality Planning & Development, 7(1).

Biran, A., & Buda, D. M. (2018). Unravelling fear of death motives in dark tourism. In P. R. Stone, R. Hartmann, A. V. Seaton, R. Sharpley, & L. White (Eds.), *The Palgrave handbook of dark tourism studies* (pp. 515–532). Palgrave Macmillan. doi:10.1057/978-1-137-47566-4_21

Bolan, P., & Williams, L. (2008). The role of image in service promotion: Focusing on the influence of film on consumer choice within tourism. *International Journal of Consumer Studies*, *32*(4), 382–390. doi:10.1111/j.1470-6431.2008.00672.x

Buda, D. M. (2015). *Affective tourism: Dark routes in conflict*. Routledge. doi:10.4324/9781315742786

Chernobyl Tour. (2021). Retrieved from: https://www.chernobyl-tour.com/

Chernobylwel.com. (2021). Retrieved from: https://www.chernobylwel.com/

Croy, W. G., & Walker, R. D. (2003). Rural tourism and film-issues for strategic regional development. In D. Hall, L. Roberts, & M. Mitchell (Eds.), *New directions in rural tourism* (pp. 115–133). Ashgate Publishing Ltd.

Cynthia, D., & Beeton, S. (2009). Supporting Independent Film Production Through Tourism Collaboration. *Tourism Review International, 13*(2), 113–119. doi:10.3727/154427209789604624

Dallen, J. T. (2011). *Cultural Heritage and Tourism – An Introduction*. Channel View Publications.

Dunkley, R., Morgan, N., & Westwood, S. (2011). Visiting the trenches: Exploring meanings and motivations in battlefield tourism. *Tourism Management, 32*(4), 860–868. doi:10.1016/j.tourman.2010.07.011

Gamma Travel. (2021). Retrieved from: https://gamma-travel.com/

Gartner, W. C. (1993). Image Formation Process. *Journal of Travel & Tourism Marketing, 2*(2-3), 191–216. doi:10.1300/J073v02n02_12

Gotham, K. F. (2007). *Dark Tourism and Disaster Tourism. The Blackwell Encyclopedia of Sociology.*

Gotham, K. F. (2015). *Dark Tourism and Disaster Tourism*. Retrieved from: https://onlinelibrary.wiley.com/doi/abs/10.1002/9781405165518.wbeosd045.pub2

Grihault, N. (2003). Film tourism-the global picture. *Travel & Tourism Analyst, 5*, 1–22.

Gunn, C. A. (1972). *Vacationscape: Designing Tourist Regions*. University of Texas.

Hao, X., & Ryan, C. (2013). Interpretation, Film Language and Tourist Destinations: A Case Study of Hibiscus Town, China. *Annals of Tourism Research, 42*, 334–358. doi:10.1016/j.annals.2013.02.016

Helmeczi, B. (2020). *Csernobil: ezért túlzás az HBO sikersorozata* [Chernobyl: why HBO's hit series is an exaggeration]. Retrieved from: https://promotions.hu/orszagos/zene-film-kultura/2020/04/26/csernobil-katasztrofa-34-ev-evfordulo-hbo-sorozat/

Hudson, S., & Ritchie, J. R. B. (2006). Promoting Destinations via Film Tourism: An Empirical Identification of Supporting Marketing Initiatives. *Journal of Travel Research, 44*(4), 387–396. doi:10.1177/0047287506286720

Irimiás, A. (2015). *Filmturizmus. A filmek és televíziós sorozatok turisztikai szerepének és hatásainak geográfiai vizsgálata [Geographical analysis of the role and impact of films and television series in tourism]*. Akadémiai Kiadó.

Kim, H., & Richardson, S. L. (2003). Motion picture impacts on destination images. *Annals of Tourism Research, 30*(1), 216–237. doi:10.1016/S0160-7383(02)00062-2

Lennon, J., & Foley, M. (2000). *Dark tourism: The attraction of death and disaster*. Continuum.

Michalkó, G. (2012). *Turizmológia: elméleti alapok [Turismology: theoretical foundations]*. Akadémiai Kiadó.

Mitchell, M. (2019). *Chernobyl on HBO location: Where is Chernobyl filmed? Where's it set?* Retrieved from: https://www.express.co.uk/showbiz/tv-radio/1123559/Chernobyl-HBO-Sky-location-Where-is-Chernobyl-filmed-set-Lithuania-Moscow-Pripyat-Ukraine?fbclid=IwAR1Jq_pplRcSiiDUjx9nLT-W8TZZPZor3XI57r5-NYBqlWBiz7z_vOqlSdJo

Morgan, N., & Pritchard, A. (1998). *Tourism promotion and power: Creating images, creating identities*. John Wiley & Sons Ltd.

Morten, R., Stone, P. R., & Jarratt, D. (2018). Dark tourism as psychogeography: An initial exploration. In The Palgrave Handbook of Dark Tourism Studies (pp. 227–255). London: Palgrave Macmillan.

Oviedo-García, M. Á., Castellanos-Verdugo, M., Trujillo-García, M. A., & Mallya, T. (2016). Film-induced tourist motivations. The case of Seville (Spain). *Current Issues in Tourism*, *19*(7), 713–733. doi:10.1080/13683500.2013.872606

Palancsa, A. (2012). Mi a katasztrófaturizmus? [What is disaster tourism?]. *Korunk*, *3*(23).

Papp-Váry, Á. (2009). Országimázs a moziban [Country image in the movies]. *Marketing & Media*, *4*(9), 13–26.

Papp-Váry, Á. (2019). *Országmárkázás. Versenyképes identitás és imázs teremtése [Country branding. Creating a competitive identity and image]*. Akadémiai Kiadó.

Peters, M., Schuckert, M., Chon, K., & Schatzmann, C. (2011). Empire and Romance: Movie-Induced Tourism and the Case of the Sissi Movies. *Tourism Recreation Research*, *36*(2), 169–180. doi:10.1080/02508281.2011.11081317

Pratt, S. (2015). The Borat Effect: Film-Induced Tourism Gone Wrong. *Tourism Economics*, *21*(5), 977–993. doi:10.5367/te.2014.0394

Puczkó, L., & Rátz, T. (2005). A turizmus hatásai [The impact of tourism]. Aula kiadó.

Puczkó, L., & Rátz, T. (2011). *Az attrakciótól az élményig [From attraction to experience]*. Akadémiai Kiadó.

Riley, R., Baker, D., & Doren, C. S. V. (1998). Movie induced tourism. *Annals of Tourism Research*, *25*(4), 919–935. doi:10.1016/S0160-7383(98)00045-0

Riley, R. W., & Van Doren, C. S. (1992). Movies as tourism promotion. *Tourism Management*, *13*(3), 267–274. doi:10.1016/0261-5177(92)90098-R

Rucińska, D., & Lechowicz, M. (2014). Natural Hazard and Disaster Tourism. *Miscellanea Geographica*, *18*(1), 17–25. doi:10.2478/mgrsd-2014-0002

Ryan, C., Page, J. S., & Aicken, M. (2005). *Taking Tourism to the Limits: Issues, Concepts and Managerial Perspectives*. Elsevier.

Schwartz, D. (2019). *Craig Mazin's Years-Long Obsession with Making' Chernobyl' Terrifyingly Accurate*. Retrieved from: https://www.vice.com/en/article/j5wbq4/craig-mazin-interview-about-chernobyl-hbo-miniseries-on-how-accurate-and-what-really-happened

Seaton, A. (1996). Guided by the dark: From thanatopsis to thanatourism. *International Journal of Heritage Studies*, *2*(4), 234–244. doi:10.1080/13527259608722178

Seaton, T. (2009). Thanatourism and its discontents: An appraisal of a decade's work with some future issues and directions. In T. Jamal & M. Robinson (Eds.), *The SAGE handbook of tourism studies* (pp. 522–544). SAGE. doi:10.4135/9780857021076.n29

Sharpley, R. (2009). Shedding Light on Dark Tourism: An Introduction. In The Dark Side of Travel: The Theory and Practice of Dark Tourism. Channel View Publications.

Sharpley, R. (2005). *Travels to the Edge of Darkness: Towards a Typology of Dark Tourism*. Academic Press.

Stachowiak, K., & Stryjakiewicz, T. (2018). The rise of film production locations and specialised film services in European semi-peripheries. *Hungarian Geographical Bulletin*, *67*(3), 223–237. doi:10.15201/hungeobull.67.3.2

Stone, P. R. (2006). A dark tourism spectrum: Towards a typology of deathand macabre related tourist-sites, attractions and exhibitions. *Tourism Management*, *54*(2), 145–160.

Sulyok, J. (2009). Filmturizmus, avagy a filmekben rejlő lehetőségek a turisztikai desztinációk számára [Film tourism or the potential of films for tourist destinations]. *Turizmus Bulletin*, *2009*(1).

The Globe Post. (2018). *Numbers of Chernobyl Disaster Zone Visitors Boom*. Retrieved from: https://theglobepost.com/2018/04/25/chernobyl-tourism/

Tóth, R. A. (2016). Széles vászon, nagy haszon [Wide canvas, big benefits]. *Turizmus Trend – Vendég & Hotel*, *2016*(5).

Tucker, H., Shelton, E. J., & Bae, H. (2017). Post-disaster tourism: Towards a tourism of transition. *Tourist Studies*, *17*(3), 306–327. doi:10.1177/1468797616671617

Turizmus.com. (2019). *Csernobil már hivatalosan is turisztikai látnivaló* [Chernobyl is now officially a tourist attraction]. Retrieved from: https://turizmus.com/utazas-kozlekedes/csernobil-mar-hivatalosan-is-turisztikai-latnivalo-1165288

Turp-Balazs, C. (2018). *HBO's Chernobyl highlights Vilnius as a go-to filming location*. Retrieved from: https://emerging-europe.com/after-hours/hbos-chernobyl-highlights-vilnius-as-a-go-to-filming-location/?fbclid=IwAR2KWgdAph0Y-pscG_UDpNvJlN1HK4GDJpVnBiXsESaifa0yNZC7PlbanJ8

Vilkos, Y. (2006). Travel Agencies Cashing in on Trips to Chernobyl. *Kyiv Post*. Retrieved from: www.kyivpost.com/nation/24356/

Wright, D. (2018). Terror park: A future theme park in 2100. *Futures*, *96*, 1–22. doi:10.1016/j.futures.2017.11.002

Wright, D., & Sharpley, R. (2018). Local community perceptions of disaster tourism: The case of L'Aquila, Italy. *Current Issues in Tourism*, *21*(14), 1569–1585. doi:10.1080/13683500.2016.1157141

ADDITIONAL READING

Goatcher, J., & Brunsden, V. (2011). Chernobyl and the Sublime Tourist. *Tourist Studies*, *11*(2), 115–137. doi:10.1177/1468797611424956

Hryhorczuk, N. (2018). Radioactive Heritage: The Universal Value of Chernobyl as a Dark Heritage Site. *Qualitative Inquiry*, *25*(9-10), 1047–1055. doi:10.1177/1077800418787553

Light, D. (2017). Progress in dark tourism and thanatourism research: An uneasy relationship with heritage tourism. *Tourism Management*, *61*, 275–301. doi:10.1016/j.tourman.2017.01.011

Nistoreanu, P., & Tuclea, C. (2011). How film and television programs can promote tourism and increase the competitiveness of tourist destinations. *Cactus Tourism Journal*, *2*(2), 25–30.

Sun, J., & Lv, X. (2021). Feeling dark, seeing dark: Mind–body in dark tourism. *Annals of Tourism Research*, *86*, 103087. doi:10.1016/j.annals.2020.103087

Yankovska, G., & Hannam, K. (2013). Dark and toxic tourism in the Chernobyl exclusion zone. *Current Issues in Tourism*, *17*(10), 929–939. doi:10.1080/13683500.2013.820260

KEY TERMS AND DEFINITIONS

Chernobyl: Chernobyl is a name used for both the city and the nuclear power facility. The nuclear power plant was located approximately 14.5 kms from the city of Chernobyl. The city was evacuated about 30 hours after the nuclear explosion and remains within the Chernobyl Exclusion Zone. Before the disaster, it was home to an estimated 14,000 people.

Dark Tourism: All travel related to death, disasters or gruesome events, visiting places associated with natural disasters, tragedies, battles, where the focus is on satisfying individual curiosity.

Disaster Tourism: A common point that emerges from the definition of dark tourism and disaster tourism is that potential destinations are places with a large number of deaths and/or tragedies resulting in severe natural disasters. The difference between the two concepts lies in their impact. Whereas dark tourism has no immediate impact, disaster tourism has an impact on people in the present day and age, even decades later.

Exclusion Zone: The Chernobyl Exclusion Zone is the 30-km radius around the power plant that was evacuated in the wake of the disaster. The zone remains in place today and is largely uninhabited. In total, about 116,000 people were evacuated from the exclusion zone.

Film Tourism: The cultural tourism sector includes trips where the primary motivation of the traveller is to visit a film location. However, it also includes film business trips, visits to film theme parks and film studios.

Place Placement: A marketing communication tool where the focus is on location placement. Filmed location placement brings the image and name of a destination to potential visitors who could only be reached through a high-cost marketing campaign.

Postmodern Tourism: The individual is in the focus, mass tourism is rejected. The most important elements are the other motivations of visitors (studying, shopping). Heterogeneous demand is typical, visitors' consumption patterns are completely different. There are few return visits to familiar, established places.

Postmodern Tourist: The postmodern tourist is well-informed, does not accept the opinions of others or mass products, wants to organise his/her own travel programme and wants to actively participate in the activities. The tourist of the post-industrial society "participates."

APPENDIX

Marketing professionals from the three travel agencies surveyed (Gamma Travel, Chernobylwel.come and Chernobyl Tour) participated in the research. The interview was conducted in writing, via e-mail. The main questions of the interviews:

1. Which tour is the most popular from the offer?
2. Where do the most visitors come from? What is the gender distribution?
3. How did the number of participants in the tours change between 2018 January and December 2020?
4. Was the impact of the Chernobyl series felt on the number of visitors?
5. What tourism developments have taken place in the Chernobyl zone in the last 5 years?
6. Do they receive any support from the national tourism organisation?

Chapter 16
Marine Life Features and Their Transmissibility Through Broadcasting:
Tourism Capitalization

Jorge Ramos
https://orcid.org/0000-0002-5115-7919
CinTurs, University of the Algarve, Portugal

ABSTRACT

Avid readers find in literature real or imaginary scenarios, which are written in such a way that make them visit the portrayed places. The same goes for films, or documentary series, where the narrator plays a fundamental role in captivating the viewer. Both means are inductors of cultural and/or nature tourism, regardless of everyone's personal motivations. The transmission of values is important and varies over time. In the past some marine animals were perceived negatively by humans, which has faded over time. Ecotourism with activities related to the sea is a relatively recent tourism niche. Nature films or series are often boosters of impulses in tourists for the search for natural, unexplored places or places with little human pressure. Some tourist activities in the nature tourism line developed all over the world—from this induction—such as the cases of diving and observation of marine animals.

INTRODUCTION

Coastal areas and the sea are places that attract people, whether for relaxation or leisure. Summer holidays are often chosen by households to tour along the beach, to enjoy the natural beauty of the coast. One of the sources to captivate people in many of these tourist choices is the fact that suggestions made by films and documentaries or by literature related to the sea and coastal areas (Hofman & Hughes, 2018). Certain places become iconic for their natural beauty or for certain peculiarities that characterize them. Some species of living beings are particularly important and serve as calling cards for these places (Mazzoldi et al., 2019).

DOI: 10.4018/978-1-7998-8262-6.ch016

In this way, these literary pieces, documentaries, or films thus have a certain influence on the decisions of individuals and amplify tourist visits to many of these places (Hoppen et al., 2014). When there is an increase in tourist flow to a location after a film has been shot there, it usually means that tourism is induced by the film (Beeton, 2006). Visiting certain places can also occur incidentally (Macionis & Sparks, 2009).

These works must have concepts that people grasp in a positive way, and thus become effective in the development of sustainable tourism related to the sea and coastal areas. In this aspect there is always the reverse of the issue. On the one hand, it is important that people spend their free time or holidays in leisure activities related to coastal areas, thus increasing demand, and making the tourism sector more robust. On the other hand, the great tourist demand of these areas puts their carrying capacity and the species that inhabit them to the test (Cater, 2010).

Literature and films play a fundamental role in people's emotions, learning, acquisition of knowledge, influencing human beings from an early age (Frevert et al., 2014). Literature and films in their various aspects are two cultural elements of learning in people as citizens who interact with their surroundings (Salomon, 2012). The ability of each person to interact positively with their surroundings is therefore greatly influenced by the information they receive from the sensory side and the ability to assimilate and transform daily. Literature and films are inductors of this sensory capacity (van Leeuwen, 2011). The environment, such as marine life in its immensity of organisms, biodiversity, ecosystem balance and its particularities, are key elements for a balanced and sustainable planet. The human being in all its aspects and its dignity contribute to this balance (Attenborough, 2020).

Tourism has for some time been a growing industry with considerable weight in the world economy (Mak, 2003). However, the pandemic crisis reshaped the tourism economy (Yang et al., 2020). Regardless of adverse factors, tourist activities generate jobs, qualify labor, and contribute positively to the trade balance of many developing countries. Tourism activities – in general – improve the standard of living of many communities, whether coastal, inland or mountain (Denny et al., 2019). Tourism is also an activity that cuts across all human beings and its enjoyment – at least at some point in their lives – demonstrates everyone's right to leisure and dignity (Bélanger & Jolin, 2011).

Sustainable tourism is increasingly important because it is governed by concepts of resilience and intergenerational legacy (Vu et al., 2020). Tourism – whether triggered by literature or induced by cinematography –, particularly about the marine world, will be even more sustainable and with a future legacy, the more conscious and responsible it is (Sakellari, 2014).

The stereotype for today's tourist is that they are increasingly aware of the problems plaguing the planet – namely microplastics, overfishing, climate change, pollution – and as a tourist they will tend to act more sustainably, reflecting on their daily routine and reducing its carbon footprint (Rico et al., 2019). The transferability of knowledge of the living and natural world derived from literature and cinematography plays a fundamental role in influencing citizens and their proactivity (Cynthia & Beeton, 2009). At sea, diving tourism (Ramos et al., 2006) and marine animal observation (Bentz et al., 2016) are the most sought after in all latitudes. In general, the tourism sector capitalizes on all these synergies (Steiner & Reisinger, 2006).

In this chapter, after introducing the theme, the goals consisted of reviewing the literature on literary tourism induced by films, namely related to marine life. It is also explored how the releases of documentaries and films with action at sea have their influence on people's perception. The importance of films and documentaries narrated by famous conservation presenters is mentioned and their contribution in inducing marine nature tourism is discussed, namely for the observation of large mammals, fish, and underwater diving. Afterwards there are presented examples of awareness of some unique marine (and aquatic) species that contribute to the enrichment of natural and cultural heritage in certain places and serve to stimulate tourist activities. Then, the discussion of all these subjects is made. Finally, the conclusions on this theme are presented.

LITERARY AND FILM-INDUCED TOURISM

Film-Induced Tourism and Marine Life

Film-induced tourism is a somewhat limited way of understanding the relationship that exists between film production, tourism, and place. The identities of places as tourist destinations are fundamental for their branding as spaces of production (Ward & O'Regan, 2009). For instance, Australian tourist authorities are generally very keen to build relationships with the film industry to promote places. A particular case of this aspect was the collaboration with Disney in the film "Finding Nemo" where the city of Sydney entered the film's plot (Hudson & Ritchie, 2006). This is a particular case where responsible authorities recognized the importance of marketing location and film-induced tourism.

Assuming that tourism is a symbolic break from everyday routines and places, traveling to discover new places or for fun are common decisions in people's lives. A simple trip to the cinema to see a premiere can trigger a series of tourist induced events. Some successful films induce tourism in different ways, namely in theme and/or amusement parks. An example of this has to do with the success of the 2003 animated film "Finding Nemo" which prompted the construction of a franchised amusement park in 2007 in Anaheim, California. For the construction of this park, a lot of marine environment consultancy had to be developed, to provide visitors with scientific knowledge and enhancement related to the film, and offer adequate amusement to people (Garner, 2019).

There are authors who have verified that there is a relationship between the increase in the observation of marine mammals (e.g., dolphin and whale) and the popularity of movies or series like "Free Willy". In these cases, nature-related films influence certain tourism niches, particularly those prone to ecotourism (Wearing et al., 2011).

Phenomena of Pilgrimage

The book "Jaws" was relatively successful in terms of sales when it was published in 1974, however nothing comparable to more recent phenomena such as "Harry Potter", where fans line up outside bookstores on the day of the release of the works and millions of books are sold around the world (Benchley, 2008).

The film "The Beach" (starring Leonardo diCaprio and Tilda Swinton) was shot on Ko Phi Phi Le Island in Thailand and premiered in 2000 (Rattanaphinanchai & Rittichainuwat, 2018). The tiny island – which includes the paradisiacal Maya Bay beach – soon became an example of a film-driven tourist destination and culminated in over-tourism. The film triggered in enthusiasts the desire to see nature in the magnificent places where the footage took place, generating phenomena of cinematic pilgrimage to the islet (Law et al., 2007). At the time, it was not possible to timely manage the large number of tourists visiting the small island. Despite generating immense income, as the place is idyllic and belongs to a national marine park, this type of tourism caused immense damage to the local fauna and flora of that islet, namely in the disappearance of about 80% of the corals, loss of biodiversity and in the large production of garbage daily. All of this has generated immense controversy and more recently there have been conservation measures to close the area for some time and open with a limitation on the daily number of visitors (Koh & Fakfare, 2020).

The characters, the real scenarios, the natural landscape, and some other factors, whether in literature or filmography, induce sensations in book readers and film viewers (Iwashita, 2006). The avidity in the consumption of these products somehow shapes people's mundane life to something more fictional, but which triggers their interest in rediscovering not only the (unreal) characters, but also the places where the actions happened (real). When looking for these real places, people go as tourists (Lovell, 2019). A search for these filming (or literature) locations is triggered. Visiting phenomena then occur, and consequently there is – usually a posteriori – a response from the supply side that will consequently try to correspond with visitation and souvenir goods and services, embodied in marketing and branding (Folker, 2006). Tourism in general will capitalize on these processes.

The Bad and the Good: Movies With Sharks and Whales

The thriller "Jaws" released in 1975 was a huge success and probably one of the most profitable films of all time. However, many criticisms emerged from the film. Somehow, the idea was created that white sharks ate bathers. Although most people are aware of fiction and reality, some subconscious ideas were created that put some negativity into the sharks (Jøn & Aich, 2015). The "Sharknado" saga, which combines shark with tornado, goes even further than "Jaws", as its fundamental point is sharks as the villains of humanity. It portrays a combination between the immense destructive power of tornadoes associated with the insatiable appetite of sharks – coming from inside the tornadoes – and that they eat everything that comes their way. Although highly fanciful, this type of film refers in a way to a negativity associated with white sharks. There are studies considering the negative interaction between people and white sharks – namely reports of attacks –, but which only consider the analysis of discourse from information sources such as newspapers and news where the real animal appears, discarding films and television series involving words related to sharks (e.g., Hardiman et al., 2019).

Moving from fiction to more realistic footage, some television series about the nature of the sea have attracted the attention of many groups of people, namely those more prone to environmental and sustainability issues. The idea of mistreating sharks because of their "bad behavior" towards humans is currently seen as simply nonsense. A practical example, unfortunately real, is the case of many fisheries, where sharks have been hunted for some time with the sole aim of removing their fins - releasing the animal later - but at the cost of leaving the fish in great suffering and eventually dying (Campana et al., 2006). The growing awareness of this problem has fostered some protection measures against illegal shark catching practices (Wilcox et al., 2021). Currently, the capture of sharks is prohibited, thanks to evidence - mostly produced through images - provided to authorities. The movie "Jaws" created a negative image about sharks, whose bites on humans were intentional, and so at the time, portraying them somewhat as enemies. However, some collective awareness - eventually stemming from this somewhat skewed and derogatory image derived from the film - led to the creation of movements aimed at protecting sharks (Pepin-Neff, 2019).

As another example, the dramatic film "Free Willy" released in 1993 was also a great success. However, despite having doubled the budget of the movie "Jaws", it achieved less than a third of the box office profit. In this plot, a scenario is created where a trapped killer whale - its common name is somewhat pejorative - is sent to an amusement park to become a tourist attraction. At the end of the movie, the whale is finally released into the sea and humans are happy again with the whale's return to its natural habitat. Marine megafauna has always triggered contrasting feelings in humans (Mazzoldi et al., 2019). The general image conveyed in comparison to "Jaws" is not so negative for humans. However, it seems to be reasonable and interesting to see whether this film induced or influenced tourism in pursuit of whale watching (Wearing et al., 2011; O'Connor et al., 2011).

Whaling, Literature, and Tourism

Another example involving the sea, whales and whalers is the novel "Stormy Isles" by Vitorino Nemésio, which portrays Azorean life in the central islands with a realistic and poetic language (Seixo, 2003). It was written during World War II, but referring to the period between 1917 and 1919, which covered World War I and the flu pandemic. Although nostalgic, it describes the surroundings of the islands of the Azores archipelago, Portugal, and the maritime culture. This realism can attract some types of cultural tourism that are passionate about the culture of the sea and that show interest in getting to know the ancient work of whalers and their livelihoods (Beier & Kramer, 2018). Tourism in the Azores islands can benefit from this superbly written book. Currently, due to the Covid-19 pandemic, the Azores are in a way being revisited by certain historical frameworks portrayed in the book Stormy Isles.

Figure 1. A) Hot spring in the village of Furnas (Island of São Miguel, Azores, Portugal); B) Horizontal tail fin of a whale seen in the ocean (Azores, Portugal). [Sources: A) Author's own; B) Adapted from a frame of the documentary "Blue Azores Expedition: A New Way Forward", 2018].

The iconic archipelago of the Azores has nine islands and numerous islets. The islands, located in the North Atlantic midway between Europe and America, are located on three distinct tectonic plates – Eurasian, African and American (Pimentel et al., 2020). Cattle raising – namely cows – and fishing have been the predominant productive activities. In the past, whale hunting was also bequeathed a lot of material and immaterial cultural heritage derived from this activity (as portrayed in Nemesio's book). In 2018, the islands of the Azores were ranked second in the world for sustainable tourism by the National Geographic Traveler magazine (just behind the Faroe Islands, Denmark) (Castanho et al., 2020). The reason for this award was to be considered a paradisiacal place, with well-kept buildings, esteemed nature, and inhabitants with experience of having lived in other places outside the archipelago. Documentaries from leading international television channels on marine life and ethnography in the Azores have contributed to tourism induction (Baldacchino, 2015). This can be found in visiting the rich island culture that is reflected in the diversity of handicraft, music, festivities, gastronomy, and nature mainly at geological (tectonics and volcanology – Figure 1A) and marine (fish and marine mammals – Figure 1B) levels.

MARINE LIFE BROADCASTS INDUCING TOURISM

Attraction and Demystification

There are curious facts regarding stories of nature and species that have somehow become of tourist interest. For example, romantic poet William Wordsworth popularized in English literature the iconic daffodil fields, which are inspiring to artists and tourists alike. David Attenborough made famous the story of his narrative with the gorillas. There are those who try to develop tourist activities related to cockroaches (Westwood & Moss, 2015). The diffusion of paradisiacal scenarios in coastal locations – by the great communicators and naturalists – has encouraged the tourist industry to develop travel packages to these locations and establish their interconnection with local communities (Tabani, 2010).

A moment of danger and adrenaline broadcast on television is something that attracts a lot of viewers. Seeing predatory animals and moving images are captivating and globalizing. In this sense, regarding the marine world, even close to places where sharks were sighted, there are adventurers who refuse to leave the water. This example demonstrates an opposite attitude towards the perception of danger, where the shark goes from being a malevolent person-eating creature to being seen almost as a docile and domesticated being. Another similar example is referred to by Taylor (2009), who describes that some people have been canoeing in wetlands where saltwater crocodiles thrive, being rammed by these dangerous animals, and only ultimately managed to escape between muddy riverbanks like if it was something trivial.

Some criticism is placed in somewhat opposite positions when it comes to public pedagogy of environmental media and conservationists. On the one hand, the conservationist vision of the presenters, who make the public think about using resources more carefully, protecting them and giving them a sense of a certain proximity and cordiality; on the other hand, the traditional view of the tourism industry that is based on improving and increasing the number of visitors to maintain the profit and growth of related activities (Blewitt, 2011). The protection of marine resources widespread by celebrity conservationists has been powerful and has induced much tourism development with capital to be applied in a more sustainable way (Brockington, 2008a).

Conservationists and Marine Life Awareness

There are authors who argue that in certain movies and TV series there are marine animals (and others) that provide a huge amount of free publicity for nature tourism, and that sometimes raise some doubts about their real conservationism (Wunder & Sheil, 2013). There are also films about travel diaries dealing with the conception of "wild", which generally raise questions related to the environment. Traveling to places not much visited by tourists, going in search of the unexplored, also marks a return to the wild, a cultural confrontation. These types of films headed by famous actors sometimes show an autobiographical side (Argod, 2018).

Wildlife broadcast on television does not always have a large audience. This audience has to do with the timing of the broadcast and who presents. Only good science and wildlife communicators can captivate large audiences (Gouyon, 2019). Making the communication of wildlife through the media a tourist attraction has several particularities. The way the media convey tourist destinations has to be appealing. There are celebrity presenters who are seen as intelligent, informed, and respected making them true audience champions. As conservationists they have broadcast wildlife and marine nature programs and films and have shown how they evolve concurrently with humanity (Brockington, 2008b). Narrators generally play a key role in documentaries and must demonstrate knowledge and power, use a speech to describe and interpret the nature, places, and people they pass through to invite the sofa audience to follow in their footsteps.

Contemporary celebrity conservationists who have built their reputations on television and other audiovisual media have a fundamental role today, about calling the attention of the public to the threats that hover over the planet, namely in marine life. Two of these key figures are the English naturalist David Attenborough and the French marine adventurer Jacques-Yves Cousteau (Huggan, 2019).

Jacques-Yves Cousteau Legacy

Captain Jacques-Yves Cousteau carried out several activities related to the exploration of the sea. Among the most notable activities are explorer, conservationist, scientist, oceanographer, inventor, filmmaker, photographer, writer. Associated with his trips with the Calypso crew, with its hydroplane, and the development of scientific submarines to explore the great depths stand out (Crylen, 2018). The inventions include the co-invention of autonomous breathing apparatus under water (Olmstead, 2008), submersible film cameras and other systems useful in diving (Munro, 2005).

The book "The Silent World: A Story of Undersea Discovery and Adventure", published in 1953, appeared to show people – in a pioneering way – what was happening under water (Cousteau & Dumas, 1953). In 1956 the film was released, which won the Palme d'Or and the Oscar for best documentary that year. The film was about the underwater expeditions of Commander Cousteau and his crew aboard the ship Calypso (Duncan, 2018). Cousteau remained for decades the only person to win the Palme d'Or for a documentary, only dethroned in 2004 by Michael Moore with "Fahrenheit 9/11".

Cousteau was one of the greatest contributors of all time to raising interest in marine life and diving adventures within everyone's reach (Cousteau & Cousteau, 1970). The developments of aqualung and diving and their broadcast induced adventurers and others to replicate surface and underwater discoveries (Cousteau & Paccalet, 1987). This knowledge transfer has allowed tour operators to emerge all over the world, fueled by these ideas of observing marine life (Oliveira et al., 2015). Tour companies have been created and profit from this initial idea, while tourists enjoy these gifts of nature.

The film that portrays Cousteau's life is also revealing of this adventurer's passion for the sea. Enthusiasts of Cousteau's unique life path want to try the spaces and experiences lived by the Calypso crew (Orams, 2002). For example, the international diving expeditions of Commander Cousteau and the Calypso crew sparked the development of diving tourism (Dimmock & Cummins, 2013).

However, environmental concern has increased in recent decades. Thus, the relatively recent phenomenon of ecotourism (or nature tourism) derived from broadcasting documentaries about nature and wildlife. Ecotourism usually has a disadvantage when compared to economic results: its massification. Natural tourism should not be massified at the cost of putting pressure on the carrying capacity of the wild places and scattering other types of disturbance (e.g., noise, agglomeration effects, etc.). The broadcast of naturalistic series has been important in creating public awareness of wildlife and environmental issues and has helped to trigger nature tourism activities. Cousteau has spread ideas that are as disturbing as they are true, appealing to global human consciousness in the way humans behave and act globally. Among these iconic phrases is one referring to water and air as the two essential fluids on which all life depends, and which have become the global dustbin. He added that plastic has also become an extraordinary unifier, and concluded that the sea, as a great unifier, is humanity's only hope (Siegle, 2018).

David Attenborough Influence

Sir David Attenborough is a BBC natural history announcer and has won BAFTAs for documentaries made in black and white and in newer formats (Collins, 2017). Broadcast documentaries have been important tools for educating people about nature. There are some who have become the most influential of all times, such as one that portrays the orca – seen as the biggest predator of the seas – and that after the documentary becomes fond of people. David Attenborough has taught us that no one will care what they are not interested in, and that no one will be interested in what they have never experienced (Colby, 2018).

It has been important for some time to direct collective awareness of environmental causes (Attenborough & Wright, 1987). The testimony of the renowned presenter of "Blue Planet" contributed to generate more and better awareness among citizens regarding the preservation of the ocean, not as a source of infinite resources, but as a sensitive world that needs human care. Investment in education - especially for the younger generations - is vitally important to face the phenomena that need to be eradicated, such as microplastics, climate change, overfishing, the greenhouse effect, and ice retreats. Attenborough is an advocate of restoring planetary biodiversity, supporter of limiting human population growth, follower of renewable energy, admirer of reduced meat consumption and devotee of applying a greater effort in natural preservation (Attenborough, 2011). In his 2020 documentary - "A Life on Our Planet" -, Attenborough considers his personal statement a testimony to the future (Attenborough, 2020).

The phenomenon of persuading the sofa audience instilling the film/television power and tourism is described by the "Attenborough Effect" (Wearing et al., 2019). The "Attenborough Effect" also known as the "Blue Planet Effect" is influencing the way people deal with plastic, as it was in the 2017 "Blue Planet 2" series that the issue of plastic recycling was placed at the top of the political agenda. The BBC program broadcast in 2017 praised humanity's impact on the oceans and the growing problem of pollution caused by the recurrent use of plastic in our daily lives. This effect has led people to consume less plastic and choose to reuse returnable packaging instead of single use. It is the perception of the concept of sustainability to be passed on to society. The marine life series "The Blue Planet" and "Blue Planet 2" captured the attention of huge audiences and certainly were shaping and educating people of various generations on reflection and awareness of trends in scientific thinking and environmental issues (Gouyon, 2019). This educational way of transmitting messages passes to the citizen in everyday life and hopefully is reflected in the citizen as a tourist consumer and in the industry as a producer of goods and services.

SINGULAR SPECIES

The Concept of Sustainability and Tourism

The term "sustainability" is relatively recent, having appeared less than 70 years ago. However, there were already terms with a similar meaning. The concept of biological sustainability is even older and derives from Malthusian ideas regarding population growth; or even Darwin's theory of evolution on intraspecific competition (Du Pisani, 2006). However, the origin of the term sustainability is debatable, and some authors argue that it essentially derives from Adam Smith's economic thinking about efficiency and competition from market forces (Stafford, 2019).

The basis for analyzing and managing marine recreational activities, including tourism, should be to ensure the sustainability of the resource on which they depend. In the film "Free Willy" spectators are given clues in this direction (Orams, 2002).

Some ways of sustainably managing marine living resources have been studied. For example, in the case of sharks there have been some studies with relational values, including the derogatory demystification of "man eater" (Skubel et al., 2019). Shark watching tourism at aggregation sites is becoming a common activity among marine life enthusiasts and beyond (Zemah-Shamir et al., 2019). This type of wildlife tourism promotes conservation and economic development, but generally lacks sustainable management practices. Some authors state that a vision for the future is needed, involving different stakeholders, to ensure the sustainability of this practice (Meza-Arce et al., 2020).

Learning is often an important element of tourism. The tourist can actively learn by having a specific purpose, and passively by comparing values, norms, and customs. There are authors who argue that travel supports active learning and is positive for sustainability, particularly in conservation. However, these learning processes regarding environmental sustainability and sustainable lifestyles remain insufficiently understood. As tourism has great economic, social, and cultural importance; tourism also has global implications for environmental sustainability, and planning is needed to enhance the effects of sustainable learning (Gössling, 2017).

Threatened Species

There is pressure on biodiversity in the ocean, but the degree and risk of extinction remain poorly characterized. However, charismatic species are more likely to gain support for conservation. For example, regarding the movie "Finding Nemo", McClenachan et al. (2012) studied the extinction risk and barriers to successful conservation of more than 15 hundred species in 16 families of marine animals that appeared in the film. They found that the species that were threatened ranged from few bony fish to most sea turtles. They found that the risk of extinction among various groups of marine living beings was greater than that of most terrestrial groups, alluding that at sea there must be even greater concern with conservation. It was verified by other authors – in relation to the same film – that some species that had spoken parts, were endemic species threatened for various reasons, including destructive fishing practices and phenomena somehow related to tourism (Arthington et al., 2016). The species represented by the characters Nemo (the warm waters clownfish also called anemonefish *Amphiprion ocellaris*) and Dory (the Indo-Pacific surgeonfish *Paracanthurus hepatus*) even though they are not threatened, have experienced a large increase in demand for the aquarium market since the movie premiered (Figure 2A and B, respectively). After the film's premiere, it became common to find aquariums with fish representing these two characters from the animated film in public spaces for tourist visits. Sometimes details such as placing the anemones that are related to the clownfish are not neglected.

Figure 2. A) Clownfish/anemonefish Amphiprion ocellaris that inspired the character Nemo in the film "Finding Nemo"; B) Indo-Pacific surgeonfish Paracanthurus hepatus who inspired the character Dory in the same film. [Sources: Author's own].

Maritime tourism allows for new recreation alternatives in coastal destinations with natural and cultural attractions that are found there. Creating a segmentation of demand for a maritime destination in terms of tourists' motivations is important and must somehow be done. In this sense, Carvache-Franco et al. (2020) found in a coastal city in Ecuador that the greater the frequency of visits, the greater the degree of motivation and satisfaction of tourists. These results provide a basis for developing tourism products and services tailored to customer demand. Among these products and services, it is important to include marine species threatened with extinction, such as the leatherback turtle or the Pacific seahorse.

Iconic Species and Tourism

In 2019 a group of students from an English elementary school asked Sir David Attenborough what his favorite animal was. The famous presenter replied that it was the weedy sea dragon (*Phyllopteryx taeniolatus*). This species belongs to the seahorse family, lives on the southern coasts of Australia and Tasmania, and is in a non-threatened status. In some ways this species is iconic and becomes even more so when someone prominent in the marine wildlife broadcasting says so.

Ornamental marine species (OMS) provide valuable income for the developing nations of the Coral Triangle that lies in the Indo-Pacific region, from where most specimens are exported. OMS aquaculture can help diversify livelihoods in the region and enables better management and conservation to reduce fishing or destructive practices in places that threaten coral reef and marine plant ecosystems. It was found that e.g., seahorses can play a role in management and conservation, complementing and diversifying fisheries and enabling livelihoods in developing nations (Williams et al., 2014). In cases like this, tourism has a responsibility and should prevent the demand for native species when they are rare or threatened, as in the case of places where there are seahorse breeding banks (Goodwin, 2011). There are cases in other world locations where the reduction of seahorses has been dramatic with a decrease above 90% of what had been reported a decade ago (Correia et al., 2014). Publicizing these issues in schools and promoting actions with children and adolescents help to create collective awareness (Figure 3A). Public constructions can serve as reminders of species to be protected as well as tourism promoters with a specific identity (Figure 3B).

Figure 3. A) Graffiti of a stylized seahorse in a public school (author: SEN; Olhão, Algarve, Portugal); B) Sculpture of a seahorse made from broken ceramic pieces located in a roundabout in a tourist area (authors: Isa Fernandes and Alex Groza; Olhão, Algarve, Portugal). [Sources: Author's own].

Choosing a species as a stylized, promotional icon can suitably promote a location, but it can also demonstrate a sustainable way of travelling. An example in this sense, taking the pink flamingo as an example, was studied in the Mexican market by Córdoba Azcárate (2011). Another case with similar birds was the case of the Andean flamingo that was threatened by pressure in its natural habitat. The creation of a reserve to protect this species made this site a tourist destination for recreational divers and professional film crews (Drumm, 2006). In these cases, there tends to be an adjustment between the value of the species for economies and tourism, so that the entire system is sustainable.

Individual distances between greater flamingos (*Phoenicopterus roseus*) are used as an indicator to determine tourist pressure (Yosef, 2000). This species is iconic and attracts the attention of tourists who like to visit natural areas located particularly in southern Europe. It's kind of such an iconic way that it serves as the inspiration for a prep school mural (Figure 4A). Despite being a species somewhat sensitive to human presence (Scarton, 2017), it adapts with some ease to artificialized areas such as lagoons in saline zones (Ramos et al., 2021) (Figure 4B).

Figure 4. A) Graffiti of a stylized flamingo in a public school (author: SEN; Olhão, Algarve, Portugal); B) Greater flamingos (Phoenicopterus roseus) in the evaporators of a traditional saltpan (Belamandil, Algarve, Portugal). [Sources: Author's own].

DISCUSSION

For some time now, there have been phenomena of literature and filmography that attract immense fans, who follow a line of cultural tourism and visits to places that inspired the work (Pascoal, 2019). Literature with an ethnographic nature and films about the natural world and the life of species, generally not generating as many followers as the fictional ones, have somehow functioned as tourist attractors to a more real world that until then could have been forgotten. This type of literary and filmographic works has stimulated and induced a tourism niche towards nature, particularly in recent years. The observation of marine life species is increasingly accessible. There is much more competition for those who provide these types of services for tourist activities (Garrod & Wilson, 2004).

With constant human evolution, there are values that change over time. Sometimes they are changed or even no longer make sense, as in the case of objects that undergo technological evolution and obsolete objects are being disregarded.

A particular case is the case of the use of plastics in excess and their reduction and reuse. The user may be increasingly aware of the problem because he receives information – e.g., from the media – and eventually adopts a more sustainable behavior, not only daily but also as a tourist. However, the perception of effectiveness as a consumer may prove to be difficult to achieve, as it is not just an individual who has an appropriate behavior that will change a problem that affects the entire society.

Another case in this line of thought happens when adopting behaviors related to the use of plastics in excess and their reduction and reuse. The user may be increasingly aware of the problem because they receive information – e.g., from the media – and eventually adopt a more sustainable behavior, not only daily but also as a tourist. However, the perception of effectiveness as a consumer may prove to be difficult to achieve, as it is not just an individual who has an appropriate behavior that will change a problem that affects the entire society (Hanss & Doran, 2020). There are also perceptions and feelings that change with the times; for example, sharks were once the bad guys, and today this negative image is somehow practically eradicated (Skubel, 2019).

Relatively recent phenomena of marine mammal watching result in a way from the release of blockbuster movies. Similarly, there have been series that have glued expedition enthusiasts to the sofa to the most secluded and enchanting places, and who wish to somehow replicate these experiences in places where they can enjoy a beautiful boat tour or a scuba diving trip in paradise waters full of biodiversity and colorful species.

Broadcast documentaries with charismatic narrators create in people a sense of genuine beauty found in nature (Wearing et al., 2019). Individuals in general - and young people in particular - are increasingly aware of the need to protect the treasures that nature offers to humanity. For this enjoyment, people look for tourist activities related to large spaces, in the open air, away from the urban hustle and often along the coast or even at sea.

In several coastal cities around the world, there are firms that provide services on boats, and include some natural features and marine life, which can be found in these places and shared as a tourist offer. In these places, it is common to go by boat to observe certain whales or dolphins, or even to find the big and fast predators of the sea such as sharks, or in the opposite direction the docile and slow turtles, to name just a few examples (Zemah Shamir et al., 2019).

The growing demand for these tourism services is important for firms to generate revenue. Therefore, it is essential for maritime tourism firms to operate in places with crystal clear water, ideally without pollution. The balance to be struck here is to be able to attract as many tourists as possible without putting resources under pressure. Such an occurrence can jeopardize the entire system.

A growing awareness of the protection of living marine beings has not only fallen on sharks, but also on other organisms. Films and documentaries have contributed to this. For example, the Archipelago of the Azores (Portugal) has been increasingly visited by tourists looking for the traits of nature. Diving and whale watching are among the tourist activities that generate the most income on the islands (Bentz et al., 2016).

Coastal and marine tourism, despite being well established, still has room for growth (Ramos & Drakeford, 2021). Many firms linked to the sea tourism industry have been established by marine life-loving entrepreneurs or biologists. The simple boat trips to watch animals seen as "charismatic" such as dolphins, whales, turtles, seahorses, flamingos is an interesting and important tourist niche to explore, as tourists are sensitive to the sighting of these animals and have references of what they watched in the media. Likewise, these activities also have a share of co-responsibility for the preservation of these species and the sustainability of marine life.

In the case of diving tourism, essentially inspired by the programs and films of naturalists such as Commander Cousteau, has led to the exploration and visitation of places all over the world. This fact often puts natural and coral reefs under pressure. An alternative that has been developed to not only diversify the offer of dive sites, but also reduce pressure, is the deployment of artificial reefs, which include the sinking of structures, namely ships (Oliveira et al., 2015). Thus, only professional divers or tourist diving constitute the main target audience for this type of tourism. For example, the 155 m long wreck of the USS "Spiegel Grove", Florida, USA is one of the largest artificial reefs in the world (Leeworthy et al., 2006) and attracts a lot of divers every year. Science and awareness of the relative fragility of the seabed means that the negative effect eventually caused by underwater mass tourism due to its enormous pressure, is mitigated.

CONCLUSION

Literature, films, and nature documentaries about marine life are sources of learning for people in general, and with particular importance for the younger ones. In everyday life these educational entertainment pieces serve to fill some free time. During the holidays, they can serve as inspirations or motivators for tourist visits to places that for some reason remained in the memory positively after reading a book or watching the media. The threat to which some marine species are subject and sustainability issues are increasingly on the agenda. Tourists can develop an affection and behavior in favor of sustainability when visiting these places. The tourism industry is adapting to market trends and fulfilling the demands of different tourist niches. The existence of singular and iconic species – which appear both in literature and in nature films and documentaries – allows attracting the attention of tourism and becomes a driving force in branding and sales of objects commonly used in everyday life, whether by children (e.g., toys, clothing) or by adults (e.g., household items). These species also have the particularity that they can be introduced in an educational way (e.g., in schools). This type of action makes it possible to draw attention to the younger generations about the importance of preserving these species locally and their cultural, social, economic, and environmental enhancement for a given location.

ACKNOWLEDGMENT

I am grateful to two anonymous referees for their helpful comments and suggestions that contributed immensely to improve the manuscript. This paper is financed by Portuguese National Funds provided by FCT - Foundation for Science and Technology through project UID/SOC/04020/2019.

REFERENCES

Argod, P. (2018). From the Road Movie to Wilderness Travel: The Quest for Adventure on the Road and the Myth of the Heroic Traveller. *Via. Tourism Review*, 14.

Arthington, A. H., Dulvy, N. K., Gladstone, W., & Winfield, I. J. (2016). Fish conservation in freshwater and marine realms: Status, threats and management. *Aquatic Conservation*, *26*(5), 838–857. doi:10.1002/aqc.2712

Attenborough, D. (2011). This heaving planet. *New Statesman (London, England)*, *140*, 28–32.

Attenborough, D. (2020). *A Life on Our Planet: My Witness Statement and a Vision for the Future*. Random House.

Attenborough, D., & Wright, D. (1987). *The First Eden: The Mediterranean world and man*. Collins.

Baldacchino, G. (2015). Small island states and territories: Vulnerable, resilient, but also doggedly perseverant and cleverly opportunistic. In *Entrepreneurship in small island states and territories* (pp. 23–50). Routledge. doi:10.4324/9781315764528-7

Beeton, S. (2006). Understanding film-induced tourism. *Tourism Analysis*, *11*(3), 181–188. doi:10.3727/108354206778689808

Beier, R., & Kramer, J. (2018). A portrait of the Azores: From natural forces to cultural identity. In *Volcanoes of the Azores* (pp. 3–26). Springer. doi:10.1007/978-3-642-32226-6_2

Bélanger, C. É., & Jolin, L. (2011). The International Organisation of Social Tourism (ISTO) working towards a right to holidays and tourism for all. *Current Issues in Tourism*, *14*(5), 475–482. doi:10.1080/13683500.2011.568056

Benchley, P. (2008). *Shark Life: True Stories About Sharks & the Sea*. Yearling.

Bentz, J., Lopes, F., Calado, H., & Dearden, P. (2016). Enhancing satisfaction and sustainable management: Whale watching in the Azores. *Tourism Management*, *54*, 465–476. doi:10.1016/j.tourman.2015.11.016

Blewitt, J. (2011). Critical practice and the public pedagogy of environmental and conservation media. *Environmental Education Research*, *17*(6), 719–734. doi:10.1080/13504622.2011.618625

Brockington, D. (2008a). Celebrity conservation: Interpreting the Irwins. *Media International Australia*, *127*(1), 96–108. doi:10.1177/1329878X0812700113

Brockington, D. (2008b). Powerful environmentalisms: Conservation, celebrity and capitalism. *Media Culture & Society*, *30*(4), 551–568. doi:10.1177/0163443708030004070l

Busby, G., Huang, R., & Jarman, R. (2013). The Stein effect: An alternative film-induced tourism perspective. *International Journal of Tourism Research*, *15*(6), 570–582. doi:10.1002/jtr.1875

Campana, S. E., Marks, L., Joyce, W., & Kohler, N. E. (2006). Effects of recreational and commercial fishing on blue sharks (Prionace glauca) in Atlantic Canada, with inferences on the North Atlantic population. *Canadian Journal of Fisheries and Aquatic Sciences*, *63*(3), 670–682. doi:10.1139/f05-251

Carvache-Franco, M., Carvache-Franco, W., Carvache-Franco, O., Hernández-Lara, A. B., & Buele, C. V. (2020). Segmentation, motivation, and sociodemographic aspects of tourist demand in a coastal marine destination: A case study in Manta (Ecuador). *Current Issues in Tourism*, *23*(10), 1234–1247. doi:10.1080/13683500.2019.1600476

Castanho, R. A., Couto, G., Pimentel, P., Carvalho, C., Sousa, Á., & Velarde, J. G. (2020). Assessing the impacts of public policies over tourism in Azores Islands. A research based on tourists and residents perceptions. *Sustainable Development, 17*(18), 19-20.

Cater, C. (2010). Any closer and you'd be lunch! Interspecies interactions as nature tourism at marine aquaria. *Journal of Ecotourism*, *9*(2), 133–148. doi:10.1080/14724040903125039

Colby, J. M. (2018). *Orca: How We Came to Know and Love the Ocean's Greatest Predator*. Oxford University Press. doi:10.1093/oso/9780190673093.001.0001

Collins, K. (2017). Calls of the wild? 'Fake' sound effects and cinematic realism in BBC David Attenborough nature documentaries. *The Soundtrack*, *10*(1), 59–77. doi:10.1386/ts.10.1.59_1

Córdoba Azcárate, M. (2011). 'Thanks God, this is not Cancun!' Alternative tourism imaginaries in Yucatan (Mexico). *Journal of Tourism and Cultural Change*, *9*(3), 183–200. doi:10.1080/14766825.2011.620118

Correia, M., Palma, J., Koldewey, H., & Andrade, J. P. (2014). The use of a non-invasive tool for capture–recapture studies on a seahorse Hippocampus guttulatus population. *Journal of Fish Biology*, *84*(4), 872–884. doi:10.1111/jfb.12304 PMID:24502668

Cousteau, J. Y., & Cousteau, P. (1970). *The shark: splendid savage of the sea* (No. Sirsi a442417). Academic Press.

Cousteau, J. Y., & Dumas, F. (1953). The silent world. Academic Press.

Cousteau, J. Y., & Paccalet, Y. (1987). *La mer blessée: La Méditerranée [Spoilt sea: The Mediterranean]*. Flammarion.

Crylen, J. (2018). Living in a World without Sun: Jacques Cousteau, Homo aquaticus, and the Dream of Dwelling Undersea. *JCMS: Journal of Cinema and Media Studies*, *58*(1), 1–23. doi:10.1353/cj.2018.0068

Cynthia, D., & Beeton, S. (2009). Supporting independent film production through tourism collaboration. *Tourism Review International*, *13*(2), 113–119. doi:10.3727/154427209789604624

Denny, L., Shelley, B., & Ooi, C. S. (2019). Education, jobs and the political economy of tourism: Expectations and realities in the case of Tasmania. *Australasian Journal of Regional Studies*, *25*(2), 282–305.

Dimmock, K., & Cummins, T. (2013). History of scuba diving tourism. *Scuba Diving Tourism*, 14-28.

Drumm, A. (2006). Tourism-based revenue generation for conservation. In *Tourism and Protected Areas: Benefits beyond Boundaries*. The Vth IUCN World Parks Congress.

Duncan, P. D. (2018). (Science) Fiction: Genre Hybridization in Jacques-Yves Cousteau and Louis Malle's The Silent World (1956). *The Journal of Popular Film and Television, 46*(2), 108–117. doi:10.1080/01956051.2018.1465886

Folker, B. (2006). *Harry Potter and international relations*. Rowman & Littlefield.

Frampton, A. (2014). *Cross-border film production: The neoliberal recolonization of an exotic island by Hollywood pirates* (Doctoral dissertation). Bowling Green State University.

Frevert, U., Eitler, P., Olsen, S., Jensen, U., Pernau, M., Brückenhaus, D., & Häberlen, J. C. (2014). *Learning how to Feel: Children's Literature and Emotional Socialization, 1870-1970*. OUP Oxford.

Garner, R. (2019). Finding Nemo's spaces: Defining and exploring transmedia tourism. *JOMEC Journal, 0*(14), 11–32. doi:10.18573/jomec.195

Garrod, B., & Wilson, J. C. (2004). Nature on the edge? Marine ecotourism in peripheral coastal areas. *Journal of Sustainable Tourism, 12*(2), 95–120. doi:10.1080/09669580408667227

Goodwin, H. (2011). Taking responsibility for tourism. Woodeaton: Goodfellow Publishers Limited.

Gössling, S. (2018). Tourism, tourist learning and sustainability: An exploratory discussion of complexities, problems and opportunities. *Journal of Sustainable Tourism, 26*(2), 292–306. doi:10.1080/09669582.2017.1349772

Gouyon, J. B. (2019). *BBC wildlife documentaries in the age of Attenborough*. Springer Nature.

Gouyon, J. B. (2019). Wildlife Television, Empathy and the End of the British Empire. In *BBC Wildlife Documentaries in the Age of Attenborough* (pp. 69–95). Palgrave Macmillan. doi:10.1007/978-3-030-19982-1_4

Hanss, D., & Doran, R. (2020). Perceived consumer effectiveness. *Responsible Consumption and Production*, 535-544.

Hardiman, N., Burgin, S., & Shao, J. (2019). News media portrayal of attributed stakeholder attitudes to shark management in Australia. *Human Dimensions of Wildlife, 24*(6), 548–563. doi:10.1080/10871209.2019.1663455

Hofman, K., & Hughes, K. (2018). Protecting the Great Barrier Reef: Analysing the impact of a conservation documentary and post-viewing strategies on long-term conservation behaviour. *Environmental Education Research, 24*(4), 521–536. doi:10.1080/13504622.2017.1303820

Hoppen, A., Brown, L., & Fyall, A. (2014). Literary tourism: Opportunities and challenges for the marketing and branding of destinations? *Journal of Destination Marketing & Management, 3*(1), 37–47. doi:10.1016/j.jdmm.2013.12.009

Hudson, S., & Ritchie, J. B. (2006). Promoting destinations via film tourism: An empirical identification of supporting marketing initiatives. *Journal of Travel Research, 44*(4), 387–396. doi:10.1177/0047287506286720

Huggan, G. (2019). *Nature's Saviours: Celebrity Conservationists in the Television Age*. Routledge.

Iwashita, C. (2006). Media representation of the UK as a destination for Japanese tourists: Popular culture and tourism. *Tourist Studies, 6*(1), 59–77. doi:10.1177/1468797606071477

Jøn, A. A., & Aich, R. S. (2015). Southern shark lore forty years after Jaws: the positioning of sharks within Murihiku, New Zealand. *Australian Folklore, 30.*

Koh, E., & Fakfare, P. (2020). Overcoming" over-tourism": The closure of Maya Bay. *International Journal of Tourism Cities*, *6*(2), 279–296. doi:10.1108/IJTC-02-2019-0023

Law, L., Bunnell, T., & Ong, C. E. (2007). The Beach, the gaze and film tourism. *Tourist Studies*, *7*(2), 141–164. doi:10.1177/1468797607083499

Leeworthy, V. R., Maher, T., & Stone, E. A. (2006). Can artificial reefs alter user pressure on adjacent natural reefs? *Bulletin of Marine Science*, *78*(1), 29–38.

Lovell, J. (2019). Fairytale authenticity: Historic city tourism, Harry Potter, medievalism and the magical gaze. *Journal of Heritage Tourism*, *14*(5-6), 448–465. doi:10.1080/1743873X.2019.1588282

Macionis, N., & Sparks, B. (2009). Film-induced tourism: An incidental experience. *Tourism Review International*, *13*(2), 93–101. doi:10.3727/154427209789604598

Mak, J. (2003). *Tourism and the economy: Understanding the economics of tourism*. University of Hawaii Press. doi:10.1515/9780824844776

Mazzoldi, C., Bearzi, G., Brito, C., Carvalho, I., Desiderà, E., Endrizzi, L., Freitas, L., Giacomello, E., Giovos, I., Guidetti, P., Ressurreição, A., Tull, M., & MacDiarmid, A. (2019). From sea monsters to charismatic megafauna: Changes in perception and use of large marine animals. *PLoS One*, *14*(12), e0226810. doi:10.1371/journal.pone.0226810 PMID:31891944

McClenachan, L., Cooper, A. B., Carpenter, K. E., & Dulvy, N. K. (2012). Extinction risk and bottlenecks in the conservation of charismatic marine species. *Conservation Letters*, *5*(1), 73–80. doi:10.1111/j.1755-263X.2011.00206.x

Meza-Arce, M. I., Malpica-Cruz, L., Hoyos-Padilla, M. E., Mojica, F. J., Arredondo-García, M. C., Leyva, C., Zertuche-Chanes, R., & Santana-Morales, O. (2020). Unraveling the white shark observation tourism at Guadalupe Island, Mexico: Actors, needs and sustainability. *Marine Policy*, *119*, 104056. doi:10.1016/j.marpol.2020.104056

Munro, C. (2005). Diving systems. *Methods for the study of marine benthos*, 112-159.

O'Connor, N., Macionis, N., Wearing, S., Buchmann, A., & Jobberns, C. (2011). Free Willy: The whale-watching legacy. *Worldwide Hospitality and Tourism Themes.*

Oliveira, M. T., Ramos, J., & Santos, M. N. (2015). An approach to the economic value of diving sites: Artificial versus natural reefs off Sal Island, Cape Verde. *Journal of Applied Ichthyology*, *31*, 86–95. doi:10.1111/jai.12953

Olmstead, K. (2008). *Jacques Cousteau: a life under the sea*. Sterling Publishing Company, Inc.

Orams, M. (2002). *Marine tourism: development, impacts and management*. Routledge. doi:10.4324/9780203197110

Pascoal, S. (2019). Smart tech para a otimização do turismo literário: O projeto TheRoute. *Dos Algarves: A Multidisciplinary e-Journal*, (35), 57-72.

Pepin-Neff, C. L. (2019a). The Rise of Shark "Attack" Discourse. In Flaws (pp. 13-38). Palgrave Macmillan. doi:10.1007/978-3-030-10976-9_2

Pepin-Neff, C. L. (2019b). Considering Sharks from a Post-Jaws Perspective. In *Flaws* (pp. 163–188). Palgrave Macmillan. doi:10.1007/978-3-030-10976-9_8

Pimentel, P., Oliveira, A., Couto, G., Ponte, J. C., & Castanho, R. (2020). The Azores Archipelago as a Region with Vast Potential for the Development of Adventure and Slow Tourism. In Peripheral Territories, Tourism, and Regional Development. IntechOpen.

Price, J. (1999). The plastic pink flamingo: A natural history. *The American Scholar*, *68*(2), 73–88.

Ramos, J., & Drakeford, B. (2021). Blue Economy and Tourism: Is There a Potential to Create Sustainable Jobs? In Handbook of Research on Human Capital and People Management in the Tourism Industry (pp. 243-264). IGI Global.

Ramos, J., Pinto, P., Pintassilgo, P., Resende, A., & Cancela da Fonseca, L. (2021). Activating an artisanal saltpan: Tourism crowding in or waterbirds crowding out? *International Journal of Culture, Tourism and Hospitality Research*. Advance online publication. doi:10.1108/IJCTHR-04-2021-0101

Ramos, J., Santos, M. N., Whitmarsh, D., & Monteiro, C. C. (2006). The usefulness of the analytic hierarchy process for understanding reef diving choices: A case study. *Bulletin of Marine Science*, *78*(1), 213–219.

Rattanaphinanchai, S., & Rittichainuwat, B. N. (2018). Film–induced tourism in Thailand: An influence of international tourists' intention to visit film shooting location. *International Journal of Tourism Sciences*, *18*(4), 325–332. doi:10.1080/15980634.2018.1551317

Rico, A., Martínez-Blanco, J., Montlleó, M., Rodríguez, G., Tavares, N., Arias, A., & Oliver-Solà, J. (2019). Carbon footprint of tourism in Barcelona. *Tourism Management*, *70*, 491–504. doi:10.1016/j.tourman.2018.09.012

Sakellari, M. (2014). Film tourism and ecotourism: Mutually exclusive or compatible? *International Journal of Culture, Tourism and Hospitality Research*, *8*(2), 194–202. doi:10.1108/IJCTHR-09-2013-0064

Salomon, G. (2012). *Interaction of media, cognition, and learning: An exploration of how symbolic forms cultivate mental skills and affect knowledge acquisition*. Routledge. doi:10.4324/9780203052945

Scarton, F. (2017). Environmental characteristics of shallow bottoms used by Greater flamingo (Phoenicopterus roseus) in a northern Adriatic lagoon. *Acrocephalus*, *38*(174-175), 161–169. doi:10.1515/acro-2017-0010

Seixo, M. A. (2003). "Mau Tempo no Canal": An Iridescent Metaphor. *Portuguese Literary and Cultural Studies*, 129-169.

Siegle, L. (2018). *Turning the Tide on Plastic: How Humanity (And You)*. Can Make Our Globe Clean Again.

Skubel, R. A., Shriver-Rice, M., & Maranto, G. M. (2019). Introducing relational values as a tool for shark conservation, science, and management. *Frontiers in Marine Science*, *6*, 53. doi:10.3389/fmars.2019.00053

Stafford, R. (2019). Sustainability: A flawed concept for fisheries management? *Elementa*, *7*, 7. doi:10.1525/elementa.346

Steiner, C. J., & Reisinger, Y. (2006). Ringing the fourfold: A philosophical framework for thinking about wellness tourism. *Tourism Recreation Research*, *31*(1), 5–14. doi:10.1080/02508281.2006.11081242

Tabani, M. (2010). The carnival of custom: Land dives, millenarian parades and other spectacular ritualizations in Vanuatu. *Oceania*, *80*(3), 309–329. doi:10.1002/j.1834-4461.2010.tb00088.x

Taylor, B. (2009). Globalization with Predators and Moving Pictures. In *Dark Green Religion* (pp. 127–154). University of California Press. doi:10.1525/9780520944459

van Leeuwen, T. (2011). Multimodality. In *The Routledge handbook of applied linguistics* (pp. 688–702). Routledge.

Vu, D. V., Tran, G. N., Nguyen, H. T. T., & Nguyen, C. V. (2020). Factors Affecting Sustainable Tourism Development in Ba Ria-Vung Tau, Vietnam. *The Journal of Asian Finance, Economics, and Business*, *7*(9), 561–572. doi:10.13106/jafeb.2020.vol7.no9.561

Ward, S., & O'Regan, T. (2009). The film producer as the long-stay business tourist: Rethinking film and tourism from a Gold Coast perspective. *Tourism Geographies*, *11*(2), 214–232. doi:10.1080/14616680902827175

Wearing, S., Buchmann, A., & Jobberns, C. (2011). Free Willy: The whale-watching legacy. *Worldwide Hospitality and Tourism Themes*, *3*(2), 127–140. doi:10.1108/17554211111122998

Wearing, S., Schweinsberg, S., & Johnson, P. (2019). Flâneur or Choraster: A Review of the Travel Narrator In the Formation of the Tourist Experience. *Tourism Analysis*, *24*(4), 551–562. doi:10.3727/108354219X15652651367415

Westwood, B., & Moss, S. (2015). Natural histories: 25 extraordinary species that have changed our world. Hachette UK.

Wilcox, C., Mann, V., Cannard, T., Ford, J., Hoshino, E., & Pascoe, S. (2021). *A review of illegal, unreported and unregulated fishing issues and progress in the Asia-Pacific Fishery Commission region.* Academic Press.

Williams, S. L., Janetski, N., Abbott, J., Blankenhorn, S., Cheng, B., Crafton, R. E., Hameed, S. O., Rapi, S., & Trockel, D. (2014). Ornamental marine species culture in the Coral Triangle: Seahorse demonstration project in the Spermonde Islands, Sulawesi, Indonesia. *Environmental Management*, *54*(6), 1342–1355. doi:10.100700267-014-0343-6 PMID:25082298

Wunder, S., & Sheil, D. (2013). On taxing wildlife films and exposure to nature. *Oryx*, *47*(4), 483–485. doi:10.1017/S0030605312001469

Yang, Y., Altschuler, B., Liang, Z., & Li, X. (2020). Monitoring the global COVID-19 impact on tourism: The COVID19tourism index. *Annals of Tourism Research*, 103120. PMID:34566202

Yosef, R. (2000). Individual distances among greater flamingos as indicators of tourism pressure. *Waterbirds*, *23*, 26–31. doi:10.2307/1522143

Zemah Shamir, Z., Zemah Shamir, S., Tchernov, D., Scheinin, A., & Becker, N. (2019). Shark aggregation and tourism: Opportunities and challenges of an emerging phenomenon. *International Journal of Sustainable Development and World Ecology*, *26*(5), 406–414. doi:10.1080/13504509.2019.1573769

Compilation of References

Afshardoost, M., & Eshaghi, M. (2020). Destination image and tourist behavioural intentions: A meta-analysis. *Tourism Management, 81*, 104154. doi:10.1016/j.tourman.2020.104154

Agarwal, S., & Shaw, G. (2017). *Heritage, screen and literary tourism.* Channel View Publications.

Akkaya, B., Karataş, E., Sarı, O., Seten, S., & Zilif, S. (Eds.). (2015). Kültür rotaları planlama rehberi. Çekul Vakfı Tarihi Kentler Birliği Yayınları.

Albuquerque, H., Costa, C., & Martins, F. (2018). The use of Geographical Information Systems for Tourism Marketing purposes in Aveiro region (Portugal). *Tourism Management Perspectives, 26*, 172-178. doi:10.1016/j.tmp.2017.10.009

Aliagaoglu, A. (2008). Savaş alanları turizmi için tipik bir yer: Gelibolu Yarımadası Tarihi Milli Parkı [A typical place to battlefield tourism: Gallipoli Peninsula Historical National Park]. *Milli Folklor, 20*(78), 88–104.

Ali-Knight, J. (2011). *The role of niche tourism products in destination development* (PhD Thesis). Edinburgh Napier University.

Almeida, M. M. (2016). "Viagens na minha serra": Percorrer a região do Caramulo na companhia de Branquinho da Fonseca ["Viagens na minha serra": Travel in the Caramulo region in the company of Branquinho da Fonseca]. In R. Baleiro, S. Quinteiro & I. D. Santos (Eds.), Viagens, relatos e itinerários [Travels, narratives and itineraries] (pp. 61-78). Universidade do Algarve.

Alper, S. (2006). Tevfik Fikret'in kırk sekiz yılı. In A. Yaraman (Ed.), *Biyografya 7: Tevfik Fikret* (pp. 11–16). Bağlam Publishing.

Alves, I. M. F. (2016). A interioridade ou o itinerário transmontano de A.M. Pires Cabral [A.M. Pires Cabral's interiority or the Trás-os-Montes itinerary]. In R. Baleiro, S. Quinteiro & I. D. Santos (Eds.), Viagens, relatos e itinerários [Travels, narratives and itineraries] (pp. 47-60). Universidade do Algarve.

Alves, M. T. A. (1999). Almeida Garrett: O texto, o intertexto e as margens do texto na representação do escritor [Almeida Garrett: the text, the intertext and the text margins in the writer representation]. *Scripta, Belo Horizonte, 3*(5), 142–154.

Amirou, R. (1999). Les nouvelles mythologies du tourisme. *Sciences Humaines, 90*, 22–25.

Amorim. (1881-1884). *Garrett: Memórias Biographicas.* Lisboa: Imprensa Nacional.

Amujo, O. C., & Otubanjo, O. (2012). Leveraging rebranding of 'unattractive' nation brands to stimulate post-disaster tourism. *Tourist Studies, 12*(1), 87–105. doi:10.1177/1468797612444196

Anar, T. (2020). *Huzur atlası – Ahmet Hamdi Tanpınar'ın Huzur romanını okuma kılavuzu.* Kapı Yayınları.

Anderson, R. (2007). *Thematic Content Analysis (TCA).* Retrieved from www.rosemarieanderson.com/wp-content/uploads/2014/08/ThematicContentAnalysis.pdf

Andriessen, D. (2004). *Making sense of intellectual capital: Designing a method for the valuation of intangibles.* Routledge. doi:10.4324/9780080510712

Andriotis, K. (2011). Genres of heritage authenticity: Denotations from a pilgrimage landscape. *Annals of Tourism Research, 38*(4), 1613–1633. doi:10.1016/j.annals.2011.03.001

Andsager, J. L., & Drzewiecka, J. A. (2002). Desirability of differences in destinations. *Annals of Tourism Research, 29*(2), 401–421. doi:10.1016/S0160-7383(01)00064-0

Araújo Vila, N., Fraiz Brea, J. A., & de Carlos, P. (2021). Film tourism in Spain: Destination awareness and visit motivation as determinants to visit places seen in TV series. *European Research on Management and Business Economics, 27*(1), 100135. Advance online publication. doi:10.1016/j.iedeen.2020.100135

Arcila Garrido, M., & López Sanchez, J. A. (2011). La cartografía cultural como instrumento para la planificación y gestión cultural. Una perspectiva geográfica. *Periférica Internacional. Revista Para El Análisis de La Cultura y El Territorio, 1*(12), 15–36. https://revistas.uca.es/index.php/periferica/article/view/1705

Arcos-Pumarola, J., Osácar Marzal, E., & Llonch-Molin, N. (2018). Literary urban landscape in a sustainable tourism context. *Human Geographies – Journal of Studies and Research in Human Geography, 12*(2), 175-189.

Arcos-Pumarola, J. (2019). Assessing Literary Heritage Policies in the Context of Creative Cities. *Journal of Spatial and Organizational Dynamics, 7*(4), 275–290.

Arcos-Pumarola, J., Marzal, E. O., & Llonch-Molina, N. (2018). Literary urban landscape in a sustainable tourism context. *Human Geographies, 12*(2), 175–189. doi:10.5719/hgeo.2018.122.3

Argod, P. (2018). From the Road Movie to Wilderness Travel: The Quest for Adventure on the Road and the Myth of the Heroic Traveller. *Via. Tourism Review*, 14.

Arnold, N. (2019). *More than Middle-Earth: How Film Tourism Changed New Zealand.* Discovery. Retrieved 14 March 2021, from https://discovery.cathaypacific.com/middle-earth-film-tourism-changed-new-zealand/

Arthington, A. H., Dulvy, N. K., Gladstone, W., & Winfield, I. J. (2016). Fish conservation in freshwater and marine realms: Status, threats and management. *Aquatic Conservation, 26*(5), 838–857. doi:10.1002/aqc.2712

Árva, L., & Deli-Gray, Zs. (2011). Posztmodern turizmus, a posztmodern marketing [Postmodern tourism, postmodern marketing]. In P. Róbert (Ed.), The socio-economic situation in Hungary in the first decades of the 21st century (pp. 1–11). Academic Press.

Árva, L., & Deli-Gray, Zs. (2011). Posztmodern turizmus, a posztmodern marketing [Postmodern tourism, postmodern marketing]. *Magyarország társadalmi-gazdasági helyzete a 21. század első évtizedeiben*, 1-11.

Ascherson, N. (2002). *Stone voices: The search for Scotland.* Hill and Wang.

Ashworth, G. J. (2009). Do Tourists Destroy the Heritage They Have Come to Experience? *Tourism Recreation Research, 34*(1), 79–83. doi:10.1080/02508281.2009.11081577

Atay, L. (2008). Alan Kılavuzluğunun Çanakkale ilindeki uygulaması ve ilgili turizm mevzuatı açısından değerlendirilmesi. *Anatolia: Turizm Araştırmaları Dergisi, 19*(2), 169–176.

Atsız, O., Cifci, I., & Rasoolimanesh, S. M. (2021). Exploring the components of meal-sharing experiences with local foods: A netnography approach. *Current Issues in Tourism*, 1–18.

Attenborough, D. (2011). This heaving planet. *New Statesman (London, England)*, *140*, 28–32.

Attenborough, D. (2020). *A Life on Our Planet: My Witness Statement and a Vision for the Future*. Random House.

Attenborough, D., & Wright, D. (1987). *The First Eden: The Mediterranean world and man*. Collins.

Augé, M. (1995). *Non-places: Introduction to an anthropology of Supermodernity*. Verso.

Augros, J., & Kitsopanidou, K. (2009). *L'économie du cinéma américain: histoire d'une industrie culturelle et de ses stratégies*. Armand Collin.

Ayvazoğlu, B. (2019). *Fikret: Kendi cevvim, kendi eflâkimde kendim tâirim*. Everest Publishing.

Bahaire, T., & Elliott-White, M. (1999). The Application of Geographical Information Systems in Sustainable Tourism Planning. *Journal of Sustainable Tourism*, *7*(2), 159–174. doi:10.1080/09669589908667333

Baird, M. (2020). *Television and Me: The Memoirs of John Logie Baird*. Berlin Ltd.

Baker, B. (2012). Branding and the opportunities of movies: Australia. In N. Morgan, A. Pritchard, & R. Pride (Eds.), *Destination Brands* (pp. 239–250). Routledge.

Bakiewicz, J., Leask, A., Barron, P., & Rakic, T. (2016). Exploring film-induced tourism. In M. Kozak & N. Kozak (Eds.), *Destination Marketing: An international perspective* (pp. 97–109). Routledge.

Bąkiewicz, J., Leask, A., Barron, P., & Rakić, T. (2017). Management Challenges at Film-Induced Tourism Heritage Attractions. *Tourism Planning & Development*, *14*(4), 548–566. doi:10.1080/21568316.2017.1303540

Baldacchino, G. (2015). Small island states and territories: Vulnerable, resilient, but also doggedly perseverant and cleverly opportunistic. In *Entrepreneurship in small island states and territories* (pp. 23–50). Routledge. doi:10.4324/9781315764528-7

Baldwin, F., & Sharpley, R. (2009). Battlefield tourism: Bringing organized violence back to life. In R. Sharpley & P. R. Stone (Eds.), *The darker side of travel: The theory and practice of dark tourism* (pp. 186–206). Channel View Publications. doi:10.21832/9781845411169-011

Baleiro, R., & Quinteiro, S. (2018). *Key concepts in literature and tourism studies*. Academic Press.

Baleiro, R., & Quinteiro, S. (2018). *Key concepts in literature and tourism studies*. Faculty of Arts and Humanities, University of Lisbon, Centre for Comparative Studies.

Baleiro, R., & Quinteiro, S. (2019). A small place, by Jamaica Kincaid: Envisioning literary tourism in Antigua. *Journal of Tourism and Cultural Change*, *17*(6), 676–688. doi:10.1080/14766825.2018.1529772

Baloglu, S., & McCleary, K. (1999). A model of destination image formation. *Annals of Tourism Research*, *26*(4), 868–897. doi:10.1016/S0160-7383(99)00030-4

Bambi, G., Iacobelli, S., Rossi, G., Pellegrini, P., & Barbari, M. (2019). Rural tourism to promote territories along the ancient roads of communication: Case study of the rediscovery of the St. Francis's ways between Florence and La Verna. *European Countryside*, *11*(3), 462–474. doi:10.2478/euco-2019-0025

Barsanti, C. (2017). *Film Review: Paterson*. http://fj.webedia.us/reviews/film-review-paterson

Basa, B., & Názer, Á. (2014). *Itt forgott - Főszerepben Budapest [Filmed here - Starring Budapest]*. Akadémiai Kiadó.

Bassano, C., Barile, S., Piciocchi, P., Spohrer, J., Iandolo, F., & Fisk, R. (2019). Storytelling about places: Tourism marketing in the digital age. *Cities (London, England)*, *87*, 10–20. doi:10.1016/j.cities.2018.12.025

Basto, A. M. (1940). O Porto Medieval [The Medieval Porto]. *Memórias de Comunicações apresentadas ao Congresso de História Medieval, II Congresso do Mundo Português, 2, 675-708*

Basto, A. M. (1962). Desenvolvimento topográfico da cidade do Porto [Topographical development in Porto]. *História da cidade do Porto, 1*, 116–158.

Bausinger, H. (1990). *Folk culture in a world of technology* (E. Dettmer, Trans.). Indiana University.

Bec, A., Moyle, B., Schaffer, V., & Timms, K. (2021). Virtual reality and mixed reality for second chance tourism. *Tourism Management, 83*, 104256. Advance online publication. doi:10.1016/j.tourman.2020.104256

Beck, J., & Egger, R. (2018). Emotionalise Me: Self-reporting and Arousal Measurements in Virtual Tourism Environments. In B. Stangl & J. Pesonen (Eds.), *Information and Communication Technologies in Tourism 2018*. Springer. doi:10.1007/978-3-319-72923-7_1

Beck, J., Rainoldi, M., & Egger, R. (2019). Virtual reality in tourism: A state-of-the-art review. *Tourism Review, 74*(3), 586–612. doi:10.1108/TR-03-2017-0049

Beeton, S. (2010). The Advance of Film Tourism. Tourism and Hospitality Planning & Development, 7(1).

Beeton, S. (2005). *Film-induced tourism*. Channel View Publications. doi:10.21832/9781845410162

Beeton, S. (2006). Understanding film-induced tourism. *Tourism Analysis, 11*(3), 181–188. doi:10.3727/108354206778689808

Beeton, S. (2010). The Advance of Film Tourism. *Tourism and Hospitality Planning and Development, 7*(1), 1–6. doi:10.1080/14790530903522572

Beeton, S. (2016). *Film-Induced Tourism* (2nd ed.). Channel View Publications. doi:10.21832/9781845415853

Beier, R., & Kramer, J. (2018). A portrait of the Azores: From natural forces to cultural identity. In *Volcanoes of the Azores* (pp. 3–26). Springer. doi:10.1007/978-3-642-32226-6_2

Beigi, H., Hossein, M., & Azani, M. (2015). Literary tourism and place identity of the central plateau of Iran. *Journal of Biodiversity and Environmental Sciences, 5*(6), 309–406.

Bélanger, C. É., & Jolin, L. (2011). The International Organisation of Social Tourism (ISTO) working towards a right to holidays and tourism for all. *Current Issues in Tourism, 14*(5), 475–482. doi:10.1080/13683500.2011.568056

Beltrán, gersón. (2016). *Geolocalización online: la importancia del dónde*. Editorial UOC. https://www.researchgate.net/publication/293824412_Geolocalizacion_online_la_importancia_del_donde

Benchley, P. (2008). *Shark Life: True Stories About Sharks & the Sea*. Yearling.

Bendix, R. (1997). *In search of authenticity: The formation of folklore studies*. The University of Wisconsin.

Benjamin, W. (1994). O Flâneur [The Flaneur] (J. C. M. Barbosa & H. A. Baptista, Trans.). In Obras escolhidas III (pp. 33-65). Brasiliense.

Bentz, J., Lopes, F., Calado, H., & Dearden, P. (2016). Enhancing satisfaction and sustainable management: Whale watching in the Azores. *Tourism Management, 54*, 465–476. doi:10.1016/j.tourman.2015.11.016

Beric, D., Kovacevic, M., Simat, K., & Bozic, S. (2013). Film tourism: A contemporary resource for promoting Serbia. *Turizam, 17*(1), 18–28. doi:10.5937/Turizam1301018B

Bhandari, K. (2008). Touristification of cultural resources: A case study of Robert Burns. *TOURISM Preliminary Communication, 56*(3), 283–293.

Bigné, J. E., Sánchez, M. I., & Sánchez, J. (2001). Tourism image, evaluation variables and after purchase behaviour: Inter-relationship. *Tourism Management, 22*(6), 607–616. doi:10.1016/S0261-5177(01)00035-8

Biran, A., & Buda, D. M. (2018). Unravelling fear of death motives in dark tourism. In P. R. Stone, R. Hartmann, A. V. Seaton, R. Sharpley, & L. White (Eds.), *The Palgrave handbook of dark tourism studies* (pp. 515–532). Palgrave Macmillan. doi:10.1057/978-1-137-47566-4_21

Birtane, Y. (2020). Sözlü ve yazılı kültür bağlamında Çanakkale muharebe anlatıları ve alan kılavuzları. In B. Borlat & B. Sayılır (Eds.), 105. Yılında Çanakkale Muharebelerine Bakış (pp. 475-483). T.C. Kültür ve Turizm Bakanlığı Çanakkale Savaşları Gelibolu Tarihi Alan Başkanlığı.

Blewitt, J. (2011). Critical practice and the public pedagogy of environmental and conservation media. *Environmental Education Research, 17*(6), 719–734. doi:10.1080/13504622.2011.618625

Boje, D. M., & Hillon, M. (2008). Transorganizational development. In T. Cummings (Ed.), *Handbook of organization development* (pp. 651–654). Sage.

Bolan, P., Boy, S., & Bell, J. (2011). We've seen it in the movies, let's see if it's true. *Worldwide Hospitality and Tourism Themes, 3*(2), 102–116. doi:10.1108/17554211111122970

Bolan, P., & Williams, L. (2008). The role of image in service promotion: Focusing on the influence of film on consumer choice within tourism. *International Journal of Consumer Studies, 32*(4), 382–390. doi:10.1111/j.1470-6431.2008.00672.x

Bonarou, C. (2021). The poetics of travel through unravelling visual representations on postcards: A critical semiotics analysis. *Journal of Tourism. Heritage & Services Marketing, 7*(1), 44–53. doi:10.5281/zenodo.4519317

Boorstin, D. J. (1961). *The image: A guide to pseudo-events in America.* Atheneum.

Booth, A. (2008). Revisiting the homes and haunts of Mary Russell Mitford. *Nineteenth-Century Contexts, 30*(1), 39–65. doi:10.1080/08905490801945538

Borrich, B. (2020). Creative Nonfictions. In S. Belasco, T. Gaul, L. Johnson, & M. Soto (Eds.), A Companion to American Literature (Vol 3, pp. 448-464). John Wiley and Sons. doi:10.1002/9781119056157.ch90

Bouvier, J.-C. (2014). Toponymie des côtes insulaires. Les îles du Frioul à Marseille (Pomègues et Ratonneau). *Nouvelle Revue d'Onomastique, 56*(1), 5–26. doi:10.3406/onoma.2014.1799

Braun, V., & Clarke, V. (2006). Using thematic analysis in psychology. *Qualitative Research in Psychology, 3*(2), 77–101. doi:10.1191/1478088706qp063oa

Brent Ritchie, J. R., Wing Sun Tung, V., & Ritchie, R. J. (2011). Tourism experience management research: Emergence, evolution and future directions. *International Journal of Contemporary Hospitality Management, 23*(4), 419–438. doi:10.1108/09596111111129968

Brockington, D. (2008a). Celebrity conservation: Interpreting the Irwins. *Media International Australia, 127*(1), 96–108. doi:10.1177/1329878X0812700113

Brockington, D. (2008b). Powerful environmentalisms: Conservation, celebrity and capitalism. *Media Culture & Society, 30*(4), 551–568. doi:10.1177/0163443708030004070l

Brown, L. (2016). Tourism and pilgrimage: Paying homage to literary heroes. *International Journal of Tourism Research, 18*(2), 167–175. doi:10.1002/jtr.2043

Bruno, G. (2002). *Atlas of Emotion – Journeys in Art, Architecture and Film.* Verso.

Buchholz, S., & Jahn, M. (2005). Space in narrative. In D. Herman, M. Jahn, & M. L. Ryan (Eds.), *Routledge encyclopedia of narrative theory* (pp. 551–555). Routledge.

Buchmann, A., & Frost, W. (2011). Wizards everywhere? Film tourism and the imagining of national identity in New Zealand. In E. Frew & L. White (Eds.), *Tourism and national identities – An international perspective* (pp. 52–64). Routledge.

Buchmann, A., Moore, K., & Fisher, D. (2010). Experiencing film tourism: Authenticity & fellowship. *Annals of Tourism Research*, *37*(1), 229–248. doi:10.1016/j.annals.2009.09.005

Buda, D. M. (2015). *Affective tourism: Dark routes in conflict*. Routledge. doi:10.4324/9781315742786

Buescu, H. C. (1990). *Incidências do olhar: Percepção e representação* [Directions of the gaze: Perception and representation]. Editorial Caminho.

Busby, G., & Devereux, H. (2015). Dark tourism in context: The diary of Anne Frank. *European Journal of Tourism. Hospitality and Recreation*, *6*(1), 27–38.

Busby, G., Ergul, M., & Eng, J. (2013). Film tourism and the lead actor: An exploratory study of the influence on destination image and branding. *Anatolia*, *24*(3), 395–404. doi:10.1080/13032917.2013.783874

Busby, G., & Haines, C. (2013). Doc Martin and film tourism: The creation of destination image. *Tourism (Zagreb)*, *61*(2), 105–120.

Busby, G., & Haines, C. (2013). Doc Martin and film tourism: The creation of destination image. *Tourism (Zagreb)*, *61*, 338–348.

Busby, G., Huang, R., & Jarman, R. (2013). The Stein effect: An alternative film-induced tourism perspective. *International Journal of Tourism Research*, *15*(6), 570–582. doi:10.1002/jtr.1875

Busby, G., & Klug, J. (2001). Movie-induced tourism: The challenge of measurement and other issues. *Journal of Vacation Marketing*, *7*(4), 316–332. doi:10.1177/135676670100700403

Busby, G., & Laviolette, P. (2006). Narratives in the net: Fiction and cornish tourism. *Cornish Studies*, *14*(1), 142–163. doi:10.1386/corn.14.1.142_1

Busby, G., & Shetliffe, E. (2013). Literary tourism in context: Byron and Newstead Abbey. *European Journal of Tourism. Hospitality and Recreation*, *4*(3), 5–45.

Butler, R. (1986). Literature as an influence in shaping the image of tourist destinations. In J. S. Marsh (Ed.), *Canadian studies of parks, recreation and tourism in foreign lands* (pp. 111–132). Trent University.

Butler, R. (1990). The Influence of the Media in Shaping International Tourist Patterns. *Tourism Recreation Research*, *15*(2), 46–53. doi:10.1080/02508281.1990.11014584

Butler, R. (2011). It's only make believe: The implications of fictional and authentic locations in films. *Worldwide Hospitality and Tourism Themes*, *3*(2), 91–101. doi:10.1108/17554211111122961

Butler, R., & Suntikul, W. (2013). Tourism and war: An ill wind. In R. Butler & W. Suntikul (Eds.), *Tourism and war* (pp. 1–11). Routledge. doi:10.4324/9780203107706

Bye, S. G. (2018). Flights by Olga Tokarczuk [Review of the book *Flights*]. *World Literature Today*. https://www.worldliteraturetoday.org/2018/july/flights-olga-tokarczuk

Cabrera, M. J., Arteaga, D. R., Carla, M. S., & Quijano, R. (2009). *Manual para hacer mapas turísticos urbanos, a partir del diagnóstico de áreas gravitacionales y focos urbanos*. Cicyt.

Cacilhas, N. (2020). Roteiro literário Levantado do Chão [Raised from the Ground Literary Roadmap]. *Dos Algarves: A Multidisciplinary eJournal, 37*, 106-122. . doi:10.18089/DAMeJ.2020.37.7

Caldeira, S. P. (2007). A cidade-personagem: Pina Bausch e Ítalo Calvino [The city-character: Pina Bausch and Italo Calvino]. *Revista de Ciências Humanas, 7*(2), 147-162. https://periodicos.ufv.br/RCH/article/view/3536

Campana, S. E., Marks, L., Joyce, W., & Kohler, N. E. (2006). Effects of recreational and commercial fishing on blue sharks (Prionace glauca) in Atlantic Canada, with inferences on the North Atlantic population. *Canadian Journal of Fisheries and Aquatic Sciences, 63*(3), 670–682. doi:10.1139/f05-251

Canary, R., & Kozicki, H. (Eds.). (1978). *The Writing of History: Literary Form and Historical Understanding*. The University of Wisconsin Press.

Candido, A. (1995). *On literature and society* (H. S. Becker, Trans.). Princeton University Press. (Original work published 1976)

CAOP. (2018). *Carta Administrativa Oficial de Portugal* [Official Administrative Map of Portugal]. Available at https://www.dgterritorio.pt/cartografia_e_geodesia/cartografia/carta_administrativa_oficial_de_portugal_caop/

Cardoso, L., Estevão, C., Fernandes, C., & Alves, H. (2017). Film induced tourism: A systematic literature review. *Tourism & Management Studies, 13*(3), 23–30. doi:10.18089/tms.2017.13303

Carey, P. (2000). *True History of the Kelly Gang*. University of Queensland Press.

Carl, D., Kindon, S., & Smith, K. (2007). Tourists' experiences of film locations: New Zealand as 'Middle Earth. *Tourism Geographies, 9*(1), 49–63. doi:10.1080/14616680601092881

Cartwright, W. (2012). Neocartography: Opportunities, Issues and Prospects. *South African Journal of Geomatics, 1*(1), 77–91. doi:10.4314/SAJG.V1I1

Carvache-Franco, M., Carvache-Franco, W., Carvache-Franco, O., Hernández-Lara, A. B., & Buele, C. V. (2020). Segmentation, motivation, and sociodemographic aspects of tourist demand in a coastal marine destination: A case study in Manta (Ecuador). *Current Issues in Tourism, 23*(10), 1234–1247. doi:10.1080/13683500.2019.1600476

Cashman, R. (2000). The heroic outlaw in Irish Folklore and Popular Culture. *Folklore, 111*(2), 191–205. doi:10.1080/00155870020004602

Castanho, R. A., Couto, G., Pimentel, P., Carvalho, C., Sousa, Á., & Velarde, J. G. (2020). Assessing the impacts of public policies over tourism in Azores Islands. A research based on tourists and residents perceptions. *Sustainable Development, 17*(18), 19-20.

Castle, A. (n.d.). *Harry Potter.* Retrieved 13 March 2021, from https://www.alnwickcastle.com/explore/on-screen/harry-potter

Cater, C. (2010). Any closer and you'd be lunch! Interspecies interactions as nature tourism at marine aquaria. *Journal of Ecotourism, 9*(2), 133–148. doi:10.1080/14724040903125039

Çevik, S. (2018b). Sarı yazma diyarında bir edebiyat turizmi mekanı: Rıfat Ilgaz Kültür ve Sanat Evi. In *II. International Congress of Tourism and Cultural Heritage* (pp. 166-178). Şanlıurfa.

Çevik, S. (2017). Kültürel miras kapsamında edebiyat turizmi deneyimi: Sait Faik Abasıyanık'ın izinde otoetnografik bir çalışma. *Turkish Studies, 12*(29), 51–168. doi:10.7827/TurkishStudies.12404

Çevik, S. (2018). Kurgusal Mekânlarda Edebiyat Turizmi Deneyimi: "Masumiyet Müzesi" Örneği. *Art-e Sanat Dergisi, 11*(22), 281–311. doi:10.21602duarte.466075

Çevik, S. (2018a). Interpretation of author-related literary tourism sites: Evidence from three sites in Turkey. *Journal of Tourism Research, 21*, 118–136.

Çevik, S. (2020). Literary tourism as a field of research over the period 1997-2016. *European Journal of Tourism Research, 24*, 2407.

Cevik, S. (2020). Literary Tourism as a field of research over the period 1997-2016. *European Journal of Tourism Research, 24*, 2407. https://ejtr.vumk.edu/index.php/about/article/view/409

Ceylan, N.B. (Director). (2018). *The wild pear tree* (Movie).

Chang, D. Y. (2015). A Study of TV Drama Series, Cultural Proximity and Travel Motivation: Moderation Effect of Enduring Involvement. *International Journal of Tourism Research, 18*(4), 399–408. doi:10.1002/jtr.2058

Charapan, N., & Mikulich, H. (2019). 'Magic realism' of Vilnius: Conceptualization of literary tourist gaze based on Tales of the Old Vilnius by Max Frei. In I. Jenkins & L. A. Lund (Eds.), *Literary tourism: theories, practice and case studies* (pp. 15–27). CABI. doi:10.1079/9781786394590.0015

Charles, R. (2019, 10 October). The Swedish Academy took a year off to fix the Nobel Prize in literature. It's still broken. *The Washington Post*. https://www.washingtonpost.com/entertainment/books/the-swedish-academy-took-a-year-off-to-fix-the-nobel-prize-in-literature-its-still-broken/2019/10/10/23f1b6da-eb7d-11e9-9306-47cb0324fd44_story.html

Chen, C. (2018). Influence of celebrity involvement on place attachment: Role of destination image in film tourism. *Asia Pacific Journal of Tourism Research, 23*(1), 1–14. doi:10.1080/10941665.2017.1394888

Chen, H., & Rahman, I. (2018). Cultural tourism: An analysis of engagement, cultural contact, memorable tourism experience and destination loyalty. *Tourism Management Perspectives, 26*, 153–163. doi:10.1016/j.tmp.2017.10.006

Chernobyl Tour. (2021). Retrieved from: https://www.chernobyl-tour.com/

Chernobylwel.com. (2021). Retrieved from: https://www.chernobylwel.com/

Christou, P. A. (2020). Tourism experiences as the remedy to nostalgia: Conceptualizing the nostalgia and tourism nexus. *Current Issues in Tourism, 23*(5), 612–625. doi:10.1080/13683500.2018.1548582

Coelho, M. H. C. (2003). Bispos e reis: oposições em torno de bens e jurisdições temporais [Bishops and kings: oppositions over temporal goods and jurisdictions]. Lusitania Sacra, 15, 279-287.

Coffin, T., & Cohen, H. (1978). *The Parade of Heroes*. Doubleday.

Coghlan, A., Buckley, R., & Weaver, D. (2012). A framework for analysing awe in tourism experiences. *Annals of Tourism Research, 39*(3), 1710–1714. doi:10.1016/j.annals.2012.03.007

Cohen, E. (1979). A phenomenology of tourist experiences. *Sociology, 13*(2), 179–201. doi:10.1177/003803857901300203

Cohen, E., & Cohen, S. A. (2012). Authentication: Hot and cool. *Annals of Tourism Research, 39*(3), 1295–1314. doi:10.1016/j.annals.2012.03.004

Cohen, R. (1986). History and genre. *New Literary History, 17*(2), 203–218. doi:10.2307/468885

Colby, J. M. (2018). *Orca: How We Came to Know and Love the Ocean's Greatest Predator*. Oxford University Press. doi:10.1093/oso/9780190673093.001.0001

Collins, K. (2017). Calls of the wild? 'Fake' sound effects and cinematic realism in BBC David Attenborough nature documentaries. *The Soundtrack, 10*(1), 59–77. doi:10.1386/ts.10.1.59_1

Collot, M., & Alves, I. (2012). Rumo a uma geografia literária [Pour une géographie littéraire] (I. Alves, Trans.). *Gragoatá, 17*(33), 17–31. doi:10.22409/gragoata.v17i33.33006

Compagnon, A. (1999). *O demônio da teoria: literatura e senso comum* [Le démon de la Théorie: littérature et sens commun] (C. P. B. Mourão, Trans.). Ed. UFMG.

Conarroe, J. (1970). *William Carlos Williams' "Paterson": Language and landscape.* University of Pennsylvania Press. http://www.jstor.org/stable/j.ctv4t80tm

Conley, T. (2006). *Cartographic cinema.* University of Minnesota Press.

Connell, J. (2005). Toddlers, tourism and Tobermory: Destination marketing issues and television-induced tourism. *Tourism Management, 26*(5), 763–776. doi:10.1016/j.tourman.2004.04.010

Connell, J. (2005a). What's the story in Balamory: The impacts of a Children's TV Programme on small tourism enterprises on the Isle of Mull, Scotland. *Journal of Sustainable Tourism, 13*(3), 228–251. doi:10.1080/01434630508668555

Connell, J. (2012). Film Tourism – Evolution, progress and prospects. *Tourism Management, 33*(5), 1007–1029. doi:10.1016/j.tourman.2012.02.008

Connell, J., & Meyer, D. (2009). Balamory revisited: An evaluation of the screen tourism destination-tourist nexus. *Tourism Management, 30*(2), 194–207. doi:10.1016/j.tourman.2008.06.001

Córdoba Azcárate, M. (2011). 'Thanks God, this is not Cancun!' Alternative tourism imaginaries in Yucatan (Mexico). *Journal of Tourism and Cultural Change, 9*(3), 183–200. doi:10.1080/14766825.2011.620118

Corral, C. (2009). La ciudad para el peatón: las ciudades históricas en la movilidad urbana. *Ingeniería y Territorio, 86,* 58–67.

Correia, M., Palma, J., Koldewey, H., & Andrade, J. P. (2014). The use of a non-invasive tool for capture–recapture studies on a seahorse Hippocampus guttulatus population. *Journal of Fish Biology, 84*(4), 872–884. doi:10.1111/jfb.12304 PMID:24502668

Coshall, J. T. (2000). Measurement of tourists' images: The repertory grid approach. *Journal of Travel Research, 39*(1), 85–89. doi:10.1177/004728750003900111

Coudounaris, D. N., & Sthapit, E. (2017). Antecedents of memorable tourism experience related to behavioral intentions. *Psychology and Marketing, 34*(12), 1084–1093.

Couldry, N. (1998). The view from inside the 'simulacrum': Visitors' tales from the set of Coronation Street. *Leisure Studies, 17*(2), 94–107. doi:10.1080/026143698375178

Cousteau, J. Y., & Cousteau, P. (1970). *The shark: splendid savage of the sea* (No. Sirsi a442417). Academic Press.

Cousteau, J. Y., & Dumas, F. (1953). The silent world. Academic Press.

Cousteau, J. Y., & Paccalet, Y. (1987). *La mer blessée: La Méditerranée [Spoilt sea: The Mediterranean].* Flammarion.

Coutinho, B. X. (1962). Arquitetura militar e religiosa [Military and religious architecture]. In História da cidade do Porto [Porto city History]. Academic Press.

Creative England. (2015). *Quantifying film and television tourism in England.* Report for Creative England in association with VisitEngland by Olsberg, SPI.

Cresswell, T. (2006). *On the move*. Routledge.

Croft, J. (2019, 10 October). The Nobel Prize was made for Olga Tokarczuk. *The Paris Review*. https://www.theparisreview.org/blog/2019/10/10/the-nobel-prize-was-made-for-olga-tokarczuk/

Crompton, J. (1979). Motivations of pleasure vacations. *Annals of Tourism Research*, *6*(4), 408–424. doi:10.1016/0160-7383(79)90004-5

Croy, W. G., & Walker, R. D. (2003). Rural tourism and film - Issues for strategic regional development. In D. Hall, L. Roberts, & M. Mitchell (Eds.), New Directions in Rural Tourism (1st ed., pp. 115-133). Ashgate Publishing Limited. doi:10.4324/9781315248097-22

Croy, G. W. (2011). Film Tourism: Sustained economic contributions to destinations. *Worldwide Hospitality and Tourism*, *3*(2), 159–164. doi:10.1108/17554211111123014

Croy, W. G., & Walker, R. D. (2003). Rural tourism and film-issues for strategic regional development. In D. Hall, L. Roberts, & M. Mitchell (Eds.), *New directions in rural tourism* (pp. 115–133). Ashgate Publishing Ltd.

Cruz, J. M. (2009). @-Turismo ativo no Parque Natural da Serra da Estrela. Criação de rotas temáticas [Active tourism in Serra da Estrela Natural Park. Development of tematic routes] [Master Dissertation]. Universidade da Beira Interior.

Crylen, J. (2018). Living in a World without Sun: Jacques Cousteau, Homo aquaticus, and the Dream of Dwelling Undersea. *JCMS: Journal of Cinema and Media Studies*, *58*(1), 1–23. doi:10.1353/cj.2018.0068

Csizmadia, N. (2016). *Geopillanat - A 21. század megismerésének térképe* [Geomoment - Map for understanding the 21st century]. Budapest: L'Harmattan Kiadó.

Culler, J. (1981). The semiotics of tourism. *American Journal of Semiotics*, *1*(1), 127–140. doi:10.5840/ajs198111/25

Culler, J. (1997). *Literary Theory: A Very Short Introduction*. Oxford University Press.

Cynthia, D., & Beeton, S. (2009). Supporting Independent Film Production Through Tourism Collaboration. *Tourism Review International*, *13*(2), 113–119. doi:10.3727/154427209789604624

Dalcastagnè, R. (2003). Sombras da cidade: O espaço na narrativa brasileira contemporânea [City shadows: space in contemporary Brazilian narrative]. *Estudos de Literatura Brasileira Contemporânea*, (21), 33–53. https://periodicos.unb.br/index.php/estudos/article/view/8928

Dallen, J. T. (2011). *Cultural Heritage and Tourism – An Introduction*. Channel View Publications.

Dann, G. (1999). Writing out the tourist in space and time. *Annals of Tourism Research*, *26*(1), 159–187. doi:10.1016/S0160-7383(98)00076-0

Dann, G. M. S. (1977). Anomie, Ego-enhancement and Tourism. *Annals of Tourism Research*, *4*(4), 184–194. doi:10.1016/0160-7383(77)90037-8

Dann, G. M. S. (1981). Tourist Motivation: An Appraisal. *Annals of Tourism Research*, *8*(2), 187–219. doi:10.1016/0160-7383(81)90082-7

Datta, S. (2000). Globalisation and representations of women in Indian cinema. *Social Scientist*, *28*(3,4), 71-82. http://www.jstor.org/stable/3518191

David, S. N. (2007). *Cartas de amor à Viscondeza da Luz* [Love letters to the Viscountess of Luz]. Famalicão: Edições Quasi.

Davies, O. (2018). *A supernatural war: Magic, divination, and faith during the First World War*. Oxford University Press.

De Laget, P. (1954). *Le Château d'If*. Imprimerie du Midi.

de Miranda, A. C. (2013). *Pensar o local: gênero e espaço urbano na narrativa brasileira contemporânea* [Doctoral Thesis, Universidade de Brasília]. https://repositorio.unb.br/bitstream/10482/12962/1/2013_AdelaideCalhmanMiranda.pdf

del Bosque, I. A. R., Martín, H. S., Collado, J., & de los Salmones, M. M. G. (2009). A framework for tourist expectations. *International Journal of Culture, Tourism and Hospitality Research, 3*(2), 139–147. doi:10.1108/17506180910962140

Delac, C. (Producer) & Pouctal, H. (Director). (1918). *Le Comte de Monte-Cristo* [Film]. Pathé frères.

Delahaye, J. (2020). *Magical Harry Potter themed breaks in the UK that all muggles need on their radars*. Mirror. Retrieved 14 March 2021, from https://www.mirror.co.uk/travel/uk-ireland/harry-potter-themed-breaks-uk-21658845

Demirer, D. (2015). *Kültürel mirasın sürdürülebilirliği kapsamında edebiyat turizminin yönetici ve tüketici perspektifinden değerlendirilmesine yönelik bir araştırma* [Unpublished master dissertation]. Düzce University, Akçakoca.

Denny, L., Shelley, B., & Ooi, C. S. (2019). Education, jobs and the political economy of tourism: Expectations and realities in the case of Tasmania. *Australasian Journal of Regional Studies, 25*(2), 282–305.

Derrett, R. (2001). Special interest tourism: Starting with the individual. In N. Douglas & R. Derret (Eds.), *Special interest tourism* (pp. 1–28). Wiley.

Dimmock, K., & Cummins, T. (2013). History of scuba diving tourism. *Scuba Diving Tourism*, 14-28.

DMPU. (2018). Pessoas e Atividades. Dinâmicas Económicas. Relatório de Caracterização e Diagnóstico [People and activities. Economic Dynamics. Diagnosis and caracpterizazion report]. In *Revisão do Plano Diretor Municipal do Porto* [Revision of the Porto Municipal Master Plan]. Direção Municipal de Urbanismo, Departamento Municipal de Planeamento Urbano.

Domínguez-Azcue, J., Almeida-García, F., Pérez-Tapia, G., & Cestino-González, E. (2021). Films and Destinations—Towards a Film Destination: A Review. *Information (Basel), 12*(1), 39. doi:10.3390/info12010039

Domínguez-Mujica, J. (2007). La cartografía en la promoción turística de Canarias (1880-1970). *Boletín de La Asociación de Geógrafos Españoles*. https://www.researchgate.net/publication/28202137_La_cartografia_en_la_promocion_turistica_de_Canarias_1880-1970

Donaire, J. A. (2008). *Turisme cultural. Entre l'experiència i el ritual*. Edicions Vitel·la.

Donaldson, C., Gregory, I., & Murrieta-Flores, P. (2015). Mapping 'Wordsworthshire': A GIS Study of Literary Tourism in Victorian Lakeland. *Journal of Victorian Culture: JVC, 20*(3), 287–307. doi:10.1080/13555502.2015.1058089

Dore, L. (2006). Gallipoli: A visitor profile. *The Historic Environment, 9*(2), 46–51.

Drumm, A. (2006). Tourism-based revenue generation for conservation. In *Tourism and Protected Areas: Benefits beyond Boundaries*. The Vth IUCN World Parks Congress.

Dubois, L. E., & Gibbs, C. (2018). Video game–induced tourism: A new frontier for destination marketers. *Tourism Review, 73*(2), 186–198. doi:10.1108/TR-07-2017-0115

Dubois, L.-E., Griffin, T., Gibbs, C., & Guttentag, D. (2021). The impact of video games on destination image. *Current Issues in Tourism, 24*(4), 554–566. doi:10.1080/13683500.2020.1724082

DumasA. (1861). *Bric à brac*. https://www.dumaspere.com/pages/bibliotheque/chapitre.php?lid=m5&cid=7

Dumas, A. (1981). L'état civil du Comte de Monte-Cristo. In *Le Comte de Monte-Cristo* (pp. 1429–1436). Gallimard. (Original work published 1867)

Dumas, A. (2008). *The Count of Monte-Cristo* (D. Coward, Trans.). Oxford University Press. (Original work published 1846) doi:10.1093/owc/9780199219650.001.0001

Duncan, P. D. (2018). (Science) Fiction: Genre Hybridization in Jacques-Yves Cousteau and Louis Malle's The Silent World (1956). *The Journal of Popular Film and Television*, *46*(2), 108–117. doi:10.1080/01956051.2018.1465886

Dunkley, R., Morgan, N., & Westwood, S. (2011). Visiting the trenches: Exploring meanings and motivations in battlefield tourism. *Tourism Management*, *32*(4), 860–868. doi:10.1016/j.tourman.2010.07.011

Dunn, C. (2006). *The cultural phenomenon of Harry Potter.* The Stanford Daily. Retrieved 13 March 2021, from https://www.stanforddaily.com/2016/10/24/the-cultural-phenomenon-of-harry-potter/

Dunne, G. (2009). *Motivation and Decision Making in City Break Travel: The Case of Dublin.* VDM Verlag Dr Muller Aktiengesellschaft & Co.

Earl, B. (2008). Literary tourism: Constructions of value, celebrity and distinction. *International Journal of Cultural Studies*, *11*(4), 401–417. doi:10.1177/1367877908096003

Edensor, T. (2001). Performing tourism, staging tourism: (Re)producing tourist space and practice. *Tourist Studies*, *1*(1), 59–81. doi:10.1177/146879760100100104

Edensor, T. (2005). Mediating William Wallace: audio-visual technologies in tourism. In D. Crouch, R. Jackson, & F. Thompson (Eds.), *The media and the tourist imagination. Converging cultures* (pp. 105–118). Routledge.

Edensor, T., & Sumartojo, S. (2015). Designing atmospheres: Introduction to Special Issue. *Visual Communication*, *14*(3), 251–265. doi:10.1177/1470357215582305

Ehrat, J. (2005). *Cinema and semiotic: Peirce and film aesthetics, narration, and representation.* University of Toronto Press.

Elkin, L. (2016). *Flâneuse: women walk the city in Paris.* Farrar, Straus and Giroux.

Elliott, A., & Radford, D. (2015). Terminal experimentation: The transformation of experiences, events and escapes at global airports. *Environment & Planning*, *33*(6), 1063–1079. doi:10.1177/0263775815595407

Engelmo-Moriche, A., Nieto-Masot, A., Horcajo-Romo, A., & Cárdenas-Alonso, G. (2021). Tourist geoportal of a rural region as a means of disseminating its offer and resources. Case study: Extremadura. *Revista de Estudios Andaluces*, *41*, 125–143. doi:10.12795/rea.2021.i41.07

Engin, V. (2006). Tevfik Fikret'in Mekteb-i Sultani müdürlüğü ve eğitimci kişiliği. In A. Yaraman (Ed.), *Biyografya 7: Tevfik Fikret* (pp. 237–252). Bağlam Publishing.

Eom, T., Han, H., & Song, H. (2020). Discovering the perceived attributes of CBT destination travelers in South Korea: A mixed method approach. *Tourism Management*, *77*, 104013. doi:10.1016/j.tourman.2019.104013

Erkoçi, I. (2016). Ismail Kadare as tourist attraction. *Dos Algarves: A Multidisciplinary e-Journal*, *28*, 1-12.

Erll, A. (2008). Literature, film, and the mediality of cultural memory. In A. Erll & A. Nünning (Eds.), *Media and cultural memory* (pp. 389–398). Walter de Gruyter.

Erll, A. (2018). Homer: A relational mnemohistory. *Memory Studies*, *11*(3), 274–286. doi:10.1177/1750698018771858

Espais Escrits. (2018). *Mapa Literari Català.* Retrieved from: http://www.mapaliterari.cat

Fagence, M. (2019). Creating heritage for tourism: 'consuming history', 'prosthetic memories' and the popularization of a folk hero's story. In C. Palmer & J. Tivers (Eds.), *Creating Heritage for Tourism* (pp. 101–114). Routledge.

Farmaki, A., Altinay, L., & Yaşarata, M. (2016). Rhetoric versus the realities of sustainable tourism: The case of Cyprus. In P. Modica & M. Uysal (Eds.), *Sustainable island tourism: Competitiveness, and quality-of-life* (pp. 35–50). CABI. doi:10.1079/9781780645421.0035

Farrelly, F., Kock, F., & Josiassen, A. (2019). Cultural heritage authenticity: A producer view. *Annals of Tourism Research, 79*, 102770. doi:10.1016/j.annals.2019.102770

Farsari, Y., & Prastacos, P. (2008). GIS Applications in the Planning and Management of Tourism. In A. Lew, M. Hall, & A. Williams (Eds.), *A Companion to Tourism* (pp. 596–607). Blackwell.

Faullant, R., Matzler, K., & Mooradian, T. A. (2011). Personality, basic emotions, and satisfaction: Primary emotions in the mountaineering experience. *Tourism Management, 32*(6), 1423–1430. doi:10.1016/j.tourman.2011.01.004

Fawcett, C., & Cormack, P. (2001). Guarding authenticity at literary tourism sites. *Annals of Tourism Research, 28*(3), 686–704. doi:10.1016/S0160-7383(00)00062-1

Ferraz, M. (2020). Apresentação. In Mrs. Dalloway (pp. 11-13). Antofágica.

Ferreira, A., Alén, E., Liberato, P., & Liberato, D. (2020). Literary Tourism: A Cultural Trip? *Advances in Tourism. Technology and Smart Systems, 171*, 505–515. doi:10.1007/978-981-15-2024-2_44

Ferreira, M. G. (2008). Teatro ao gosto de Almeida Garrett [Theatre in the style of Almeida Garrett]. *Teatro do Mundo, 03*, 53–60.

Figyelő. (2018). *születésnapját ünnepli a Hatoslottó* [The Six Number Lottery celebrates its 30th birthday]. Retrieved from: https://figyelo.hu/hirek/30-szuletesnapjat-unnepli-a-hatoslotto-21927/

Foley, M., & Lennon, J. (1996). Heart of darkness. *International Journal of Heritage Studies, 2*(4), 195–197. doi:10.1080/13527259608722174

Folker, B. (2006). *Harry Potter and international relations*. Rowman & Littlefield.

Fonseca, A. M. (2020). Dentro do Segredo: Representações da Coreia da Norte na narrativa de viagens de José Luís Peixoto [Inside the Secret: Representations of North Korea in José Luís Peixoto's travel narrative]. *Dos Algarves: A Multidisciplinary e-Journal, 37*, 22-35. doi:10.18089/DAMeJ.2020.37.2

Fontal Merillas, O. (2007). El patrimonio cultural del entorno próximo : Un diseño de sensibilización para secundaria. *Enseñanza de Las Ciencias Sociales, 6*(31), 31–47.

Forsyth, A. (2015). What is a walkable place? The walkability debate in urban design. *URBAN DESIGN International, 20*(4), 274–292. doi:10.1057/udi.2015.22

Fowler, A. (1982). *Kinds of Literature*. Clarendon Press.

Frampton, A. (2014). *Cross-border film production: The neoliberal recolonization of an exotic island by Hollywood pirates* (Doctoral dissertation). Bowling Green State University.

Frevert, U., Eitler, P., Olsen, S., Jensen, U., Pernau, M., Brückenhaus, D., & Häberlen, J. C. (2014). *Learning how to Feel: Children's Literature and Emotional Socialization, 1870-1970*. OUP Oxford.

Gamír, A. (2019). El giro espacial en las humanidades digitales y sus productos cartográficos. *Biblio 3w. Revista Bibliográfica de Geografía y Ciencias Sociales, 0*(1), 1–27. doi:10.1344/b3w.0.2019.28482

Gamma Travel. (2021). Retrieved from: https://gamma-travel.com/

Gammage, B. (1998). Kelly, Edward 'Ned. In G. Davison, J. Hirst, & S. Macintyre (Eds.), *The Oxford Companion to Australian History* (pp. 362–363). Oxford University Press.

García-Hernández, M., de la Calle-Vaquero, M., & Yubero, C. (2017). Cultural heritage and urban tourism: Historic city centres under pressure. *Sustainability (Switzerland)*, *9*(8), 1346. doi:10.3390u9081346

Gariepy, P. (2014). *Gardens of Hell: Battles of the Gallipoli Campaign.* University of Nebraska Press. doi:10.2307/j.ctt1d9nkq4

Garner, R. (2019). Finding Nemo's spaces: Defining and exploring transmedia tourism. *JOMEC Journal*, *0*(14), 11–32. doi:10.18573/jomec.195

Garrett, A. (2011). *O Arco de Sant'Ana* [The Arch of Sant'Ana]. Porto Editora.

Garrison, S., & Wallace, C. (2021). Media Tourism and Its Role in Sustaining Scotland's Tourism Industry. *Sustainability, 13*(11), 6305. doi:10.3390/su13116305

Garrod, B., & Wilson, J. C. (2004). Nature on the edge? Marine ecotourism in peripheral coastal areas. *Journal of Sustainable Tourism*, *12*(2), 95–120. doi:10.1080/09669580408667227

Gartner, W. C. (1993). Image Formation Process. *Journal of Travel & Tourism Marketing*, *2*(2-3), 191–216. doi:10.1300/J073v02n02_12

Gentile, R., & Brown, L. (2015). A Life as a work of art: Literary tourist's motivations and experiences at il vittoriale degli Italiani. *European Journal of Tourism, Hospitality and Recreation*, *6*(2), 25–47.

Gentile, R., & Brown, L. (2015). A life as a work of art: Literary tourists' motivations and experiences at Il Vittoriale degli Italiani European. *Journal of Tourism, Hospitality and Recreation*, *6*(2), 25–47.

Gentile, R., & Brown, L. (2015). A life as a work of art: Literary tourists' motivations and experiences at Il Vittoriale Degli Italiani. *European Journal of Tourism, Hospitality and Recreation*, *6*(2), 25–47.

Gil, A. C. (2008). *Métodos e Técnicas de Pesquisa Social* [Methods and techniques of social research] (6th ed.). Editora Atlas.

Gilmore, J. H., & Pine, B. J. Jr. (2007). *Authenticity: What consumers really want?* Harvard Business School Press.

Gjorgievski, M., & Trpkova, S. (2012). Movie Induced Tourism: A New Tourism Phenomenon. *UTMS Journal of Economics (Skopje)*, 3.

Goodwin, H. (2011). Taking responsibility for tourism. Woodeaton: Goodfellow Publishers Limited.

Gorrini, A., & Bertini, V. (2018). Walkability assessment and tourism cities: The case of Venice. *International Journal of Tourism Cities*, *4*(3), 355–368. doi:10.1108/IJTC-11-2017-0072

Gössling, S. (2018). Tourism, tourist learning and sustainability: An exploratory discussion of complexities, problems and opportunities. *Journal of Sustainable Tourism*, *26*(2), 292–306. doi:10.1080/09669582.2017.1349772

Gotham, K. F. (2015). *Dark Tourism and Disaster Tourism.* Retrieved from: https://onlinelibrary.wiley.com/doi/abs/10.1002/9781405165518.wbeosd045.pub2

Gotham, K. F. (2007). *Dark Tourism and Disaster Tourism. The Blackwell Encyclopedia of Sociology.*

Gottschall, J. (2013). *The storytelling animal: How stories make us human.* Mariner.

Goulding, C. (2000). The commodification of the past, postmodern pastiche, and the search for authentic experiences at contemporary heritage attractions. *European Journal of Marketing*, *34*(7), 835–853. doi:10.1108/03090560010331298

Gouyon, J. B. (2019). *BBC wildlife documentaries in the age of Attenborough*. Springer Nature.

Gouyon, J. B. (2019). Wildlife Television, Empathy and the End of the British Empire. In *BBC Wildlife Documentaries in the Age of Attenborough* (pp. 69–95). Palgrave Macmillan. doi:10.1007/978-3-030-19982-1_4

Graham, B. (2002). Heritage as knowledge: Capital or culture? *Urban Studies (Edinburgh, Scotland)*, *39*(5/6), 1003–1017. doi:10.1080/00420980220128426

Gramsci, A. (2011). *Prison notebooks* (Vol. 2; J. A. Buttigieg, Trans.). Columbia University Press. (Original work published 1950)

Griffin, T., Giberson, J., Lee, S. H., Guttentag, D., Kandaurova, M., Sergueeva, K., & Dimanche, F. (2017). Virtual Reality and Implications for Destination Marketing. *Travel and Tourism Research Association: Advancing Tourism Research Globally*, 29.

Griffin, T. (2002). An optimistic perspective on tourism's sustainability. In R. Harris, T. Griffin, & P. William (Eds.), *Sustainable Tourism* (pp. 24–32). Butterworth Heinmann. doi:10.1016/B978-0-7506-8946-5.50006-X

Grihault, N. (2003). Film tourism-the global picture. *Travel & Tourism Analyst*, *5*, 1–22.

Grönroos, C. (1990). *Service management and marketing: Managing the moments of truth in service competition*. Lexington Books.

Guide@Hand. (2021). *Film Destination Budapest*. Retrieved from: https://guideathand.com/hu/downloads/guide@hand

Gunn, C. A. (1972). *Vacationscape: Designing Tourist Regions*. University of Texas.

Gupta, S. B., & Gupta, S. (2013). Representation of social issues in cinema with specific reference to Indian cinema: Case study of Slumdog Millionaire. *The Marketing Review*, *13*(3), 271–282. doi:10.1362/146934713X13747454353619

Güzel, Ö., & Güzel, B. (2017). The Influence of Films on Destination Attributes (DAs) for Arabian Tourists in Turkey. *International Journal of Business & Information*, *12*(3), 209–242. doi:10.6702/ijbi.2017.12.3.1

Haggarty, L. (1996). What is content analysis? *Medical Teacher*, *18*(2), 99–101. doi:10.3109/01421599609034141

Hahm, J., & Wang, Y. (2011). Film-Induced Tourism as a Vehicle For Destination Marketing: Is it Worth the Efforts? *Journal of Travel & Tourism Marketing*, *28*(2), 165–179. doi:10.1080/10548408.2011.546209

Hall, C. M. (2003). *Tourism Issues, Agenda Setting and the Media, e-review of Tourism Research*. Academic Press.

Hall, C. M., & Weiler, B. (1992). Introduction: What's so special about Special Interest Tourism? In B. Weiler & M. C. Hall (Eds.), *Special interest tourism* (pp. 1–14). Belhaven Press.

Hall, J., & Basarin, V. J. (2009). Motives for attendance at Anzac Day commemoration in Turkey in tourist experiences. In *Meanings, motivations, behaviours conference*. The University of Central Lancashire Press.

Han, B.-C. (2013). Im Schwarm, Ansichten des Digitalen [In the swarm: Views of the digital]. Matthes & Seitz Berlin Verlag.

Hanefors, M., & Mossberg, L. (2002). TV travel shows – A pre-taste of the destination. *Journal of Vacation Marketing*, *8*(3), 235–246. doi:10.1177/135676670200800303

Hanss, D., & Doran, R. (2020). Perceived consumer effectiveness. *Responsible Consumption and Production*, 535-544.

Hao, X., & Ryan, C. (2013). Interpretation, Film Language and Tourist Destinations: A Case Study of Hibiscus Town, China. *Annals of Tourism Research*, *42*, 334–358. doi:10.1016/j.annals.2013.02.016

Happer, C., & Philo, G. (2013). The Role of the Media in the Construction of Public Belief and Social Change. *Journal of Social and Political Psychology*, *1*(1), 321–336. doi:10.5964/jspp.v1i1.96

Hardiman, N., Burgin, S., & Shao, J. (2019). News media portrayal of attributed stakeholder attitudes to shark management in Australia. *Human Dimensions of Wildlife*, *24*(6), 548–563. doi:10.1080/10871209.2019.1663455

Harrison, D., & Sharpley, R. (2017). Introduction. In D. Harrison & R. Sharpley (Eds.), *Mass tourism in a small world* (pp. 1–14). CABI. doi:10.1079/9781780648545.0001

Hart, P. (2011). Gallipoli. Profile Books.

Hartmann, R. (2013). The Anne Frank House in Amsterdam: A museum and literary landscape goes virtual reality. *Journalism and Mass Communication*, *3*(10), 625–644.

Harvey, D. (1989). From managerialism to entrepreneurialism: The transformation in urban governance in late capitalism. *Geografiska Annaler*, *71*(1), 3–17. doi:10.1080/04353684.1989.11879583

Harvey, D. (2007). (Re)creating culture through tourism: Black heritage sites in New Jersey. In M. Smith (Ed.), *Tourism, culture, and regeneration* (pp. 59–68). CABI.

Heinich, N. (1983). L'aura de Walter Benjamin. *Actes de la Recherche en Sciences Sociales*, *49*(1), 107–109. doi:10.3406/arss.1983.2201

Hellman, P. T. (2005). *Historical Gazetteer of the United States*. Routledge.

Helmeczi, B. (2020). *Csernobil: ezért túlzás az HBO sikersorozata* [Chernobyl: why HBO's hit series is an exaggeration]. Retrieved from: https://promotions.hu/orszagos/zene-film-kultura/2020/04/26/csernobil-katasztrofa-34-ev-evfordulo-hbo-sorozat/

Hendrix, H. (2014), Literature and Tourism: Explorations, Reflections, and Challenges. In S. Quinteiro & R. Baleiro (Orgs.), Lit&Tour: Ensaios sobre literatura e turismo [Lit&Tour: Essays on literature and tourism] (pp. 19-30). Húmus.

Hendrix, H. (2014). Literature and tourism: Explorations, reflections, and challenges. In S. Quinteiro & R. Baleiro (Eds.), Lit & Tour: Ensaios sobre literatura e turismo [Lit & Tour: Essays on literature and tourism] (pp. 19-29). Húmus.

Hendrix, H. (2020). The early musealization of writers' and artists' houses through guidebooks. *Nordic Museology*, *1*, 8–22. doi:10.5617/nm.7962

Hendrix, H. (2007). The early-modern invention of literary tourism: Petrarch's houses in France and Italy. In H. Hendrix (Ed.), *Writer's houses and the making of memory* (pp. 15–30). Routledge.

Hendrix, H. (2008). *Writers' houses and the making of memory*. Routledge.

Hendrix, H. (2009). From Early Modern to Romantic Literary Tourism: A Diachronical Perspective. In N. J. Watson (Ed.), *Literary tourism and nineteenth-century culture* (pp. 13–24). Palgrave. doi:10.1057/9780230234109_2

Hendrix, H. (2014). Literature and tourism: explorations, reflections and challenges. In S. Quinteiro & R. Baleiro (Eds.), *Lit&tour: Ensaios sobre literatura e turismo* (pp. 19–30). Húmus.

Herbert, D. (1995). *In heritage tourism and society tourism leisure and recreation*. Mansell Publishing.

Herbert, D. (2001). Literary places, tourism and the heritage experience. *Annals of Tourism Research*, *28*(2), 312333. doi:10.1016/S0160-7383(00)00048-7

Herbert, D. T. (1996). Artistic and literary places in France as tourist attractions. *Tourism Management, 17*(2), 77–85. doi:10.1016/0261-5177(95)00110-7

Hertzog, A. (2012). War, battlefields, tourism, and imagination. *Tourist Imaginaries, 1*, 1-13.

Hesmondhalgh, D. (2007). *The Cultural Industries*. SAGE Publications.

Hladnik, M., & Fridl, J. (2012). Space and its geographical presentation in Slovene historical narratives. *Slavistična Revija, 60*(3), 443–455.

Hofman, K., & Hughes, K. (2018). Protecting the Great Barrier Reef: Analysing the impact of a conservation documentary and post-viewing strategies on long-term conservation behaviour. *Environmental Education Research, 24*(4), 521–536. doi:10.1080/13504622.2017.1303820

Holak, S. L., & Havlena, W. J. (1998). Feelings, fantasies, and memories: An examination of the emotional components of nostalgia. *Journal of Business Research, 42*(3), 217–226. doi:10.1016/S0148-2963(97)00119-7

Hoppen, A., Brown, L., & Fyall, A. (2014). Literary tourism: Opportunities and challenges for the marketing and branding of destinations? *Journal of Destination Marketing & Management, 3*(1), 37–47. doi:10.1016/j.jdmm.2013.12.009

Horzum, S. (2016). Reflections on the Gallipoli Campaign in Turkish literature. In M. Gurcan & R. Johnson (Eds.), *The Gallipoli Campaign The Turkish perspective* (pp. 206–216). Routledge.

Hosany, S., Buzova, D., & Sanz-Blas, S. (2019). The Influence of Place Attachment, Ad-Evoked Positive Affect, and Motivation on Intention to Visit: Imagination Proclivity as a Moderator. *Journal of Travel Research, 59*(3), 477–495. doi:10.1177/0047287519830789

Hovi, T. (2016). *Finding heritage through fiction in Dracula tourism*. Suomalainen Tiedeakatemia/Academia Scientiarum Fennica.

Howkins, J. (2004). *Az alkotás gazdagít: Bátorítás kreatív embereknek [The creative economy. How People Make Money from Ideas]*. HVG.

Hsieh, H.-K., & Shannon, S. (2005). Three approaches to qualitative analysis. *Qualitative Health Research, 15*(9), 1277–1288. doi:10.1177/1049732305276687 PMID:16204405

Hsu, H. C. C., & Huang, S. (2008). Travel Motivation: A Critical Review of the Concept's Development. In A. G. Woodside & D. Martin (Eds.), *Tourism Management: Analysis, Behaviour and Strategy* (pp. 14–27). CABI. doi:10.1079/9781845933234.0014

Hudson, S. (2020). *COVID-19 and Travel: Impacts, Responses and Outcomes*. Goodfellow Publishers Limited. doi:10.23912/9781911635703-4387

Hudson, S., & Ritchie, J. R. B. (2006). Promoting Destinations via Film Tourism: An Empirical Identification of Supporting Marketing Initiatives. *Journal of Travel Research, 44*(4), 387–396. doi:10.1177/0047287506286720

Hudson, S., Wang, Y., & Gil, S. M. (2011). The Influence of a Film on Destination Image and the Desire to Travel: A Cross-Cultural Comparison. *International Journal of Tourism Research, 13*, 177–190.

Hudson, S., Wang, Y., & Gil, S. M. (2011). The influence of a film on destination image and the desire to travel: A cross-cultural comparison. *International Journal of Tourism Research*. Advance online publication. doi:10.1002/jtr.808

Huggan, G. (2019). *Nature's Saviours: Celebrity Conservationists in the Television Age*. Routledge.

Hughes, G. (1995). Authenticity in tourism. *Annals of Tourism Research, 22*(4), 781–803. doi:10.1016/0160-7383(95)00020-X

Hunt, N. (2010). *Memory, war, and trauma.* Cambridge University Press. doi:10.1017/CBO9780511845017

ICOMOS. (1964). *The Venice Charter 1964.* https://www.icomos.org/charters/venice_e.pdf

ICOMOS. (2008). *Québec Declaration on the Preservation of the Spirit of Place.* Available at: https://whc.unesco.org/uploads/activities/documents/activity-646-2.pdf

Iles, J. (2006). Recalling the ghosts of war: Performing tourism on the battlefields of the western front. *Text and Performance Quarterly, 26*(2), 162–180. doi:10.1080/10462930500519374

Imbert-Bouchard Ribera, D., Llonch Molina, N., Martín Piñol, C., & Osácar Marzal, E. (2013). Turismo cultural y apps. Un breve panorama de la situación actual. *Heritage & Museography, 0*(13), 44-54–54.

Inankur, Z. (2016). Ressam Tevfik Fikret. In A. Yersu (Ed.), *Ölümünün 100. Yıldönümünde Tevfik Fikret Sempozyumu* (pp. 33–48). Istanbul Metropolitan Municipality Directorate of Libraries and Museums.

INE. (2021). *Annual Provisional Resident Population Estimates, according to the administrative territorial organisation.* Available at: https://www.ine.pt/xportal/xmain?xpid=INE&xpgid=ine_indicadores&contecto=pi&indOcorrCod=0008273&selTab=tab0&xlang=en

Innovációs és Technológiai Minisztérium. (2019). *Együttműködési megállapodást kötött az ITM és a Moholy-Nagy Művészeti Egyetem* [Cooperation agreement between ITM and Moholy-Nagy University of the Arts]. Retrieved from: https://www.kormany.hu/hu/innovacios-es-technologiai-miniszterium/gazdasagstrategiaert-es-szabalyozasert-felelos-allamtitkar/hirek/egyuttmukodesi-megallapodast-kotott-az-itm-es-a-moholy-nagy-muveszeti-egyetem

Instituto Camões. (2021). *Dia Mundial da Língua Portuguesa, 5 de maio de 2021* [World day of the Portuguese language, May, 5th, 2021]. Available at https://www.instituto-camoes.pt/images/pdf_noticias/Dados_sobre_a_l%C3%ADngua_portuguesa-2021.pdf

Irimiás, A., Papp-Váry, Á., & Szűts, Z. (2013). Filmturizmus [Film tourism]. *Figyelő Hetilap, 2013*(7).

Irimiás, A. (2015). *Filmturizmus. A filmek és televíziós sorozatok turisztikai szerepének és hatásainak geográfiai vizsgálata [Film tourism. Geographical analysis of the role and impact of films and television series on tourism].* Akadémiai Kiadó.

Irimiás, A. (2015). *Filmturizmus. A filmek és televíziós sorozatok turisztikai szerepének és hatásainak geográfiai vizsgálata [Geographical analysis of the role and impact of films and television series in tourism].* Akadémiai Kiadó.

Ismail, M. N. I., Sumarjan, N., Hadi Khan, N. F. A., & Hanafiah, M. H. (2017). Movie Touring: The Influence of Film on Viewers' Travel Intention. In A. Saufi, I. Andilolo, N. Othman, & A. Lew (Eds.), *Balancing Development and Sustainability in Tourism Destinations.* Springer. doi:10.1007/978-981-10-1718-6_11

Iso-Ahola, E. (1982). Towards a social psychology theory of tourism motivation: A rejoinder. *Annals of Tourism Research, 9*(2), 256–262. doi:10.1016/0160-7383(82)90049-4

İstanbul'un edebiyat coğrafyası, Ahmet Hamdi Tanpınar. (2020, May 20). *İbb Tv.* https://ibb.tv/haberler/detay/evine-tabela-asildi

Iternatura. (2021). *Literapolis BCN.* Retrieved from: https://www.iternatura.eu/proyectos/literapolis/

Iwashita, C. (2006). Media representation of the UK as a destination for Japanese tourists: Popular culture and tourism. *Tourist Studies, 6*(1), 59–77. doi:10.1177/1468797606071477

Jain, J., & Rai, S. (2002). *Films and feminism*. Rawat Publications.

Jancewicz, K., & Borowicz, D. (2017). Tourist maps – definition, types and contents. *Polish Cartographical Review, 49*(1), 27–41. doi:10.1515/pcr-2017-0003

Janiskee, R. L. (1996). Historic houses and special events. *Annals of Tourism Research, 23*(2), 398–414. doi:10.1016/0160-7383(95)00069-0

Jansen-Verbeke, M. (2016). Tourismification of cultural landscapes: synergies between tangible and intangible heritage resources. *Proceedings of TCL2016 Conference*, 276-281.

Jarmusch, J. (Director). (2016). *Paterson* [Film]. K5 International, Le Pacte, Animal Kingdom, Inkjet Productions.

Jenkins, I. (2019). Practical applications of literature to the tourist experience and tourist industry. In I. Jenkins & L. A. Lund (Eds.), *Literary tourism: theories, practice and case studies* (pp. 71–72). CABI. doi:10.1079/9781786394590.0000

Jenkins, O. H. (1999). Understanding and measuring tourist destination images. *International Journal of Tourism Research, 1*(1), 1–15. doi:10.1002/(SICI)1522-1970(199901/02)1:1<1::AID-JTR143>3.0.CO;2-L

Jewell, B., & McKinnon, S. (2008). Movie Tourism—A New Form of Cultural Landscape? *Journal of Travel & Tourism Marketing, 24*(2), 153–162. doi:10.1080/10548400802092650

Jia, H. (2009). The construction of literary tourism site. *Tourism (Zagreb), 57*(1), 69–83.

Jiang, L., & Xu, H. (2016). Reading, tourism, and geography consumption in literary places. *Tourism Geographies, 18*(5), 483–502. doi:10.1080/14616688.2016.1217033

Jiang, L., & Xu, H. (2017). The growth of literary places in ancient town tourism destinations: Based on the theories of Bourdieu, Danto and Dickie. *Journal of Tourism and Cultural Change, 15*(3), 213–228. doi:10.1080/14766825.2016.1148153

Jiang, L., & Yu, L. (2019). Consumption of a literary tourism place: A perspective of embodiment. *Tourism Geographies, 22*(1), 127–150. doi:10.1080/14616688.2019.1586985

Jøn, A. A., & Aich, R. S. (2015). Southern shark lore forty years after Jaws: the positioning of sharks within Murihiku, New Zealand. *Australian Folklore, 30*.

Jones, D., & Smith, K. (2005). Middle-earth meets New Zealand: Authenticity and location in the making of the Lord of the Rings. *Journal of Management Studies, 42*(5), 923–945. doi:10.1111/j.1467-6486.2005.00527.x

Jones, I. (2008). *Ned Kelly: a short life*. Lothian Books.

Jones, M. (2007). What should historians do with heroes? Reflections on nineteenth- and twentieth-century Britain. *History Compass, 5*(2), 439–454. doi:10.1111/j.1478-0542.2007.00390.x

Jurkus, E., Taminskas, J., Povilanskas, R., Kontautiene, V., Baltranaitė, E., Dailidė, R., & Urbis, A. (2021). Delivering Tourism Sustainability and Competitiveness in Seaside and Marine Resorts with GIS. *Journal of Marine Science and Engineering, 9*(3), 312. doi:10.3390/jmse9030312

Kaczmarek, J. (2020). The aleatoric aspect of literary tourism: An example from Havana. *Tourism (Zagreb), 30*(1), 31–42.

Kałamucki, K., & Czerny, A. (2015). Role of Cartographic Products in the Promotion of Tourism Based on the Example of Roztocze. *Barometr Regionalny. Analizy i Prognozy, 1*(39), 145–151.

Kaplan, M. (1962). Bir şairin romanı: Huzur. *İstanbul Üniversitesi Edebiyat Fakültesi Türk Dili ve Edebiyatı Dergisi, 12,* 33-86.

Kaplan, C. (1996). *Questions of Travel: Postmodern Discourses of Displacement*. Duke University press.

Kaplan, M. (1987). *Tevfik Fikret: Devir-Şahsiyet-Eser* (2nd ed.). Dergâh Publishing.

Kassabova, K. (2017, 3 June). Flights by Olga Tokarczuk review: The ways of wanderers. *The Guardian*. https://www.theguardian.com/books/2017/jun/03/flights-by-olga-tokarczuk-review

Kassean, H., & Gassita, R. (2013). Exploring tourists push and pull motivations to visit Mauritius as a tourist destination. *African Journal of Hospitality, Tourism and Leisure*, *2*(3), 1–13.

Keegan, W., Moriarty, S., & Duncan, T. (1992). *Marketing*. Prentice Hall.

Keresnyei, K., & Egedy, T. (2015). Adalékok a kreatív gazdaság elméletéhez [Contributions to the theory of the creative economy]. *Földrajzi Közlemények*, 30–42.

Kerstetter, D., Confer, J., & Graefe, A. (2001). An exploration of the specialization concept within the context of heritage tourism. *Journal of Travel Research*, *39*(3), 267–274. doi:10.1177/004728750103900304

Keyes, D. (2020). Amazon has surpassed 150 million Prime subscribers globally. *Business Insider*. https://www.businessinsider.com/amazon-surpasses-150-million-prime-subscribers-2020-2?r=US&IR=T

Kim, H., & Richardson, S. (2003). Motion picture impacts on destination images. *Annals of Tourism Research*, *30*(1), 216–237. doi:10.1016/S0160-7383(02)00062-2

Kim, J. (2014). The antecedents of memorable tourism experiences: The development of a scale to measure the destination attributes associated with memorable experiences. *Tourism Management*, *44*, 34–45. doi:10.1016/j.tourman.2014.02.007

Kim, J. H., Ritchie, J. B., & McCormick, B. (2012). Development of a scale to measure memorable tourism experiences. *Journal of Travel Research*, *51*(1), 12–25. doi:10.1177/0047287510385467

Kim, S. (2010). Extraordinary Experience: Re-enacting and Photographing at Screen Tourism Locations. *Tourism and Hospitality Planning & Development*, *7*(1), 59–75. doi:10.1080/14790530903522630

Kim, S. (2012). Audience involvement and film tourism experiences: Emotional places, emotional experiences. *Tourism Management*, *33*, 387–396.

Kim, S. (2012). Audience involvement and film tourism experiences: Emotional places, emotional experiences. *Tourism Management*, *33*, 387–396. doi:10.1016/j.tourman.2011.04.008

Kim, S. S., Agrusa, J., Lee, H., & Chon, K. (2007). Effects of Korean TV dramas on the flow of Japanese tourists. *Tourism Management*, *28*(5), 1340–1353. doi:10.1016/j.tourman.2007.01.005

Kim, S. S., Kim, S. S., & Petrick, J. (2019). The Effect of Film Nostalgia on Involvement, Familiarity, and Behavioral Intentions. *Journal of Travel Research*, *58*(2), 283–297. doi:10.1177/0047287517746015

Kim, S. S., Lee, C., & Klenosky, D. (2003). The influence of push and pull factors at Korean national parks. *Tourism Management*, *24*(2), 169–180. doi:10.1016/S0261-5177(02)00059-6

Kim, S., & Kim, S. (2017). Segmentation of potential film tourists by film nostalgia and preferred film tourism program. *Journal of Travel & Tourism Marketing*, *35*(3), 285–305. doi:10.1080/10548408.2017.1284705

Kim, S., Long, P., & Robinson, M. (2009). Small Screen, Big Tourism: The role of popular Korean television dramas in South Korean Tourism. *Tourism Geographies*, *11*(3), 308–333. doi:10.1080/14616680903053334

Kim, S., & O'Connor, N. (2011). A cross-cultural study of screen-tourists' profiles. *Worldwide Hospitality and Tourism Themes*, *3*(2), 141–158. doi:10.1108/17554211111123005

King, R. S. (2018). *Writing to the World.* Johns Hopkins University Press.

Kinsella, M. (2011). *Legend-tripping online: Supernatural folklore and the search for Ong's hat.* University of Mississippi Press. doi:10.14325/mississippi/9781604739831.001.0001

Kirshenblatt-Gimblett, B. (1998). *Destination culture: Tourism, museums, and heritage.* University of California Press.

Klamer, A. (2013). The values of cultural heritage. In I. Rizzo & A. Mignosa (Eds.), *Handbook on the economics of cultural heritage* (pp. 421–437). Edward Elgar. doi:10.4337/9780857931009.00035

Klein, L. (2003). Creating virtual product experiences: The role of telepresence. *Journal of Interactive Marketing*, *17*(1), 41–55. doi:10.1002/dir.10046

Klenosky, D. (2002). The "Pull" of Tourism Destinations: A Means-End Investigation. *Journal of Travel Research*, *40*(4), 396–403. doi:10.1177/004728750204000405

Knudsen, D. (2015). Landscape studies and tourism research. In P. Howard, I. Thompson, & E. Waterton (Eds.), *The Routledge companion to landscape studies* (pp. 286–295). Routledge.

Kocak, A. (2013). Savaşı şiirden okumak: Çanakkale zaferinin Türk şirinde akisleri, *Uluslararası Gelibolu Sempozyumu*, 467-479.

Koh, E., & Fakfare, P. (2020). Overcoming" over-tourism": The closure of Maya Bay. *International Journal of Tourism Cities*, *6*(2), 279–296. doi:10.1108/IJTC-02-2019-0023

Konecny, M. (2011). Review: Cartography: challenges and potential in the virtual geographic environments era. Annals of GIS. *Annals of GIS*, *17*(3), 135–146. doi:10.1080/19475683.2011.602027

Korda Filmstúdió. (2021). *About us.* Retrieved from: https://kordastudio.hu/

Korez-Vide, R. (2017). Storytelling in sustainable tourism management: Challenges and opportunities for Slovenia. *Journal of Advanced Management Science*, *5*(5), 380–386. doi:10.18178/joams.5.5.380-386

Kork, Y. (2016). How Film Tourists Experience Destinations. In V. Katsoni & A. Stratigea (Eds.), *Tourism and Culture in the Age of Innovation: Second International Conference IACuDiT, Athens 2015* (pp. 145-56). Springer International Publishing AG. 10.1007/978-3-319-27528-4_10

Koroglu, E. (2007). *Ottoman propaganda and Turkish identity: Literature in Turkey during world war I.* I.B.Tauris.

Krotscheck, C., & Fend, M. (2014). The CCC method in brief. In Cultural Resources Strategy for Sustainable Regional Development-The CCC Method. Verlag.

Kültür ve Turizm Bakanlığı. (2007). *Türkiye turizm stratejisi 2023.* http://www.sp.gov.tr/upload/xSPTemelBelge/files/J4WC8+20070302-17-1.pdf

Kuran, N. (2016). Tevfik Fikret'in aile hayatı ve hayatında sanatın önemi. In A. Yersu (Ed.), *Ölümünün 100. Yıldönümünde Tevfik Fikret Sempozyumu* (pp. 7–22). Istanbul Metropolitan Municipality Directorate of Libraries and Museums.

Kuran, Ö. (2016). Mimar Tevfik Fikret. In A. Yersu (Ed.), *Ölümünün 100. Yıldönümünde Tevfik Fikret Sempozyumu* (pp. 77–90). Istanbul Metropolitan Municipality Directorate of Libraries and Museums.

Landsberg, A. (2004). *Prosthetic Memory: The transformation of American remembrance in the age of mass culture.* Columbia University Press.

Lappin, L. (2015). *The Soul of Place.* Travellers' Tales.

Lastécouères, C. (2014). *"L'Île-prison": insularité, enfermement et pouvoir dans Le Comte de Monte-Cristo d'Alexandre Dumas (1844).* https://journals.openedition.org/criminocorpus/2866

Law, L., Bunnell, T., & Ong, C. E. (2007). The Beach, the gaze and film tourism. *Tourist Studies*, *7*(2), 141–164. doi:10.1177/1468797607083499

Lawless, J. W., & Silva, K. D. (2017). Towards an integrative understanding of 'authenticity' of cultural heritage: An analysis of World Heritage Site Designations in the Asian context. *Journal of Heritage Management*, *1*(2), 148–159. doi:10.1177/2455929616684450

Laws, E., Scott, N., & Parfitt, N. (2002). Synergies in destination image management: A case study and conceptualisation. *International Journal of Tourism Research*, *4*(1), 39–55. doi:10.1002/jtr.353

Le Blanc, G. (2013). Ce qu'imaginer veut dire. *Entrelacs, 10.* https://journals.openedition.org/entrelacs/489

Lee, S., & Bai, B. (2016). Influence of popular culture on special interest tourists' destination image. *Tourism Management*, *52*, 161–169. doi:10.1016/j.tourman.2015.06.019

Lee, S., Scott, D., & Kim, H. (2008). Celebrity fan involvement and destination perceptions. *Annals of Tourism Research*, *35*(3), 809–832. doi:10.1016/j.annals.2008.06.003

Leeworthy, V. R., Maher, T., & Stone, E. A. (2006). Can artificial reefs alter user pressure on adjacent natural reefs? *Bulletin of Marine Science*, *78*(1), 29–38.

Lefebvre, H. (1991). *The production of space.* Blackwell.

Lennon, J., & Foley, M. (2000). *Dark tourism: The attraction of death and disaster.* Continuum.

Lichrou, M., O'Malley, L., & Patterson, M. (2008). Place-product or place narratives(s)? Perspectives in the marketing of tourism destinations. *Journal of Strategic Marketing*, *16*(1), 27–39. doi:10.1080/09652540701794429

Liew, J. F., Sulaiman, Z., Zakwan, N., & Awang, S. R. (2020). Memorable adventure tourism experience: Constructs and consequences. *Asian Journal of Arts. Culture and Tourism*, *2*(3), 14–22.

Li, S., Li, H., Song, H., Lundberg, C., & Shen, S. (2017). The economic impact of on-screen tourism: The case of The Lord of the Rings and the Hobbit. *Tourism Management*, *60*, 177–187.

Litlong. (2021). *Lit Long: Edinburgh.* Retrieved from: https://litlong.org

Liu, Z. (2003). Sustainable tourism development: A critique. *Journal of Sustainable Tourism*, *11*(6), 459–475. doi:10.1080/09669580308667216

Lopes, O. (1981). De «O Arco de Sant'ana» a «Uma Família Inglesa» [From "O Arco de Sant'ana" to "An English Family"]. Separata da «Revista de História», 4, 9.

Lovell, J. (2019). Fairytale authenticity: Historic city tourism, Harry Potter, medievalism and the magical gaze. *Journal of Heritage Tourism*, *14*(5-6), 448–465. doi:10.1080/1743873X.2019.1588282

Lowenthal, D. (2009). *The heritage crusade and the spoils of history* (7th ed.). Cambridge University.

MacCannell, D. (1973). Staged authenticity: Arrangements of social space in tourist settings. *American Journal of Sociology*, *79*(3), 589–603. doi:10.1086/225585

MacCannell, D. (1976). *The tourist: A new theory of the leisure class.* Schocken Books Inc.

MacCannell, D. (1999). *The Tourist: A New Theory of the Leisure Class.* University of California Press.

MacCannell, D. (2003). *El turista. Una nueva teoría de la clase ociosa*. Melusina.

MacFarlane, A. (2012). *The Kelly Gang Unmasked*. Oxford University Press.

Machado, A., Machado, C., Moreira, F. J., Reis, J., Aurindo, M. J., Almeida, M. M., & Faria, R. (2020). Story Map Caminhos d'O conspirador: Um percurso literário na vila de Marvão com Branquinho da Fonseca [Story Map Caminhos d'O conspirador: A literary journey through the town of Marvão with Branquinho da Fonseca]. *Dos Algarves: A Multidisciplinary e-Journal, 37*, 88-105. . doi:10.18089/DAMeJ.2020.37.6

Macionis, N. (2004). Understanding the film induced tourist. In *Proceedings of the international tourism and media conference*. Tourism Research Unit, Monash University.

Macionis, N. (2004). Understanding the Film-Induced Tourist. In W. Frost, G. Croy & S. Beeton (Eds.), *International Tourism and Media Conference Proceedings* (pp. 86-97). Tourism Research Unit, Monash University.

Macionis, N., & Sparks, B. (2009). Film-induced Tourism: An Incidental Experience. *Tourism Review International, 13*(2), 93-101.

Macionis, N. (2004). Understanding the Film-Induced Tourist. In *International Tourism and Media Conference Proceedings* (pp. 24-26). La Trobe University, Monash University and University of Otago.

Macionis, N., & O'Connor, N. (2011). How can the film-induced tourism phenomenon be sustainably managed? *Worldwide Hospitality and Tourism Themes, 3*(2), 87–90. doi:10.1108/17554211111123032

Macionis, N., & Sparks, B. (2009). Film-Induced Tourism: An Incidental Experience. *Tourism Review International, 13*(2), 93–101. doi:10.3727/154427209789604598

Mackie, C. (2017). Further investigation: Gallipoli's rich history of conflict started well before 1915 (conversation). In Gallipoli, Anzac and the Great War (pp. 36-37), LaTrobe University Press.

MacLeod, N. (2020). A faint whiff of cigar': The literary tourists' experience of visiting writers' homes. *Current Issues in Tourism*, 1–16.

MacLeod, N. (2021). 'A faint whiff of cigar': The literary tourist's experience of visiting writers' homes. *Current Issues in Tourism, 24*(9), 1211–1226. doi:10.1080/13683500.2020.1765996

MacLeod, N., Shelley, J., & Morrison, A. M. (2018). The touring reader: Understanding the bibliophile's experience of literary tourism. *Tourism Management, 67*, 388–398. doi:10.1016/j.tourman.2018.02.006

Madariaga, I. (2004). *Urbanismo con perspectiva de género* [Urbanism with a gender perspective]. Instituto Andaluz de La Mujer. https://www.juntadeandalucia.es/institutodelamujer/institutodelamujer/ugen/sites/default/files/documentos/98.pdf

Magadán Díaz, M. & Rivas García, J. (2012). *Estructura, economía y política turística* [Structure, economy and tourism policy]. Septem Ediciones.

Mak, J. (2003). *Tourism and the economy: Understanding the economics of tourism*. University of Hawaii Press. doi:10.1515/9780824844776

Marasco, A. (2020). Beyond virtual cultural tourism: History-living experiences with cinematic virtual reality. *Tourism and Heritage Journal, 2*, 1–16. doi:10.1344/THJ.2020.2.1

Marine-Roig, E. (2015). Identity and authenticity in destination image construction. *Anatolia, 26*(4), 574–587. doi:10.1080/13032917.2015.1040814

Marques, A. M. S. (2012). *O Anacronismo no Romance Histórico Português Oitocentista* [Anachronism in the 19th century Portuguese historical novel]. Edições Afrontamento.

Marques, L., & Cunha, C. (2013). Literary rural tourism entrepreneurship: Case study evidence from Northern Portugal. *Journal of Policy Research in Tourism, Leisure & Events*, *5*(3), 289–303. doi:10.1080/19407963.2013.801158

Martínez Gil, T., López Benito, V., Asensio Brouard, M., & Santacana Mestre, J. (2018). ¿Entretener o aprender? La evaluación de las tecnologías móviles en la Didáctica del Patrimonio [Entertainment or learning? Mobile technologies evaluation for heritage education]. *ENSAYOS. Revista de La Facultad de Educación de Albacete*, *33*, 39–52.

Maslow, A. H. (1970). *Motivation and Personality* (2nd ed.). Harper & Row.

Massey, D. (2009). *Space, place and gender*. University of Minnesota Press.

Mazzoldi, C., Bearzi, G., Brito, C., Carvalho, I., Desiderà, E., Endrizzi, L., Freitas, L., Giacomello, E., Giovos, I., Guidetti, P., Ressurreição, A., Tull, M., & MacDiarmid, A. (2019). From sea monsters to charismatic megafauna: Changes in perception and use of large marine animals. *PLoS One*, *14*(12), e0226810. doi:10.1371/journal.pone.0226810 PMID:31891944

Mcadam, D. (1999). The Value and Scope of Geographical Information Systems in Tourism Management. *Journal of Sustainable Tourism*, *7*(1), 77–92. doi:10.1080/09669589908667327

McClenachan, L., Cooper, A. B., Carpenter, K. E., & Dulvy, N. K. (2012). Extinction risk and bottlenecks in the conservation of charismatic marine species. *Conservation Letters*, *5*(1), 73–80. doi:10.1111/j.1755-263X.2011.00206.x

McClinchey, K. A. (2015). Travel books, place experience and the 'Je ne sais quoi' of Paris. *Tourism Geographies*, *17*(5), 701–718. doi:10.1080/14616688.2015.1066843

McDermott, A. (2001). *The Jerilderie Letter*. Faber and Faber.

McGinn, C. (2004). *Mindsight: Image, dream, meaning*. Harvard University Press.

McIntosh, R. W., Goeldner, C. R., & Ritchie, J. R. B. (1995). *Tourism: Principles, practices, philosophies* (7th ed.). John Wiley and Sons.

McQuilton, J. (1979). *The Kelly Outbreak: The geographical dimensions of social banditry*. Melbourne University Press.

McWha, M. R., Frost, W., Laing, J., & Best, G. (2015). Writing for the anti-tourist? Imagining the contemporary travel magazine reader as an authentic experience seeker. *Current Issues in Tourism*, *19*(1), 85–99. doi:10.1080/13683500.2015.1013525

McWha, M., & Beeton, S. (2018). Engaging with popular media through tourism: A brief introduction. *Tourist Studies*, *18*(3), 257–260. doi:10.1177/1468797618782254

Medynska-Gulij, B. (2003). The effect of cartographic content on tourist map users. *Cartography*, *32*(2), 49–54. doi:10.1080/00690805.2003.9714252

Medynska-Gulij, B. (2008). Point symbols: Investigating principles and originality in cartographic design. *The Cartographic Journal*, *45*(1), 62–67. doi:10.1179/000870408X276602

Meinig, D. W. (2002). O olhar que observa: Dez versões da mesma cena [The gaze that observes: Ten versions of the same scene]. *Espaço e cultura, 13*, 35-46.

Menezes, J. (2020). Casa do Rio Vermelho: From place of memory to literary place - a possibility of sustainable cultural tourism in Salvador, Bahia, Brazil. *Dos Algarves: A Multidisciplinary e-Journal*, *37*, 71-87.

Mesquita, A. C. (2020). Toda a vida em um só dia [All life in one day]. In Mrs. Dalloway (pp. 373-387). Antofágica.

Meza-Arce, M. I., Malpica-Cruz, L., Hoyos-Padilla, M. E., Mojica, F. J., Arredondo-García, M. C., Leyva, C., Zertuche-Chanes, R., & Santana-Morales, O. (2020). Unraveling the white shark observation tourism at Guadalupe Island, Mexico: Actors, needs and sustainability. *Marine Policy*, *119*, 104056. doi:10.1016/j.marpol.2020.104056

Michalkó, G. (2012). *Turizmológia: elméleti alapok [Turismology: theoretical foundations]*. Akadémiai Kiadó.

Midford, S. (2010). From Achilles to Anzac: Heroism in the Dardanelles from antiquity to The Great War. *ASCS 31 Proceedings*, 1-12.

Midford, S. (2017). Gallipoli, Anzac and the Great War. LaTrobe University Press. doi:10.26826/1002

Midford, S. (2011). Constructing the Australian Iliad: Ancient heroes and Anzac diggers in the Dardanelles. *Melbourne Historical Journal*, *2011*(Special Issue), 59–79.

Miles, S. (2017). Sensorial engagement in tourism experiences on the Western Front. In N. J. Saunders & P. Cornish (Eds.), Modern conflict and the senses. Routledge. doi:10.4324/9781315682228-6

Ministry of Forestry and Water Affairs. (2012). *Simulation and information center for the legend of Gallipoli*. Author.

Mintel. (2011). *Literary tourism – international – September 2011*. Mintel International Group.

Misztal, B. A. (2003). *Theories of social remembering*. Open University Press.

Mitchell, M. (2019). *Chernobyl on HBO location: Where is Chernobyl filmed? Where's it set?* Retrieved from: https://www.express.co.uk/showbiz/tv-radio/1123559/Chernobyl-HBO-Sky-location-Where-is-Chernobyl-filmed-set-Lithuania-Moscow-Pripyat-Ukraine?fbclid=IwAR1Jq_pplRcSiiDUjx9nLTW8TZZPZor3XI57r5-NYBqlWBiz7z_vOqlSdJo

Mochocki, M. (2021). Heritage sites and video games: Questions of authenticity and immersion. *Games and Culture*. Advance online publication. doi:10.1177/15554120211005369

Mokras-Grabowska, J. (2013). Folk culture resources as a component of tourism space. *Tourism (Zagreb)*, *23*(2), 43–47.

Molina, G. (2018). Cities literary factory: The mobilization of literature in urban planning in Europe. *Cities (London, England)*, *77*, 4–12. Advance online publication. doi:10.1016/j.cities.2017.08.016

Molony, J. (1980). *Ned Kelly*. Melbourne University Press.

Moon, H., & Han, H. (2018). Destination attributes influencing Chinese travelers' perceptions of experience quality and intentions for island tourism: A case of Jeju Island. *Tourism Management Perspectives*, *28*, 71–82. doi:10.1016/j.tmp.2018.08.002

Morettoni, M. M. (2020). A modernidade como destino e como questão: A cidade em Max Weber e Georg Simmel [Modernity as a destination and as a question: The city in Max Weber and Georg Simmel]. *Revista Ensaios*, *16*, 156–175. doi:10.22409/revistaensaios.v16.47766

Morgan, N., & Pritchard, A. (1998). *Tourism promotion and power: Creating images, creating identities*. John Wiley & Sons Ltd.

Morris, C. (2017). *Secular relics: Narrative objects and material biography in the museums of Darwin, Elgar & Holmes* (PhD Thesis), University of the West of England.

Morris, J. (2019). Screen tourism: how much money is this tourism trend making in the UK? *Tourism Review News*. Retrieved 10 July from https://www.tourism-review.com/film-related-inbound-tourism-in-uk-news11231

Morten, R., Stone, P. R., & Jarratt, D. (2018). Dark tourism as psychogeography: An initial exploration. In The Palgrave Handbook of Dark Tourism Studies (pp. 227–255). London: Palgrave Macmillan.

Moryson, F. (1617). *An itinerary Written by Fynes Moryson Gent* (Vol. 1). Internet Archive. https://archive.org/details/anitinerarycont17morygoog/page/n422/mode/2up?q=petrarch

Moustakas, C. (1990). Heuristic Research: Design, methodology, and applications. *Sage (Atlanta, Ga.)*.

Mukhametzyanova, L. R., Yao, M. K., Emanova, J. G., & Pozdnyakova, K. G. (2017). Aesthetic and semanticaccents of the neo-gothic interpretation in Russia. *Journal of Fundamental and Applied Sciences*, *9*, 7.

Müller, D. K. (2006). Unplanned development of literary tourism in two municipalities in rural Sweden. *Scandinavian Journal of Hospitality and Tourism*, *6*(3), 214–228. doi:10.1080/15022250600667433

Munro, C. (2005). Diving systems. *Methods for the study of marine benthos*, 112-159.

Mura, P., & Sharif, S. (2017). Narrative Analysis in Tourism: A critical review. *Scandinavian Journal of Hospitality and Tourism*, *17*(2), 194–207. doi:10.1080/15022250.2016.1227276

Murrieta-Flores, P., Donaldson, C., & Gregory, I. (2017). GIS and Literary History: Advancing Digital Humanities research through the Spatial Analysis of historical travel writing and topographical literature. *Digital Humanities Quarterly*, *11*(1). Available at https://www.digitalhumanities.org/dhq/vol/11/1/000283/000283.html

Murrieta-Flores, P., & Howell, N. (2017). Towards the Spatial Analysis of Vague and Imaginary Place and Space: Evolving the Spatial Humanities through Medieval Romance. *Journal of Map & Geography Libraries*, *13*(1), 29–57. doi:10.1080/15420353.2017.1307302

Najemy, J. (1993). *Between Friends*. Princeton University Press.

Narlı, M., & Aliağaoğlu, A. (2012). Edebi miras turizmi ve Türkiye'de edebî mekânlar. *Bilig*, *60*, 1–18.

Neace, C. (2015). *Literary Tourism: American Author Museums*. Available at: https://bookriot.com/literary-tourism-american-author-museums/

Neal, J. D., Sirgy, J. M., & Uysal, M. (1999). The role of satisfaction with leisure travel/tourism services and experience in satisfaction with leisure life and overall life. *Journal of Business Research*, *44*(3), 153–163. doi:10.1016/S0148-2963(97)00197-5

Nemzeti Filmintézet. (2017). *Megéri hollywoodi produkciókat az országba csábítani* [It is worth attracting Hollywood productions to the country]. Retrieved from: https://nfi.hu/hu/filmipar/magyarorszagnak-megeri-hollywoodi-produkciokat-az-orszagba-csabitani

Nemzeti Filmintézet. (2019). *110,1 milliárd forintból készültek filmek Magyarországon 2018-ban* [HUF 110.1 billion spent on films in Hungary in 2018]. Retrieved from https://nfi.hu/hu/filmipar/1101-milliard-forintbol-keszultek-filmek-magyarorszagon-2018-ban

Nemzeti Filmintézet. (2021). *Rekordot döntött a magyar filmipar* [The Hungarian film industry sets a record]. Retrieved from: https://nfi.hu/hu/filmipar/rekordot-dontott-a-magyar-filmipar-1

Netflix. (2021). *Q4 2020 Financial Statements*. https://ir.netflix.net/financials/financial-statements/default.aspx

Nezhad, S. F., Eshrati, P., & Eshrati, D. (2015). A definition of authenticity concept in conservation of cultural landscapes. *Archnet-IJAR*, *9*(1), 93–107. doi:10.26687/archnet-ijar.v9i1.473

Nihad, M. (2012). *Büyük harpte Çanakkale seferi [The Gallipoli Campaign in the Great War]*. Türk Şehitlikleri İmar Vakfı.

O'Connor, N., Flanagan, S., & Gilbert, D. (2008). The integration of film-induced tourism and destination branding in Yorkshire, UK. *International Journal of Tourism Research*, *10*(5), 423–437. doi:10.1002/jtr.676

O'Connor, N., Macionis, N., Wearing, S., Buchmann, A., & Jobberns, C. (2011). Free Willy: The whale-watching legacy. *Worldwide Hospitality and Tourism Themes*.

Ocak, A. (1998). *Hıdırellez. In TDV İslâm ansiklopedisi* (Vol. 17). TDV İslâm Araştırmaları Merkezi.

Ofcom. (2020). *Media Nations 2020 UK Report*. https://www.ofcom.org.uk/__data/assets/pdf_file/0010/200503/media-nations-2020-uk-report.pdf

Okay, O. (1998). Huzur. In *TDV İslam Ansiklopedisi* (Vol. 18, pp. 439–441). Türkiye Diyanet Vakfı Yayınları.

Okay, O. (2010). Ahmet Hamdi Tanpınar. In *TDV İslam Ansiklopedisi* (Vol. 39, pp. 567–570). Türkiye Diyanet Vakfı Yayınları.

Okay, O. (2012). *Bir hülya adamının romanı*. Dergah Yayınları.

Oliveira, L. L. (2002). Apresentação. In L. L. Oliveira (Org.), *Cidade: história e desafios* (pp 8-13). Editora FGV. http://cpdoc.fgv.br/producao_intelectual/arq/1264.pdf

Oliveira, M. (2016). Lisboa, Livro de Bordo, de José Cardoso Pires: A construção de um guia cultural [Lisboa, Livro de Bordo, by José Cardoso Pires: The construction of a cultural guide]. In S. Quinteiro, R. Baleiro & I. D. Santos (Eds.), Turistas, viajantes e lugares literários [Tourists, travellers and literary places] (pp. 85-102). Universidade do Algarve.

Oliveira, R. (2013). *Duas muralhas, duas cidades. A História Militar do Porto Medieval* [Two walls, two cities. The militar history of Medieval Porto] [Master Dissertation]. University of Porto.

Oliveira, S. (2017). *Um Porto de encontro entre Turismo e Literatura* [A meeting point between Tourism and Literature] [Master Dissertation]. University of Porto.

Oliveira, M. T., Ramos, J., & Santos, M. N. (2015). An approach to the economic value of diving sites: Artificial versus natural reefs off Sal Island, Cape Verde. *Journal of Applied Ichthyology*, *31*, 86–95. doi:10.1111/jai.12953

Olmstead, K. (2008). *Jacques Cousteau: a life under the sea*. Sterling Publishing Company, Inc.

Olsberg SPI. (2015). *Quantifying Film and Television Tourism in England. Report for Creative England in association with VisitEngland*. Olsberg•SPI.

Onfray, M. (2007). *La theorie du voyage: Poétique de la géographie* [The theory of travel: Poetics of geography]. Livre de Poche.

Onfray, M. (2007). *Théorie du voyage: Poétique de la géographie* [Theory of travel: Poetics of Geography]. Le livre de Poche.

Orams, M. (2002). *Marine tourism: development, impacts and management*. Routledge. doi:10.4324/9780203197110

Orr, A. (2018). Plotting Jane Austen: Heritage sites as fictional worlds in the literary tourist's imagination. *International Journal of Heritage Studies*, *24*(3), 243–255. doi:10.1080/13527258.2017.1378911

Oviedo-García, M., Castellanos-Verdugo, M., Trujillo-García, M., & Mallya, T. (2014). Film-induced tourist motivations. The case of Seville (Spain). *Current Issues in Tourism*, *19*(7), 713–733. doi:10.1080/13683500.2013.872606

Pagotto, A. M. de C. (2010). *A modernidade no ensaísmo político de Octávio Paz* [Masters Dissertation, Universidade Estadual Paulista]. http://hdl.handle.net/11449/93187

Palancsa, A. (2012). Mi a katasztrófaturizmus? [What is disaster tourism?]. *Korunk*, *3*(23).

Papp-Váry, Á., & Marthy, R. (2009). Magyar "Borat"-ok: Budapest és Magyarország imázsa a hollywoodi filmekben [Hungarian "Borats": the image of Budapest and Hungary in Hollywood films]. In "Új marketing kihívások a 21. században – Fenntartható fogyasztás", a Magyar Marketing Szövetség Marketing Oktatók Klubja, Kaposvár. Academic Press.

Papp-Váry, Á. (2009). Országimázs a moziban [Country image in the cinema]. *Marketing & Media*, *4*(9), 13–26.

Papp-Váry, Á. (2009). Országimázs a moziban [Country image in the movies]. *Marketing & Media*, *4*(9), 13–26.

Papp-Váry, Á. (2019). *Országmárkázás. Versenyképes identitás és imázs teremtése [Country branding. Creating a competitive identity and image]*. Akadémiai Kiadó.

Park, H., & Stephenson, M. (2007). *A Critical Analysis of the Symbolic Significance of Heritage Tourism*. Retrieved from https://repository.usp.ac.fj/8692/1/Article_49_HBMSU.pdf

Parramon, P., & Medina, F. X. (2017). ¿Son Las Grandes Producciones Televisivas Un Atractivo Turístico De Base Cultural? Percepciones En Relación Con El Caso De Juego De Tronos En Girona. *International Journal of Scientific Management and Tourism*, *3*, 3–403.

Pascoal, S. (2019). Smart tech para a otimização do turismo literário: O projeto TheRoute. *Dos Algarves: A Multidisciplinary e-Journal*, (35), 57-72.

Patricio Mulero, M., & Rius-Ulldemolins, J. (2017). From creative city to generative governance of the cultural policy system? The case of Barcelona's candidature as UNESCO City of Literature. *City. Cultura e Scuola*, *10*(January), 1–10. doi:10.1016/j.ccs.2017.05.001

Pauls, A. (2017). Prefácio: O fio tênue da ficção [Preface: The fine thread of fiction]. In Mrs. Dalloway. Companhia das Letras.

Paz, O. (1984). *Os Filhos do Barro: do Romantismo à Vanguarda* (O. Savary, Trans.). Nova Fronteira.

Pearce, P. (2005). *Tourist Behaviour. Themes and Conceptual Schemes*. Channel View Publications. doi:10.21832/9781845410247

Pearce, P., & Caltabiano, M. (1983). Inferring Travel Motivation from Travellers Experiences. *Journal of Travel Research*, *22*(2), 16–20. doi:10.1177/004728758302200203

Pearce, P., Morrison, A. M., & Rutledge, J. L. (1998). *Tourism: Bridges across continents*. McGraw-Hill.

Pearce, P., Morrison, A., & Moscardo, G. (2003). Individuals as tourist icons: A developmental and marketing analysis. *Journal of Hospitality & Leisure Marketing*, *10*(1-2), 63–85. doi:10.1300/J150v10n01_05

Pearce, P., & Moscardo, G. (1986). The Concept of Authenticity in Tourist Experiences. *Journal of Sociology (Melbourne, Vic.)*, *22*(1), 121–132. doi:10.1177/144078338602200107

Pepin-Neff, C. L. (2019a). The Rise of Shark "Attack" Discourse. In Flaws (pp. 13-38). Palgrave Macmillan. doi:10.1007/978-3-030-10976-9_2

Pepin-Neff, C. L. (2019b). Considering Sharks from a Post-Jaws Perspective. In *Flaws* (pp. 163–188). Palgrave Macmillan. doi:10.1007/978-3-030-10976-9_8

Pereiro, X. P. (2009). Turismo cultural. Uma visão antropológica [Cultural tourism: An anthropological perspective]. *Pasos (El Sauzal)*.

Pérez Isasi, S. (2018). Hacia un mapa digital de las relaciones literarias ibéricas (1870-1930): Algunas reflexiones teóricas y metodológicas. *Artnodes. Revista de Arte. Ciencia y Tecnología*, *22*, 93–101.

Pérez, Y., & Nello, M. (2008). Los mapas turísticos disponibles a través de Internet. El caso de la Costa Dorada, Tarragona. *Ar@cne: Revista Electrónica de Recursos En Internet Sobre Geografía y Ciencias Sociales*. https://www.raco.cat/index.php/Aracne/article/view/114766/143621

Pessoa, F. (1982/2002). *The book of disquiet* (R. Zenith, Trans.). Penguin Classics.

Peters, M., Schuckert, M., Chon, K., & Schatzmann, C. (2011). Empire and Romance: Movie-Induced Tourism and the Case of the Sissi Movies. *Tourism Recreation Research*, *36*(2), 169–180. doi:10.1080/02508281.2011.11081317

Peterson, R. A. (2005). In search of authenticity. *Journal of Management Studies*, *42*(5), 1083–1098. doi:10.1111/j.1467-6486.2005.00533.x

Petroman, I., Petroman, C., & Brătulescu, M. (2008). Virtual tourism: Balzac's house in Paris (France). *Research Journal of Agricultural Science*, *40*(3), 303–306.

Piaceski, D. P. F. (2019). Maria Valéria Rezende: Colorindo invisíveis por meio da literatura [Maria Valéria Rezende: coloring the invisible through literature]. *Revista Crioula*, *24*(24), 250–267. doi:10.11606/issn.1981-7169.crioula.2019.160624

Pike, S. (2002). Destination image analysis: A review of 142 papers from 1973 to 2000. *Tourism Management*, *23*(5), 541–549. doi:10.1016/S0261-5177(02)00005-5

Pike, S., & Ryan, C. (2004). Destination Positioning Analysis through a Comparison of Cognitive, Affective, and Conative Perceptions. *Journal of Travel Research*, *42*(4), 333–342. doi:10.1177/0047287504263029

Pimentel, P., Oliveira, A., Couto, G., Ponte, J. C., & Castanho, R. (2020). The Azores Archipelago as a Region with Vast Potential for the Development of Adventure and Slow Tourism. In Peripheral Territories, Tourism, and Regional Development. IntechOpen.

Planells Costa, M., & Crespi Vallbona, M. (2014). *Sistemas y servicios de información turística*. Sintesis.

Pocock, D. (1992). Catherine Cookson Country: Tourist expectation and experience. *Geography (Sheffield, England)*, *77*(3), 236–243. https://www.jstor.org/stable/40572194

Pollock, G. (1988). *Vision and difference: femininity and the histories of art*. Routledge.

Potočnik-Topler, J., & Špenko, T. (2019). Film tourism as a tool of tourism development: The Representation of Scotland in the Outlander TV series. *TIMS. Acta*, *13*(2), 79–88. doi:10.5937/timsacta1902079P

Pratt, S. (2009). A movie map conversion study: A case study of Pride & Prejudice. *Advances In Tourism Destination Marketing: Managing Networks*, *22*, 59–73.

Pratt, S. (2015). The Borat Effect: Film-Induced Tourism Gone Wrong. *Tourism Economics*, *21*, 977–993.

Prayag, G., & Ryan, C. (2012). Antecedents of Tourists' Loyalty to Mauritius. *Journal of Travel Research*, *51*(3), 342–356. doi:10.1177/0047287511410321

Price, J. (1999). The plastic pink flamingo: A natural history. *The American Scholar*, *68*(2), 73–88.

Prideaux, S. (2007). Echoes of war: Battlefield tourism. In C. Ryan (Ed.), *Battlefield tourism: History, place, and interpretation* (pp. 17–27). Elsevier. doi:10.1016/B978-0-08-045362-0.50007-5

Puczkó, L., & Rátz, T. (2005). A turizmus hatásai [The impact of tourism]. Aula kiadó.

Puczkó, L., & Rátz, T. (2011). *Az attrakciótól az élményig [From attraction to experience]*. Akadémiai Kiadó.

Qiao, H., Peterson, A. T., Campbell, L. P., Soberón, J., Ji, L., & Escobar, L. E. (2016). Niche: A Creating virtual species and ecological niches in multivariate environmental scenarios. *Ecography*, *39*(8), 805–813. doi:10.1111/ecog.01961

Quaglieri Domínguez, A., & Paolo Russo, A. (2010). Paisajes urbanos en la época post-turística. Propuesta de un marco analítico. *Scripta Nova, 14*(323).

Quinteiro, S., Carreira, V., & Rodrigues Gonçalves, A. (2020). Coimbra as a literary tourism destination: Landscapes of literature. *International Journal of Culture, Tourism and Hospitality Research, 14*(3), 361–372. doi:10.1108/IJCTHR-10-2019-0176

Radomskaya, V. (2018). *Popular culture as a powerful destination marketing tool: an Australian study* (Doctoral dissertation). James Cook University, https://researchonline.jcu.edu.au/58966

Ramos, J., & Drakeford, B. (2021). Blue Economy and Tourism: Is There a Potential to Create Sustainable Jobs? In Handbook of Research on Human Capital and People Management in the Tourism Industry (pp. 243-264). IGI Global.

Ramos, J., Pinto, P., Pintassilgo, P., Resende, A., & Cancela da Fonseca, L. (2021). Activating an artisanal saltpan: Tourism crowding in or waterbirds crowding out? *International Journal of Culture, Tourism and Hospitality Research.* Advance online publication. doi:10.1108/IJCTHR-04-2021-0101

Ramos, J., Santos, M. N., Whitmarsh, D., & Monteiro, C. C. (2006). The usefulness of the analytic hierarchy process for understanding reef diving choices: A case study. *Bulletin of Marine Science*, *78*(1), 213–219.

Ramsay, J. (2008). *Cultivating Landscape spirits to create an identity.* Retrieved from https://openarchive.icomos.org/id/eprint/126 on 31/01/2021

Ram, Y., & Hall, C. M. (2018). Walking tourism in cities: Introducing the special issue. *International Journal of Tourism Cities*, *4*(3), 281–284. doi:10.1108/IJTC-09-2018-098

Rattanaphinanchai, S., & Rittichainuwat, B. N. (2018). Film–induced tourism in Thailand: An influence of international tourists' intention to visit film shooting location. *International Journal of Tourism Sciences*, *18*(4), 325–332. doi:10.1080/15980634.2018.1551317

Reijnders, S. (2011). *Places of the imagination: media, tourism, culture.* Ashgate Publishing Limited.

Reijnders, S. (2011). *Places of the imagination: Media, tourism, culture.* Ashgate.

Reijnders, S. (2011). Stalking the count: Dracula, fandom and tourism. *Annals of Tourism Research*, *38*(1), 231–248. doi:10.1016/j.annals.2010.08.006

Reijnders, S. (2016). Stories that move: Fiction, imagination, tourism. *European Journal of Cultural Studies*, *19*(6), 672–689. doi:10.1177/1367549415597922

Rennaissance. (2019). *What kids are reading.* Rennaissance. https://www.renaissance.com/learning-analytics/wkar/

Reynaud, D. (2014). National versions of the Great War: Modern Australian Anzac cinema. In M. Loschnigg & M. Sokolowska-Paryz (Eds.), *The Great War in post-memory literature and film* (pp. 289–304). De Gruyter.

Rezende, M. V. (2014a). Bastidores: Há 10 dias uma conhecida ligou no meio da noite [Backstage: 10 days ago an acquaintance called in the middle of the night]. *Pernambuco*, (102), 3. http://www.suplementopernambuco.com.br/images/pdf/PE_102_web.pdf

Rezende, M. V. (2014b). *Quarenta dias* [Forty days]. Objetiva.

Ribeiro, J. M. (2020). Liberalismo e Revolução Liberal de 1820. Novas interpretações [Liberalism and the Liberal Revolution of 1820. New interpretations]. *História. Revista da FLUP*, *10*(2), 3–8. doi:10.21747/0871164X/hist10_2a1

Richards, G. (1996). *Cultural tourism in Europe*. CAB International.

Richards, G. (2001). The market for cultural attractions. In G. Richards (Ed.), *Cultural attractions and European tourism* (pp. 31–54). CAB International. doi:10.1079/9780851994406.0031

Richards, G. (2017). Sharing the new localities of tourism. In D. Dredge & S. Gyimóthy (Eds.), *Collaborative economy and tourism: Perspectives, politics, policies and prospects* (pp. 169–184). Springer. doi:10.1007/978-3-319-51799-5_10

Rickly-Boyd, J. M. (2012). Authenticity & aura: A Benjaminian approach to tourism. *Annals of Tourism Research*, *39*(1), 269–289. doi:10.1016/j.annals.2011.05.003

Rickly, J. M., & McCabe, S. (2017). Authenticity for Tourism Design and Experience. In D. Fesenmaier & Z. Xiang (Eds.), *Design Science in Tourism. Tourism on the Verge*. Springer. doi:10.1007/978-3-319-42773-7_5

Rico, A., Martínez-Blanco, J., Montlleó, M., Rodríguez, G., Tavares, N., Arias, A., & Oliver-Solà, J. (2019). Carbon footprint of tourism in Barcelona. *Tourism Management*, *70*, 491–504. doi:10.1016/j.tourman.2018.09.012

Riesman, C. (2008). Narrative Methods for the Human Sciences. *Sage (Atlanta, Ga.)*.

Riley, R., Baker, D., & Doren, C. S. V. (1998). Movie induced tourism. *Annals of Tourism Research*, *25*, 919–935.

Riley, R., & van Doren, C. (1992). Movies as Tourism Promotion: A 'Pull' Factor' in a 'Push' location. *Tourism Management*, *13*(3), 267–274. doi:10.1016/0261-5177(92)90098-R

Rittichainuwat, B., & Rattanaphinanchai, S. (2015). Applying a mixed method of quantitative and qualitative design in explaining the travel motivation of film tourists in visiting a film-shooting destination. *Tourism Management*, *46*, 136–147. doi:10.1016/j.tourman.2014.06.005

Roberson, R., & Grady, M. (2015). The "Shawshank Trail": A Cross Disciplinary Study in Film Induced Tourism and Fan Culture. *Almatourism*, *6*(4), 47–66. doi:10.6092/issn.2036-5195/4953

Robertson, J. P., & Radford, L. A. (2009). The private uses of quiet grandeur: A mediation on literary pilgrimage. *Changing English*, *16*(2), 203–209. doi:10.1080/13586840902863186

Robinson, M. & Andersen H. C. (Orgs.). (n.d.). *Literature and tourism: Essays in the reading and writing of tourism*. Thomson Learning.

Robinson, M. (2002/2011). Between and beyond the pages: Literature - Tourism Relationships. In M. Robinson & H. C. Andersen (Orgs.), Literature and tourism: Essays in the reading and writing of tourism (pp. 39-79). Thomson Learning.

Robinson, M., & Novelli, M. (2005). Niche tourism: An introduction. *Niche tourism: Contemporary issues, trends and cases,* 1-11.

Robinson, M. (2002). Reading between the lines: Literature and the creation of touristic spaces. *Current Writing: Text and Reception in Southern Africa*, *14*(1), 1–28. doi:10.1080/1013929X.2002.9678112

Robinson, M. (2004). Narratives of being elsewhere: Tourism and travel writing. In A. A. Lew, C. M. Hall, & A. M. Williams (Eds.), *A Companion to tourism* (pp. 303–315). Blackwell Publishing. doi:10.1002/9780470752272.ch24

Robinson, M., & Andersen, H. C. (2002). Literature and tourism: Reading and writing tourism. *Continuum.*

Robinson, M., & Andersen, H.-C. (2002). Literature and tourism: Reading and writing tourism texts. *Continuum.*

Robinson, M., & Novelli, M. (2005). Niche tourism: An introduction. In M. Novelli (Ed.), *Niche tourism: Contemporary issues, trends and cases* (pp. 1–11). Elsevier. doi:10.1016/B978-0-7506-6133-1.50007-X

Rodaway, P. (1994). *Sensuous Geographies*. Routledge. doi:10.4324/9780203306864

Rodrigues, J. (2020). Literatura de turismo sobre o Algarve: O caso dos romances de Rolf Osang [Tourist literature about the Algarve: The case of Rolf Osang's novels]. *Dos Algarves: A Multidisciplinary e-Journal*, *37*, 36-50. doi:10.18089/DAMeJ.2020.37.3

Rodrigues, M. F. (Interviewer) & Rezende, M. V. (Interviewer). (2014, May 2). *Maria Valéria Rezende viveu na rua para escrever romance* [Maria Valéria Rezende lived on the street to write a novel]. Estadão. https://cultura.estadao.com.br/noticias/geral,maria-valeria-rezende-viveu-na-rua-para-escrever-romance,1161541

Roesch, S. (2009). *The Experience of Film Location Tourists*. Channel View Publications. doi:10.21832/9781845411220

Roesch, S. (2018). Film Tourism in the Golden Age of Television. In *The Routledge Handbook of Popular Culture and Tourism*. Routledge. doi:10.4324/9781315559018-8

Roginsky, D. (2007). Folklore, folklorism, and synchronization: Preserved created folklore in Israel. *Journal of Folklore Research*, *44*(1), 41–66. doi:10.2979/JFR.2007.44.1.41

Rosselló, V. (2008). *Cartografia, paisatge i territori. Treballs de La Societat Catalana de Geografia*, 35-49.

Rucińska, D., & Lechowicz, M. (2014). Natural Hazard and Disaster Tourism. *Miscellanea Geographica*, *18*(1), 17–25. doi:10.2478/mgrsd-2014-0002

Ryan, C., Page, J. S., & Aicken, M. (2005). *Taking Tourism to the Limits: Issues, Concepts and Managerial Perspectives*. Elsevier.

Sağıroğlu, M. (2019). Turistik mekanda gerçekliğin algisal değişimi: Modern ve post-modern turizm deneyimlerinde özgünlük. *Planlama*, *29*(2), 90–101. doi:10.14744/planlama.2019.42204

Sakellari, M. (2014). Film tourism and ecotourism: Mutually exclusive or compatible? *International Journal of Culture, Tourism and Hospitality Research*, *8*(2), 194–202. doi:10.1108/IJCTHR-09-2013-0064

Salomon, G. (2012). *Interaction of media, cognition, and learning: An exploration of how symbolic forms cultivate mental skills and affect knowledge acquisition*. Routledge. doi:10.4324/9780203052945

Sampaio, S. (2014). Watching narratives of travel-as-transformation in *The Beach* and *The Motorcycle Diaries*. *Journal of Tourism and Cultural Change*, *12*(2), 184–199. doi:10.1080/14766825.2014.915094

Santana, M. H. (2004). *Edição crítica das Obras de Almeida Garrett* [Critical edition of the Works of Almeida Garrett]. Imprensa Nacional – Casa da Moeda.

Santesso, A. (2004). The birth of the birthplace: Bread Street and literary tourism before Stratford. *English Literary History*, *71*(2), 377–403. doi:10.1353/elh.2004.0032

Santos, C., & Kim, S. (2017). Promoting Tourism, Projecting Power: The Role of Television Commercials. *Journal of Travel & Tourism Marketing*, *34*(2), 192–208. doi:10.1080/10548408.2016.1156610

Sarikanonp, C., & Sahachaisaereep, N. (2010). Graphical design features responding to tourist mapping need: A case of Bangkok's maps for foreign tourists. *Procedia: Social and Behavioral Sciences*, *5*, 1226–1231. doi:10.1016/j.sbspro.2010.07.265

Scarton, F. (2017). Environmental characteristics of shallow bottoms used by Greater flamingo (Phoenicopterus roseus) in a northern Adriatic lagoon. *Acrocephalus*, *38*(174-175), 161–169. doi:10.1515/acro-2017-0010

Schmitz, H. (2016). Atmospheric Spaces. *Ambiances*. https://journals.openedition.org/ambiances/711

Schofield, P. (1996). Cinematographic images of a city: Alternative heritage tourism in Manchester. *Tourism Management*, *17*(5), 333–340. doi:10.1016/0261-5177(96)00033-7

Schofield, P. (2000). Deciphering day trip destination choice using a tourist expectation/satisfaction construct. In A. G. Woodside, G. I. Crouch, J. A. Mazanec, M. Oppermann, & M. Y. Sakai (Eds.), *Consumer Psychology of Travel, Hospitality and Leisure* (pp. 269–293). CAB International.

Schorch, P. (2014). Cultural feelings and the making of meaning. *International Journal of Heritage Studies*, *20*(1), 22–35. doi:10.1080/13527258.2012.709194

Schwartz, D. (2019). *Craig Mazin's Years-Long Obsession with Making' Chernobyl' Terrifyingly Accurate*. Retrieved from: https://www.vice.com/en/article/j5wbq4/craig-mazin-interview-about-chernobyl-hbo-miniseries-on-how-accurate-and-what-really-happened

Seal, G. (1996). *The Outlaw Legend*. Cambridge University Press.

Seal, G. (2001). *Encyclopaedia of Folk Heroes*. ABC-CLIO.

Seal, G. (2004). *Inventing Anzac: The digger and national mythology*. University of Queensland Press.

Seal, G. (2010). Folklore, history, and myth at an Anzac memorial. *Australian Folklore*, *25*, 171–181.

Seal, G. (2011). *Outlaw Heroes in Myth and History*. Anthem Press.

Seaton, A. (1996). Guided by the dark: From thanatopsis to thanatourism. *International Journal of Heritage Studies*, *2*(4), 234–244. doi:10.1080/13527259608722178

Seaton, T. (2009). Thanatourism and its discontents: An appraisal of a decade's work with some future issues and directions. In T. Jamal & M. Robinson (Eds.), *The SAGE handbook of tourism studies* (pp. 522–544). SAGE. doi:10.4135/9780857021076.n29

Secchi, B. (2009). *A cidade do século vinte* [The twentieth century city]. Perspectiva.

Sedikides, C., Wildschut, T., Routledge, C., Arndt, J., Hepper, E. G., & Zhou, X. (2015). To nostalgize: Mixing memory with affect and desire. *Advances in Experimental Social Psychology*, *51*, 189–273. doi:10.1016/bs.aesp.2014.10.001

Seixo, M. A. (2003). "Mau Tempo no Canal": An Iridescent Metaphor. *Portuguese Literary and Cultural Studies*, 129-169.

Sellick, M. C. (2004). Discovery, connection, nostalgia: Key travel motives within the senior market. *Journal of Travel & Tourism Marketing*, *17*(1), 55–71. doi:10.1300/J073v17n01_04

Sétaműhely. (2021). *Sétakínálat* [Offer of Walks]. Retrieved from: https://setamuhely.hu/setak/

Seyfi, S., Hall, C. M., & Rasoolimanesh, S. M. (2020). Exploring memorable cultural tourism experiences. *Journal of Heritage Tourism*, *15*(3), 341–357. doi:10.1080/1743873X.2019.1639717

Shaheer, I. (2021). The nexus of video games and heritage attractions. *Current Issues in Tourism*, 1–5. Advance online publication. doi:10.1080/13683500.2021.1915254

Sharpley, R. (2005). *Travels to the Edge of Darkness: Towards a Typology of Dark Tourism*. Academic Press.

Sharpley, R. (2009). Shedding Light on Dark Tourism: An Introduction. In The Dark Side of Travel: The Theory and Practice of Dark Tourism. Channel View Publications.

Shishmanova, M. V. (2015). Cultural tourism in cultural corridors, itineraries, areas and cores networked. *Procedia: Social and Behavioral Sciences*, *188*, 246–254. doi:10.1016/j.sbspro.2015.03.382

Shi, Y., Bettache, K., Zhang, N., & Xue, L. (2021). Constructing nostalgia in tourism: A comparison analysis of genuine and artificial approaches. *Journal of Destination Marketing & Management*, *19*, 100488. doi:10.1016/j.jdmm.2020.100488

Siegle, L. (2018). *Turning the Tide on Plastic: How Humanity (And You)*. Can Make Our Globe Clean Again.

Signorielli, N., & Morgan, M. (2001). Television and the Family: The Cultivation Perspective. In J. Bryant & J. A. Bryant (Eds.), *Television and the American Family* (2nd ed.). Lawrence Erlbaum Associates, Inc.

Simancas Cruz, M. (2016). *La planificación y gestión territorial del turismo*. Síntesis. https://dialnet.unirioja.es/servlet/libro?codigo=674345

Simmell, G. (1973). [1903]. A metrópole e a vida mental. In *O fenômeno urbano* (pp. 10–24). Zahar Editores.

Simmell, G. (2005). A grande cidade e a vida do espírito (L. Waizbort, Trans.). *Mana*, *11*(2), 577–591. doi:10.1590/S0104-93132005000200010

Skubel, R. A., Shriver-Rice, M., & Maranto, G. M. (2019). Introducing relational values as a tool for shark conservation, science, and management. *Frontiers in Marine Science*, *6*, 53. doi:10.3389/fmars.2019.00053

Slade, P. (2003). Gallipoli thanatourism: The meaning of ANZAC. *Annals of Tourism Research*, *30*(4), 779–794. doi:10.1016/S0160-7383(03)00025-2

Smith, K. A. (2003). Literary enthusiasts as visitors and volunteers. *International Journal of Tourism Research*, *5*(2), 83–95. doi:10.1002/jtr.419

Smith, P. (2017). How the mind works. In *Devotion*. Yale University Books. doi:10.2307/j.ctv1bvnfgr.3

Smith, S. (2013). *And they loved not their lives unto death: The history of Worstead and Westwick's war memorial and war dead*. Tommies Guides.

Smith, V. L. (1998). War and tourism: An American ethnography. *Annals of Tourism Research*, *25*(1), 202–227. doi:10.1016/S0160-7383(97)00086-8

Sontag, S. (1979). *On photography*. Penguin.

Sousa, A. (1994). Tempos medievais [Medieval Times]. In L. A. Oliveira Ramos (Ed.), *História do Porto*. Porto Editora.

Sousa, P. R., & Fernandes, S. B. (2007). Aplicação dos sistemas de informação geográfica no turismo [Application of geographical information systems in tourism]. *Finisterra*, *42*(84), 105–118.

Spengemann, W. (1980). *The Forms of Autobiography*. Yale University Press.

Spigel, L. (2013). *Make Room for TV: Television and the Family Ideal in Postwar America*. The University of Chicago Press.

Squire, S. J. (1994). The cultural values of literary tourism. *Annals of Tourism Research*, *21*(1), 103–120. doi:10.1016/0160-7383(94)90007-8

Squire, S. J. (1996). Literary tourism and sustainable tourism: promoting 'Anne of Green Gables' in Prince Edward Island. *Journal of Sustainable Tourism*, *4*(3), 119–134. doi:10.1080/09669589608667263

Stachowiak, K., & Stryjakiewicz, T. (2018). The rise of film production locations and specialised film services in European semi-peripheries. *Hungarian Geographical Bulletin*, *67*(3), 223–237. doi:10.15201/hungeobull.67.3.2

Stachowiak, K., & Stryjakiewicz, T. (2018). The rise of film production locations and specialized film services in European semi-peripheries. *Hungarian Geographical Bulletin*, *67*(3), 223–237.

Stadler, J. (2015). Conceptualizing and Mapping Geocultural Space. *International Journal of Humanities and Arts Computing*, *9*(2), 133–141. doi:10.3366/ijhac.2015.0145

Stafford, R. (2019). Sustainability: A flawed concept for fisheries management? *Elementa*, *7*, 7. doi:10.1525/elementa.346

Stankova, M., & Vassenska, I. (2015). Raising cultural awareness of local traditions through festival tourism. *Tourism & Management Studies*, *11*(1), 120–127.

Stanley, L. (2015). The Death of the Letter? Epistolary Intent, Letterness and the Many Ends of Letter-writing. *Cultural Sociology*, *9*(2), 240–255. doi:10.1177/1749975515573267

Steiner, C. J., & Reisinger, Y. (2006). Ringing the fourfold: A philosophical framework for thinking about wellness tourism. *Tourism Recreation Research*, *31*(1), 5–14. doi:10.1080/02508281.2006.11081242

Stewart, J. (2016). *Seeking the 'Sites/Sights' Seen on the Small Screen! An investigation of tourist motivation in the context of television film-induced tourism* [Unpublished master thesis]. Dublin Institute of Technology, Dublin, Ireland.

Sthapit, E. (2017). Exploring tourists' memorable food experiences: A study of visitors to Santa's official hometown. *Anatolia*, *28*(3), 404–421. doi:10.1080/13032917.2017.1328607

Stiebel, L. (2004). Hitting the hot spots: Literary tourism as a research field with particular reference to KwaZulu-Natal South Africa. *Critical Arts: South-North Cultural and Media Studies*, *18*(2), 31–44. doi:10.1080/02560240485310151

Stiebel, L. (2007). Going on (literary) pilgrimage: Constructing literary trails with particular reference to KwaZulu-Natal. *Scrutiny*, *12*(1), 93–106. doi:10.1080/18125440701398596

Stiebel, L. (2013). 'When in Rome…?': Literary tourism in Rome from a South African perspective. *Journal of Human Ecology (Delhi, India)*, *43*(1), 93–99. doi:10.1080/09709274.2013.11906615

Stoll, D. S. (2020). A flânerie de uma andarilha urbana [The Flânerie of an Urban Walker]. *Revista Estudos Feministas*, *28*(2), e57230. doi:10.1590/1806-9584-2020v28n157230

Stone, P. (2005). Dark Tourism: An old concept in a new world. *Tourism Magazine*, *4*(25), 20.

Stone, P. R. (2006). A dark tourism spectrum: Towards a typology of deathand macabre related touristsites, attractions and exhibitions. *Tourism Management*, *54*(2), 145–160.

Stone, P., & Sharpley, R. (2008). Consuming dark tourism: A thanatological perspective. *Annals of Tourism Research*, *35*(2), 574–595. doi:10.1016/j.annals.2008.02.003

Sturken, M. (2007). *Tourists of history. Memory, kitsch, and consumerism from Oklahoma City to Ground Zero*. Duke University Press.

Sulyok, J. (2009). Filmturizmus, avagy a filmekben rejlő lehetőségek a turisztikai desztinációk számára [Film tourism or the potential of films for tourist destinations]. *Turizmus Bulletin*, *2009*(1).

Sunsets and Suitcases. (2017). *A trip to Hogwarts (Alnwick Castle)*. Retrieved 15 March 2021, from https://sunsetsandsuitcases.com/trip-hogwarts-alnwick-castle/

Suntikul, W. (2017). Nostalgia-motivated Thai domestic tourism at Amphawa, Thailand. *Asia Pacific Journal of Tourism Research, 22*(10), 1038–1048. doi:10.1080/10941665.2017.1363064

Szerdahelyi, K. (2020). *Jön a Budapest Brand Zrt., új turisztikai stratégia a fővárosban* [Budapest Brand Zrt. is coming; new tourism strategy for the capital]. Retrieved from: https://turizmus.com/szabalyozas-orszagmarketing/jon-a-budapest-brand-zrt-uj-turisztikai-strategia-a-fovarosban-1171906

Tabani, M. (2010). The carnival of custom: Land dives, millenarian parades and other spectacular ritualizations in Vanuatu. *Oceania, 80*(3), 309–329. doi:10.1002/j.1834-4461.2010.tb00088.x

Tally, R. Jr. (2013). *Spatiality*. Routledge.

Tandem Budapest. (2021). *Budapesti túrák* [Budapest Walks]. Retrieved from: http://tandembudapest.hu/budapest/

Tanpınar Edebiyat Araştırmaları ve Uygulama Merkezi. (2017). *Huzur'un İstanbul'u* [Data set]. http://www.tanpinarmerkezi.com/roman-haritalari/

Tanpınar Merkezi. (2013). https://twitter.com/tanpinarmerkezi/media

Tanpınar, A. H. (2001). *Beş Şehir*. Dergah Yayınları.

Tanpınar, A. H. (2013). *Huzur*. Dergah Yayınları.

Taşkıran, C. (2009). 18 Mart Canakkale deniz savaşı. *Atatürk araştırma Merkezi Dergisi Çanakkale Savaşı Özel Sayısı, 73*(25), 25-38.

Taylor, B. (2009). Globalization with Predators and Moving Pictures. In *Dark Green Religion* (pp. 127–154). University of California Press. doi:10.1525/9780520944459

Taylor, J. E., & Pouliot, A. (2020). Introduction: Placing the author in ecologies of literary tourism. *Nineteenth-Century Contexts, 42*(4), 381–389. doi:10.1080/08905495.2020.1782014

Telex.hu. (2021) *Készüljünk fel Hollybudra? Filmes bérgyártásban Magyarország a globális verseny élmezőnyébe tart* [Shall we prepare for Hollybud? Hungary is heading to the top in the global competition in contract film production]. Retrieved from: https://telex.hu/gazdasag/2021/08/04/budapest-kornyeke-mar-ma-is-vezeto-filmgyartasi-kozpont-de-a-nagy-boom-miatt-ujabb-meretes-beruhazasok-indulnak

Teng, H. Y., & Chen, C. Y. (2020). Enhancing celebrity fan-destination relationship in film-induced tourism: The effect of authenticity. *Tourism Management Perspectives, 33*(Jan), 1–11. doi:10.1016/j.tmp.2019.100605

Teng, H.-Y. (2020). Can film tourism experience enhance tourist behavioural intentions? The role of tourist engagement. *Current Issues in Tourism*. Advance online publication. doi:10.1080/13683500.2020.1852196

The Globe Post. (2018). *Numbers of Chernobyl Disaster Zone Visitors Boom*. Retrieved from: https://theglobepost.com/2018/04/25/chernobyl-tourism/

The Inn Collection. (n.d.). *The Hog's Head Inn, Alnwick*. Retrieved 14 March 2021, from https://www.inncollection-group.com/hogs-head-inn/

The Numbers. (2021). *All Time Worldwide Box Office*. Retrieved 9 January 2021, from https://www.the-numbers.com/box-office-records/worldwide/all-movies/cumulative/all-time

Thelen, T., Kim, S., & Scherer, E. (2020). Film tourism impacts: A multi-stakeholder longitudinal approach. *Tourism Recreation Research, 45*(3), 291–306. doi:10.1080/02508281.2020.1718338

Thomas, J. (2012). *Shakespeare's shrine: The bard's birthplace and the invention of Stratford-upon-Avon*. University of Pennsylvania Press. doi:10.9783/9780812206623

Thompson, C. (2011). *Travel writing: The new critical idiom*. Routledge. doi:10.4324/9780203816240

Throsby, D. (2001). *Economics and Culture*. Cambridge University Press.

Throsby, D. (2010). *The economics of cultural policy*. Cambridge University. doi:10.1017/CBO9780511845253

Tian, Y., Lu, D., & Wu, T. (2015). Impacts of awe emotion and perceived value on tourist satisfaction and loyalty—The case of Tibet. *East China Economic Management*, *29*(10), 79–85.

Tiberghien, G., Bremner, H., & Milnec, S. (2020). Authenticity and disorientation in the tourism experience. *Journal of Outdoor Recreation and Tourism*, *30*, 100283. Advance online publication. doi:10.1016/j.jort.2020.100283

Tilden, F. (1977). *Interpreting our heritage* (3rd ed.). The University of North Carolina Press.

Tkalec, M., Zilic, I., & Recher, V. (2017). The effect of film industry on tourism: Game of Thrones and Dubrovnik. *International Journal of Tourism Research*, *19*(6), 705–714. doi:10.1002/jtr.2142

Tokarczuk, O. (2007/2017). *Flights* (J. Croft, Trans.). Fitzcarraldo Editions.

Tolbert, J. (2016). Introduction. In M. D. Foster & J. A. Tolbert (Eds.), *The folkloresque: Reframing folklore in a popular culture world* (pp. 37–41). University Press of Colorado.

Tooke, N., & Baker, M. (1996). Seeing is believing: The effect of film on visitor numbers to screened locations. *Tourism Management*, *17*(2), 87–94. doi:10.1016/0261-5177(95)00111-5

Topuz, H. (2012). *Elbet Sabah Olacaktır* (10th ed.). Remzi Publishing.

Torchin, L. (2002). Location, location, location: The destination of the Manhattan TV tour. *Tourist Studies*, *2*(3), 247–266. doi:10.1177/14687976020023002

Törenek, M. (2011). Tevfik Fikret ve Aşiyan. *Yeni Türk Edebiyatı Araştırmaları*, *3*(5), 7–27.

Torun, F., Doğan, Y., & Gülüm, K. (2020). Seyit Onbaşı'nın Çanakkale savaşlarındaki rolüne yönelik öğrencilerin ve öğretmen adaylarının metaforik algıları: Boylamsal bir araştırma. *Balıkesir Üniversitesi Sosyal Bilimler Enstitüsü Dergisi*, *23*(43), 63–86.

Tóth, R. A. (2016). Széles vászon, nagy haszon [Wide canvas, big benefits]. *Turizmus Trend – Vendég & Hotel, 2016*(5).

Tóth, R. A. (2016). Széles vászon, nagy haszon [Wide screen, big profits]. *Turizmus Trend – Vendég & Hotel, 2016*(5).

Tours Weta Workshop. (n.d.). *The Weta Cave | Weta Workshop Tours*. Retrieved 15 March 2021, from https://tours.wetaworkshop.com/wellington/wetacave/

Trimurti, C. P., & Utama, I. G. B. R. (2020). The examination of the Bali destination marketing chain model based on tourist motivation. *International Journal of Scientific and Technology Research*, *9*(3), 4491–4495.

TripAdvisor. (2021). *About us*. Available at: https://tripadvisor.mediaroom.com/us-about-us

Tuan, Y.-F. (1977/2001). *Space and place: The perspective of experience* (8th ed.). University of Minnesota Press.

Tucker, H., Shelton, E. J., & Bae, H. (2017). Post-disaster tourism: Towards a tourism of transition. *Tourist Studies*, *17*(3), 306–327. doi:10.1177/1468797616671617

Tujaka, D. (2010). La e-cartografía turística, una TIC infrautilizada. In J. L. C. H. A. Jesús Guevara Plaza & A. Luis Aguayo Maldonado (Eds.), *Turitec 2010: VIII Congreso Nacional Turismo y Tecnologías de la Información y las Comunicaciones* (pp. 301–316). https://dialnet-unirioja-es.sire.ub.edu/servlet/articulo?codigo=3814712

Tung, V., & Ritchie, J. (2011). Exploring the essence of memorable tourism experiences. *Annals of Tourism Research*, *38*(4), 1367–1386. doi:10.1016/j.annals.2011.03.009

Turismo Ciudad de Granada. (n.d.) *Ruta de Federico García Lorca*. Retrieved from: http://www.granadatur.com/rutas-tematicas/10-ruta-de-federico-garcia-lorca

Turizmus.com. (2019). *Csernobil már hivatalosan is turisztikai látnivaló* [Chernobyl is now officially a tourist attraction]. Retrieved from: https://turizmus.com/utazas-kozlekedes/csernobil-mar-hivatalosan-is-turisztikai-latnivalo-1165288

Turner, A. J. (2006). Introduction to Neogeography. In A. Turner (Ed.), *OReilly Short Cuts*. OReilly. https://www.oreilly.com/library/view/introduction-to-neogeography/0596529953/index.html

Turp-Balazs, C. (2018). *HBO's Chernobyl highlights Vilnius as a go-to filming location*. Retrieved from: https://emerging-europe.com/after-hours/hbos-chernobyl-highlights-vilnius-as-a-go-to-filming-location/?fbclid=IwAR2KWgdAph0Y-pscG_UDpNvJlN1HK4GDJpVnBiXsESaifa0yNZC7PlbanJ8

Uccella, F. R. (2008). La patrimonialització literària a Catalunya: Una aproximació. *Mnemosine (Rio de Janeiro)*, *5*, 61–72.

Uçman, A. (2016). Bir Servet-i Fünun şairi olarak Tevfik Fikret. In A. Yersu (Ed.), *Ölümünün 100. Yıldönümünde Tevfik Fikret Sempozyumu* (pp. 23–32). Istanbul Metropolitan Municipality Directorate of Libraries and Museums.

UNESCO Barcelona Ciutat de la Literatura. (n.d.). *Mapa Literari*. Retrieved from: https://ajuntament.barcelona.cat/ciutatdelaliteratura/ca/mapa-literari

UNESCO. (1994). *Nara Document on Authenticity*. https://whc.unesco.org/archive/nara94.htm

UNESCO. (1994). *Routes as part of our Cultural Heritage*. Convention concerning the protection of the world cultural and natural heritage, World Heritage Comittee, 18th session. WHC-94/CONF.003/INF.13. Available at: https://whc.unesco.org/archive/routes94.htm

UNESCO. (2003). *Convention for the Safeguarding of the Intangible Cultural Heritage*. Available at: https://ich.unesco.org/en/convention

UNESCO. (2005). Convention on the Protection and Promotion of the Diversity of Cultural Expressions. UNESCO. doi:10.1163/ej.9789004164543.1-0.38

UNESCO. (2017). *Tangible Cultural Heritage*. https://webarchive.unesco.org/20170130173132/http://www.unesco.org/new/en/cairo/culture/tangible-cultural-heritage/

UNESCO. (2018). *Basic Texts of the 2003 Convention for the Safeguarding of the Intangible Cultural Heritage*. https://ich.unesco.org

UNESCO. (n.d.). *Spring celebration, Hıdrellez*. https://ich.unesco.org/en/RL/spring-celebration-hdrellez-01284

United Nations. (2018). *United Nations Conference on Trade Development*. Retrieved from: Creative Economy Outlook: https://unctad.org/en/PublicationsLibrary/ditcted2018d3_en.pdf

UNWTO. (2017). *Definitions*. Committee on Tourism and Competitiveness. Available at http://cf.cdn.unwto.org/sites/all/files/docpdf/ctcdefinitionsenweb.pdf

UNWTO. (2018). *Tourism and culture synergies*. UNWTO.

Urbain, J.-D. (1991). *l'idiot du voyage, histoires de touristes*. Petite Bibliothèque Payot.

Urry, J. (1995). *Consuming places*. Routledge.

Urry, J. (1990). The tourist gaze: Leisure and travel in contemporary societies. *Sage (Atlanta, Ga.)*.

Urry, J. (1990). The Tourist gaze: Leisure and travel in contemporary societies. *Sage (Atlanta, Ga.)*.

Urry, J. (1990). *The tourist gaze: leisure and travel in contemporary societies*. Sage Publications Ltd.

Urry, J. (1990). *Tourist gaze: Leisure and travel in contemporary societies*. Sage Publications.

Urry, J. (1992). The tourist gaze "revisited". *The American Behavioral Scientist*, *36*(2), 172–186. doi:10.1177/0002764292036002005

Urry, J. (2002). The Tourist Gaze 2.0. *Sage (Atlanta, Ga.)*.

Urry, J., & Larsen, J. (1990/2011). The tourist gaze 3.0. *Sage (Atlanta, Ga.)*.

Urry, J., & Larson, J. (2011). *The Tourist Gaze 3.0*. SAGE Publications Ltd. doi:10.4135/9781446251904

van Es, N., & Reijnders, S. (2018). Making sense of capital crime cities: Getting underneath the urban facade on crime-detective fiction tours. *European Journal of Cultural Studies*, *21*(4), 502–520. doi:10.1177/1367549416656855

van Leeuwen, T. (2011). Multimodality. In *The Routledge handbook of applied linguistics* (pp. 688–702). Routledge.

Varley, P., Schilar, H., & Rickly, J. M. (2020). Tourism non-places: Bending airports and wildscapes. *Annals of Tourism Research*, *80*, 102791. Advance online publication. doi:10.1016/j.annals.2019.102791

Veasna, S., Wu, W., & Huang, C. (2013). The impact of destination source credibility on destination satisfaction: The mediating effects of destination attachment and destination image. *Tourism Management*, *36*, 511–526. doi:10.1016/j.tourman.2012.09.007

Vilkos, Y. (2006). Travel Agencies Cashing in on Trips to Chernobyl. *Kyiv Post*. Retrieved from: www.kyivpost.com/nation/24356/

Virzi, P. (Director). (2017). *The leisure seeker* [Film]. Indiana Production Company.

Vitić-Ćetković, A., Jovanović, I., & Potočnik Topler, J. P. (2020). Literary tourism: the role of Russian 19th century travel literature in the positioning of the smallest European royal capital – Cetinje. *Annales. Series historia et sociologia*, *30*(1).

Volgger, M. (2020). Staging *Genius Loci*: Atmospheric Interventions in Tourism Destinations. In M. Volgger & D. Pfister (Eds.), *Atmospheric Turn in Culture and Tourism* (pp. 139–151). Emerald Publishing.

Von Maltzahn, K. E. (1994). *Nature as landscape: Dwelling and understanding*. McGill-Queen's University Press.

Vu, D. V., Tran, G. N., Nguyen, H. T. T., & Nguyen, C. V. (2020). Factors Affecting Sustainable Tourism Development in Ba Ria-Vung Tau, Vietnam. *The Journal of Asian Finance, Economics, and Business*, *7*(9), 561–572. doi:10.13106/jafeb.2020.vol7.no9.561

Waitt, G. (2000). Consuming heritage: Perceived historical authenticity. *Annals of Tourism Research*, *27*(4), 835–862. doi:10.1016/S0160-7383(99)00115-2

Wallace, C. (2009). Yeats's Country and 'Yeats Country': Conceptualizing literary spaces. *Journal of Tourism and Cultural Change*, *7*(1), 48–60. doi:10.1080/14766820902803424

Walt Disney Company. (2021). *The Walt Disney company reports first quarter earnings for fiscal 2021*. https://thewaltdisneycompany.com/app/uploads/2021/02/q1-fy21-earnings.pdf

Wang, H. J., & Zhang, D. (2017). Comparing literary tourism in mainland China and Taiwan: The Lu Xun native place and the Lin Yutang house. *Tourism Management, 59*, 234–253. doi:10.1016/j.tourman.2016.08.008

Wang, N. (1999). Rethinking authenticity in tourism experience. *Annals of Tourism Research, 26*(2), 349–370. doi:10.1016/S0160-7383(98)00103-0

Ward, S., & O'Regan, T. (2009). The film producer as the long-stay business tourist: Rethinking film and tourism from a Gold Coast perspective. *Tourism Geographies, 11*(2), 214–232. doi:10.1080/14616680902827175

Watson, C., & Saunders, R. (2004). The production of literary landscapes. In M. Robinson & D. Picard (Eds.), *Conference proceedings tourism and literature: travel, imagination and myth*. Centre for Tourism and Cultural Change.

Watson, N. (Ed.). (2009). Literary tourism and nineteenth century culture. Palgrave.

Watson, C., & Saunders, R. (2004). The production of literary landscapes. In M. Robinson, & D. Picard (Eds.), *Conference Proceedings Tourism and Literature: Travel, Imagination and Myth*. Centre for Tourism and Cultural Change.

Watson, N. J. (2006). *The literary tourist: Readers and places in Romantic and Victorian Britain*. Palgrave. doi:10.1057/9780230584563

Watson, N. J. (2009). *Literary tourism and nineteenth-century culture*. Palgrave. doi:10.1057/9780230234109

Wearing, S. L., & Foley, C. (2017). Understanding the tourist experience of cities. *Annals of Tourism Research, 65*, 97–107. doi:10.1016/j.annals.2017.05.007

Wearing, S., Buchmann, A., & Jobberns, C. (2011). Free Willy: The whale-watching legacy. *Worldwide Hospitality and Tourism Themes, 3*(2), 127–140. doi:10.1108/17554211111122998

Wearing, S., Schweinsberg, S., & Johnson, P. (2019). Flâneur or Choraster: A Review of the Travel Narrator In the Formation of the Tourist Experience. *Tourism Analysis, 24*(4), 551–562. doi:10.3727/108354219X15652651367415

Weigelt, K., & Camerer, C. (1988). Reputation and corporate strategy: A review of recent theory and applications. *Strategic Management Journal, 9*(5), 443–454. doi:10.1002mj.4250090505

Wen, H., Josiam, B., Spears, D., & Yang, Y. (2018). Influence of movies and television on Chinese Tourists perception toward international tourism destinations. *Tourism Management Perspectives, 28*, 211–219. doi:10.1016/j.tmp.2018.09.006

Westover, P. (2012). *Necromanticism: Travelling to meet the dead, 1750-1860*. Palgrave Macmillan. doi:10.1057/9780230369498

Westwood, B., & Moss, S. (2015). Natural histories: 25 extraordinary species that have changed our world. Hachette UK.

Whang, H., Yong, S., & Ko, E. (2016). Pop culture, destination images, and visit intentions: Theory and research on travel motivations of Chinese and Russian tourists. *Journal of Business Research, 69*(2), 631–641. doi:10.1016/j.jbusres.2015.06.020

Wheeler, K. S. (2015). The Relationships Between Television Viewing Behaviors, Attachment, Loneliness, Depression, and Psychological Well-Being. *University Honors Program Theses*, Paper 98. https://digitalcommons.georgiasouthern.edu/honors-theses/98

White, S. (2017). Tourism guru off to 'Star Wars' site. *Otago Daily Times*. https://www.odt.co.nz/news/dunedin/tourism-guru-star-wars-site

Wilcox, C., Mann, V., Cannard, T., Ford, J., Hoshino, E., & Pascoe, S. (2021). *A review of illegal, unreported and unregulated fishing issues and progress in the Asia-Pacific Fishery Commission region*. Academic Press.

Wildemuth, B. (Ed.). (2018). *Applications of Social Research Methods* (2nd ed.). ABC-CLIO.

Wilkins, H. (2010). Souvenirs: What and Why We Buy. *Journal of Travel Research, 50*(3), 239–247. doi:10.1177/0047287510362782

Williams, S. L., Janetski, N., Abbott, J., Blankenhorn, S., Cheng, B., Crafton, R. E., Hameed, S. O., Rapi, S., & Trockel, D. (2014). Ornamental marine species culture in the Coral Triangle: Seahorse demonstration project in the Spermonde Islands, Sulawesi, Indonesia. *Environmental Management, 54*(6), 1342–1355. doi:10.100700267-014-0343-6 PMID:25082298

Wolff, J. (1985). The invisible flâneuse: Women and the literature of modernity. *Theory, Culture & Society, 2*(3), 34–50. doi:10.1177/0263276485002003005

Wood, J. (2012). *Como funciona a ficção* [How Fiction Works] (D. Bottmann, Trans.). Cosac Naify.

Woolf, V. (2021). Mrs. Dalloway. Penguin Books. (Original publication 1925)

World Travel and Tourism Council. (2020). *Travel and tourism economic impact and trends 2020*. Author.

Wright, D. (2018). Terror park: A future theme park in 2100. *Futures, 96*, 1–22. doi:10.1016/j.futures.2017.11.002

Wright, D., & Sharpley, R. (2018). Local community perceptions of disaster tourism: The case of L'Aquila, Italy. *Current Issues in Tourism, 21*(14), 1569–1585. doi:10.1080/13683500.2016.1157141

Wunder, S., & Sheil, D. (2013). On taxing wildlife films and exposure to nature. *Oryx, 47*(4), 483–485. doi:10.1017/S0030605312001469

Wylie, J. (2007). *Landscape*. Routledge. doi:10.4324/9780203480168

Yakici, A. (1994). Canakkale savaşları etrafında oluşan menkıbelerin Türk folkloru içindeki yeri [The place of the religious legends (menkabe/menkibe) formed around the Canakkale wars in Turkish folklore]. *Atatürk Araştırma Merkezi Dergisi, 10/30*, 599–608.

Yang, Y., Altschuler, B., Liang, Z., & Li, X. (2020). Monitoring the global COVID-19 impact on tourism: The COVID-19tourism index. *Annals of Tourism Research*, 103120. PMID:34566202

Yen, C., & Teng, H. (2013). Celebrity Involvement, Perceived Value, and Behavioural Intentions in Popular Media-Induced Tourism. *Journal of Hospitality & Tourism Research (Washington, D.C.), 39*(2), 225–244. doi:10.1177/1096348012471382

Yersu, A. (2016). Tevfik Fikret'in Aşiyanı ve yüzyıllık serüveni. In A. Yersu (Ed.), *Ölümünün 100. Yıldönümünde Tevfik Fikret Sempozyumu* (pp. 131–153). Istanbul Metropolitan Municipality Directorate of Libraries and Museums.

Yiannakis, J. N., & Davies, A. (2012). Diversifying rural economies through literary tourism: A review of literary tourism in Western Australia. *Journal of Heritage Tourism, 7*(1), 33–44. doi:10.1080/1743873X.2011.618538

Yilmaz, B. (2015). *Çanakkale savaşları etrafında teşekkül eden halk anlatıları–Çanakkale örneği* [Folk narratives occurred around the Gallipoli Campaign: The case of Canakkale] (PhD thesis). Çanakkale Onsekiz Mart Üniversitesi, Sosyal Bilimler Enstitüsü.

Yilmazer, T. (2020). Cabuk, Seyit Ali. *International Encyclopedia of the First World War*. https://encyclopedia.1914-1918-online.net/article/cabuk_seyit_ali

Yi, X., Fu, X., Yu, L., & Jiang, L. (2018). Authenticity and loyalty at heritage sites: The moderation effect of postmodern authenticity. *Tourism Management, 67*, 411–424. doi:10.1016/j.tourman.2018.01.013

Yi, X., Lin, V. S., Jin, W., & Luo, Q. (2017). The authenticity of heritage sites, tourists' quest for existential authenticity, and destination loyalty. *Journal of Travel Research, 56*(8), 1032–1048. doi:10.1177/0047287516675061

Yoo, C. K., Yoon, D., & Park, E. (2018). Tourist motivation: An integral approach to destination choices. *Tourism Review, 73*(2), 169–185. doi:10.1108/TR-04-2017-0085

Yoon, Y., & Uysal, M. (2005). An examination of the effects of motivation and satisfaction on destination loyalty: A structural model. *Tourism Management, 26*(1), 45–56. doi:10.1016/j.tourman.2003.08.016

Yosef, R. (2000). Individual distances among greater flamingos as indicators of tourism pressure. *Waterbirds, 23*, 26–31. doi:10.2307/1522143

Youngs, T. (2013). *The Cambridge introduction to travel writing*. Cambridge University Press.

Yung, R., & Khoo-Lattimore, C. (2017). New realities: A systematic literature review on virtual reality and augmented reality in tourism research. *Current Issues in Tourism, 22*(17), 2056–2081. doi:10.1080/13683500.2017.1417359

Yurdal, M. O. (2018). Efsanelere kurban edilen savaş [War Sacrificed to Legends]. *Tarih, 53*, 28–33.

Zachar, B. (2006). *Van egy magyar filmtörvény 1* [There is a Hungarian Film Act I]. Budapest: Elektronikus Periodika Archívum és Adatbázis. Retrieved from: http://www.epa.oszk.hu/00300/00373/00029/filmtorv.html

Zealand, N. (n.d.). *The Lord of the Rings Trilogy Experiences | 100% Pure NZ*. Retrieved 15 March 2021, from https://www.newzealand.com/int/feature/the-lord-of-the-rings-trilogy-experiences/

Zemah Shamir, Z., Zemah Shamir, S., Tchernov, D., Scheinin, A., & Becker, N. (2019). Shark aggregation and tourism: Opportunities and challenges of an emerging phenomenon. *International Journal of Sustainable Development and World Ecology, 26*(5), 406–414. doi:10.1080/13504509.2019.1573769

Zhang, H., Fu, X., Cai, L., & Lu, L. (2014). Destination image and tourist loyalty: A meta-analysis. *Tourism Management, 40*, 213–223. doi:10.1016/j.tourman.2013.06.006

Zhang, H., Wu, Y., & Buhalis, D. (2018). A model of perceived image, memorable tourism experiences and revisit intention. *Journal of Destination Marketing & Management, 8*, 326–336. doi:10.1016/j.jdmm.2017.06.004

Zhang, Q., & Xu, H. (2020). Understanding aesthetic experiences in nature-based tourism: The important role of tourists' literary associations. *Journal of Destination Marketing & Management, 10*, 100429. doi:10.1016/j.jdmm.2020.100429

Zhang, Y., & Wildemuth, B. (2018). Qualitative analysis of Content. In B. Wildemuth (Ed.), *Applications of Social Research Methods* (2nd ed., pp. 318–329). ABC-CLIO.

Zhao, L., Chen, Q., & Zhou, W. (2013). The ontology description and intention construct of tourism experience under phenomenology perspective. *Luyou Xuekan, 28*(10), 97–106.

Zhong, Y. Y. S., Busser, J., & Baloglu, S. (2017). A model of memorable tourism experience: The effects on satisfaction, affective commitment, and storytelling. *Tourism Analysis, 22*(2), 201–217. doi:10.3727/108354217X14888192562366

Zhu, Y. (2012). Performing heritage: Rethinking authenticity in tourism. *Annals of Tourism Research, 39*(3), 1495–1513. doi:10.1016/j.annals.2012.04.003

About the Contributors

Rita Baleiro holds a PhD and a Masters in English and Portuguese Studies from the New University of Lisbon, and a first degree in Modern Languages and Literatures – Portuguese and English – also from the New University of Lisbon. She is a senior lecturer of English for Specific Purposes and Portuguese Language and Culture at the School of Management, Hospitality and Tourism of the University of the Algarve. She is a member of CITUR (Centre for Research, Development and Innovation in Tourism) and she collaborates with the Tourism, Space and Urbanities Research Group of the Federal University of Rio de Janeiro. She has been co-chief-editor of Dos Algarves: A Multidisciplinary e-Journal since 2007. Her research interests include literature studies, academic writing and the intersections between literature and tourism. She has authored and co-authored various national and international scientific publications. Orcid 0000-0002-3188-5150.

Rosária Pereira has a PhD in tourism from the Faculty of Economics of the University of the Algarve and is an Associate Professor at the School of Management, Hospitality and Tourism of the University of the Algarve. She was, between 2015 and 2017, an Invited Assistant Lecturer at the School of Tourism and Hospitality of the University Europeia in Lisbon. She is a member of the Research Centre for Tourism, Sustainability and Well-being (CinTurs). Her research interests are related to tourism, tourist marketing, branding, and destinations branding and personality. She has published articles in international journals such as International Journal of Culture, Tourism and Hospitality Research, Journal of Destinations Marketing and Management and Journal of Quality Assurance in Tourism and Hospitality, among others, as well as some book chapters. She also belongs to the Scientific Commission of the journal Tourism & Management Studies, participates in international conferences in the area of tourist and tourism marketing and has been invited several times to participate in scientific committees of international conferences on tourism. Orcid 0000-0002-2335-5694

* * *

Helena Albuquerque is an Assistant Professor in the Department of Tourism, Heritage and Culture, and member of the REMIT (Research on Economics, Management, and Information Technologies) of the Portucalense University, being responsible for the line Tourism, Hospitality and Heritage. She has participated in several research projects in the areas of sustainable tourism and environmental conservation. Her main research interests are Sustainable Tourism, Geographic Information Systems applied to Tourism and Nature Tourism.

Jordi Arcos-Pumarola holds a PhD in Education and Society from the University of Lleida, a Master Degree in Innovation in Tourism Management, specialization in Cultural and Natural Heritage Management by Barcelona, School of Tourism, Hospitality and Gastronomy CETT – UB, and a Bachelor in Philosophy at the University of Barcelona. He is director of the Research Group on Tourism, Culture, and Territory at CETT-UB, where he also works as a lecturer on cultural tourism and tourism research methods. His two main research lines are cultural and literary tourism, and ethics in tourism.

Ozan Atsız holds a PhD degree and a Master's degree in Tourism Management (Turkey). Currently, he is a research assistant in the Faculty of Tourism at Yozgat Bozok University. His research interests include destination marketing, heritage experience and food experience.

Márcia Cavalcanti Moreira completed her degree studies in Tourism from the Federal Rural University of Rio de Janeiro (UFRRJ) in 2014 and Librarianship and Documentation graduated from the Fluminense Federal University (UFF) in 2019. Work as a Tourism teacher/tutor at the CECIERJ Foundation/Cederj, since 2015; and as a collaborating researcher of the Geography, Tourism and Literature Research Network (Rede Entremeio) in Brazil. Her main research interests are in Literary Tourism and Library Tourism.

Samet Çevik has a PhD in Tourism Management from Anadolu University. He is an Associate Professor in the Department of Tourism and Hotel Management at Bandırma Onyedi Eylül University, Turkey. His research interests include destination marketing, cultural tourism, literary tourism.

Maria Elicher has a PhD in Geography from Universidade Federal Fluminense (UFF) - Niterói/Brasil (2014). Founder and coordinator of Entremeio – Research network in Geography, Tourism and Literature. Since 2010, is a professor of Geography of Tourism, Tourism and Literature and Tourism and Production of Space at the Department of Tourism and Patrimony at Universidade Federal do Estado do Rio de Janeiro (Unirio), Rio de Janeiro/Brasil. Researches in the areas of the production of space, tourism in relation with art, and in particular, the interdisciplinary relation between Tourism, Literature and Arts.

Michael Thomas Fagence has linked his training in 'thinking geographically' with elements of tool kits from history and from semiotics in order to forge investigative and interpretive processes that can be applied to cases of history-linked and heritage-based tourism. His most recent appointment was as Honorary Research Fellow at the University of Queensland.

Adriana Coelho Florent has held the position of maître de conférences at the University of Aix-Marseille (AMU) since 2010. Her Master's works (Master 1 and 2) deal with the influence of Balzac in the work of Eça de Queiroz and the representation of dreams in the realistic novels of the same author. His thesis on Graciliano Ramos and the writer's function in Brazilian society from 1930 to 1945 was published by Terceira margem publishing house in 2011, under the title Graciliano Ramos em seu tempo - o meio literário na era Vargas. She has published articles in France and Brazil regarding Brazilian, Portuguese, Angolan and Mozambican works of fiction from a perspective linked to sociocriticism and comparative literature. Her current research revolves around literary tourism in Lusophone countries, particularly in its relations with memorial tourism linked to migrations and diasporas between Europe, Brazil and Africa.

Núria Guitart Casalderrey graduated in Tourism and Official Master Degree in Innovation in Tourism Management, specialization in Tourism Management of Urban Destinations by Barcelona, School of Tourism, Hospitality and Gastronomy CETT – UB. PhD Student in Geography, Territorial Planning and Environmental Management from the University of Barcelona. She is a researcher of the Research Group on Tourism, Culture, and Territory at CETT-UB. Her main research lines are urban tourism management and tourism governance.

Erol Gülüm works as an Assistant Professor at the Department of Turkish Language and Literature, in Bilecik Şeyh Edebali University, Turkey. He graduated from Hacettepe University, Department of Turkish Language and Literature in 2010. He also got his M.A (2012) and PhD (2017) from Hacettepe University, Department of Turkish Folklore. He studied as a visiting doctoral researcher at The State University of New York at Binghamton in the U.S. A for a full year between 2015-2016. Currently, he is a postdoctoral visiting fellow at Frankfurt Memory Studies Platform (FMSP). His recent interests include digital culture studies, cultural studies, cultural creative industries and intangible cultural heritage.

Daniel Imbert-Bouchard Ribera holds a PhD in Geography, graduated in Tourism and official master's degree in Territorial Planning and Environmental Management from the University of Barcelona. Coordinator of official master's degree in Innovation of the Tourism Management of Urban Destinations at the CETT-UB, where he also works as a lecturer and researcher. His field of study is tourist mobility and the sustainable management of tourist spaces. He is the author of various publications in national and international journals.

Eleni (Elina) Michopoulou is an Associate Professor in Business Management at University of Derby, UK. Her research interests include information systems in tourism, online consumer behaviour and technology acceptance. She is particularly interested in the field of accessible and wellness tourism, which she has actively been researching for over fifteen years. Eleni has published over 60 academic journal articles, book chapters and conference papers. She is the co-founder of THEINC International Conference, sits on the editorial board of over 10 high impact academic journals and is the European Editor for the International Journal of Spa and Wellness.

Delia Gabriela Moisa is a former student at the University of Derby, where she obtained a degree in Travel and Tourism Management. From there, she continued her master studies in International Hospitality Management at the Manchester Metropolitan University. Her working experience in multicultural environments across various countries including China, Mexico and the United Kingdom together with her passion for cultures and tourism motivated her to continue her research on a variety of tourism and hospitality related topics.

Zeliha Öztürk received her Ph.D. degree in Modern Turkish Literature (Turkish Language and Literature) at the University of Erciyes in 2017. Dr. Ozturk is an assistant professor in the Department of Turkish Language and Literature at the Bilecik Seyh Edebali University, Turkey. Her teaching and research interests focus on Turkish Literature, text analysis, feminism, orientalism, comparative literature, modernism, city culture, narratology, myth criticism, and archetypal criticism. Dr. Ozturk has been publishing in the field of modern Turkish Literature since 2010. She has published articles in international journals. In addition, his work has appeared in several books and the proceedings of conferences scientific meetings of her specialty.

Árpád Ferenc Papp-Váry is the dean of the Faculty of Tourism, Business and Communication at the Budapest Metropolitan University, Hungary. He is also the head of the Commerce and Marketing BSc programme and vocational programme, the Marketing MSc programme (running from 2019), and the Digital Marketing executive MBA postgraduate programme. At the same time he is research associate of the Urban Marketing and Geostrategy Centre of John von Neumann University, Kecskemét, Hungary. Besides university education, Árpád regularly holds training sessions and provides branding consultancy for cities, companies and professionals. He is serving as Vice President of the Hungarian Marketing Association. His teaching and research areas are country branding, city branding, personal branding, sports branding and branded entertainment. Árpád is the author of six books and several hundred publications, most of which are available online at www.papp-vary.hu.

Jorge Ramos is an Integrated Researcher of the Research Centre for Tourism, Sustainability and Well-being (CinTurs) at the University of Algarve, Portugal. He has worked on socioeconomics related to coastal issues, particularly ecotourism, small-scale fisheries and offshore aquaculture. His research interests include aspects related to ecotourism, stakeholder analysis, artificial reefs, among other areas. He holds a PhD in Fisheries Economics and Management, a PgC in Social Research Methods, a MSc in Fisheries Economics (University of Portsmouth, UK) and a graduate degree in Marine Biology and Fisheries (University of Algarve, Portugal).

Aleksandra Siurnicka is currently studying for the MSc Tourism Management at the University of Derby, UK. Her research interests focus on the role of media and particularly of films in tourism, and the implications for destinations and tourism providers.

Jennifer Stewart is an Assistant Lecturer with Technological University Dublin, in the School of Hospitality Management and Tourism, College of Arts and Tourism. She lectures primarily in tourism and marketing in the hospitality, tourism, leisure, and events area. Jennifer teaches modules on Sustainable Tourism Futures, Tourism Policy Planning and Regulation, Tourism and Technology, Cultural Tourism, Destination Marketing, and Marketing Strategy. Jennifer has undertaken research projects in the field of film tourism for large scale film production studios and has contributed to marketing plans for various tourism businesses at Local Authority and Industry level.

Selman Temiz holds a Master's degree in Business Administration at Yozgat Bozok University and PhD candidate in Business Administration at Yozgat Bozok University. Currently, he is a lecturer in the Vocational School of Social Sciences at Yozgat Bozok University. His research interests include tourism marketing, alternative tourism, and tourism destinations.

Tímea Zsófia Tóth is a Ph.D. student at the Department of International and Regional Economics, University of Sopron. In April 2019 she won third place in the Economics Section of the National Scientific Students' Associations Conference with her research on film tourism. Since November 2019 she has been a student member of the Scientific Students' Associations of the University of Sopron Alexandre Lamfalussy Faculty of Economics. Between 2020 and 2021 she was a Student Member of the Executive

Board of the XXXV National Students' Scientific Conference. Her educational work includes practical and theoretical courses for tourism and catering BA and MA students. Her field of research includes cultural tourism, film-induced tourism, special tourism products, travel motivation and mental mapping.

Isabel Vaz de Freitas is a Full Professor at Portucalense University, PhD in History, research member of REMIT (Research on Economics, Management, and Information Technologies) and researcher member of LAB2PT at Minho University, Correspondent of the Portuguese Academy of History, Head of the Department of Tourism, Heritage and Culture.

Index

www.ingramcontent.com/pod-product-compliance
Lightning Source LLC
LaVergne TN
LVHW062120300926
849161LV00010B/2502
9781799882626